Near-infrared Nanomaterials
Preparation, Bioimaging and Therapy Applications

RSC Nanoscience & Nanotechnology

Editor-in-Chief:
Professor Paul O'Brien FRS, *University of Manchester, UK*

Series Editors:
Professor Ralph Nuzzo, *University of Illinois at Urbana-Champaign, USA*
Professor Joao Rocha, *University of Aveiro, Portugal*
Professor Xiaogang Liu, *National University of Singapore, Singapore*

Honorary Series Editor:
Sir Harry Kroto FRS, *University of Sussex, UK*

Titles in the Series:
1: Nanotubes and Nanowires
2: Fullerenes: Principles and Applications
3: Nanocharacterisation
4: Atom Resolved Surface Reactions: Nanocatalysis
5: Biomimetic Nanoceramics in Clinical Use: From Materials to
   Applications
6: Nanofluidics: Nanoscience and Nanotechnology
7: Bionanodesign: Following Nature's Touch
8: Nano-Society: Pushing the Boundaries of Technology
9: Polymer-based Nanostructures: Medical Applications
10: Metallic and Molecular Interactions in Nanometer Layers, Pores and
    Particles: New Findings at the Yoctolitre Level
11: Nanocasting: A Versatile Strategy for Creating Nanostructured Porous
    Materials
12: Titanate and Titania Nanotubes: Synthesis, Properties and Applications
13: Raman Spectroscopy, Fullerenes and Nanotechnology
14: Nanotechnologies in Food
15: Unravelling Single Cell Genomics: Micro and Nanotools
16: Polymer Nanocomposites by Emulsion and Suspension
17: Phage Nanobiotechnology
18: Nanotubes and Nanowires, 2nd Edition
19: Nanostructured Catalysts: Transition Metal Oxides
20: Fullerenes: Principles and Applications, 2nd Edition
21: Biological Interactions with Surface Charge Biomaterials
22: Nanoporous Gold: From an Ancient Technology to a High-Tech Material
23: Nanoparticles in Anti-Microbial Materials: Use and Characterisation
24: Manipulation of Nanoscale Materials: An Introduction to
    Nanoarchitectonics
25: Towards Efficient Designing of Safe Nanomaterials: Innovative Merge of
    Computational Approaches and Experimental Techniques
26: Polymer–Graphene Nanocomposites

How to obtain future titles on publication:
A standing order plan is available for this series. A standing order will bring delivery of each new volume immediately on publication.

For further information please contact:
Book Sales Department, Royal Society of Chemistry, Thomas Graham House, Science Park, Milton Road, Cambridge, CB4 0WF, UK
Telephone: +44 (0)1223 420066, Fax: +44 (0)1223 420247
Email: booksales@rsc.org
Visit our website at www.rsc.org/books

Near Infrared Nanomaterials
Preparation, Bioimaging, and Therapy Applications

Edited by

Fan Zhang
Fudan University, Shanghai, China
Email: zhang_fan@fudan.edu.cn

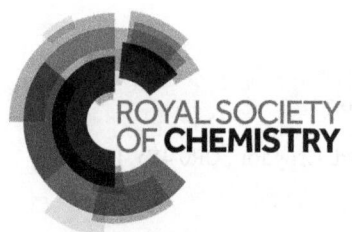

THE QUEEN'S AWARDS
FOR ENTERPRISE:
INTERNATIONAL TRADE
2013

RSC Nanoscience & Nanotechnology No. 40

Print ISBN: 978-1-78262-319-9
PDF eISBN: 978-1-78262-393-9
EPUB eISBN: 978-1-78262-831-6
ISSN: 1757-7136

A catalogue record for this book is available from the British Library

Published by The Royal Society of Chemistry,
Thomas Graham House, Science Park, Milton Road,
Cambridge CB4 0WF, UK

Registered Charity Number 207890

For further information see our web site at www.rsc.org

Printed in the United Kingdom by CPI Group (UK) Ltd, Croydon, CR0 4YY, UK

Foreword

In the last decade, bioimaging and therapy based on near infrared (NIR) nanomaterials have played an important role in biotechnology due to their intrinsic advantages over traditional imaging probes and medicines, such as greater penetration depth, low detection threshold concentration, and better targeted performance. Nanomaterials based on organic dyes, lanthanides, carbon, quantum dots (QDs), and noble metals are major components in this big family of NIR bionanomaterials. Exciting developments have been made at a very fast pace by many research groups. The vast literature published about NIR nanomaterials over the past two decades is a clear witness to this; the number of papers has increased exponentially, with most of the activity and development happening in the last 10 years.

NIR materials have absorption/excitation/emission maxima falling in the region of minimal tissue absorbance/autofluorescence between 650 and 1700 nm, an "imaging window." In tissues, light absorbance and scattering is minimal in this wavelength range, allowing light to penetrate more deeply. This enables animal bioimaging and therapy with high sensitivity in real time without the need for dissection or invasive procedures. In the past two decades, related theories, methods, and techniques have been explored. As a consequence, novel NIR materials are increasingly emerging, and their applications extend from traditional fields such as optical communication amplifiers and solid-state lasers to high-tech fields including biosensors, bioimaging, and disease therapy. Researchers in this field can therefore give a deep insight into synthesis strategies and behavior of NIR nanomaterials, and in particular, establish structure–function–synthesis relationships.

This book contains 11 chapters. Chapter 1 describes some distinctive characteristics of lanthanide-based NIR nanomaterials (upconversion and downconversion) which are directly related to their bioimaging applications,

RSC Nanoscience & Nanotechnology No. 40
Near Infrared Nanomaterials: Preparation, Bioimaging and Therapy Applications
Edited by Fan Zhang
© The Royal Society of Chemistry 2016
Published by the Royal Society of Chemistry, www.rsc.org

such as color tunability, energy transfer principles, and some strategies for enhancing luminescent efficiency. Bioimaging based on NIR QDs has advantages including lower absorption and relatively low autofluorescence, resulting in deeper penetrating depth and lower background. Chapter 2 summarizes developments in non-toxic QDs, especially for synthesis and *in vivo* bioimaging. Chapter 3 introduces the fabrication and fundamental properties of carbon dots (CDs) and nanodiamonds (NDs) and then focuses on their recent bioapplications in bioimaging. Challenges and perspectives for future developments are also briefly discussed. We hope this chapter will provide critical insights to inspire further exciting research on CDs/NDs for biological imaging applications, to better realize the potential of these intriguing materials in the near future. Chapter 4 summarizes recent progress in the synthesis and *in vivo* behavior of NIR-emitting gold nanoparticles, and discusses future challenges and opportunities for them.

Besides the inorganic NIR nanomaterials already mentioned, nanomaterials based on organic molecules are a novel type of NIR imaging agents. Chapter 5 focuses on recent progress in this area, including major NIR organic chromophores, luminescent principles, and construction methods, as well as biomedical applications and challenges. Photodynamic therapy (PDT) is a treatment for cancer that uses the reactive oxygen species generated by a photosensitizer drug following irradiation at a specific wavelength to destroy cancerous tissue. Chapter 6 provides an overview of the main principles and mechanisms for biosensing based on NIR QDs and the use of QDs for simultaneous diagnostics and therapy of disease. Chapter 7 focuses on state-of-the art use of NIR nanomaterials for PDT, including both *in vitro* and *in vivo* applications. Chapter 8 introduces NIR light-triggered drug and gene delivery platforms, including photoresponsive nanocarriers, photocaging of bioactive cargos, and photothermal transduction for NIR-triggered nanocarriers. It has been suggested that, due to the differences in cell-killing mechanisms, synergetic tumor responses may be achieved if two modalities are combined in an appropriate sequence. Therefore, multifunctional nanocarriers that enable combination cancer therapy with different therapeutic mechanisms in one system may play increasingly important roles in the fight against cancer due to their unique advantages such as minimal side effects and high efficacies. In Chapter 9, we discuss the most significant progress made in the field of NIR-responsive nanotheranostics for synergistic cancer therapy. NIR-induced photothermal ablation therapy (PAT) has attracted increasing interest as a minimally invasive and potentially effective treatment technology for cancer. A prerequisite for the development of NIR-induced PAT is to obtain low-cost and biocompatible photothermal agents with high photothermal conversion efficiency. In Chapter 10, we first introduce the measurement method for photothermal conversion efficiency, and then summarize the research progress of these photothermal agents as well as the combination of PAT with other nanobiotechnology techniques.

Numerous and extensive studies have been devoted to the design, synthesis, and development of various NIR nanomaterials for biomedical imaging and imaging-guided therapy of cancers; however, so far there is little information on the toxicological properties of NIR nanomaterials and their long-term toxicity or health effects. Moreover, most of the present data are either conflicting or not in the public domain, preventing the scientific community from properly evaluating the effect of nanomaterials on human health and environment. Therefore, Chapter 11 focuses primarily on recent progress in toxicity studies of NIR nanomaterials (including carbon-based materials, QDs, noble metal-based nanoparticles, upconversion nanoparticles, and narrow-bandgap semiconductors), discussing in detail how the biophysicochemical properties of NIR nanomaterials influence their toxicity, and finally presenting a broad overview of the available *in vitro* and *in vivo* toxicity assessments.

Research on NIR nanomaterials has developed rapidly in the past decade. A comprehensive review is thus necessary, and it is the main purpose of this book. Chapters are organized along the following lines: (1) following the forefront of current research, and striving to reflect the latest progress and developments; (2) comprehensive review focusing on basic fundamental research; and (3) practical research experience in methodology, experimental skills, and data analysis. Each chapter also includes understanding, induction, and summaries from the authors. We have taken care to include fundamental information about NIR nanomaterials, making the book especially suitable for beginners and graduate students who have just entered this field. We hope that, through reading this book, they can fully understand the chemistry of photon upconversion nanomaterials, grasp the skills required for their synthesis, obtain high-quality materials, and therefore learn to deeply appreciate the chemical and physical properties of these materials and their applications.

This book is a distillation of the authors' knowledge and hard work. We want it to help and inspire researchers who are working in the fields of chemistry and materials science, especially in nanobiology. We also hope it can provide a reference source or serve as a textbook for undergraduate and graduate students majoring in chemistry, chemical engineering, physics, materials science, and biology, as well as readers who are already interested in NIR nanomaterials.

Galen Stucky
University of California, Santa Barbara

Contents

RSC Nanoscience & Nanotechnology No. 40
Near Infrared Nanomaterials: Preparation, Bioimaging and Therapy Applications
Edited by Fan Zhang
© The Royal Society of Chemistry 2016
Published by the Royal Society of Chemistry, www.rsc.org

CHAPTER 1

Lanthanide-Based Near Infrared Nanomaterials for Bioimaging

RUI WANG[a] AND FAN ZHANG*[a]

[a]Fudan University, Department of Chemistry, 220 Handan Rd, Shanghai, 200433, P. R. China
*E-mail: zhang_fan@fudan.edu.cn

1.1 Introduction

Lanthanide elements are spectroscopically rich species, a property that facilitates their use as optical codes in a spectral window distinct from fluorescent dyes used for labeling biological samples. The lanthanide 4f orbitals are buried beneath the 6s, 5p, and 5d orbitals; hence, spectra arising from f–f transitions are narrow and insensitive to their environment, unlike transition metal (3d) spectra.[1] Most importantly, this gives rise to a rich energy-level structure in the near infrared (NIR), visible (VIS), and ultraviolet (UV) spectral range (Figure 1.1). Triply ionized lanthanide ions in solid hosts typically have emission line widths of ~10–20 nm (FWHM, full width at half maximum), which is about half that observed for quantum dots (QDs, ~25–40 nm) and much narrower than that observed for organic dyes (~30–50 nm) or transition metal ions (~100 nm).[2,3] This feature allows more resolvable bands to be packed into the same spectral bandwidth, which enables a larger

RSC Nanoscience & Nanotechnology No. 40
Near Infrared Nanomaterials: Preparation, Bioimaging and Therapy Applications
Edited by Fan Zhang
Published by the Royal Society of Chemistry, www.rsc.org

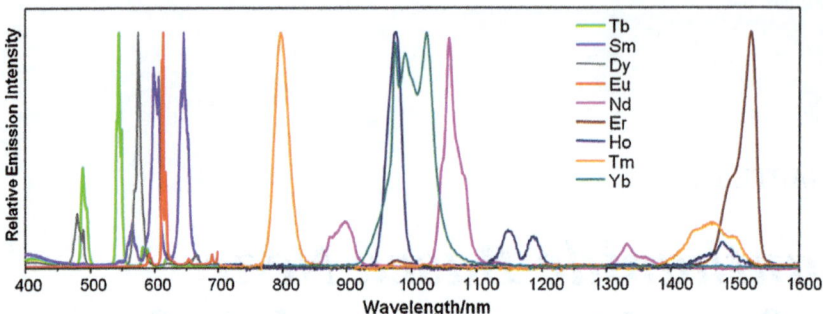

Figure 1.1 Normalized emission spectra of luminescent lanthanide complexes in solution, illustrating the sharp emission bands and minimal overlap of lanthanide luminescence.[4-6] (Reproduced with permission from S. Petoud, *et al.*, *J. Am. Chem. Soc.*, 2003, **125**, 13324–13325.[5] Copyright (2003) American Chemical Society and from ref. 6 with permission from John Wiley and Sons. Copyright © 2005 Wiley-VCH Verlag GmbH & Co. KGaA, Weinheim.)

number of distinct combinations. Because lanthanide emissions involve only atomic transitions, they are extremely resistant to photobleaching. The energy-level structure in lanthanide ions also creates the possibility for large shifts between the excitation and emission bands. This shift can be several hundred nanometers, containing discrete gaps with zero absorption. By comparison, the HOMO–LUMO (highest occupied molecular orbital–lowest unoccupied molecular orbital) transition in organic dyes typically results in overlapping excitation and emission bands and a Stokes shift of only 10–30 nm between the absorption and emission maxima. The large variety of absorption and emission wavelengths, the independence on host materials, and low vibration energy losses make lanthanides ions be ideal for spectral conversion. Lanthanide ions can be doped in a variety of solids such as crystals, fibers, or glass ceramics to give them the desired downconversion and upconversion optical properties. Thus, their remarkable luminescence properties have been widely applied in lasers, solar cells, analytical sensors, photodynamic therapy, and optical imaging.[4-6] In this chapter, we focus on some distinct characteristics of lanthanide-based NIR nanomaterials that are closely related to their bioimaging applications, such as color tunability, energy transfer principles, and some strategies for enhancing luminescent efficiency. In addition, we systematically introduce the most recent bioimaging work based on lanthanide NIR nanomaterials.

1.2 Upconversion Nanoparticles (UCNPs)

Upconversion materials, which emit high-energy photons under excitation by the NIR light (anti-Stokes shift) were first discovered in the 1960s,[7] but have primarily been exploited for the development of some remarkably effective optical devices such as infrared quantum counter detectors,[8,9] temperature sensors,[10,11] and compact solid state lasers.[12–15] Thus, for more than 30 years the use of the upconversion effect has been limited to bulk glass or crystalline materials.[16–19] Because of their suitable size (small enough to go in and out through many biological host materials, such as the cytoplasm or nucleus of a cell) and their unique properties, such as high chemical stability, low cytotoxicity, and high signal-to-noise ratio, the biological applications of UCNPs for analytical assays and bioimaging were readily recognized.[20–23] The upconversion process proceeds by different mechanisms: excited state absorption (ESA), energy transfer upconversion (ETU), and photon avalanche (PA).[24] ESA and ETU are based on the sequential absorption of two or more photons by metastable, long-lived energy states. In the case of ESA, the ground state activator absorbs at least two photons of suitable energy sequentially. In ETU, activator and neighboring sensitizer both absorb one photon at first, then energy is transferred between sensitizer and activator, resulting in a population of emitting ions in a highly excited state.[25] PA was first reported by Chivian *et al.* in 1979.[26] They found that when Pr^{3+}-doped $LaCl_3$ or $LaBr_3$ crystal is exposed to laser-pump radiation slightly in excess of a certain critical intensity, the fluorescence of Pr^{3+} increases by orders of magnitude.[13,26] PA-induced UC features an unusual pump mechanism that requires a pump intensity above a certain threshold value and always responds slowly to excitation (up to several seconds). The quantum yield (QY) is differs considerably among these three mechanisms: in theory, ESA < ETU < PA, but PA always needs a rather high excitation energy and suffers from slow response to excitation. The QY of ETU is two orders of magnitude higher than that of ESA, making it well understood and widely applied in many fields.[7]

Efficient UCNPs are composed of three components: a host matrix, a sensitizer, and an activator. An ideal host matrix needs to be optically transparent and have low lattice phonon energy, in order to minimize non-radiative losses and maximize radiative emission. $NaYF_4$,[27–29] $NaGdF_4$,[30,31] $NaLuF_4$,[32] LaF_3,[33] and CaF_2,[34] among others, have proven to be ideal candidates. With the development of nanotechnology, several methods have been used to fabricate uniform monodispersed UCNPs with controlled crystalline phases and sizes, including coprecipitation,[35–42] thermal decomposition,[43–47] hydro(solvo)-thermal synthesis,[48–62] sol–gel process,[63–66] and combustion synthesis,[67] which have been reviewed in many papers.[68–72] A proper choice of synthesis method enables the development of UCNPs whose properties match the need for the applications envisioned. On the other hand, the rational choice of different lanthanide ions as sensitizer and activator is also very important. Typically, Yb^{3+} is always chosen as sensitizer because of its large absorption cross-section of around 980 nm. Er^{3+}, Tm^{3+}, and Ho^{3+} feature ladder-like arranged energy levels which favor multiphoton process, and thus are frequently used as activators.[25] Besides these three components, rational design of core–shell structure to minimize surface quenching effects and improve the luminescence efficiency of UCNPs is also very crucial.[25,70–72]

1.2.1   UCNPs Excited at 980 nm

The upconversion materials were first used in tissue imaging in 1999, when Zijlmans and coworkers reported the first upconversion bioimaging based on submicron-sized Y_2O_2S:Yb^{3+}/Tm^{3+} particles (0.2–0.4 μm).[73] Upon excitation at 980 nm, they observed a low autofluorescence signal and no bleaching even after continuous exposure to high excitation energy levels. After that, the idea of upconversion bioimaging was realized by employing other oxysulfide or oxide nanomaterials (*e.g.*, Y_2O_3:Yb^{3+}/Er^{3+} and Gd_2O_3:Yb^{3+}/Er^{3+}).[74,75] However, the size of the oxysulfide or oxide particles was at the submicron level, which limited applications. The necessary requirements for material selection in practical bioimaging are small size, bright luminescence, and biological safety. Recently, with the rapid development of synthesis techniques, fluoride-based UCNPs with smaller size but high-quality luminescence have been explored extensively and widely used in cell, tissue, and animal imaging. In contrast to oxysulfides or oxides, fluorides are considered better host materials for the doping of lanthanide ions to achieve intense UC emissions, owning to their low phonon energies and the resulting minimization of quenching of the excited state of the lanthanide ions. One of the first reports of *in vivo* imaging of UCNPs in small animals reported spot measurements of UCNPs injected subcutaneously in rats by Zhang *et al.* in 2008, which showed much deeper penetration under 980 nm excitation compared to commercial green-emitting QDs excited by UV.[76] Although the NIR excitation light for UC materials has strong penetration ability, the UV/VIS UC emissions are still easily absorbed by biological samples, which definitely limits their further applications in the observation of deep biological tissues. For small-animal *in vivo* imaging, highly sensitive NIR–NIR systems have attracted increasing attention, because both the excitation and emission light is located in the NIR region. For this purpose, Tm^{3+} ions are frequently chosen as dopants since they can exhibit strong NIR emission between 750 and 850 nm due to the transition from $^3H_4$ to $^3H_6$ under CW laser excitation at 980 nm. Nyk *et al.* reported the use of 20–30 nm $NaYF_4$ NPs doped with Tm^{3+} and Yb^{3+}, which has an emission around 800 nm for both *in vitro* and *in vivo* imaging (Figure 1.2).[77] High-contrast photoluminescence (PL) imaging was possible in cells and small animals due to the better tissue penetration properties achieved since both the excitation and emission are in the NIR region. After that, *in vivo* whole-body imaging of small animals has been successfully realized based on Tm^{3+}-doped $NaYF_4$,[78] $NaGdF_4$,[79] $NaLuF_4$,[80] and $NaYbF_4$ NPs.[81]

Besides 750–850 nm emission from Tm^{3+}, the red emission (625–690 nm, centered at 660 nm) from Er^{3+} or Ho^{3+} is also ideal for bioimaging. Therefore, much effort has also been devoted to enhancing the red emission of Er^{3+} or Ho^{3+}—in other words, to achieving an enhanced red/green (R/G) ratio in the Yb/Er codoped upconversion system. In 2011, Liu *et al.*[82] described an oil-based synthetic method for the preparation of $KMnF_3$ nanocrystals with lanthanide dopants homogeneously incorporated in the host lattice. With Yb^{3+}/Er^{3+} doping, blue and green emissions of Er^{3+} disappeared completely,

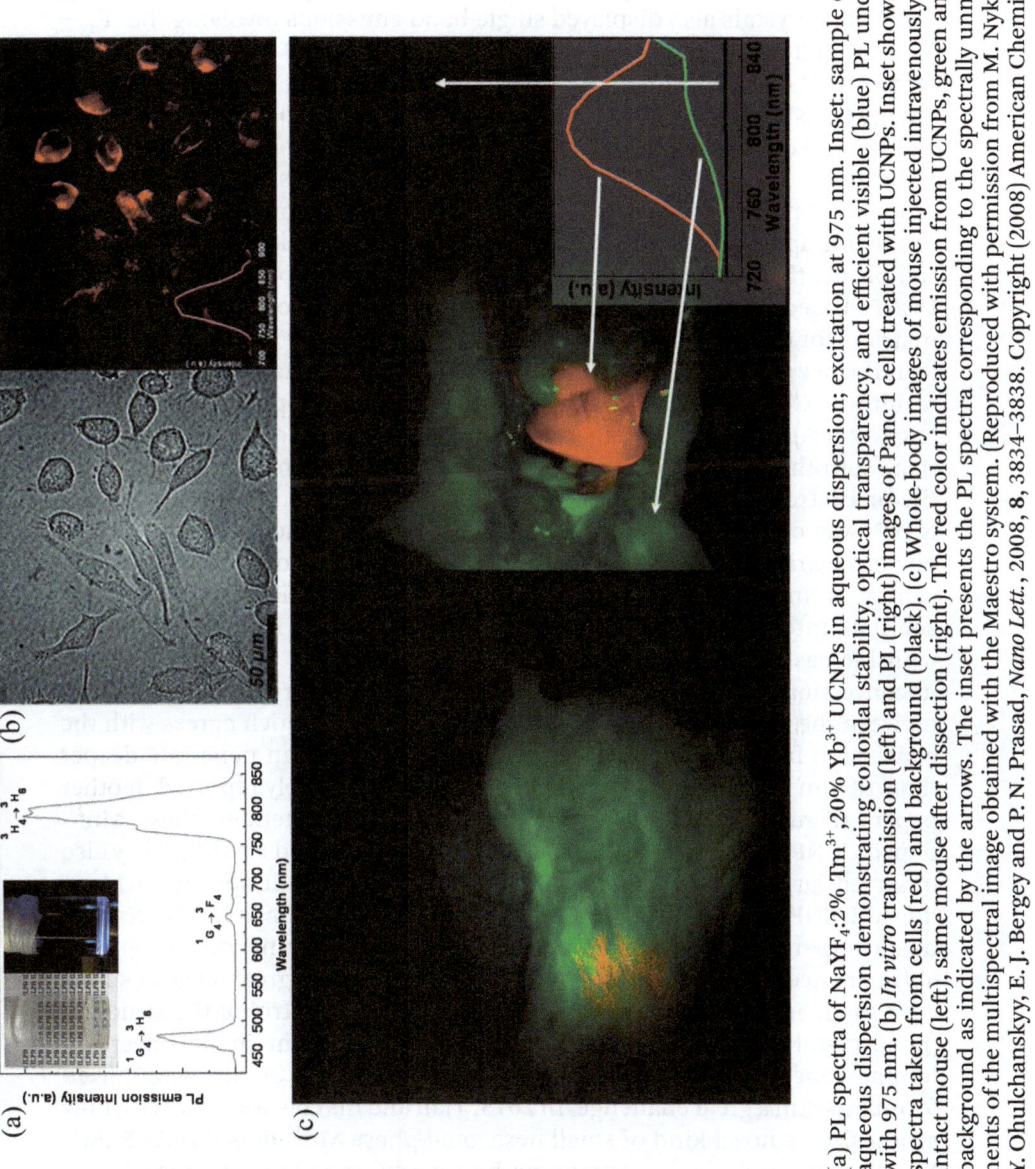

Figure 1.2 (a) PL spectra of NaYF$_4$,:2% Tm^{3+},20% Yb^{3+} UCNPs in aqueous dispersion; excitation at 975 nm. Inset: sample of the UCNPs aqueous dispersion demonstrating colloidal stability, optical transparency, and efficient visible (blue) PL under excitation with 975 nm. (b) *In vitro* transmission (left) and PL (right) images of Panc 1 cells treated with UCNPs. Inset shows localized PL spectra taken from cells (red) and background (black). (c) Whole-body images of mouse injected intravenously with UCNPs; intact mouse (left), same mouse after dissection (right). The red color indicates emission from UCNPs, green and black show background as indicated by the arrows. The inset presents the PL spectra corresponding to the spectrally unmixed components of the multispectral image obtained with the Maestro system. [Reproduced with permission from M. Nyk, R. Kumar, T. Y. Ohulchanskyy, E. J. Bergey and P. N. Prasad, *Nano Lett.*, 2008, **8**, 3834–3838. Copyright (2008) American Chemical Society.][77]

suggesting an extremely efficient exchange-energy transfer process between the Er^{3+} and Mn^{2+} ions, which can be largely attributed to the close proximity and effective mixing of wave functions of the Er^{3+} and Mn^{2+} ions in the crystal host lattices (Figure 1.3a). Besides Yb^{3+}/Er^{3+}, Yb^{3+}/Ho^{3+} and Yb^{3+}/Tm^{3+} doped $KMnF_3$ nanocrystals were synthesized, respectively. Importantly, these nanocrystals also displayed single-band emissions involving the $^5F_5 \rightarrow$ $^5I_8$ (centered at 650 nm) transition in Ho^{3+} (Figure 1.3b) and the $^3H_4 \rightarrow$ $^3H_6$ (centered at 800 nm) transition in Tm^{3+} (Figure 1.3c). As a result of efficient energy transfer between the dopant ion and host Mn^{2+} ion, remarkably pure single-band upconversion emissions were generated in the red and NIR spectral regions. The complete lack of short-wavelength emission of these lanthanide-doped nanocrystals in the visible spectral region provides a platform for promising applications in biolabeling studies, for which imaging at different sample depths is required. One year later, Zhao *et al.*[83] reported red-emission UCNPs based on $NaYF_4$ system, which is considered as one of the best host matrices for the upconversion process. In this work, Mn^{2+} served as a dopant to influence the growth dynamics of the crystalline phase and size of the resulting UCNPs, rather than host material. The Mn^{2+}-doped $NaYF_4$:Yb^{3+}/Er^{3+} (18/2 mol%) NPs were obtained by using a modified liquid–solid solution (LSS) solvothermal strategy: with the increased doping amount of Mn^{2+} ions, the phase transformation from hexagonal to cubic was obvious. Pure cubic $NaYF_4$ was obtained when the level of Mn^{2+} ions reached 5 mol% and no obvious extra diffraction peaks were detected even when the Mn^{2+} ion concentration increased to 30 mol%, indicating the formation of a Y–Mn solid solution. Interestingly, the R/G ratio gradually increased from 0.83 to 163.78 with increasing Mn^{2+} dopant content. The appearance of single-band red upconversion emission suggests that the exchange-energy transfer process between the Er^{3+} and Mn^{2+} ions is extremely efficient, which agrees with the conclusion from Liu *et al.* These red-emitting UCNPs can penetrate deeper than 10 mm when used for imaging *in vivo*, which is rarely reported in other papers (Figure 1.3d–f). Later in 2014, Hao *et al.*[84] extended these Mn^{2+}-doped UCNPs to other host materials, such as $NaLuF_4$ and $NaYbF_4$. They also observed large enhancements in overall UC luminescent spectra of Mn^{2+}-doped UCNPs (~59.1 times for the $NaLuF_4$ host, ~39.3 times for the $NaYbF_4$ host compared to the UCNPs without Mn^{2+} doping), mainly due to remarkably enhanced luminescence in the red band. Although great advances have been made in these three papers,[82–84] simultaneous control of the structure (nanocrystal size, shape, and phase) and enhancement in upconversion luminescence especially dominated by red emission in UCNPs with a fixed formula is still a great challenge. In 2015, Tian and his coworkers successfully synthesized a novel kind of small hexagonal-phase Mn^{2+}-doped $NaYbF_4$:Er^{3+} UCNPs with bright and red emission by a modified codeposition method.[85] This method was more controllable and convenient than the hydrothermal or solvothermal methods used previously[82–84] (Figure 1.3g–l). Moreover, Tian *et al.* found that the dopant Mn^{2+} ions have a negligible effect on the phase structure since all the diffraction peaks of the samples still correspond to

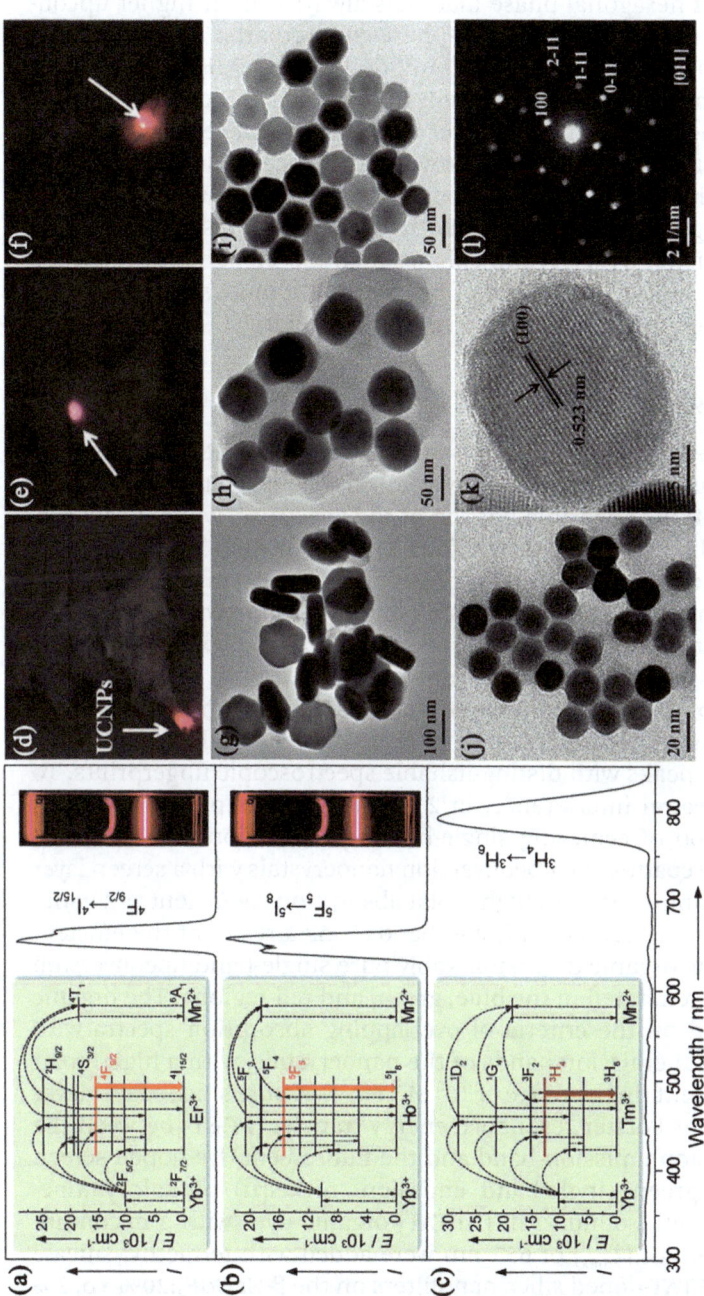

Figure 1.3 Room-temperature UC emission spectra of solutions containing: (a) KMnF₃:Yb/Er (18:2 mol%), (b) KMnF₃:Yb/Er (18:2 mol%) and (c) KMnF₃:Yb/Tm (18:2 mol%) nanocrystals in cyclohexane (insets: proposed energy transfer mechanisms and corresponding luminescent photos of the colloidal solutions). (Reproduced from ref. 82 with permission from John Wiley and Sons. Copyright © 2011 Wiley-VCH Verlag GmbH & Co. KGaA, Weinheim.) (d–f) *In vivo* upconversion luminescence animal imaging using Mn²⁺-doped NaYF₄:Yb³⁺/Er³⁺ (18:2 mol%) NPs. (Reproduced with permission from ref. 83.) (g–l) TEM images of Mn²⁺-doped NaYbF₄:Er³⁺ UCNPs obtained after heating for 1 h at 310 °C in the presence of 0, 10, 20, and 40 mol% Mn²⁺ dopant ions in (g), (h), (i), and (j), respectively. HRTEM and SEAD patterns of 40 mol% Mn²⁺ doping NaYbF₄:Er³⁺ UCNPs in (k) and (l). (Reproduced from ref. 85 with permission from the Royal Society of Chemistry.)

the pure hexagonal phase without admixture of cubic phase or other impurities. This finding was totally different from Zhao's and Hao's results, and it is well known that hexagonal-phase materials always exhibit higher upconversion efficiency relative to their cubic-phase counterparts. Therefore, Tian *et al.* concluded these hexagonal-phase Mn^{2+} doped $NaYbF_4$:Er^{3+} UCNPs may highly desirable in biomedicine, especially in bioimaging. Most recently, Rai *et al.*[86] reported significant enhancement in the red upconversion emission of Er^{3+} in $NaSc_{0.8}Er_{0.02}Yb_{0.18}F_4$ UCNPs through resonance energy transfer and plasmonic effect from Au NPs. Attachment of Au NPs on the surface of UCNPs gave two advantages: reduction in green band (through resonance energy transfer with efficiency 31.54%) and enhancement in red band (through the plasmonic effect). It gave a R/G ratio of nearly 20:1 (almost single-band red UC), which is quite promising for imaging applications.

1.2.2 Single-Band UCNPs

Single-band UCNPs have only one emission band under NIR excitation. Triply ionized lanthanide ions in UCNPs typically have emission line widths of 10–20 nm (FWHM) in the visible portion of the spectrum, which is approximately half the line width observed for QDs (25–40 nm) and much narrower than the line width observed for organic dyes (30–50 nm). This feature increases the number of distinguishable emission bands within a specific spectral bandwidth, enabling a large number of multiplexed detections. Although UCNPs have shown significant advantages over the traditional organic fluorophores or QD fluorescent biolabels, a problem remains: each lanthanide ion has a unique set of energy levels and generally exhibits a set of sharp emission peaks with distinguishable spectroscopic fingerprints. To minimize this spectral interference, in 2015 our group reported a general and simple method of achieving single-band upconversion emission with different colors by coating the upconversion nanocrystals with a screen layer containing an organic dye with a high molar absorption coefficient as a nanofilter to remove the unwanted emission bands.[87] As a result of the efficient reabsorption of the organic dye, remarkably pure single-band upconversion emissions can be generated in the blue, green, and red regions. The organic dyes were selected on the criteria of overlapping absorption spectra with only one of the dual emission bands of the nanocrystals, with a high molar absorption coefficient (in the range of $10^5\ M^{-1}\ cm^{-1}$). A pure silica spacer layer was used to prevent Förster resonance energy transfer (FRET) between the filtered upconversion emission band and the fluorescent dye-doped screen layer. To obtain green single-band emission, nickel(II) phthalocyanine-tetrasulfonic acid tetrasodium salt (NPTAT) organic dyes with a maximum absorption wavelength (λ_{max}) of 657 nm were added with tetraethyl silicate (TEOS) to form NPTAT-doped silica nanofilters on the β-$NaGdF_4$:20% Yb, 2% Er@$NaGdF_4$@SiO_2 NPs to filter the red emission band efficiently. With the β-$NaGdF_4$2:0% Yb, 2% Er@$NaGdF_4$@SiO_2@NPTAT-doped SiO_2 nanostructure, only the narrow green emission centered at 540 nm was observed, in stark contrast to the dual upconversion emission bands of β-NaGdF$_4$:20% Yb, 2% Er@NaGdF$_4$. To obtain the blue and red emission single-band UCNPs, β-NaGdF$_4$2:0% Yb, 0.2% Tm@NaGdF$_4$, and α-NaYbF$_4$1:0% Er@NaYF$_4$ nanocrystals with strong blue to red and red-to-green upconversion emission ratios were first prepared. After coating the pure SiO$_2$ layers, nanofilters doped with NPTAT and rhodamine B isothiocyanate were used to filter the red and green emissions to obtain the final blue and red single-band UCNPs, respectively. This general approach permits not only the removal of minor emission peaks away from the main peaks using appropriate nanofilters, but also the alternative removal of the main peaks to leave the minor peaks for single-band upconversion emission. For example, green emission single-band UCNPs can be obtained by coating the α-NaYbF$_4$1:0% Er@NaYF$_4$ nanocrystals with NPTAT-doped nanofilters. Besides green (550 nm) emission single-band UCNPs, remarkably pure single-band upconversion emissions can also be generated in the blue (480 nm) and red (650 nm) regions.

Significantly, in this work, we have demonstrated the use of single-band UCNPs for the multiplexed detection of three tumor biomarkers in both cultured human breast cancer cells and paraffin-embedded clinical tissue sections. The simultaneous quantification of estrogen receptor (ER), progesterone receptor (PR), and HER2 receptor expression levels in the breast cancer cell specimens correlated closely with the results of the traditional western blot method. Furthermore, the application of conjugated single-band UCNPs and quantitative spectroscopy may be more accurate than immunoenzyme-based immunohistochemical (IHC) methods for the simultaneous quantification of proteins present at low levels in cancer cells and tissue specimens. Thus, single-band UCNP-based technology may be well suited to the molecular profiling of tumor biomarkers *in vitro* and represent a clinically translational application of upconversion nanomaterials for cancer prognosis. The ability to detect multiple target proteins in small samples of cancer tissues could enable more effective therapeutic decisions when used in combination with regular IHC methods. The next step is to conduct large-scale clinical studies to establish protocols and practices for single-band UCNP-based molecular pathology.

In the work discussed above, single-band UCNPs were used for multispectral *in vitro* biodetection, but we believe that single-band UCNPs may also hold great promise in multispectral bioimaging which requires no overlapping signals between different imaging agents (Figure 1.4).

1.2.3   UCNPs Excited at Another Wavelength Range

Traditionally, highly efficient UCNPs require a NIR laser at a wavelength of about 980 nm as the excitation source since the sensitizer ion (Yb^{3+}) has a high absorption cross-section in its absorption band. Unfortunately the 980 nm laser light is strongly absorbed by water and biological specimens. Thus, 980 nm excitation has associated problems such as limited penetration depth and tissue damage due to sample overheating. Therefore,

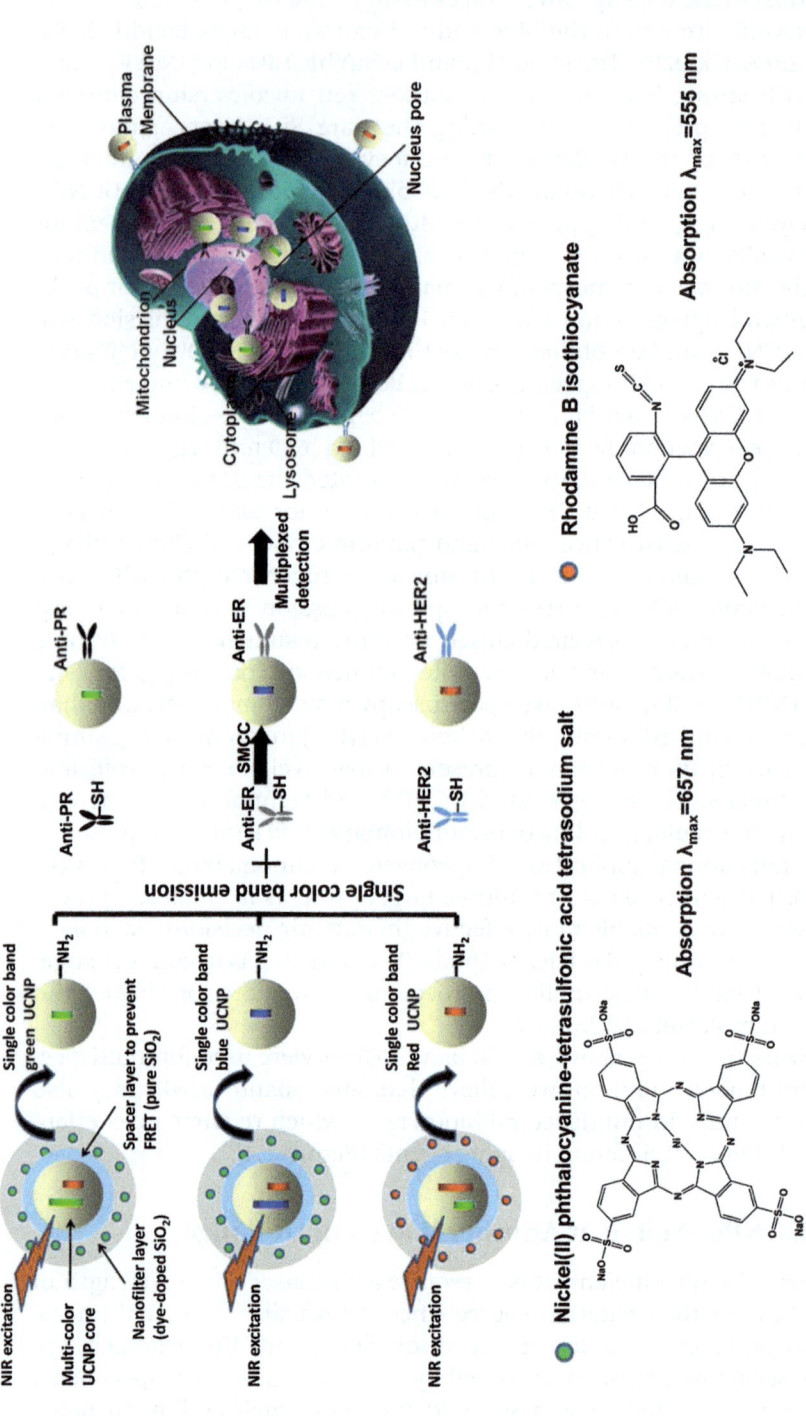

Figure 1.4 Schematic diagram of single-band UCNP fabrication for multiplexed detection. Surface amino modifications of the multilayer structure of green, blue, and red single-band UCNPs and conjugates with antibodies to the breast cancer biomarkers PR, ER, and HER2, respectively, for multiplexed *in situ* molecular mapping of breast cancer biomarkers. (Reproduced with permission from ref. 87.)

excitation band tuning of UCNPs into an appropriate range is also important for improving their performance. Zou *et al.* reported the concept of a UCNP where an organic NIR dye is used as an antenna to harvest the NIR photons within a broad band (740–850 nm) for the β-NaYF$_4$:Yb,Er NPs in which the upconversion occurs (Figure 1.5a). The overall upconversion by the dye-sensitized NPs is dramatically enhanced (by a factor of ~3300) as a result

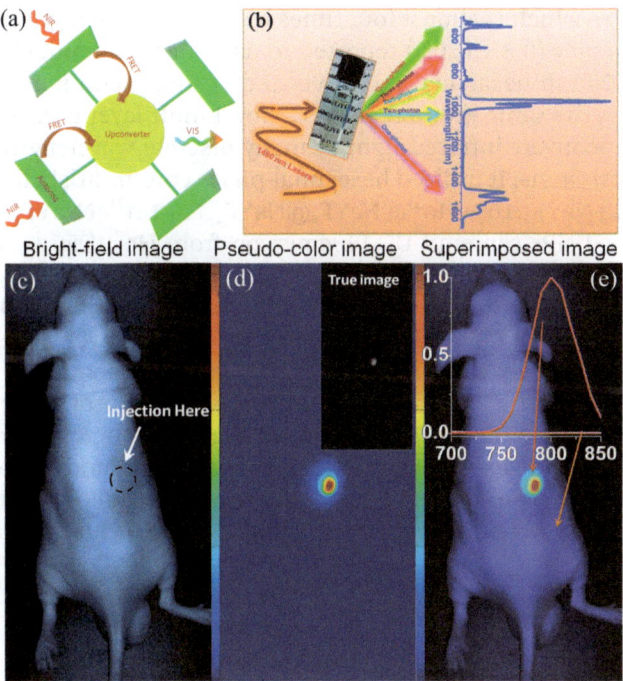

Figure 1.5 (a) Principal concept of the dye-sensitized nanoparticle. Antenna dyes (green) absorb NIR solar energy (red wavy arrows) and transfer it (brown arrows) to the nanoparticle core (in yellow), where upconversion occurs. Upconversion denotes a non-linear (on the incident radiation intensity) process in which the energies of two NIR quanta are summed to emit a quantum of higher energy in the green–yellow region (green–yellow wavy arrow). (Reproduced with permission from ref. 88.) (b) Intense visible and near infrared upconversion PL in colloidal LiYF$_4$:Er^{3+} nanocrystals under excitation at 1490 nm. (Reproduced with permission from G. Y. Chen, T. Y. Ohulchanskyy, A. Kachynski, H. Agren and P. N. Prasad, *ACS Nano*, 2011, **5**, 4981–4986. Copyright (2011) American Chemical Society.[89]) (c–e) *In vivo* whole-body image of a NaYbF$_4$:Yb^{3+}/Tm^{3+} injected nude mouse: (c) bright field image; (d) pseudocolor image obtained from true image (the inset black/white image); (e) superimposed image (bright field image and pseudocolor image) with the unmixed spectra of *in vivo* image (the inset chart) of UC signal and background as indicated by the arrows. (Reproduced with permission from Q. Q. Zhan, J. Qian, H. J. Liang, G. Somesfalean, D. Wang, S. L. He, Z. G. Zhang and S. Andersson-Engels, *ACS Nano*, 2011, **5**, 3744–3757. Copyright (2011) American Chemical Society.[91])

of increased absorptivity and overall broadening of the absorption spectrum of the upconverter.[88] Prasad's group also reported the intense upconversion PL in colloidal $LiYF_4:Er^{3+}$ nanocrystals under excitation at 1490 nm telecom wavelength (Figure 1.5b). The intensities of two- and three-photon anti-Stokes upconversion PL bands are higher than or comparable to that of the Stokes emission under excitation with low power density in the range 5–120 W cm^{-2}. The QY of the upconversion PL was measured to be as high as ~1.2 ± 0.1%, which is almost four times higher than the highest upconversion PL efficiency (0.3 ± 0.1%) reported to date for lanthanide-doped nanocrystals in 100 nm hexagonal $NaYF_4:Yb^{3+},Er^{3+}$ using excitation at 980 nm.[89] Later in 2015, the same group reported a novel multilayer core–shell design to broadly upconvert infrared light at many discrete wavelengths into visible or NIR emissions. It utilized hexagonal-phase core/multishell $NaYF_4$:10% $Er^{3+}@NaYF_4@NaYF_4$:10% $Ho^{3+}@NaYF_4@NaYF_4$:1% $Tm^{3+}@NaYF_4$ NPs. These core–multishell NPs can emit UC PL emission from Ho^{3+} ($^5F_5 \rightarrow {}^5I_8$, 625–685 nm range), Tm^{3+} ($^3H_4 \rightarrow {}^3H_6$, 760–860 nm range), and Er^{3+} ($^4S_{3/2} \rightarrow {}^4I_{15/2}$, 510–570 nm range) when excited at ~1120–1190 nm (due to Ho^{3+}), ~1190–1260 nm (due to Tm^{3+}), and ~1450–1580 nm (due to Er^{3+}), respectively. The excitation light could collectively cover a broad spectral range of about 270 nm in the NIR range.[90] Zhan *et al.* also demonstrated a new and promising excitation approach for better NIR-to-NIR UC PL *in vitro* or *in vivo* imaging employing a cost-effective 915 nm laser. This novel laser excitation method led to much less heating of the biological specimen and greater imaging depth in the animals or tissues because water absorption was quite low (Figure 1.5c).[91]

1.2.4 Nd^{3+} Sensitized UCNPs

Recently, researchers have paid more attention to the 800 nm excitation UCNPs. Notably, Nd^{3+} has multiple NIR excitation bands at wavelengths shorter than 980 nm, such as 730, 808, and 865 nm, corresponding to transitions from $^4I_{9/2}$ to $^4F_{7/2}$, $^4F_{5/2}$, and $^4F_{3/2}$, respectively. At all of these wavelengths, water absorption is lower, and the typical absorption coefficient is 0.02 cm^{-1} at 808 nm, in contrast to 0.48 cm^{-1} at 980 nm.[92] Consequently, the laser-induced heating effect, especially for biological tissues, is expected to be greatly minimized. Meanwhile, Nd^{3+} has a large absorption cross-section in the NIR region (1.2 × 10^{-19} cm^2 at 808 nm) compared to that of Yb^{3+} (1.2 × 10^{-20} cm^2 at 980 nm), which also benefits the efficiency of the Nd^{3+}-sensitized upconversion process.

Han *et al.* found when Nd^{3+}, Yb^{3+}, Er^{3+}/Tm^{3+} were codoped in $NaYF_4$, upconversion distinguishable by the naked eye can be observed upon the 800 nm excitation (Nd^{3+} has a broad excitation wavelength in the range of 790–810 nm). In these novel UCNPs, Nd^{3+} serves as an 800 nm photon sensitizer and Yb^{3+} as a bridging ion, with the energy transfer procedure performing as $Nd^{3+} \rightarrow Yb^{3+} \rightarrow Er^{3+}/Tm^{3+}$. However, the doping limit of the Nd^{3+} is only 1% because of concentration quenching.[93] Apparently, novel structure needs to be designed to increase the doping concentration of Nd^{3+}. In 2013, Liu *et al.* reported a new type of core–shell UCNPs (Figure 1.6a). Through spatially confined doping of Nd^{3+}, which can be effectively excited at 795 nm, they claimed that the active $NaYF_4$:Nd^{3+} shell layer can effectively prevent surface quenching of Yb^{3+} emission and can simultaneously promote the transfer of excitation energy to Yb^{3+} ions, which significantly enhances the

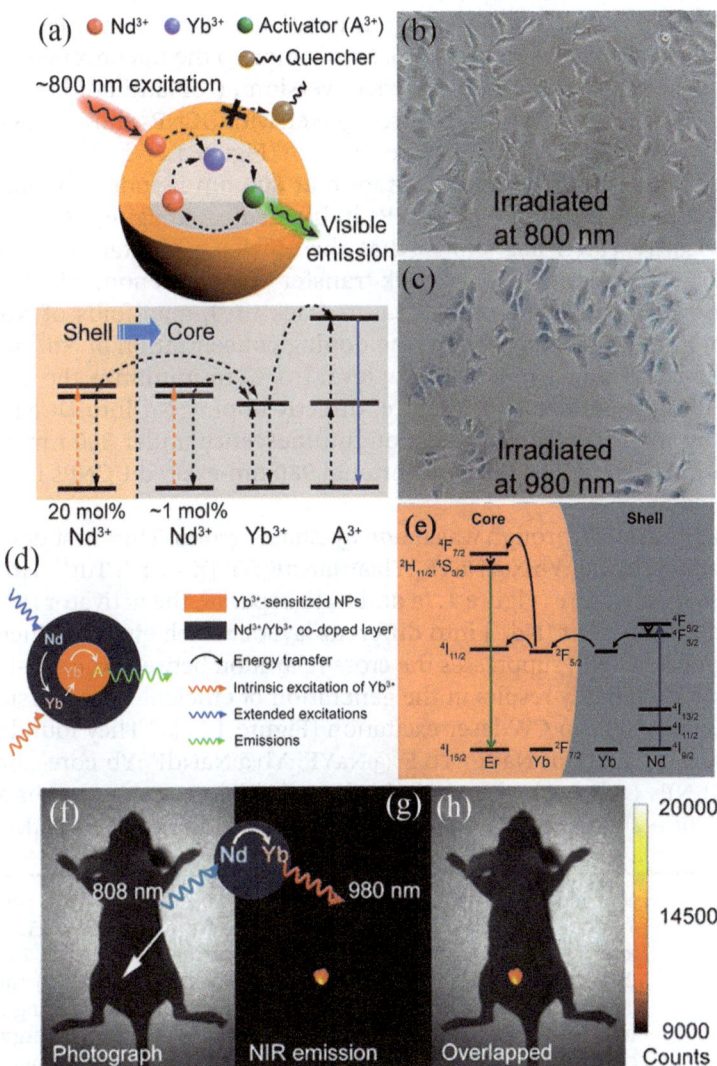

Figure 1.6 (a) Schematic design (top) and simplified energy level diagram (bottom) of a core–shell nanoparticle for photon upconversion under 800 nm excitation. Nd^{3+} ions doped in the core and shell layers serve as sensitizers to absorb the excitation energy and subsequently transfer it to Yb^{3+} ions. After energy migration from the Yb^{3+} ions to activator ions, activator emission is achieved *via* the Nd^{3+}-sensitization process. (b and c) Optical microscopy images of trypan blue-treated HeLa cells

upconversion; also, 800 nm is better than 980 nm for heat generation, as confirmed by cell irradiation experiment (Figure 1.6b and c).[94] Almost at the same time, Yan and his co-workers reported a similar core–shell structure (Figure 1.6d) with a similar design idea by doping Nd^{3+} in the shell to ensure successive $Nd^{3+} \rightarrow Yb^{3+} \rightarrow$ activator energy transfer. *In vivo* imaging of a nude mouse subcutaneously injected with UCNPs showed that comparable photon numbers can be measured when irradiated with 980 nm laser and 808 nm laser, respectively (Figure 1.6e). In addition to the upconversion process, these authors also performed downconversion PL of lanthanide-doped NPs. *In vivo* NIR imaging of a nude mouse injected with UCNPs showed high signal-to-noise ratio under 808 nm laser excitation (Figure 1.6f–h).[95]

These results indicate that excitation at 800 nm is indeed a good future direction for development of UCNPs in biomedical imaging. However, more theoretical research has shown that the efficiency limitation of Nd^{3+}-sensitized UCNPs is the "energy back-transfer" phenomenon, which can efficiently transfer energy from activators back to ^4I_J manifolds of Nd^{3+}, such as from Er^{3+} to Nd^{3+}. As a result, the doping concentration of Nd^{3+} in UCNPs must be constrained to a very low level (<1%) to minimize the quenching of the excitation energy. Therefore, directly doping Nd into UCNPs always results in much lower upconversion luminescence under 800 nm NIR laser irradiation than that of the conventional 980 nm-excited UCNPs in the same conditions.

In 2014, a breakthrough was made by Zhao's group. They first developed a well-defined $NaYF_4$:Yb,X@$NaYF_4$:Yb@$NaNdF_4$:Yb (X = Er^{3+}, Tm^{3+}, Ho^{3+}) core–shell–shell structure (Figure 1.7a and b) to separate the activator (Er^{3+}, Tm^{3+}, Ho^{3+}) and sensitizer (Nd^{3+}) into different layers, which enables efficient harvesting of NIR light, suppresses the cross-relaxation between the sensitizer and activator, and finally results in the generation of efficient upconversion emissions under 800 nm CW laser excitation (Figure 1.7c).[96] They found that the emission intensity of $NaYF_4$:Yb,Er@$NaYF_4$:Yb@$NaNdF_4$:Yb core–shell–shell (ErCSS) NPs (~16 nm) with optimized shell thickness shows enhancement factors of ~2000 (compared with the conventional $NaYF_4$:Yb,Er@$NaYF_4$

---

recorded after irradiation for 5 min at 800 and 980 nm, respectively (6 W cm^{-2}). (Reproduced with permission from X. J. Xie, N. Y. Gao, R. R. Deng, Q. Sun, Q. H. Xu and X. G. Liu, *J. Am. Chem. Soc.*, 2013, **135**, 12608–12611. Copyright (2013) American Chemical Society.[94]) (d) Integration scheme of $Nd^{3+} \rightarrow Yb^{3+}$ energy transfer process by introducing Nd^{3+}/Yb^{3+} codoped shell. The resulting $Nd^{3+} \rightarrow Yb^{3+} \rightarrow$ activator energy transfer could extend the effective excitation bands for conventional Yb^{3+}-sensitized UCNPs. (e) Energy transfer pathway from Nd^{3+} to Yb^{3+}-activated Er^{3+} upconversion emission in core–shell structured NPs under 808 nm excitation. (f–h) *In vivo* NIR imaging of a nude mouse injected with Er@ Nd NPs dispersed in water. (f) White-light photograph, (g) NIR image obtained with 808 nm excitation, and (h) overlapped image. Injection site is denoted by the white arrow. (Reproduced with permission from Y. F. Wang, G. Y. Liu, L. D. Sun, J. W. Xiao, J. C. Zhou and C. H. Yan, *ACS Nano*, 2013, **7**, 7200–7206. Copyright (2013) American Chemical Society.[95])

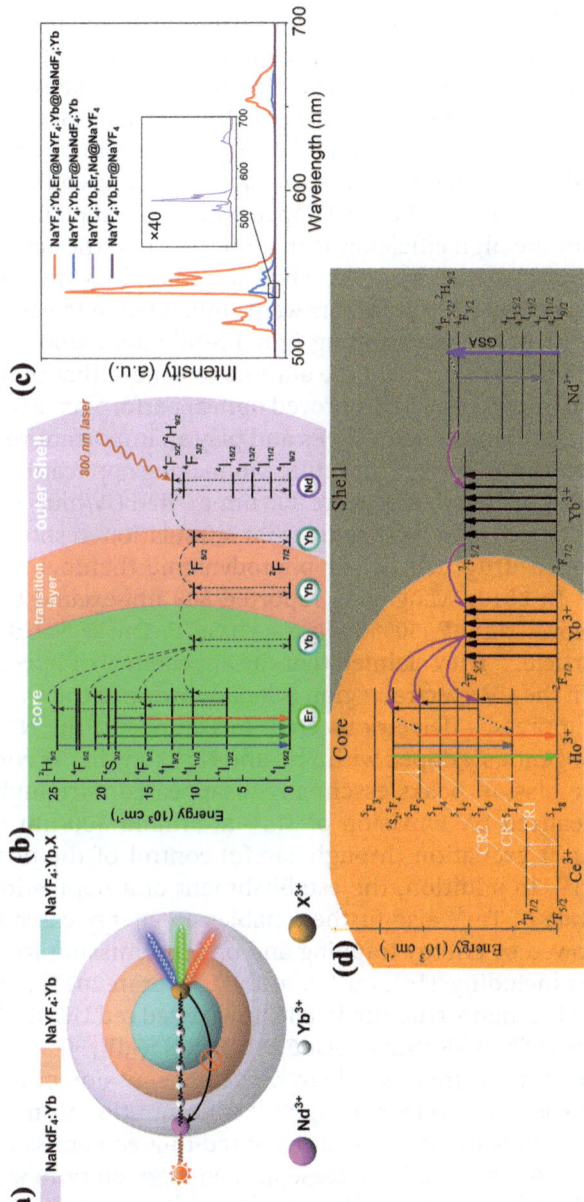

Figure 1.7 (a) Schematic illustration of the proposed energy-transfer mechanisms in the quenching-shield sandwich-structured UCNPs upon 800 nm excitation. (b) Proposed energy-transfer mechanisms in the quenching-shield sandwich NPs upon 800 nm diode-laser excitation. (c) Upconversion emission spectra of the synthesized quenching-shield sandwich NPs (red line), the Nd-coating core–shell NPs (blue line), the Nd/Yb/Er triply doped NPs (violet line), and the conventional Yb³⁺-sensitized UCNPs (dark violet line) under 800 nm excitation (0.5 W cm⁻²). (Reproduced from ref. 96 with permission from John Wiley and Sons. Copyright © 2013 Wiley-VCH Verlag GmbH & Co. KGaA, Weinheim.) (d) Energy level diagrams of Ce³⁺, Ho³⁺, Yb³⁺, and Nd³⁺ as well as the proposed mechanisms for the achievement of pure red UC luminescence under 980 or 808 nm laser excitation in Yb/Ho/Ce:NaGdF₄@Yb/Nd:NaYF₄ core–shell NPs. (Reproduced with permission from D. Chen, L. Liu, P. Huang, M. Ding, J. Zhong and Z. Ji, *J. Phys. Chem. Lett.*, 2015, 6, 2833–2840. Copyright (2015) American Chemical Society.⁹⁷)

core–shell (ErCS) NPs), ~100 (compared with the Nd/Yb/Er tri-doped UCNPs) and ~8 (compared with the Nd-coated core–shell UCNPs) under 800 nm irradiation. Notably, the upconversion emission of the 800 nm-excited ErCSS NPs (~16 nm) is ~7 times higher than that of the conventional 980 nm-excited ErCS NPs (~20 nm, hexagonal phase) at a low excitation power (0.1 W). Moreover, other lanthanide ions conventionally used for generating upconversion emissions, including Tm^{3+} and Ho^{3+}, can also serve as the activator in an Nd^{3+}-sensitized core–shell–shell system (Tm: TmCSS; Ho: HoCSS), because the efficient energy transfer from Yb^{3+} to these lanthanide ions can still be facilitated. Similar enhanced upconversion emission could be observed in Tm (~4.6 times) and Ho (~2 times) doped core–shell–shell nanocrystals. An explanation of the high efficiency is that the Nd^{3+} and the activators (Er^{3+}, Tm^{3+}, Ho^{3+}) were separated by a core–shell structure, in which the Nd^{3+} was confined in the shell and the activators were embedded in the core. Therefore, the non-radiative processes resulting from Er–Nd interactions are largely impeded. These characteristics lead the authors to believe that these 800 nm-excited core–shell–shell NPs with improved optical performance will outperform conventional 980 nm-excited UCNPs and play an important role in the development of fluorescent probes for future bioimaging applications.

Recent advances in Nd^{3+}-sensitized UCNPs are 800 nm-excited UV/blue and red light, which are considered good light sources for stimulation–response systems (such as controlled drug delivery or photodynamic therapy) and bioimaging, respectively. In 2013, Wang *et al.* reported 808 nm-excited UV/blue emission from a novel $NaYbF_4$:50% Nd@Na(Yb,Gd)F_4:Tm^{3+}@NaGdF_4 core–shell–shell nanoparticle.[98] They claimed that the appearance of upconversion emission peaks in the UV spectral region is largely owing to the core–shell structure, which suppresses deleterious cross-relaxations in the NPs. When the Tm^{3+} was homogenously doped with Yb^{3+} and Nd^{3+} ions in the core layer, UV upconversion emission peaks essentially disappeared. Strikingly, for the first time they realized UV emission of Tm^{3+} at around 300 nm in NPs by 808 nm diode laser excitation through careful control of the doping concentration of Tm^{3+}. In addition, the establishment of a population at a high-lying energy state of Tm^{3+} also further enables an energy cascade in the Gd sublattice followed by energy trapping and optical emission from common lanthanide ions including Tb^{3+}, Eu^{3+}, and Dy^{3+}. Later in 2015, they applied their core–shell–shell nanostructure in 800 nm-excited red UCNPs.[99] They have assessed a series of $NaYbF_4$:Nd@NaGdF_4:Yb/Er@NaGdF_4 NPs with varying Yb^{3+} concentrations in the inner shell layer. The red emission of Er^{3+} gradually dominated the spectra with increasing Yb^{3+} concentration from 18 to 78 mol%, which corresponds to intensity ratios of red-to-green emission from 0.5 to 0.7, 0.85, and 1.9. The steady increase in the red/green ratio was mainly induced by the $^4S_{3/2} + {}^2F_{7/2} \rightarrow {}^4I_{13/2} + {}^2F_{5/2}$ and $^4I_{13/2} + {}^2F_{5/2} \rightarrow {}^4F_{9/2} + {}^2F_{7/2}$ cross-relaxations at elevated Yb^{3+} concentrations. Further increasing Yb^{3+} concentration did not lead to noticeable improvement in red emission of Er^{3+}, probably due to the poor shell quality as a result of the fast shell deposition process in the absence of Gd^{3+} cofactors. They also examined relevant

NPs comprising high concentrations of Er^{3+} in the inner shell layer, which are known to induce $^4S_{3/2} + {}^4I_{9/2} \rightarrow {}^4F_{9/2} + {}^4F_{9/2}$ cross-relaxation between Er^{3+} ions for promoting the red emission of Er^{3+}. PL investigation showed that the emission spectra of Er^{3+} can only be marginally tuned by varying the Er^{3+} concentration, accompanied by a decrease in the overall emission intensity at high Er^{3+} content. Taken together, the optimal Yb/Er concentration was determined to be 78/2 mol%. In their *in vitro* experiment, they found the optical emission can be clearly observed when a 5 mm sample of pork muscle tissue is placed between the cells and the irradiating laser. The upconversion emission signals were still detectable after the thickness of the muscle tissue was increased to 10 mm. As a control experiment, classical $NaYF_4$:Yb/ Er (18/2 mol%)@$NaYF_4$ core–shell NPs can hardly be detected when the laser beam is blocked by a muscle tissue as thin as 5 mm. Almost at the same time, Chen *et al.* presented another strategy to achieve 808 nm-excited single-band red upconversion luminescence of Ho^{3+} *via* Ce^{3+} to change the red/green ratio in the $NaGdF_4$:Yb/Ho/Ce@Yb/Nd:$NaYF_4$ active-core@active-shell nanoarchitecture (Figure 1.7d).[97] The doping of Ce^{3+} plays a key role in the realization of Ho^{3+} single-band red luminescence *via* the efficient cross-relaxation processes between Ce^{3+} and Ho^{3+}: that is, Ho^{3+5}:$S_2/^5F_4 + Ce^{3+2}$:$F_{5/2} \rightarrow Ho^{3+5}$:$F_5 + Ce^{3+2}$:$F_{7/2}$ and Ho^{3+5}:$I_6 + Ce^{3+2}$:$F_{5/2} \rightarrow Ho^{3+5}$:$I_7 + Ce^{3+2}$:$F_{7/2}$. This nanoarchitecture enables the spatial separation between Ho^{3+} and Nd^{3+} and subsequently the high-content Nd^{3+} doping to efficiently improve upconversion luminescence, which might finally provide highly attractive luminescent biomarkers for bioimaging without the problematic overheating effect.

1.3 Lanthanide Downconversion Nanoparticles (DCNPs)

Lanthanide-based NIR downconversion nanoparticles (DCNPs) emerged a little later than UCNPs. The synthesis and surface modification procedures successfully used for UCNPs can also be well utilized on DCNPs. Ions of almost all the 15 lanthanide possess the downconversion property, and their emission wavelength can be located in the UV, visible, NIR I, NIR II and even mid-IR range.[100–102] Typically, 865–900 nm, 1060 nm, and 1300 nm from Nd^{3+} ($^4F_{3/2} \rightarrow {}^4I_{9/2}$, $^4F_{3/2} \rightarrow {}^4I_{13/2}$, and $^4F_{3/2} \rightarrow {}^4I_{15/2}$, respectively), 900–1000 nm from Yb^{3+} ($^2F_{5/2} \rightarrow {}^2F_{7/2}$), 1185 nm from Ho^{3+} ($^5I_6 \rightarrow {}^5I_8$), 1310 nm from Pr^{3+} ($^1G_4 \rightarrow {}^3H_5$), 1475 nm from Tm^{3+} ($^3H_4 \rightarrow {}^3F_4$), and 1525 nm from Er^{3+} ($^4I_{13/2} \rightarrow {}^4I_{15/2}$) are common NIR emission wavelengths in the spectral database of the lanthanide ions. However, the luminescence efficiency of these wavelengths varies greatly, and developing efficient DCNPs with proper emission wavelengths is an important issue in this field.

The first attempts to utilize DCNPs as biomedical imaging agents (NIR I or NIR II) have been made in the last decade. As early as 2002, Veggel *et al.* reported Nd^{3+}-doped LaF_3 NPs with both NIR I and NIR II emission under 514 nm laser excitation, as well as Er^{3+}- or Ho^{3+}-doped NPs. Because the interesting luminescence emission is in the telecommunication window (*i.e.*, Er^{3+} at 1530 nm, Nd^{3+} at 1330 nm, and Ho^{3+} at 966 nm and 1450 nm), they had considered them as promising materials for polymer-based optical components.[103] In 2006, inspired by Veggel's work, Wang *et al.* developed a simple method to synthesize LaF_3:Nd^{3+} in aqueous solution at low temperature, and have pointed out that this kind of DCNP would have potential application in biomedical imaging. The emission of Nd^{3+}-doped LaF_3 nanocrystals is located in NIR II under 802 nm laser excitation.[104] Later in 2008, NdF_3 NPs were synthesized in aqueous solution by a similar method as used to synthesize traditional Nd^{3+}-doped LaF_3 NPs. NdF_3 NPs showed no doping concentration quenching effect; moreover, the vibrational quenching caused by the O–H groups on the surfaces of the NdF_3 NPs can be suppressed after coating with silica shells. For deep-tissue imaging, mice were injected intramuscularly and intraperitoneally with 100 μL of NdF_3/SiO_2 NPs (1.0 μg mL^{-1}) into the thigh and abdominal cavity respectively. NIR signals (1050 nm) from the deep tissues of the thigh and abdominal cavity can both be clearly distinguished from the tissue autofluorescence under 730 nm excitation.[105] Later in 2013, Nd^{3+}-doped DCNPs based on a GdF_3 host matrix were realized by Mimun *et al.*[106] The GdF_3:Nd^{3+} NPs were small, with an average size of 5 nm, and formed stable colloids that lasted for several weeks without settling, enabling their use for several biomedical and photonic applications. Their excellent NIR properties, such as a nearly 11% QY the 1064 nm emission, make them ideal contrast agents and biomarkers for *in vitro* and *in vivo* NIR optical bioimaging. The nanophosphors, which were coated with poly(maleic anhydride-*alt*-1-octadicene) (PMAO), were implemented in cellular imaging, showing no significant cellular toxicity for concentrations up to 200 mg mL^{-1}. A proof-of-concept experiment for imaging through tissue was conducted by placing the GdF_3:Nd^{3+} NPs under varying thicknesses of pig skin, ranging from 0.67 to 5 mm. Emission spectra were collected through each thickness, and the 1064 nm emission was easily discernible even at the greatest tissue thickness of 5 mm. Furthermore, the incorporation of Gd into the nanocrystalline structure endowed these NPs with exceptional magnetic properties, making them ideal for use as magnetic resonance imaging (MRI) contrast agents. Almost at the same time, the same group reported their investigation of the downconversion absolute quantum yields measurement on the powder, PMAO-coated powder, and colloidal solution states of GdF_3:Nd^{3+} NPs.[107] The maximum total absolute downconversion QY of 10.2 ± 1.5% was measured for the GdF_3:Nd^{3+} nanophosphor powder at an excitation power density of 12.74 ± 2.0 W cm^{-2} at 800 nm excitation. Similarly, downconversion QYs of 5.02 ± 0.75% and 2.2 ± 0.33% were measured at an excitation power density of 5.3 ± 0.8 and 1.4 ± 0.2 W cm^{-2}, respectively. With the known QY 10% for IR-140 dye in the spectral range of 862–1013 nm at an excitation power of 150 mW under 800 nm excitation, a comparison method was also implemented to check the accuracy of the measurement. Comparison measurement for GdF_3:1% Nd^{3+} powder shows that the downconversion QY in GdF_3:1% Nd^{3+} is 5.8 ± 0.87% at 150 mW (4.77 W cm^{-2}) excitation under

800 nm, which is very close to the measured QY at 5.3 W cm^{-2} using integrating spheres for GdF$_3$:1% Nd^{3+} powder. Scaling of the downconversion emission spectra revealed that the 1064 nm emission from Nd^{3+} represents around 90% of the overall downconversion emission intensity. In addition, compared with the upconversion QY of 0.005 ± 0.0005% (at 150 W cm^{-2}) reported by van Veggel *et al.* for β-NaYF$_4$:20% Yb^{3+}/2% Er^{3+} of 8–10 nm sized particles, the downconversion QY for GdF$_3$:1% Nd^{3+} nanophosphor powder is 2000 times higher even at an excitation power density of 12.74 ± 2.0 W cm^{-2} at 800 nm excitation.[108] This shows that these particles have a higher QY within the biological window which yield more photon counts (information density) for bioimaging applications compared to UCNPs. Furthermore, the comparison method was implemented to measure the downconversion QY for colloidal GdF$_3$:1% Nd^{3+} with respect to the reported QY for the dye IR-140 to mimic the experimental conditions. Using the QY of 10% reported for the IR-140 at 150 mW (4.77 W cm^{-2}) excitation under 800 nm, the downconversion QY for GdF$_3$:1% Nd^{3+} at a concentration of 0.05 mg mL^{-1} was measured to be 1 ± 0.05%. Similarly, the QY of 1.5 ± 0.075% was measured for GdF$_3$:1% Nd^{3+} at 255 mW (8.28 W cm^{-2}). This verifies that the QY for colloidal GdF$_3$:1% Nd^{3+} is dependent on concentration and excitation power density, indicating that DCNPs yield more photon counts (information density) for bioimaging applications than UCNPs. The authors also have measured the NIR emission spectra obtained with and without the additional PMAO coating for the GdF$_3$:1% Nd^{3+} at an excitation power density of 12.74 ± 2.0 W cm^{-2}. Coating the NPs with PMAO does not significantly change the measured downconversion QY, which is important since the polymer coating is essential for making the particles biocompatible.

Although considerable research has been done on LaF$_3$:Nd^{3+}, GdF$_3$:Nd^{3+}, and NdF$_3$, morphology control of the NPs is still a big problem for this type of materials. Thanks to the well-developed synthesis methods for UCNPs, uniform and monodispersed NaReF$_4$:Nd^{3+} NPs (where Re = rare earth elements) have been obtained, and have attracted increasing attention in recent years.[109-111] In 2012, Prasad *et al.* reported highly efficient NaGdF$_4$:Nd^{3+}@ NaGdF$_4$ for NIR–NIR biomedical imaging.[112] Unlike LaF$_3$:Nd^{3+}, these novel DCNPs benefit from controllable morphology and even ~3 nm uniform NaGdF$_4$ shells have been successfully synthesized (Figure 1.8a). These DCNPs exhibited spectrally sharp, photostable, and large Stokes-shifted NIR PL at 900, 1050, and 1300 nm when excited at 740 nm (Figure 1.8b). The absolute QY of this NIR-to-NIR downconversion PL was evaluated to be as high as 40% for core–shell NaGdF$_4$:Nd^{3+}@NaGdF$_4$ NPs dispersed in hexane and 20% for ligand-free NaGdF$_4$:Nd^{3+}@NaGdF$_4$ NPs dispersed in water. The high luminescent efficiency in NPs was realized by effective suppression of non-radiative losses originating from surface passivation and cross-relaxation between Nd^{3+} dopants, as revealed by the PL steady-state and time-resolved studies. A facile high-contrast NIR-to-NIR imaging of HeLa cells and a nude mouse were demonstrated by the authors, utilizing excitation from an incoherent light source through observation of NIR PL at 900 nm (Figure 1.8c–e).

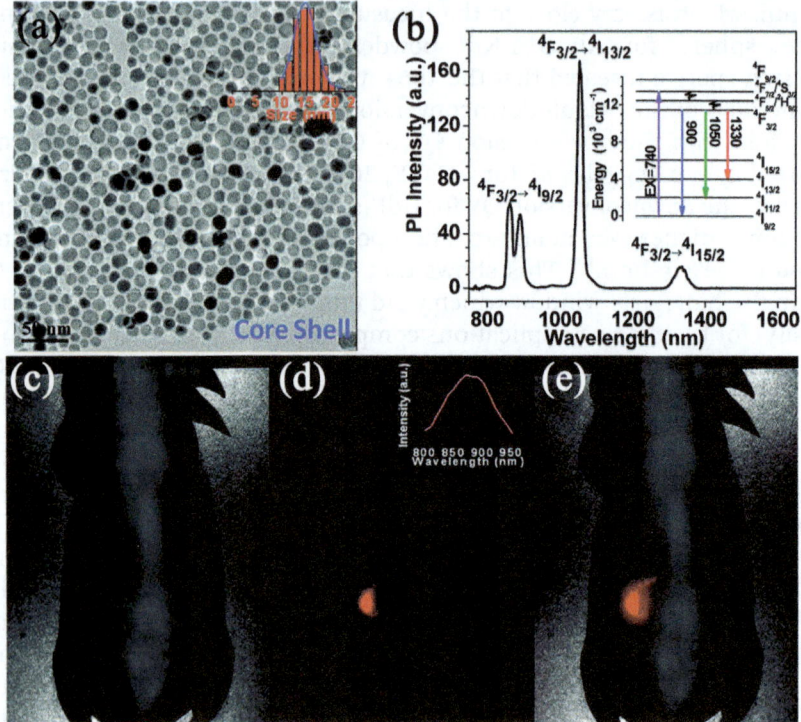

Figure 1.8   (a) TEM images of NaGdF$_4$:Nd^{3+}@NaGdF$_4$ core–shell nanocrystals. (b) PL spectrum of colloidal NaGdF$_4$:Nd^{3+} nanocrystals under laser excitation at 740 nm. The inset shows the mechanism of generation of the observed PL emissions. (c–e) *In vivo* whole-body image of a nude mouse subcutaneously injected with ligand-free NaGdF$_4$:Nd^{3+}@NaGdF$_4$ core–shell nanocrystals: (c) bright field image, (d) PL image, and (e) superimposed image (bright field image and spectrally unmixed PL image). Inset in (d) is the spectra of the NIR PL signals with a background taken from the injection site and noninjected area. (Reproduced with permission from G. Y. Chen, *et al.*, *ACS Nano*, 2012, **6**, 2969–2977. Copyright (2012) American Chemical Society.[112])

Recently, some attempts have been made to explore the downconversion optical property of the traditional UCNPs. Nagasaki *et al.* had reported the application of Y$_2$O$_3$:Yb^{3+},Er^{3+} as a UCNP biomedical imaging nanoprobe in 2008.[113] In 2011, they used the same kind of NPs for the downconversion bioimaging research. A strong NIR signal (1550 nm) can be observed 24 h after intravenous injection of DCNPs upon the irradiation of 980 nm laser.[114] Although this work is at an early stage, it also indicates that other lanthanide ions, such as Er^{3+}, may hold promise for NIR bioimaging. Er^{3+} possesses a stable energy level in the NIR range ($^4I_{13/2}$), which can emit NIR light in the wavelength range of 1450–1650 nm. In 2014, exciting work on the downconversion phenomenon of UCNPs was published by Moghe and his colleagues. They first developed a library of rare earth nanomaterials with tunable, discrete SWIR (short-wavelength infrared, 1000–2300 nm, including NIR II) emissions and proceeded to evaluate their optical performance for several clinical imaging applications including real-time, multispectral *in vivo* SWIR imaging.[115] The rare earth nanomaterials they used were $NaYF_4:Yb^{3+}$, Ln^{3+} (Ln = Er, Ho, Tm, or Pr)@$NaYF_4$ core–shell NPs. By doping the $NaYF_4$ core with Yb and one of several other rare earth elements, such as Er, Ho, Tm, or Pr, the emission properties of rare earth nanomaterials can be tailored in both the SWIR and visible ranges. The fluorescence of rare earth nanomaterials occurs following the resonant transfer of excitation energy from a sensitizer (Yb) to an activator dopant such as Er, Ho, Tm, or Pr. The relaxation from an excited state results in the generation of SWIR emissions that are unique to the specific rare earth activator. Hexagonal-phase $NaYF_4:Yb^{3+},Er^{3+}$ NPs were among the brightest SWIR-emitting rare-earth-doped phosphors (with an optical efficiency >1.1%), and therefore were chosen to illustrate not only the benefits of SWIR compared to conventional optical imaging methods but also the biomedical potential for SWIR-based imaging approaches. Compared to other SWIR emitters, the rare earth nanomaterials presented in this work are considerably more effective at generating SWIR emissions than single-walled carbon nanotubes (SWNTs) and IR-26, an organic SWIR dye. Only highly toxic lead-based QDs matched the SWIR emission power output of the rare earth nanomaterials. This group further investigated the attenuation properties of actual biological tissues in the SWIR by measuring the absorbance spectra of excised tissue obtained from a mouse exhibiting pigmented tumor lesions. The majority of tissue samples exhibit markedly low attenuation at 1000–1350 nm as well as at 1500–1650 nm, effectively extending the wavelength region of lowered attenuation within the second 'tissue-transparent window' of SWIR. Furthermore, strong absorbers in the tissue, such as melanin in tumor samples and hemoglobin in blood samples, exhibited strong attenuation in the visible range whereas attenuation in the SWIR was weak. Importantly, <0.4% of NIR I light penetrated through 0.5 cm of pigmented tumor tissue compared to ~80% transmittance achieved by SWIR, suggesting that NIR I has limited use for detecting optical probes in melanin-containing tissues. Furthermore, they determined the actual tissue penetration depth of SWIR compared to the current standard imaging wavelength region by measuring the intensity of SWIR and NIR I light through tissue phantoms composed of both scattering and absorbing agents. The intensity of the NIR I signal rapidly diminishes over increasing phantom depth, with complete signal loss occurring by 5 mm. In contrast, the high SWIR emission from the rare earth nanomaterial pellet saturates the camera's detector for phantom tissues at 5 mm or less. Notably, SWIR signal was detectable through 1 cm of phantom tissue, whereas no signal above background was seen using NIR I. In the *in vivo* SWIR imaging experiment, rare earth nanomaterials also showed good performance. Using a series of image-processing algorithms, SWIR signal intensities as a function of concentration of rare earth nanomaterials were found to exhibit a linear relationship in both tissue phantoms and subcutaneously injected mice, with a detection threshold at ~3 nM rare earth nanomaterials at an excitation power density of 0.14 W cm^{-2} and camera exposure time of ~50 ms per frame. In comparison to QDs and SWNTs with reported detection limits of ~5 nM and 6 nM respectively, rare earth nanomaterials can be detected at a lower concentration under comparable excitation. After injection, SWIR emissions were first identified in the tail vein (5 s) before clearing the vasculature to enter the heart and lungs (10 s). The beating of the heart in the chest of the mouse was visualized by pulsing SWIR emissions captured in real time (Figure 1.9). Over the course of 60 s, the SWIR signal became progressively more intense in organs such as the liver and spleen, which are part of the reticuloendothelial system that mediates nanoparticles. In contrast, the accompanying visible signal from UCNPs was notably absent, probably due to absorption and scattering losses caused by blood and tissue components.

Although great advances have been made by Moghe and his colleagues, 980 nm-excited lanthanide-based NIR nanomaterials, both UCNPs and DCNPs, suffer from the same problem: the water in biological structures

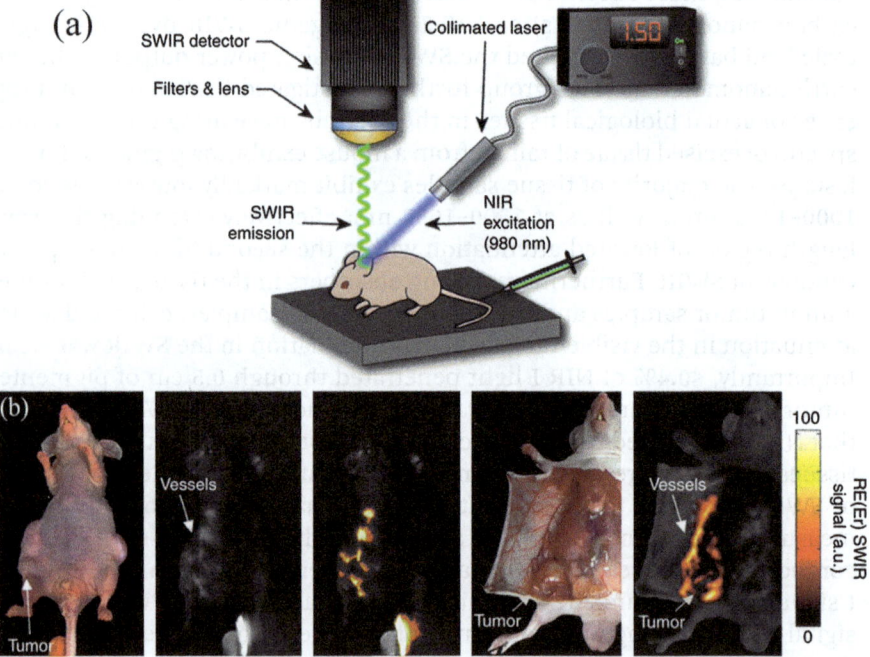

Figure 1.9   (a) Schematic of the portable SWIR imaging prototype, consisting of a room temperature-cooled InGaAs camera operating at a typical exposure time of 50 ms, adjustable filter mounts, a collimated laser with an output power density of 0.14 W cm^{-2}, and a neoprene rubber imaging surface. (b) Nude mice bearing melanoma xenografts were intravenously injected with rare earth nanomaterials and imaged near surrounding tumor regions before dissection from the ventral aspect. (Reproduced with permission from ref. 115.)

would overwhelmingly attenuate 980 nm light, and transform its energy into local heat which could damage cells and tissues. For efficient bioimaging, it is therefore essential to develop a novel lanthanide-based SWIR probe with excitation source optimization. For UCNPs, a solution has already been mentioned above: Nd^{3+}-sensitized UCNPs. The same idea can also be used for DCNPs. In 2014, our group reported a novel kind of β-NaGdF$_4$/Na(Gd,Yb) F$_4$:Er/NaYF$_4$:Yb/NaNdF$_4$:Yb C/S1/S2/S3 DCNPs as an efficient 800 nm NIR to 1525 nm SWIR probe for *in vivo* bioimaging.[116] C/S1/S2/S3 DCNPs were synthesized using the epitaxial seeded growth method, and are composed of the NaGdF$_4$ core (seed for epitaxial growth), a Na(Gd,Yb)F$_4$:Er shell (S1, the SWIR-emitting layer), a NaYF$_4$:Yb shell (S2, the energy transfer layer), and a NaNdF$_4$:Yb shell (S3, the energy absorption layer) (Figure 1.10). After absorbing 800 nm excitation energy (Nd^{3+}, $^4I_{9/2} \rightarrow {}^4F_{5/2}$), the S3 transfers its energy into the inner S2 layer ($Nd^{3+} \rightarrow Yb^{3+}$, $^2F_{7/2} \rightarrow {}^2F_{5/2}$). Energy is transferred within this layer by the codoped Yb^{3+} until the Er^{3+} in the S1 is sensitized ($Yb^{3+} \rightarrow Er^{3+}$, $^4I_{15/2} \rightarrow {}^4I_{11/2}$). This results in relaxation from the excited state of Er^{3+} *via* the release of a 1525 nm ($^4I_{13/2} \rightarrow {}^4I_{15/2}$) photon along with phonon vibration.

As the 1525 nm SWIR PL occurs following the resonant transfer of excitation energy from the Yb sensitizer to the Er activator, efficient resonant energy transfer theoretically requires a high concentration of the Yb sensitizer. Therefore, in order to realize the highly efficient 1525 nm SWIR PL, we needed to find an optimal doping concentration for the Yb sensitizer. In our

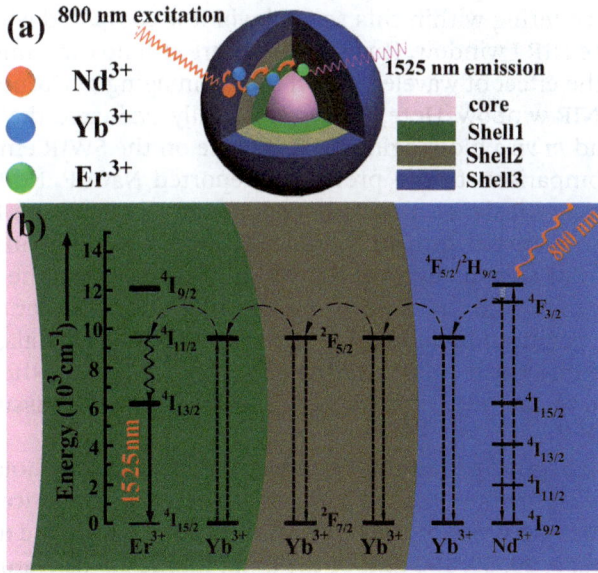

Figure 1.10   (a) Schematic design of the C/S1/S2/S3 DCNPs for 1525 nm luminescence. (b) Proposed energy transfer mechanisms in the multilayer core–shell DCNPs. (Reproduced from ref. 116 with permission from John Wiley and Sons. Copyright © 2014 Wiley-VCH Verlag GmbH & Co. KGaA, Weiheim.)

case, the doping ratio of Yb^{3+} and Er^{3+} can be tuned easily without changing the morphology due to the epitaxial seeded growth method. If we had instead used the $Na(Gd,Yb)F_4$:Er as the starting core of this core–shell nanostructure, the particle size of the $Na(Gd,Yb)F_4$:Er would have been uncontrollable when the Yb:Er doping ratio was changed, resulting in non-comparable spectral data because the PL properties depend on the size of the lanthanide-based NPs. With delicate adjustment of Yb:Er doping ratio, the best composition of the S1 layer was optimized to $NaYbF_4$:2Er to realize the most efficient SWIR emission. Similar experiments were also carried out to optimize the Yb and Nd dopant concentrations in the S2 and S3 layers. We finally found that the optimum composition for efficient 1525 nm SWIR emission under 800 nm excitation is $NaGdF_4/NaYbF_4$:2Er/$NaYF_4$:10Yb/$NaNdF_4$:10Yb.

We next studied the penetration depth of our C/S1/S2/S3 NPs. A pellet of the NPs was excited at 800 nm and the 1525 nm signal was imaged using an InGaAs camera through increasing thickness of tissue (pork slices). Notably, SWIR signals from C/S1/S2/S3 NPs were clearly detectable through 1.8 cm of pork tissue, much deeper than the 1.0 cm reported for the 1525 SWIR probe excited by a 980 nm laser under comparable excitation power density (~30 μW), further confirming that the 800 nm excitation source is more suitable for *in vivo* applications. As mentioned above, Moghe *et al.* demonstrated that 1525 nm SWIR transmits more effectively through tissue phantoms than 800 nm NIR light with the same intensity. That was the first experimental demonstration of the imaging advantages of SWIR due to the reduced tissue absorbance and scattering within this second window compared with that of an emitter in the NIR I window. However, there are still no experimental results to evaluate the effect of wavelength on the bioimaging performance around the second NIR window. Here, we systematically evaluated the dependence of *in vitro* and *in vivo* bioimaging performance on the SWIR emission wavelength in comparison to the previously reported $NaGdF_4$:Nd/$NaGdF_4$ NPs with 1060 nm SWIR signal, also upon 800 nm excitation (Figure 1.11a and b). For the 1060 nm wavelength, unlike the 1525 nm wavelength, no signal above background was seen using $NaGdF_4$:Nd/$NaGdF_4$ NPs when the tissue slices were thicker than 1 cm. It is worth mentioning that, in order to accurately compare the penetration depth of 1525 nm and 1060 nm signals, the emitted power of the rare earth pellet and the output power of the NIR source were first matched to have identical spectral intensity before the tissue phantoms were applied.

In addition to good penetration depth, low detection threshold concentration with high resolution is also an essential quality for an optical biomedical imaging agent. As a proof-of-concept experiment, we embedded different concentrations (1, 5, 20, 50, 100 nM) of the amphiphilic 1,2-distearoyl-*sn-glycero*-3-phosphoethanolamine-*N*-[carboxy-(polyethylene glycol)-2000] (DSPE-PEG-2000-COOH) modified C/S1/S2/S3 NPs and $NaGdF_4$:Nd/$NaGdF_4$ NPs into pork muscle tissue at varied depths (5, 8, 15 mm) to investigate the feasibility of bioimaging by a InGaAs camera. It can be seen that the SWIR signals of C/S1/S2/S3 NPs were detectable under 5 mm of pork tissue even when the

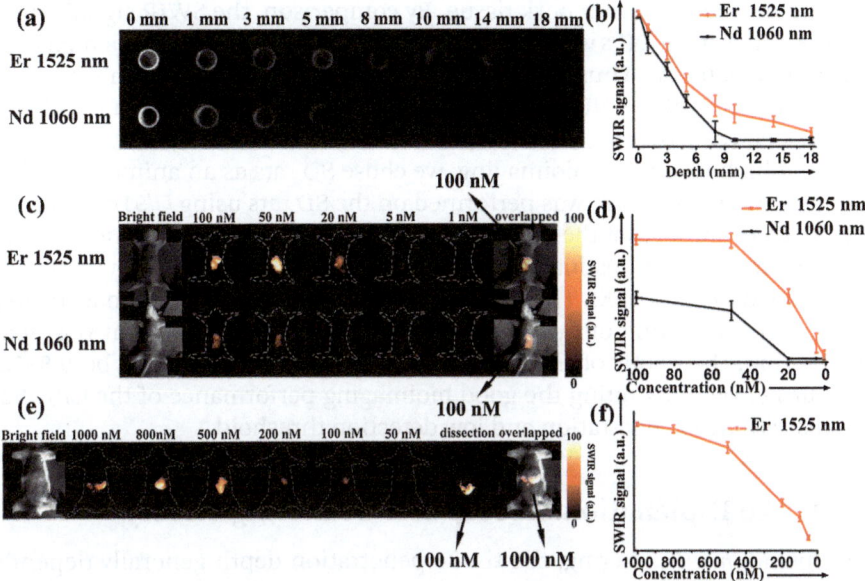

Figure 1.11 (a and c) Comparison of SWIR signal of C/S1/S2/S3 NPs (Er 1525 nm) and NaGdF$_4$:Nd/NaGdF$_4$ (Nd 1060 nm) in (a) pork muscle tissues of varying thickness and in (c) stomach of nude mice with different concentration. (e) SWIR images of C/S1/S2/S3 NCs in the stomach of SD rats with different concentration. (b, d and f) Corresponding signal intensity curve of (a), (c) and (e). Note that the intensities of the two systems at 0 mm were almost the same and appropriate filters were equipped. (Reproduced from ref. 116 with permission from John Wiley and Sons. Copyright © 2014 Wiley-VCH Verlag GmbH & Co. KGaA, Weiheim.)

concentration was as low as 1 nM. For tissue thickness 8 mm and 15 mm, the detection threshold increased to 20 nM and 50 nM, respectively. Nevertheless, the detection thresholds of NaGdF$_4$:Nd/NaGdF$_4$ NPs were 20 nM and 50 nM for 5 mm and 8 mm, respectively, much higher than those of the C/S1/S2/S3 NPs. For 15 mm thickness, little signal could be detected when NaGdF$_4$:Nd/NaGdF$_4$ NPs were embedded.

To evaluate the capability of these modified C/S1/S2/S3 NPs for *in vivo* imaging, bioimaging experiments were carried out on nude mice and SD rats. At first, we designed the bioimaging experiment with two groups of 6 week old nude mice. For group 1, 0.2 mL of different concentrations (1, 5, 20, 50, 100 nM) of water-soluble C/S1/S2/S3 NPs were introduced to the stomach by gastric syringe. For group 2, the same dose of water-soluble NaGdF$_4$:Nd/NaGdF$_4$ NPs was used instead. After 20 min digestion, SWIR images were taken of both groups upon 800 nm irradiation (0.2 W cm^{-2}) with appropriate filters. Bright 1525 nm SWIR signals were detected when C/S1/S2/S3 NPs were used (Figure 1.11c). Moreover, the 1525 nm SWIR signals were still detectable when the concentration of NPs was as low as 5 nM. Considering the depth of the stomach in nude mice (3–5 mm), these results were consistent with the detection thresholds obtained from pork tissue. By comparison, the SWIR signals from NaGdF$_4$:Nd/NaGdF$_4$ NPs were weaker and could be detected only at a high concentration (>50 nM). Thus, we demonstrated that the low detection threshold property of C/S1/S2/S3 NPs can be extended to *in vivo* imaging in a small-animal model. To further highlight the use of the C/S1/S2/S3 NPs for deep penetration and high-resolution bioimaging, we chose SD rats as an animal model. A similar gavage procedure was performed on the SD rats using C/S1/S2/S3 NPs. Figure 1.11e shows that these C/S1/S2/S3 NPs also exhibit bright fluorescence in the stomach and intestinal tract of the SD rats, with the detection threshold determined to be 100 nM. The outline of the stomach can be well imaged after dissection, suggesting our previous images were taken in the right position. In addition, the depth of the SD rat stomach was determined to be 0.8–1.2 cm, further demonstrating the good bioimaging performance of the C/S1/S2/S3 NPs with deep penetration and low detection threshold.

1.3.1   An Explanation: Absorption–Scattering Theory

For optical *in vivo* imaging, the tissue penetration depth generally depends on the absorption and scattering of the excitation and emission light. Bashkatov *et al.*[117] have proposed a theoretical model to calculate the penetration depth:

$$\delta = [3\mu_a(\mu_a + \mu_s')]^{-1/2}$$

where μ_a is the optical absorption extinction coefficient which depends on the wavelength (Figure 1.12a), μ_s' ($\sim\lambda^{-w}$) is the reduced scattering coefficient, and δ is the resulting penetration depth. The exponent (w) depends on the size and concentration of scatterers in the tissue and ranges from 0.22 to

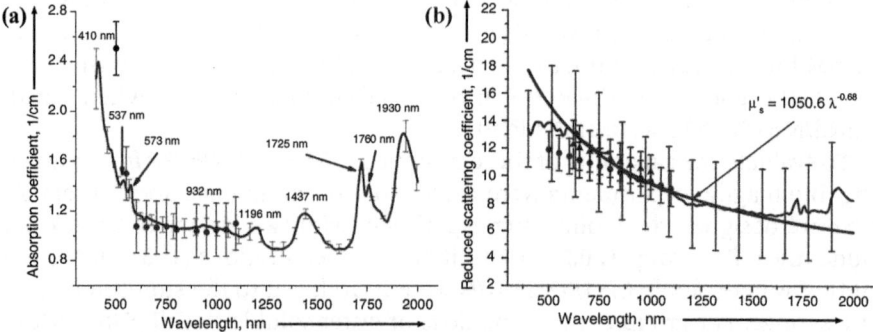

Figure 1.12   (a) The wavelength dependence of the absorption coefficients μ_a of human subcutaneous adipose tissue, showing the increased absorption of tissue in the SWIR compared to the NIR I and visible region. (b) The spectral dependence of the reduced scattering coefficient μ_s' of human subcutaneous adipose tissue, showing decreased scattering with increasing wavelength. The vertical lines show the standard deviation values.

1.68 for different tissues (Figure 1.12b). From this theoretical model and our results, we can conclude that water absorption is stronger for 1525 nm than for 1060 nm radiation, but 1060 nm radiation suffers from more tissue scatter than 1525 nm radiation. Obviously, in our case, the reduced scattering coefficient μ_s' was more important in determining the resulting penetration depth δ than the absorption extinction coefficient μ_a. We therefore suggest that the imaging performance of SWIR is significantly influenced by wavelength, which directly affects both μ_a and μ_s'. Moreover, which of the two plays the more important role cannot be ascertained on the basis of simulation analysis but depends on the application, in this case biomedical imaging.

1.3.2 NIR-IIa Window

In 2014, Dai *et al.*[118] explored a new biologically transparent sub-window in the 1.3–1.4 μm wavelength range (the NIR-IIa region) and performed non-invasive brain imaging in this window using SWNTs which emit in the whole 1000–1700 nm NIR II window. They resolved cerebral vasculature with a high spatial resolution of <10 μm at a depth of >2 mm in an epifluorescence imaging mode. Compared to previous NIR II work, they found that the 1.3–1.4 μm NIR-IIa window for *in vivo* imaging can further reduce tissue scattering by rejecting photons with a wavelength <1.3 μm. The truly non-invasive nature and dynamic capability of NIR-IIa imaging could allow cerebrovascular imaging with high spatial and temporal resolution to follow biological processes in the brain at the molecular scale.

The emission of lanthanide ions is not only abundant in the NIR-IIa, but also more spectrally pure than the emission of SWNTs. In 2015, García *et al.* demonstrated how the use of the particular emission band at 1340 nm of Nd^{3+} ions in SrF_2 NPs can be used for deep-tissue, autofluorescence-free, high-resolution *in vivo* imaging.[119] In their experiment, they found that food autofluorescence extends up to about 1100 nm upon 808 nm excitation, and this kind of infrared autofluorescence was found to exist in all the different diets available in their animal center. Nevertheless it is important to note here that both the fluorescence intensity and spatial distribution within the pellets were found to be strongly dependent on the particular pellet analyzed. Concerning the possible origin of this infrared luminescence, previous studies dealing with the visible autofluorescence of animal food have correlated it with the presence of plant components (in particular with the presence of alfalfa in the food pellets). García *et al.* also claim that the infrared food fluorescence they reported (extending up to 1100 nm) was generated by the presence of plant components, in particular by chlorophyll.

Up to this point, it is clear that high-contrast (autofluorescence-free) NIR *in vivo* imaging requires the use of luminescent materials that show intense fluorescence in the NIR-IIa region. The emission spectrum of Nd^{3+} (after 808 nm excitation) displays three emission bands centered at around 900, 1060, and 1340 nm corresponding respectively to the $^4F_{3/2} \rightarrow {}^4I_{9/2}$, $^4F_{3/2} \rightarrow {}^4I_{11/2}$ and $^4F_{3/2} \rightarrow {}^4I_{13/2}$ intra-4f electronic transitions of this ion. As mentioned above, the use of 900 nm and 1060 nm bands directly for fluorescence contrast have been reported. Nevertheless, it is clear that the use of these two bands does not ensure the complete removal of food autofluorescence, and these two bands are outside the NIR-IIa region. In fact, the removal of food autofluorescence would only be achieved if fluorescence images were recorded based on the 1340 nm fluorescence band. The principal reason why this band of Nd-doped NPs has not been previously used for *in vivo* imaging is its lower intensity when compared to the 900 and 1060 nm emission bands. Indeed, for most of the Nd^{3+}-doped crystals the fluorescence branching ratio of the $^4F_{3/2} \rightarrow {}^4I_{13/2}$ (1340 nm) transition is about three times lower than that of the $^4F_{3/2} \rightarrow {}^4I_{9/2}$ (900 nm) or $^4F_{3/2} \rightarrow {}^4I_{11/2}$ (1060 nm) transitions. In fact, generally speaking, only 15% of the radiative de-excitations generated from the $^4F_{3/2}$ metastable state of Nd ions is produced through the $^4F_{3/2} \rightarrow {}^4I_{13/2}$ fluorescence channel. This implies that using the 900 or 1060 nm bands would provide brighter images than those obtained from the 1340 nm emission. Thus, *in vivo* fluorescence imaging based on the 1340 nm weak emission would require the use of a host matrix providing large radiative de-excitation probabilities for Nd^{3+} ions. In this respect the SrF_2 host seems to be particularly interesting as, compared to other fluoride nanocrystals, it has been shown to provide the highest emission intensities for other trivalent rare earth ions. This superior performance is based on various causes. First of all, multiphonon relaxation is made inefficient in this host due to the low wavenumbers of the vibrational modes. Moreover, although the local site symmetry at the cationic sites is highly symmetric in the SrF_2 lattice (O_h), due to the necessity of charge compensation (Nd^{3+} replaces Sr^{2+}), non-centrosymmetric crystal field components are expected to be present around the Nd ions, thereby leading to partially allowed forced electric dipole transition and, hence, to bright fluorescence. Thus, in principle, $Nd:SrF_2$ NPs are expected to provide a highly intense 1340 nm emission, compared to other previously studied Nd:NPs. The QY of the as-synthesized $Nd:SrF_2$ NPs is estimated to be 0.9 ± 0.1. Indeed, this large QY can be now compared to that recently reported for Ag_2S QDs (QY = 0.15), which have also been used for high-brightness infrared fluorescence *in vivo* imaging.

García *et al.* then performed different infrared fluorescence imaging experiments by using an 808 nm diode laser as the excitation source and a Peltier-cooled InGaAs infrared camera for detection. Two different experimental configurations were adopted. In the "free-running" mode (FRM) no filters were used, such that the obtained fluorescence images accounted for the spatial distribution of emitted light integrated in the 900–1500 nm range. Fluorescence images were also obtained in the "food fluorescence free" mode (FFFM), by attaching a 1300 nm long-pass filter to the camera objective. Thus, in this mode images were constructed by recording only the 1300–1500 nm fluorescence range. The complete removal of food fluorescence by using the FFFM configuration is demonstrated in Figure 1.13. Figure 1.13a shows an optical image of a food pellet and an Eppendorf tube partially filled with the aqueous solution of $Nd:SrF_2$ NPs. As can be observed in Figure 1.13b,

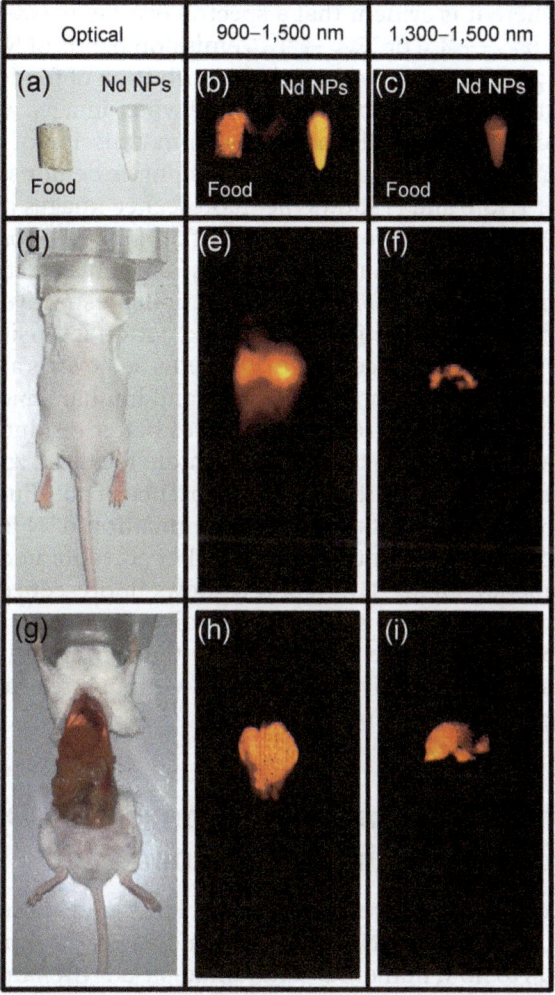

Figure 1.13 Left column: optical images of a mouse food pellet and an Eppendorf tube containing a colloidal solution of Nd:SrF$_2$ NPs (a), of a living mouse after intravenous administration through the retro-orbital venous sinus of 50 µL of a colloidal solution of Nd:SrF$_2$ NPs in phosphate buffered saline (d), and of the same mouse after being sacrificed and opened to obtain direct access to organs (g). Middle column: (b), (e) and (f) are corresponding fluorescence images of the three systems obtained under 808 nm illumination and recording the fluorescence in the 900–1500 nm range. Right column: (c), (f) and (i) are corresponding fluorescence images under 808 nm illumination but in this case the fluorescence intensity was recorded in the 1300–1500 nm range. (Reproduced from ref. 119 with kind permission from Springer Science + Business Media.)

both the food pellet and Nd:SrF$_2$ NPs appear in the FRM fluorescence image. This agrees well with the emission spectra included in an *ex vivo* imaging experiment, where it is evident that a spectral overlap between food fluorescence and the $^4F_{3/2} \rightarrow ^4I_{9/2}$ or $^4F_{3/2} \rightarrow ^4I_{11/2}$ emission bands of Nd^{3+} ions occurs. In contrast, Figure 1.13c shows how the contribution of food emission to the fluorescence image was completely removed when images were acquired in the FFFM mode. The suitability of the 1340 nm emission of Nd:SrF$_2$ NPs for autofluorescence-free *in vivo* and *ex vivo* imaging is demonstrated in Figure 1.13e–i. Figure 1.13(e and f) shows the fluorescence images of a nude mouse (Figure 1.13d) after intravenous injection of 50 µL of a phosphate buffered saline (PBS) solution containing Nd:SrF$_2$ NPs (at a concentration of 0.3 wt%), as obtained in the FRM and FFFM configurations, respectively. Both images were obtained 1 h after injection using an excitation intensity 0.5 W cm^{-2} at 808 nm. The FRM fluorescence image (Figure 1.13e) reveals a bright luminescence generated from the whole abdominal and lumbar zones. On the other hand, the FFFM image (Figure 1.13f) provides better resolution and greater contrast, showing fluorescence only from a more localized zone at the upper area of the abdomen. Due to the complete absence of autofluorescence background, Figure 1.13f suggests a strong accumulation of Nd:SrF$_2$ NPs in the liver and/or spleen. This point has been further corroborated by performing *ex vivo* experiments. Figure 1.13(h and i) shows fluorescence *ex vivo* images of the same mouse after being killed, as obtained in FRM and FFFM configurations, respectively. Figure 1.13i confirms that infrared fluorescence was mainly generated from the liver and spleen, without a relevant contribution from other organs such as lungs or kidneys.

1.4 Upconversion and Downconversion Dual-Mode Luminescence in One Nanoparticle

As is mentioned in the two previous sections, lanthanide-based NIR nanomaterials, both UCNPs and DCNPs, have become a specific topic of interest in recent years. Up to now, numerous lanthanide-based NIR nanomaterials have been developed, which can emit strong UC or DC luminescence by tuning different lanthanide dopants. However, it remains difficult to realize efficient nanoprobes with combined UC and DC dual-mode functions under a single NIR excitation and fulfill the excitation/emission wavelength requirements of *in vitro* and *in vivo* applications at the same time. Furthermore, for the UC process, because of well-established efficient UC luminescence, considerable efforts have been devoted to the synthesis of lanthanide-doped NaYF$_4$(NaGdF$_4$) NPs, where Yb^{3+}, acting as the sensitizer with a large absorption cross-section around 980 nm, is usually codoped along with the most common UC activator ions (Er^{3+}, Tm^{3+}, Ho^{3+}, Pr^{3+}, and Tb^{3+}) to produce strong visible and UV emissions. However, excitation light around 980 nm suffers from strong water absorption, as already discussed in the section on Nd^{3+}-sensitized UCNPs. Therefore, if one can engineer nanomaterial structures with both the above-mentioned Yb^{3+}–Nd^{3+}–RE^{3+} UC system and the Nd^{3+}-doped DC system, a Nd^{3+}-sensitized UC/DC dual-mode nanoprobe under a single excitation around 800 nm can be realized, with a low heat effect and highly efficient bioimaging function. In 2014, our group developed a strategy to fabricate uniform multilayer C/S1/S2/S3 β-$NGdF_4$:Nd/$NaYF_4$/$NaGdF_4$:Nd,Yb,Er/$NaYF_4$ NPs that were composed of the $NaGdF_4$:Nd core (DC) and the $NaGdF_4$:Nd,Yb,Er shell (S2) (UC) in order to achieve dual-mode luminescence under a single excitation around 800 nm.[120] This kind of NP could serve as a multiplexed luminescent biolabel for both *in vitro* and *in vivo* bioimaging applications. As shown in Figure 1.14a and b, this NP consists of four parts, each part having a specific role and working together to fulfill the dual-mode luminescence.

Nd^{3+} has an intense absorption cross-section at around 800 nm and highly efficient NIR-to-NIR DC luminescence (QY ~ 40%) could be obtained when an Nd^{3+} activator was used. Therefore, in our C/S1/S2/S3 NPs the $NaGdF_4$:Nd was constructed as a core for emitting the NIR DC luminescence under 800 nm excitation. This NIR-to-NIR DC luminescence is ideal for *in vivo* applications with deep light penetration, low light scattering, and heat effect because both the exciting and the emission wavelength are located in the NIR biological window. For the UC luminescence, unlike the typical single-sensitizer (Yb^{3+}) upconversion system, efficient 800 nm-excited NIR-to-visible UC emission of Er^{3+} is realized (in S2) by taking advantage of Nd^{3+} and Yb^{3+} ions as double sensitizers, which can be used for the efficient *in vitro* bioimaging with a low autofluorescence and reduced photo-damage effects. Furthermore, we also found that there is a competitive relationship between DC and UC due to the energy transfer, governed by the law of conservation of energy. Hence, in order to avoid energy transfer between the DC core and the UC S2, a $NaYF_4$ host layer (S1) was inserted. To ensure the photostability of the UC (S2), a $NaYF_4$ host layer (S3) was also fabricated as the outer layer of the particle to minimize sublattice defects and external deactivators, as it is well known that these energy traps can be a major deleterious factor in regard to luminescence of colloidal NPs.

Dual-mode UC/DC luminescence could be obtained by the cooperation these layers. Upon NIR excitation around 800 nm, the C/S1/S2/S3 NPs exhibit characteristic UC and DC emission peaks of Er^{3+} (525 nm ($^2H_{11/2} \rightarrow {}^4I_{15/2}$), 540 nm ($^4S_{3/2} \rightarrow {}^4I_{15/2}$), and 650 nm ($^4F_{9/2} \rightarrow {}^4I_{15/2}$)) and Nd^{3+} (862, 892 nm ($^4F_{3/2} \rightarrow {}^4I_{9/2}$)), respectively. Comparing the luminescence intensity evolution during our synthesis, the strongest UC emission was observed in $NaGdF_4$:Nd/$NaYF_4$/$NaGdF_4$:Nd,Yb,Er/$NaYF_4$ C/S1/S2/S3 NPs. This was about 20 times stronger than that of NPs without the passivation $NaYF_4$ layer (S3). In contrast, $NaGdF_4$:Nd/$NaYF_4$ has optimal DC emission, which is about 1.2 times stronger than that of C/S1/S2/S3 NPs. This result is reasonable because the 800 nm light source could be partially blocked and absorbed by the UC layer before reaching the DC core. It should be noted that $NaGdF_4$:Nd NPs can also exhibit very weak UC emission around 525 and 585 nm under 800 nm excitation, which can be attributed to the $^2K_{13/2}$, $^4G_{7/2} \rightarrow {}^4I_{9/2}$, and $^4G_{5/2}$, $^2G_{7/2} \rightarrow {}^4I_{9/2}$ transitions of Nd^{3+}. This weak UC emission of Nd^{3+} can be neglected after coating with the $NaGdF_4$:Nd,Yb,Er/$NaYF_4$ UC layers.

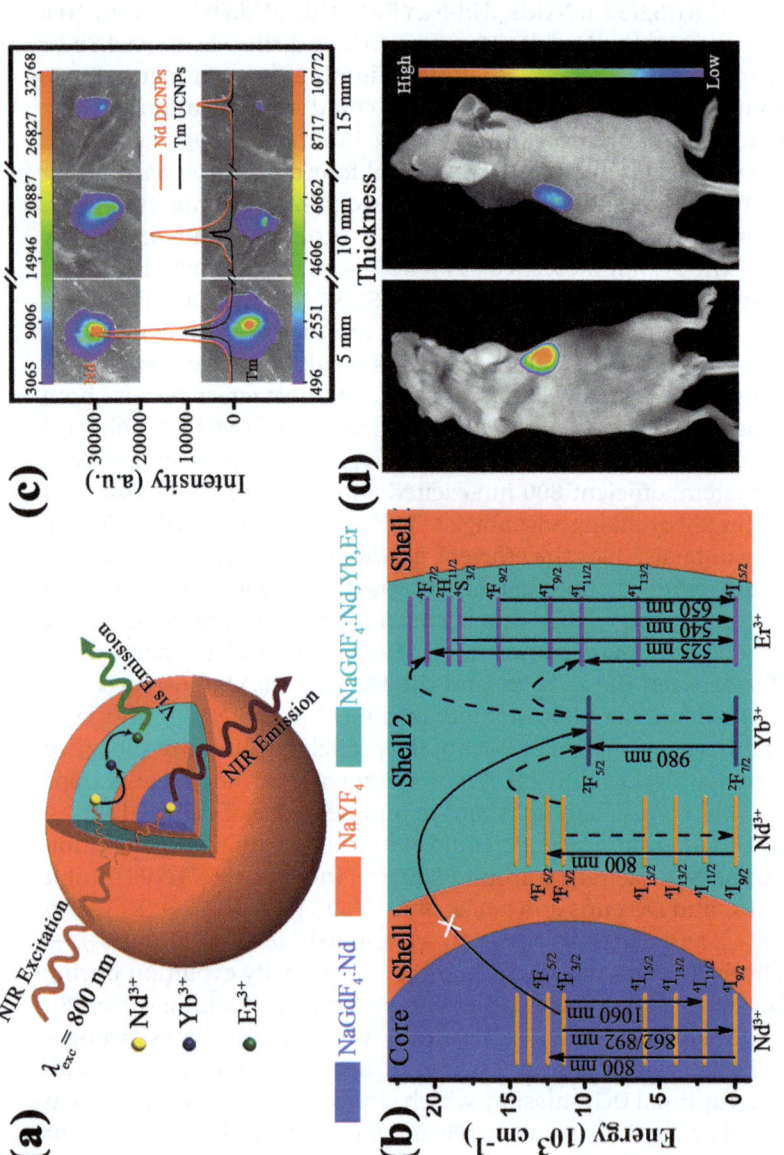

Figure 1.14 (a) General strategy to achieve the UC and DC dual-mode luminescence with multilayer C/S1/S2/S3 β-NGdF₄:Nd/NaYF₄/NaGdF₄:Nd,Yb,Er/NaYF₄ NPs. (b) Proposed energy transfer mechanisms in the multilayer core–shell NPs. (c) Merged bright-field images with NIR-to-NIR DC emission of NaGdF₄:Nd/NaYF₄/NaGdF₄:Nd,Yb,Er/NaYF₄ NPs under 800 nm excitation (top row) and UC emission of traditional NaGdF₄:Yb,Tm/NaYF₄ NPs under 980 nm excitation (bottom row). Intensities of the signals are also summarized here. (d) *In vivo* imaging of a nude mouse from the ventral aspect (left) and from the back (right) with water-soluble NaGdF₄:Nd/NaYF₄/NaGdF₄:Nd,Yb,Er/NaYF₄ NPs introduced into the stomach. (Reproduced with permission from ref. 120.)

To determine whether these C/S1/S2/S3 NPs can be used for cellular imaging by using 800 nm-excited green UC emission, we have performed *in vitro* cellular bioimaging using human lymphocytes (as a cell suspension). After incubation with 0.2 mg mL^{-1} NaGdF$_4$:Nd/NaYF$_4$/NaGdF$_4$:Nd,Yb,Er/NaYF$_4$ C/S1/S2/S3 NPs in PBS (pH 7.4) for 3 h at 37 °C, the unbound NPs were washed away, and the living cells were imaged using 800 nm excitation. Cellular uptake of the NPs can be clearly observed from the merged bright-field and UC signal of cells. Local spectral analysis of overall cell staining confirms that our NPs are the origin of the cellular luminescence signal.

As a proof-of-concept experiment, we embedded the modified NaGdF$_4$:Nd/ NaYF$_4$/NaGdF$_4$:Nd,Yb,Er/NaYF$_4$ C/S1/S2/S3 NPs into pork muscle tissue at varying depths (0–15 mm) to investigate the feasibility of bioimaging by a modified *in vivo* imaging system. As shown in Figure 1.14c, the NPs can be detected even at a depth of 15 mm under an excitation power density of approximately 1 W cm^{-2}. In identical experimental settings, however, when traditional NIR (980 nm)-to-NIR (808 nm) UC NaGdF$_4$:Yb,Tm/NaYF$_4$ was used as a biomarker, the signals are much weaker than that of the C/S1/S2/S3 NPs at the same tissue depth, especially in deep muscle tissue (>10 mm). To demonstrate the capability of the NIR (800 nm)-to-NIR (860–895 nm) DC for *in vivo* imaging, 0.2 mL of 5 mg mL^{-1} water-soluble NPs was introduced to the stomach of nude mice by gastric syringe. A clear high-contrast luminescence image was observed, with almost no autofluorescence observed elsewhere (Figure 1.14d). It is noteworthy that the DC signal from C/S1/S2/S3 NPs can be detected even from the back of the mouse, suggesting that the 800 nm-excited low-heat-effect UC and DC dual-mode nanoprobe can not only be used for NIR (800 nm)-to-visible (540 nm) *in vitro* bioimaging but also shows great penetration capability in the NIR I window.

Within a year, the same idea was also shared by other groups, such as Qin *et al.* who reported sub-10 nm BaLuF$_5$:Yb^{3+},Er^{3+}@BaLuF$_5$:Yb^{3+} NPs which can upconvert and downconvert 980 nm light simultaneously.[121] Chaudhuri *et al.* designed and synthesized NaGdF$_4$:Nd^{3+}, Yb^{3+}, Tm^{3+} nanophosphors with combined dual-mode DC and upconversion UC PL upon 800 nm excitation. A broad range of PL peaks covering NIR I/NIR II (860–900, 1000, and 1060 nm), and visible emission including blue (475 nm), green (520 and 542 nm), and yellow (587 nm) light were exhibited by NaGdF$_4$:Nd^{3+}, Yb^{3+}, Tm^{3+} NPs after excitation at 800 nm.[122] These results prove that such dual-mode luminescence NPs can open the door to engineering the excitation and emission wavelengths of UC/DC NPs and provide a new tool for a wide variety of applications in the fields of bioanalysis and bioimaging.

1.5 Conclusion

In the last decade, NIR biomedical imaging based on lanthanide nanomaterials has played an important role in biotechnology due to its intrinsic advantages over the traditional imaging probes. They benefit from high photostability, non-blinking properties, and low toxicity, but they also suffer from shortcomings. The absorption coefficient of UCNPs is quite low, which results in a low QY (<1%). As far as we know, the low absorption coefficient is limited by the low 980 nm absorption cross-section of Yb^{3+}, which is even lower in the case of Er^{3+} and Tm^{3+}. As mentioned above, many efforts have been made to improve the QY of UCNPs, such as changing the excitation wavelength. However, further improvements are still needed. There is only a handful of lanthanide DCNPs, which are not even well studied yet. Among the reported lanthanide-based DCNPs, the QY of Nd^{3+} is much higher than any other lanthanide ions such as Er^{3+}, Tm^{3+}, and Pr^{3+}. More advanced nano-structures must be designed to reduce the crystal defects or modify the crystal field of the host. What is more, in order to obtain highly efficient lanthanide nanoprobes, increasing the doping concentration without lowering the QY is also something to think about.

Acknowledgements

The work was supported by the NSFC (Grant No. 21322508, 21210004) and the China National Key Basic Research Program (973 Project) (No. 2013CB934100, 2012CB224805).

References

1. A. Poddar, S. C. Gedam and S. J. Dhoble, *J. Lumin.*, 2013, **143**, 579–582.
2. W. C. W. Chan and S. M. Nie, *Science*, 1998, **281**, 2016–2018.
3. M. Y. Han, X. H. Gao, J. Z. Su and S. Nie, *Nat. Biotechnol.*, 2001, **19**, 631–635.
4. H. Uh and S. Petoud, *C. R. Chim.*, 2010, **13**, 668–680.
5. S. Petoud, S. M. Cohen, J. C. G. Bunzli and K. N. Raymond, *J. Am. Chem. Soc.*, 2003, **125**, 13324–13325.
6. J. Zhang, P. D. Badger, S. J. Geib and S. Petoud, *Angew. Chem., Int. Ed.*, 2005, **44**, 2508–2512.
7. F. Auzel, *Chem. Rev.*, 2004, **104**, 139–173.
8. L. Esterowi, A. Schnitzl, J. Noonan and J. Bahler, *Appl. Opt.*, 1968, **7**, 2053.
9. S. T. Liu and R. B. Maciolek, *J. Electron. Mater.*, 1974, **3**, 864.
10. G. S. Maciel, L. D. Menezes, A. S. L. Gomes, C. B. deAraujo, Y. Messaddeq, A. Florez and M. A. Aegerter, *IEEE Photonics Technol. Lett.*, 1995, **7**, 1474–1476.
11. H. Berthou and C. K. Jorgensen, *Opt. Lett.*, 1990, **15**, 1100–1102.
12. E. Downing, L. Hesselink, J. Ralston and R. Macfarlane, *Science*, 1996, **273**, 1185–1189.
13. M. F. Joubert, *Opt. Mater.*, 1999, **11**, 181–203.
14. R. C. Stoneman and L. Esterowitz, *Opt. Lett.*, 1992, **17**, 816–818.
15. J. A. Hutchinson and T. H. Allik, *Appl. Phys. Lett.*, 1992, **60**, 1424–1426.
16. E. M. Pacheco and C. B. Dearaujo, *Chem. Phys. Lett.*, 1988, **148**, 334–336.

17. M. A. Chamarro and R. Cases, *J. Lumin.*, 1988, **42**, 267–274.
18. R. S. Quimby, M. G. Drexhage and M. J. Suscavage, *Electron. Lett.*, 1987, **23**, 32–34.
19. D. C. Yeh, W. A. Sibley and M. J. Suscavage, *J. Appl. Phys.*, 1988, **63**, 4644–4650.
20. S. V. Eliseeva and J. C. G. Bunzli, *Chem. Soc. Rev.*, 2010, **39**, 189–227.
21. H. Kobayashi, M. Ogawa, R. Alford, P. L. Choyke and Y. Urano, *Chem. Rev.*, 2010, **110**, 2620–2640.
22. H. S. Mader, P. Kele, S. M. Saleh and O. S. Wolfbeis, *Curr. Opin. Chem. Biol.*, 2010, **14**, 582–596.
23. L. Cheng, K. Yang, Y. G. Li, J. H. Chen, C. Wang, M. W. Shao, S. T. Lee and Z. Liu, *Angew. Chem., Int. Ed.*, 2011, **50**, 7385–7390.
24. M. Haase and H. Schafer, *Angew. Chem., Int. Ed.*, 2011, **50**, 5808–5829.
25. F. Wang and X. G. Liu, *Chem. Soc. Rev.*, 2009, **38**, 976–989.
26. J. S. Chivian, W. E. Case and D. D. Eden, *Appl. Phys. Lett.*, 1979, **35**, 124–125.
27. J. C. Boyer, M. P. Manseau, J. I. Murray and F. van Veggel, *Langmuir*, 2010, **26**, 1157–1164.
28. G. A. Hebbink, J. W. Stouwdam, D. N. Reinhoudt and F. van Veggel, *Adv. Mater.*, 2002, **14**, 1147–1150.
29. F. Wang, Y. Han, C. S. Lim, Y. Lu, J. Wang, J. Xu, H. Chen, C. Zhang, M. Hong and X. Liu, *Nature*, 2010, **463**, 1061–1065.
30. F. Wang, R. R. Deng, J. Wang, Q. X. Wang, Y. Han, H. M. Zhu, X. Y. Chen and X. G. Liu, *Nat. Mater.*, 2011, **10**, 968–973.
31. F. Wang, J. Wang and X. Liu, *Angew. Chem., Int. Ed.*, 2010, **49**, 7456–7460.
32. Q. Liu, Y. Sun, T. Yang, W. Feng, C. Li and F. Li, *J. Am. Chem. Soc.*, 2011, **133**, 17122–17125.
33. Y. W. Zhang, X. Sun, R. Si, L. P. You and C. H. Yan, *J. Am. Chem. Soc.*, 2005, **127**, 3260–3261.
34. W. Zheng, S. Zhou, Z. Chen, P. Hu, Y. Liu, D. Tu, H. Zhu, R. Li, M. Huang and X. Chen, *Angew. Chem., Int. Ed.*, 2013, **52**, 6671–6676.
35. F. Wang and X. G. Liu, *J. Am. Chem. Soc.*, 2008, **130**, 5642–5643.
36. F. Wang, D. K. Chatterjee, Z. Q. Li, Y. Zhang, X. P. Fan and M. Q. Wang, *Nanotechnology*, 2006, **17**, 5786–5791.
37. Z. Q. Li and Y. Zhang, *Angew. Chem., Int. Ed.*, 2006, **45**, 7732–7735.
38. J. H. Zeng, J. Su, Z. H. Li, R. X. Yan and Y. D. Li, *Adv. Mater.*, 2005, **17**, 2119–2123.
39. G. S. Yi and G. M. Chow, *J. Mater. Chem.*, 2005, **15**, 4460–4464.
40. G. S. Yi, H. C. Lu, S. Y. Zhao, G. Yue, W. J. Yang, D. P. Chen and L. H. Guo, *Nano Lett.*, 2004, **4**, 2191–2196.
41. S. Heer, K. Kompe, H. U. Gudel and M. Haase, *Adv. Mater.*, 2004, **16**, 2102–2105.
42. S. Heer, O. Lehmann, M. Haase and H. U. Gudel, *Angew. Chem., Int. Ed.*, 2003, **42**, 3179–3182.
43. J. C. Boyer, L. A. Cuccia and J. A. Capobianco, *Nano Lett.*, 2007, **7**, 847–852.

44. H. X. Mai, Y. W. Zhang, L. D. Sun and C. H. Yan, *J. Phys. Chem. C*, 2007, **111**, 13721–13729.
45. H. X. Mai, Y. W. Zhang, R. Si, Z. G. Yan, L. D. Sun, L. P. You and C. H. Yan, *J. Am. Chem. Soc.*, 2006, **128**, 6426–6436.
46. J. C. Boyer, F. Vetrone, L. A. Cuccia and J. A. Capobianco, *J. Am. Chem. Soc.*, 2006, **128**, 7444–7445.
47. R. Si, Y. W. Zhang, L. P. You and C. H. Yan, *Angew. Chem., Int. Ed.*, 2005, **44**, 3256–3260.
48. F. Zhang and D. Y. Zhao, *ACS Nano*, 2009, **3**, 159–164.
49. F. Zhang, J. Li, J. Shan, L. Xu and D. Y. Zhao, *Chem. Eur. J.*, 2009, **15**, 11010–11019.
50. L. W. Yang, H. L. Han, Y. Y. Zhang and J. X. Zhong, *J. Phys. Chem. C*, 2009, **113**, 18995–18999.
51. G. F. Wang, Q. Peng and Y. D. Li, *J. Am. Chem. Soc.*, 2009, **131**, 14200–14201.
52. R. F. Qin, H. W. Song, G. H. Pan, X. Bai, B. Dong, S. H. Xie, L. N. Liu, Q. L. Dai, X. S. Qu, X. G. Ren and H. F. Zhao, *Cryst. Growth Des.*, 2009, **9**, 1750–1756.
53. H. Schafer, P. Ptacek, H. Eickmeier and M. Haase, *Adv. Funct. Mater.*, 2009, **19**, 3091–3097.
54. Y. Wei, F. Q. Lu, X. R. Zhang and D. P. Chen, *J. Alloys Compd.*, 2008, **455**, 376–384.
55. Z. Q. Li and Y. Zhang, *Nanotechnology*, 2008, **19**, 345606.
56. F. Zhang, Y. Wan, T. Yu, F. Q. Zhang, Y. F. Shi, S. H. Xie, Y. G. Li, L. Xu, B. Tu and D. Y. Zhao, *Angew. Chem., Int. Ed.*, 2007, **46**, 7976–7979.
57. L. Y. Wang and Y. D. Li, *Chem. Mater.*, 2007, **19**, 727–734.
58. C. H. Liu and D. P. Chen, *J. Mater. Chem.*, 2007, **17**, 3875–3880.
59. X. Wang, J. Zhuang, Q. Peng and Y. D. Li, *Inorg. Chem.*, 2006, **45**, 6661–6665.
60. L. Y. Wang and Y. D. Li, *Nano Lett.*, 2006, **6**, 1645–1649.
61. R. X. Yan and Y. D. Li, *Adv. Funct. Mater.*, 2005, **15**, 763–770.
62. X. Wang, J. Zhuang, Q. Peng and Y. D. Li, *Nature*, 2005, **437**, 121–124.
63. C. K. Jayasankar, K. U. Kumar, V. Venkatramu, P. Babu, T. Troster, W. Sievers and G. Wortmann, *J. Lumin.*, 2008, **128**, 718–720.
64. K. S. Yang, F. Zheng, R. N. Wu, H. S. Li and X. Y. Zhang, *J. Rare Earths*, 2006, **24**, 162–166.
65. A. Patra, C. S. Friend, R. Kapoor and P. N. Prasad, *Chem. Mater.*, 2003, **15**, 3650–3655.
66. A. Patra, C. S. Friend, R. Kapoor and P. N. Prasad, *J. Phys. Chem. B*, 2002, **106**, 1909–1912.
67. X. Qin, T. Yokomori and Y. G. Ju, *Appl. Phys. Lett.*, 2007, **90**, 073104.
68. G. F. Wang, Q. Peng and Y. D. Li, *Acc. Chem. Res.*, 2011, **44**, 322–332.
69. L. H. Fischer, G. S. Harms and O. S. Wolfbeis, *Angew. Chem., Int. Ed.*, 2011, **50**, 4546–4551.
70. L. Cheng, C. Wang and Z. Liu, *Nanoscale*, 2013, **5**, 23–37.
71. F. Wang, D. Banerjee, Y. Liu, X. Chen and X. Liu, *Analyst*, 2010, **135**, 1839–1854.

72. J. Zhou, Z. Liu and F. Li, *Chem. Soc. Rev.*, 2012, **41**, 1323–1349.
73. H. Zijlmans, J. Bonnet, J. Burton, K. Kardos, T. Vail, R. S. Niedbala and H. J. Tanke, *Anal. Biochem.*, 1999, **267**, 30–36.
74. S. A. Hilderbrand, F. W. Shao, C. Salthouse, U. Mahmood and R. Weissleder, *Chem. Commun.*, 2009, 4188–4190.
75. L. J. Zhou, Z. J. Gu, X. X. Liu, W. Y. Yin, G. Tian, L. Yan, S. Jin, W. L. Ren, G. M. Xing, W. Li, X. L. Chang, Z. B. Hu and Y. L. Zhao, *J. Mater. Chem.*, 2012, **22**, 966–974.
76. D. K. Chatteriee, A. J. Rufalhah and Y. Zhang, *Biomaterials*, 2008, **29**, 937–943.
77. M. Nyk, R. Kumar, T. Y. Ohulchanskyy, E. J. Bergey and P. N. Prasad, *Nano Lett.*, 2008, **8**, 3834–3838.
78. A. Xia, Y. Gao, J. Zhou, C. Y. Li, T. S. Yang, D. M. Wu, L. M. Wu and F. Y. Li, *Biomaterials*, 2011, **32**, 7200–7208.
79. J. Zhou, Y. Sun, X. X. Du, L. Q. Xiong, H. Hu and F. Y. Li, *Biomaterials*, 2010, **31**, 3287–3295.
80. A. Xia, M. Chen, Y. Gao, D. M. Wu, W. Feng and F. Y. Li, *Biomaterials*, 2012, **33**, 5394–5405.
81. H. Y. Xing, W. B. Bu, Q. G. Ren, X. P. Zheng, M. Li, S. J. Zhang, H. Y. Qu, Z. Wang, Y. Q. Hua, K. L. Zhao, L. P. Zhou, W. J. Peng and J. L. Shi, *Biomaterials*, 2012, **33**, 5384–5393.
82. J. Wang, F. Wang, C. Wang, Z. Liu and X. G. Liu, *Angew. Chem., Int. Ed.*, 2011, **50**, 10369–10372.
83. G. Tian, Z. J. Gu, L. J. Zhou, W. Y. Yin, X. X. Liu, L. Yan, S. Jin, W. L. Ren, G. M. Xing, S. J. Li and Y. L. Zhao, *Adv. Mater.*, 2012, **24**, 1226–1231.
84. S. J. Zeng, Z. G. Yi, W. Lu, C. Qian, H. B. Wang, L. Rao, T. M. Zeng, H. R. Liu, H. J. Liu, B. Fei and J. H. Hao, *Adv. Funct. Mater.*, 2014, **24**, 4051–4059.
85. D. P. Tian, D. L. Gao, B. Chong and X. Z. Liu, *Dalton Trans.*, 2015, **44**, 4133–4140.
86. M. Rai, S. K. Singh, A. K. Singh, R. Prasad, B. Koch, K. Mishra and S. B. Rai, *ACS Appl. Mater. Interfaces*, 2015, **7**, 15339–15350.
87. L. Zhou, R. Wang, C. Yao, X. M. Li, C. L. Wang, X. Y. Zhang, C. J. Xu, A. J. Zeng, D. Y. Zhao and F. Zhang, *Nat. Commun.*, 2015, **6**, 6938.
88. W. Q. Zou, C. Visser, J. A. Maduro, M. S. Pshenichnikov and J. C. Hummelen, *Nat. Photonics*, 2012, **6**, 560–564.
89. G. Y. Chen, T. Y. Ohulchanskyy, A. Kachynski, H. Agren and P. N. Prasad, *ACS Nano*, 2011, **5**, 4981–4986.
90. W. Shao, G. Y. Chen, T. Y. Ohulchanskyy, A. Kuzmin, J. Damasco, H. L. Qiu, C. H. Yang, H. Agren and P. N. Prasad, *Adv. Opt. Mater.*, 2015, **3**, 575–582.
91. Q. Q. Zhan, J. Qian, H. J. Liang, G. Somesfalean, D. Wang, S. L. He, Z. G. Zhang and S. Andersson-Engels, *ACS Nano*, 2011, **5**, 3744–3757.
92. S. Singh, R. Smith and L. Van Uitert, *Phys. Rev. B*, 1974, **10**, 2566–2572.
93. J. Shen, G. Chen, A.-M. Vu, W. Fan, O. S. Bilsel, C.-C. Chang and G. Han, *Adv. Opt. Mater.*, 2013, **1**, 644–650.

94. X. J. Xie, N. Y. Gao, R. R. Deng, Q. Sun, Q. H. Xu and X. G. Liu, *J. Am. Chem. Soc.*, 2013, **135**, 12608–12611.

95. Y. F. Wang, G. Y. Liu, L. D. Sun, J. W. Xiao, J. C. Zhou and C. H. Yan, *ACS Nano*, 2013, **7**, 7200–7206.

96. Y. T. Zhong, G. Tian, Z. J. Gu, Y. J. Yang, L. Gu, Y. L. Zhao, Y. Ma and J. N. Yao, *Adv. Mater.*, 2014, **26**, 2831–2837.

97. D. Chen, L. Liu, P. Huang, M. Ding, J. Zhong and Z. Ji, *J. Phys. Chem. Lett.*, 2015, **6**, 2833–2840.

98. H. Wen, H. Zhu, X. Chen, T. F. Hung, B. Wang, G. Zhu, S. F. Yu and F. Wang, *Angew. Chem., Int. Ed.*, 2013, **52**, 13419–13423.

99. Q. Ju, X. Chen, F. J. Ai, D. F. Peng, X. D. Lin, W. Kong, P. Shi, G. Y. Zhu and F. Wang, *J. Mater. Chem. B*, 2015, **3**, 3548–3555.

100. T. Hirai, S. Hashimoto, S. Sakuragi and N. Ohno, *Chem. Phys. Lett.*, 2007, **446**, 138–141.

101. Q. Y. Zhang, G. F. Yang and Z. H. Jiang, *Appl. Phys. Lett.*, 2007, **91**, 051903.

102. R. Lisiecki, E. Augustyn, W. Ryba-Romanowski and M. Żelechower, *Opt. Mater.*, 2011, **33**, 1630–1637.

103. J. W. Stouwdam and F. van Veggel, *Nano Lett.*, 2002, **2**, 733–737.

104. F. Wang, Y. Zhang, X. Fan and M. Wang, *J. Mater. Chem.*, 2006, **16**, 1031–1034.

105. X.-F. Yu, L.-D. Chen, M. Li, M.-Y. Xie, L. Zhou, Y. Li and Q.-Q. Wang, *Adv. Mater.*, 2008, **20**, 4118–4123.

106. L. C. Mimun, G. Ajithkumar, M. Pokhrel, B. G. Yust, Z. G. Elliott, F. Pedraza, A. Dhanale, L. Tang, A. L. Lin, V. P. Dravid and D. K. Sardar, *J. Mater. Chem. B*, 2013, **1**, 5702–5710.

107. M. Pokhrel, L. C. Mimun, B. Yust, G. A. Kumar, A. Dhanale, L. Tang and D. K. Sardar, *Nanoscale*, 2014, **6**, 1667–1674.

108. J.-C. Boyer and F. C. J. M. van Veggel, *Nanoscale*, 2010, **2**, 1417–1419.

109. L. Xu, S. Y. Zhang and J. Q. Xu, *Laser Phys. Lett.*, 2010, **7**, 303–306.

110. A. Bednarkiewicz, D. Wawrzynczyk, M. Nyk and W. Strek, *Opt. Mater.*, 2011, **33**, 1481–1486.

111. A. Bednarkiewicz, D. Wawrzynczyk, M. Nyk and W. Strek, *Appl. Phys. B*, 2010, **103**, 847–852.

112. G. Y. Chen, T. Y. Ohulchanskyy, S. Liu, W. C. Law, F. Wu, M. T. Swihart, H. Agren and P. N. Prasad, *ACS Nano*, 2012, **6**, 2969–2977.

113. M. Kamimura, D. Miyamoto, Y. Saito, K. Soga and Y. Nagasaki, *J. Photopolym. Sci. Technol.*, 2008, **21**, 183–187.

114. M. Kamimura, N. Kanayama, K. Tokuzen, K. Soga and Y. Nagasaki, *Nanoscale*, 2011, **3**, 3705–3713.

115. D. J. Naczynski, M. C. Tan, M. Zevon, B. Wall, J. Kohl, A. Kulesa, S. Chen, C. M. Roth, R. E. Riman and P. V. Moghe, *Nat. Commun.*, 2013, **4**, 2199.

116. R. Wang, X. M. Li, L. Zhou and F. Zhang, *Angew. Chem., Int. Ed.*, 2014, **53**, 12086–12090.

117. A. N. Bashkatov, E. A. Genina, V. I. Kochubey and V. V. Tuchin, *J. Phys. D: Appl. Phys.*, 2005, **38**, 2543–2555.

118. G. S. Hong, S. Diao, J. L. Chang, A. L. Antaris, C. X. Chen, B. Zhang, S. Zhao, D. N. Atochin, P. L. Huang, K. I. Andreasson, C. J. Kuo and H. J. Dai, *Nat. Photonics*, 2014, **8**, 723–730.
119. I. Villa, A. Vedda, I. X. Cantarelli, M. Pedroni, F. Piccinelli, M. Bettinelli, A. Speghini, M. Quintanilla, F. Vetrone, U. Rocha, C. Jacinto, E. Carrasco, F. S. Rodriguez, A. Juarranz, B. del Rosal, D. H. Ortgies, P. H. Gonzalez, J. G. Sole and D. J. García, *Nano Res.*, 2015, **8**, 649–665.
120. X. Li, R. Wang, F. Zhang, L. Zhou, D. Shen, C. Yao and D. Zhao, *Sci. Rep.*, 2013, **3**, 3536.
121. Y. L. Zhang, X. H. Liu, Y. B. Lang, Z. Yuan, D. Zhao, G. S. Qin and W. P. Qin, *J. Mater. Chem. C*, 2015, **3**, 2045–2053.
122. X. W. Zhang, Z. Zhao, X. Zhang, D. B. Cordes, B. Weeks, B. S. Qiu, K. Madanan, D. Sardar and J. Chaudhuri, *Nano Res.*, 2015, **8**, 636–648.

CHAPTER 2

Near Infrared Quantum Dots for Bioimaging

YI LIN[a] AND DAI-WEN PANG*[a]

[a]Key Laboratory of Analytical Chemistry for Biology and Medicine (Ministry of Education), College of Chemistry and Molecular Sciences, State Key Laboratory of Virology, and Wuhan Institute of Biotechnology, Wuhan University, Wuhan, 430072, P. R. China
*E-mail: dwpang@whu.edu.cn

2.1 Introduction

Seeing is believing, and imaging is the most important and straightforward way to discover information hidden from our naked eye. For scientists in biological or biomedical fields, imaging is a technique for monitoring and recording both structural and functional information from subcellular organelles or interconnected organs, enabling the discovery of biomolecular mechanisms and the diagnosis of various diseases with sufficient sensitivity and accuracy.[1,2]

Quantum dots (QDs), also known as semiconductor nanocrystals, are semiconductors whose excitons are quantum confined in all three spatial dimensions and exhibit quantum mechanical properties.[3–5] QDs have attracted significant attention in biomedical research and applications during the past decade because they are characterized by broad excitation spectra, narrow and symmetric emission peaks, ultrahigh brightness, great stability, and resistance to photobleaching. Furthermore, the emissions of QDs are closely related to their size and highly tunable.[6,7]

RSC Nanoscience & Nanotechnology No. 40
Near Infrared Nanomaterials: Preparation, Bioimaging and Therapy Applications
Edited by Fan Zhang
© The Royal Society of Chemistry 2016
Published by the Royal Society of Chemistry, www.rsc.org

Near infrared (NIR) (wavelength > 650 nm) QD-based biological imaging has advantages including lower absorption and relatively low autofluorescence, resulting in greater penetrating depth and lower background.[8] With the development of synthesis techniques for NIR QDs, bioimaging based on high-quality but toxic NIR QDs, especially for deep tissue biological imaging, have been of great interest in recent years.

2.2 NIR QDs

2.2.1 Structures and Properties

Since the excitons are confined in all three spatial dimensions, the electrons of QDs are quantized to certain energies, similar to those of small molecules (Figure 2.1).[9–12] The bandgap decreases as the size of the QDs increases. As more and more atoms are bound together, the discrete energy levels merge into energy bands.

Based on the quantum confinement effect arising from their small dimensions, the bandgap of QDs can be easily tuned. Composite or structural variation can also change the bandgap. Since fluorescence occurs when an excited electron relaxes to the ground state, the excitation and emission of QDs can be easily modulated by changing the their diameters, composites, or structures, resulting in broad absorption spectra with narrow, symmetric, and tunable emission spectra spanning the NIR region.[10–14] For example, by controlling the size of synthesized CdSe QDs (blue), InP QDs (green), and InAs QDs (red), fluorescence spectra can be modulated within a range of 400 nm–2 μm (Figure 2.2).

Compared to conventional organic dyes, QDs are characterized by large Stokes shifts. This allows overlap of the excitation and the emission spectra to be effectively avoided (Figure 2.3).[15] Furthermore, multicolor QD-labeled real samples can be excited with a single light source, enabling multichannel detection or imaging.[16–18]

QDs typically have high quantum yield (QY) and high resistance to photobleaching.[15,19] For example, although Alexa dye 488 is more stable than other organic dyes, its photostability is much lower than that of QDs (Figure 2.4).[19] When 3T3 cell nuclei and microtubules were labeled either with QDs emitting at 630 nm or with dye Alexa 488, the fluorescence of the QDs remain unchanged, while the Alexa 488 signal weakened rapidly with elapsed time.

2.2.2 Classification and Preparation

For bioimaging, NIR QDs provide many alternatives to organic fluorophores. The NIR QDs are typically composed of elements from groups II–VI (*e.g.*, Cd, Zn, Se, and Te) or III–V (*e.g.*, In, P, and As). Currently, key concerns for the design and synthesis of NIR QDs available for bioimaging include:[20–22]

- Larger absorbance at a wavelength >400 nm with emission at >650 nm.
- High QY.

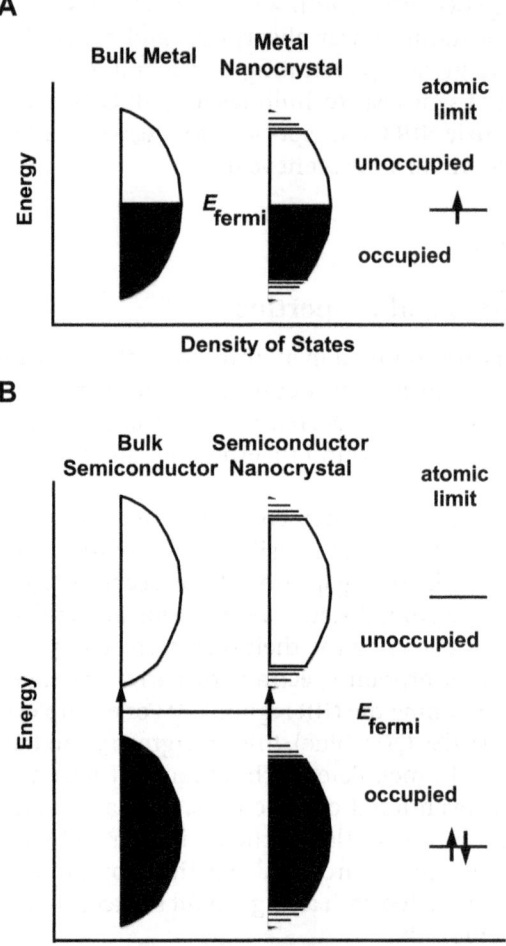

Figure 2.1 Density of states in metal (A) and semiconductor (B) nanocrystals. In each case, the density of states is discrete at the band edges. The Fermi level is in the center of a band in a metal, and so kT will exceed the level spacing even at low temperatures and small sizes. In contrast, in semiconductors, the Fermi level lies between two bands, so that the relevant level spacing remains large even at large sizes. The HOMO–LUMO gap increases in semiconductor nanocrystals of smaller size. (Reproduced with permission from A. P. Alivisatos, *J. Phys. Chem. B*, 1996, **100**, 13226. Copyright (1996) American Chemical Society.[10])

- Water solubility, or easy conversion to aqueous solution and functionalization.
- Good colloidal stability and photostability.
- Relatively small size.
- Absence of toxic metallic elements.

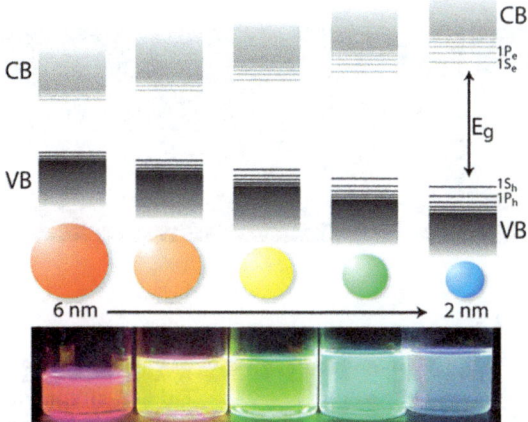

Figure 2.2 Schematic representation of the quantum confinement effect on the energy-level structure of a semiconductor material. The lower panel shows colloidal suspensions of CdSe nanocrystals of different sizes under ultraviolet excitation. (Reproduced from ref. 13 with permission from the Royal Society of Chemistry.)

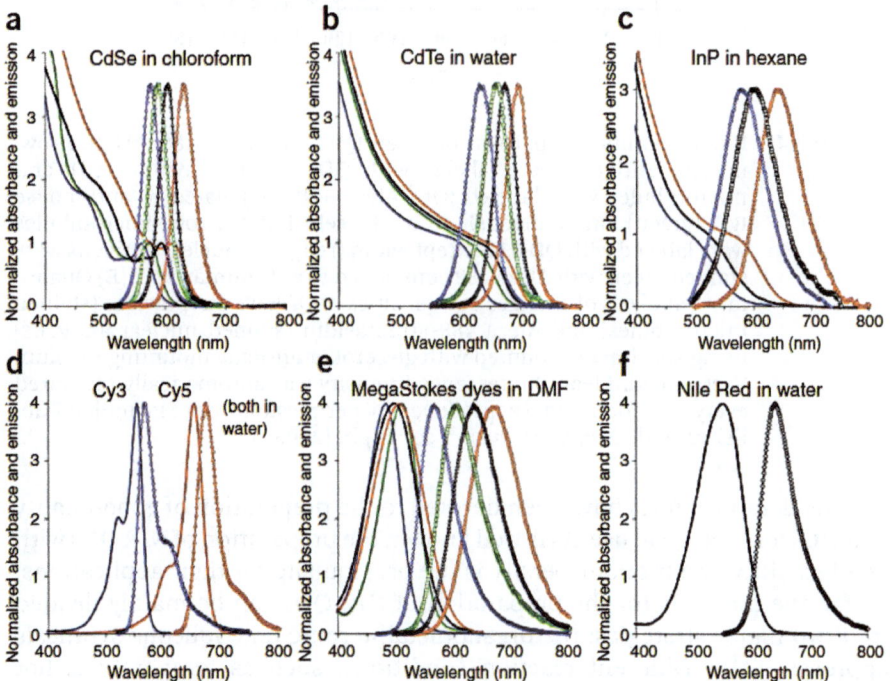

Figure 2.3 Spectra of QDs and organic dyes. (a–f) Absorption (lines) and emission (symbols) spectra of representative QDs (a–c) and organic dyes (d–f) color coded by size (blue < green < black < red). MegaStokes dyes were designed for spectral multiplexing in dimethylformamide (DMF). (Reproduced by permission from Macmillan Publishers Ltd.: *Nat. Methods*,[15] copyright (2008).)

A

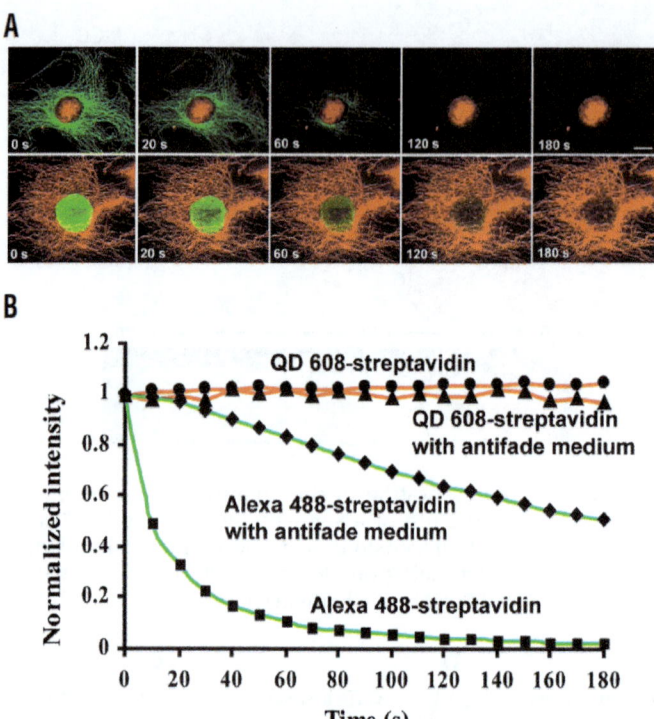

B

Figure 2.4 Photostability comparison between QDs and Alexa 488. (A) Top row: nuclear antigens were labeled with QD 630–streptavidin (red), and microtubules were labeled with Alexa 488 conjugated to anti-mouse IgG (green) simultaneously in a 3T3 cell. Bottom row: microtubules were labeled with QD 630–streptavidin (red), and nuclear antigens were stained green with Alexa 488 conjugated to anti-human IgG. (B) Quantitative analysis of changes in intensities of QD 608–streptavidin (stained microtubules) and Alexa 488–streptavidin (stained nuclear antigens) using specimens mounted with glycerol or antifade mounting medium Vectashield. Mean fluorescence intensity was automatically measured every 10 s for 3 min. (Reproduced by permission from Macmillan Publishers Ltd.: *Nat. Biotechnol.*,[19] copyright (2003).)

Considerable efforts have been devoted to the preparation of good-quality NIR QDs for biological use. As is well known, the preparation of NIR QDs with good stability and monodispersity is the prerequisite for their applications. So far, the methods for the preparation of NIR QDs can be mainly divided into two major strategies: the organometallic route and aqueous synthesis approaches.[23–25] Different reaction conditions, such as temperature, hot injection time, precursor reagents, and exchange ligands, have been modulated to prepare NIR QDs with diverse components.[25] As QDs synthesized in the organic phase usually have very poor water solubility and biocompatibility, the synthesis of QDs in aqueous solution has become an alternate option, which also has the advantages of convenience, easy control, reproducibility, low cost, and large-scale preparation. However, because of the poor crystal quality with non-uniform particle size, broad spectra, and low QY of the resulting QDs, this method still requires improvement.

Many problems arise during the synthesis of NIR QDs. Take the hot injection method, for example. Although the synthesis looks extremely simple and is used for the majority of QDs,[20–22,26–29] the hot injection method is based on the concept of adding one of the reactants very quickly into a hot reaction mixture that contains all the other reactants, stabilizers, coordinating and non-coordinating solvents, thus creating a supersaturated solution which leads to the formation of a burst of nuclei that then grow over time, preferably all at the same rate, to the desired size. However, even the LaMer model[30,31] that is the basis of this approach has recently been challenged.[32,33] The reaction time, temperature, concentration, and stabilizer play a role in controlling the size and emission of the QDs. Theoretically, at a specific time, the reaction is quickly quenched and purified to isolate the QDs. However, the injection speed, the purity of the reactants, and other chemicals used in the synthesis play a crucial role in the formation of QDs.[34] Furthermore, small amounts of reagents remain, changing the outcome. This small-scale method presents other challenges with respect to reproducibility if one wants to scale it up to grams, tens of grams, and beyond. Fast injection obviously becomes problematic in large-scale reactions, but controlling the temperature of large reaction flasks or reactors is also far from trivial, as is the quenching of the reaction mixture to stop the further growth of the QDs.

The major classification and synthesis methods of NIR QDs are summarized as follows.

2.2.2.1 Group II–VI NIR QDs

CdTe QDs were among the earliest NIR QDs to emerge. Although the emission of CdTe QDs can be tuned to the NIR region, their QYs were rather low.[35] By referring to bandgap engineering theory and simultaneously adjusting the components and structure of the QDs, small-size CdTe core material was prepared. Cu^{2+} was introduced to prepare novel CdTe/CdS:Cu QDs, with both the wavelength and the lifetime regulated. Singh *et al.* synthesized tunable NIR Ni-doped CdTeSe/CdS QDs (MNIR-QDs). The resulting composites were converted into aqueous solution and functionalized with folic acid, and subsequently used for NIR fluorescence imaging of cancer cells.[36] NIR $CdTe_xSe_{1-x}$/ZnS QDs with high brightness (QY ~80%), controllable morphology, and tunable wavelength (650–870 nm) were prepared by doping. Results showed that emission wavelengths were closely related to the ratio of Te and Se contents, especially the Te content. The resulted QDs were rather stable after surface ligand exchange. In the presence of salt at high ionic strength (2 M NaCl) or different pH (4–10), the stability was still excellent. Gd ions with magnetic properties were adsorbed on the QD surface, enabling the construction of NIR QD-based probes for dual-mode fluorescence imaging and magnetic resonance imaging (MRI). The probe was also used for *in vivo* imaging of lymph node.[37]

Some core–shell QDs with CdSe or CdTe as cores or alloys emit high-efficiency NIR fluorescence. It is reported that by capping QDs with a semiconductor material of the same crystal structure and wider bandgap, NIR QDs can be prepared with better fluorescence and fewer defects.[36,38-40] Core–shell NIR QDs CdTe/CdSe were synthesized by successive ionic layer adsorption reactions, and used for the *in vivo* imaging analysis of living cells by modification with double thiol polymers.[23]

In addition, by combining cation exchange reaction and Hg doping on small CdTe QDs, the emission wavelength of the QDs was adjusted to the NIR range. With ZnS and a polymer package, increased stability and decreased biological toxicity of the QDs was realized. NIR fluorescence/MRI dual-modality imaging was also achieved by chelating with Gd particles.[35] However, both the doping and the wrapping with a shell further complicated the synthesis process. Furthermore, the doping ions may easily leach out of the QDs during bioimaging, and the size of core–shell QDs is significantly greater.

Functional CdTe and CdHgTe NIR QDs were also synthesized by using biological peptides as ligands, and were used for biological analysis and live cell imaging.[41] There are also a few relatively new II–VI NIR QDs. Chalcogenide Hg QDs, such as HgS QDs and HgTe QDs, have a suitable bandgap and emit relatively strong NIR fluorescence.[42,43] The HgTe colloidal QDs were prepared *via* a simple two-step injection method and were tunable between 1.3 and 5 μm.[42] The HgS QDs, prepared with a simple two-step route in bovine serum albumin (BSA), showed a broader emission with the maximum wavelength at 730 nm and a QY of 4–5%.[43]

2.2.2.2   Group IV–VI NIR QDs

Generally, group IV–VI QDs, especially PbX (X = S, Se, Te), have very high QYs. However, both the toxicity of these elements and the large size (100 nm) of the QDs hinders their use in bioimaging. Core–shell QDs based on PbS QDs or PbSe QDs can greatly increase their stability and QYs. The QYs remained at 20–30% after conversion into aqueous solution. The disadvantage, again, is the large full width half maximum (FWHM > 200 nm).[44,45]

NIR PbS QDs were prepared with reagents such as Pb(II) acetate and bis(trimethylsilyl) sulfide in oleic acid and octadecene,[46] which gave exciton absorption features from about 800 to 1800 nm and a QY of ~20%. Moreels *et al.*[47] synthesized PbS QDs with a QY of 90% in the organic phase. Ozin and coworkers showed the extinction coefficient to be size dependent.[48] Kanatzidis *et al.* reported the use of Sn(II) in a $Pb_{2-x}Sn_xS_2$ ternary compound in a mixture of trioctylphosphine (TOP), oleyl amine, oleic acid, and octadecene,[49] enhancing the NIR absorption. These ternary compounds are stable up to ~300 °C, at which point they decompose into two binary compounds. Weiss and coworkers showed that PbS QDs can be oxidized with tetracyanoquinodimethane (TCNQ).[50] Rao *et al.* synthesized NIR PbS QDs based on a luciferin template.[25] The FWHM depends on QD size, thus is highly tunable. The fluorescent lifetime was as long as milliseconds.

The first, seminal, report of a wet chemical synthesis of colloidal PbSe QDs, in 2001, was by injection of a solution of Pb oleate and TOP-Se in TOP into stirred diphenyl ether at 150 °C.[51] These authors tuned the reaction conditions to get QD sizes of 3.5–15 nm. This size range gives exciton absorption from ~2200 nm to ~1200 nm. Guyot-Sionnest *et al.*[52] and Kraus *et al.*[53] reported ultrasmall PbSe QDs with exciton absorptions down to ~700 nm and a relative QY of up to 90%.

Pb(II) acetate and TOP-Se seem to be the precursors most often used for PbSe QDs.[54–57] Koole *et al.* published a detailed study on the optical confinement properties of PbSe QDs.[57] Ternary Pb chalcogenide QDs, $PbSe_xTe_{1-x}$, PbS_xTe_{1-x}, and $PbSe_xSe_{1-x}$, have been reported by Smith *et al.*[58]

Less work has been reported on PbTe QDs than on PbS QDs and PbSe QDs. Liu *et al.*[59] reported the synthesis of PbTe QDs in an analogous way to the PbSe QDs by reacting Pb(II) acetate with TOP-Te. Superlattices also formed based on the prepared PbTe QDs through self-assembly because of the small size dispersion (<5%). The QDs sizes were tuned from 4 to 14 nm, with exciton absorption being observed from 1900 nm to almost 2400 nm. Progress has been made by Soriano *et al.* on the incorporation of Sb.[60] They used TOP-Se, Pb(II) acetate, and Sb acetate in a mixture of oleic acid and oleyl amine to synthesize $Pb_mSb_{2n}Te_{m+3n}$ QDs with m = 2, 3, 4, 6, 8, and 10 and n = 1 and 2, but the size dispersion is still modest. A novel synthesis route has been proposed by Chubilleau *et al.*, using laser fragmentation to make PbTe QDs, but the size dispersion was large.[61]

2.2.2.3 Group III–V NIR QDs

NIR QDs without toxic elements are promising bioimaging probes and their controlled synthesis has been explored. Group III–V NIR QDs are generally used as optoelectronic devices since the QDs have poor stability and low emission efficiency, and it is very hard to control their size and their synthesis.[62,63]

In 2006, Bawendi *et al.* synthesized InAs/ZnSe core–shell QDs. By carefully controlling the core sizes, the shell thicknesses, and their composites, the emission wavelength were tuned from 750 nm to 920 nm, with the absolute QY in hexane being 7–10%. Although this QY is low, researchers have found that QDs with a QY of 1–2% were good enough for bioimaging.[63] Furthermore, three-layered InAs/InP/ZnSe QDs emitting in the NIR region were also synthesized.[64] Notably, Peng *et al.* proposed a method to synthesize ZnSe-coated Cu-deposited InP QDs (Cu:InP/ZnSe) with a high QY of 40%. The emission wavelength was tuned from 630 nm to 1100 nm. Since they contain no toxic heavy metal element, these QDs were very promising for NIR bioimaging.[65]

2.2.2.4 Group I–VI NIR QDs

AgX_2 (X = S, Se, Te) is an ideal class of materials for NIR QDs because of their small bandgap. However, since they all have a very low solubility product, e.g., $K_{sp}(AgS_2)$ = 6.3 × 10^{-50}, the crystal growth rates are generally so fast that it is hard to get QDs.[66–72]

Most NIR Ag$_2$S QDs are synthesized in the organic phase, with the result that as-prepared Ag$_2$S QDs cannot be directly dispersed in the aqueous phase. In 2010, Wang *et al.* synthesized NIR AgS$_2$ QDs based on a single precursor with maximum emission at 1058 nm.[66] Our group synthesized tunable NIR Ag$_2$S QDs in the organic phase, but with low QY.[68] We also synthesized NIR Ag$_2$Se QDs[69] and NIR Ag$_2$S QDs[70] in aqueous solution, using the newly proposed two-step method to synthesize these low-toxicity NIR QDs. The QDs were transferred to the aqueous phase, and the fluorescence QY was maintained.

A method of preparing water-soluble Ag$_2$S QDs based on AgNO$_3$ and mercaptopropionic acid (MPA) was also developed.[68] By adjusting the QD growth time, Ag$_2$S QDs with an emission wavelength in the NIR region were realized. Results of NIR imaging in nude mice show that the NIR fluorescence of Ag$_2$S QDs can successfully penetrate the body and skin of mice. Thus the background interference is small, showing that the Ag$_2$S QDs prepared by this method have the potential to be used in biological imaging.

Ag$_2$Se QDs are a new kind of low-toxicity NIR QDs. Since the solubility product in aqueous solution is really small (K_{sp} = 2.0 × 10^{-64}), it is difficult to control particle sizes of Ag$_2$Se. Our group proposed a "quasi-biological" system. By using two biochemical pathways—enzymatic reduction of Na$_2$SeO$_3$ and Ag$^+$ coordination with alanine—low-toxicity Ag$_2$Se QDs were synthesized in aqueous solution, with the sizes kept <3 nm. Two kinds of cell activity experiments and the toxicity of Ag$_2$Se QDs to various normal cells and cancer cells were studied. Results showed that at least 96% of the cells maintained good cell activity.[69]

Ag$_2$Te QDs are another type of NIR QDs. Ag$_2$Te NIR QDs with a particle size of 2.5 nm were synthesized with a NIR-II (1000–1700 nm) fluorescence window emission.[73] An amphipilic polymer coating was used to transfer the water-soluble Ag$_2$Te QDs to the aqueous phase where they were then modified with functionalities. The emission intensity of Ag$_2$Te QDs in the NIR-II range was 32 times higher than that of HiPCO, a very stable carbon nanotube emitting in the NIR-II window. The application of PEG-modified Ag$_2$Te QDs in brain and hind limbs of rats was investigated. Results showed that the Ag$_2$Te QDs could be used to achieve continuous ultrafast imaging of the brain with an exposure time of 5 ms.

2.2.2.5 Group I–III–VI NIR QDs

Triple-component QDs are emerging as a new class of NIR fluorophores. Typical examples are CuInS$_2$, CuInSe$_2$, AgInSe$_2$, and core–shell QDs based on them. Hot injection is the main synthesis method for these QDs.[74] Bawendi *et al.* proposed a hot injection method and synthesized CuInSe QDs that can be tuned from 600 nm to 1000 nm.[75] In 2010, Klimov *et al.* proposed a highly efficient large batch method to synthesize CuInS$_2$ QDs, with emission from 650 nm to 800 nm.[76] With a ZnS coating, the QY increased from 5–10% to >80%. With its long lifetime of 500 ns, this is a suitable bioimaging probe.

Su *et al.* synthesized water-soluble CuInS$_2$ QDs with a wavelength of 660 nm *via* a hydrothermal synthesis method.[77] Although the QY was low, the method offers new opportunities for the development of QD synthesis. An novel one-pot preparation technique was developed by Zhong *et al.* to prepare low-toxicity I–III–VI NIR CuInSe$_2$ and CuInS$_2$ QDs. The size, composition, and surface chemistry of the QDs was varied to modulate their emission wavelength. CuInSe$_2$ QDs with an emission wavelength of 700–900 nm and a QY of >60% were obtained.[78–80]

2.2.2.6 Group IV NIR QDs

Group IV NIR QDs contain only one element such as Si, Ge, or C. Related research has been going on for many years. For example, Si QDs are a typical zero-dimensional nanomaterial. Synthesis methods include electrochemical corrosion, laser ablation, mechanical grinding, and plasma sputtering. The emission of Si QDs stabilized with different ligands can be tuned from the visible to the NIR range, making them an ideal material for the preparation of bioprobes. However, many factors affect their application, including size, surface passivation, surface oxidation and defects, and aggregation.

In 2005, Mangolini *et al.*[81] proposed a method to synthesize Si QDs that emit at 800 nm, based on plasma sputtering. The resulting QY was ~60%. However, the synthesis was achieved in the organic phase, causing the resulting Si QDs to be unstable in the aqueous phase and thus not available for biological applications. Many researchers have made extensive improvements to the synthesis method. Erogbogbo *et al.*[82] and He *et al.*[83] reported the conversion of organic ligand-capped Si QDs into aqueous solution and used them for cell imaging. Sailor *et al.*[84] synthesized multiporous Si nanoparticles with a diameter of 130 nm and emission at 700–900 nm. The QY was 10%. Interestingly, porous Si nanoparticles decomposed into soluble silicate when used in drug delivery, and rapidly degraded in mice. This approach is promising in molecular imaging and targeted treatment of cancer.

Ge QDs can be synthesized with both the solvent method and the sol–gel method. The maximum emission varies from 800 nm to 1200 nm and the QY from 0.05% to 8% according to the QD size or synthesis methods. The emission spectra of Ge QDs are rather wide. Furthermore, they are easily oxidized and a core–shell structure is not yet available, so as yet there are no reports on the conversion of Ge QDs into the aqueous phase or their application in bioimaging.[85–87]

For C-based QDs, nanodiamond[88] and carbon dots[89] are most extensively reported. Synthesis techniques include surface modification, oxidation with concentrated acid, electrochemistry, laser ablation, organic carbonization, and template methods.[90–92]

In summary, the synthesis of high-quality and low-toxicity QDs is a prerequisite for the use of NIR QDs in bioimaging. Current research focuses on the development of synthesis methods in aqueous solution for QDs without toxic elements. More effort should be focused on novel methods that take advantage of biological systems. The development of new methods or novel chemical reactions for the synthesis of NIR QDs is also valuable.

2.3 NIR QDs for Bioimaging

Although NIR QDs are superior to conventional organic dyes, many key problems must be solved before they can be used in bioimaging and several criteria for the water solubilization and functionalization of QDs must be satisfied. First, the NIR QDs must be made water soluble while maintaining their fluorescence properties as far as possible, since bioimaging has to be done in aqueous solution. Second, the structure of NIR QDs must not be affected by coatings or conjugations. Third, QDs have to be stable colloids and maintain their optical properties as much as possible in biological media such as phosphate buffered saline (PBS) or cell culture media. Furthermore, smaller sizes are much preferred, to eliminate effects of QDs on biological systems. Last but not least, functional groups such as amine, carboxyl, and thiol groups are often required on the surface of the QDs for further bioconjugation.[93]

2.3.1 Surface Chemistry

For NIR QDs synthesized in the organic phase, common strategies for changing or capping the stabilizing layer after synthesis and purification are illustrated in Figure 2.5. These methods are generally based on silica coating, ligand exchange, or hydrophobic interaction.[94] Colvin *et al.* used 11-mercaptoundecanoic acid to transfer PbSe QDs to water.[95] Hyun and coworkers exchanged the surface oleates with carboxylic acids bearing a terminal thiol or amine group, to arrive at water-dispersible PbS and PbSe QDs.[96] Sargent and coworkers used 1-mercaptoundecyl-tetra(ethylene glycol) to transfer oleate-capped PbS QDs to HEPS, TRIS, and PBS buffers with a phospholuminescence (PL) QY of 26%.[97] Lifshitz and coworkers used 2-aminoethanol to make PbSe QDs water-dispersible.[98] Lin *et al.* used poly(acrylic acid) to transfer PbS QDs to water.[99] Modification based on hydrophobic interaction between oleate monolayer and alkyl chains of an amphiphilic polymer was realized by Ma *et al.*[100–102] Robinson *et al.* stripped off the ligands from PbS and CdSe QDs with Na_2S, leading to very closely packed water-dispersible hexagonal superstructures.[103] All the above approaches have some disadvantages, such as shift of the emission peak upon transfer, decreased QY, or colloidal instability in aqueous solution, especially in buffers of varying pH.

In vivo targeting and imaging is very challenging because of the relatively large overall size and short circulation time of QD conjugates. Usually, polymers such as poly(ethylene glycol) (PEG) can be incorporated on to the QDs to reduce non-selective accumulation in the reticuloendothelial system, thereby enhancing circulation times, increasing biocompatibility, and improving targeting efficiency. Furthermore, phospholipid micelle encapsulation of QDs

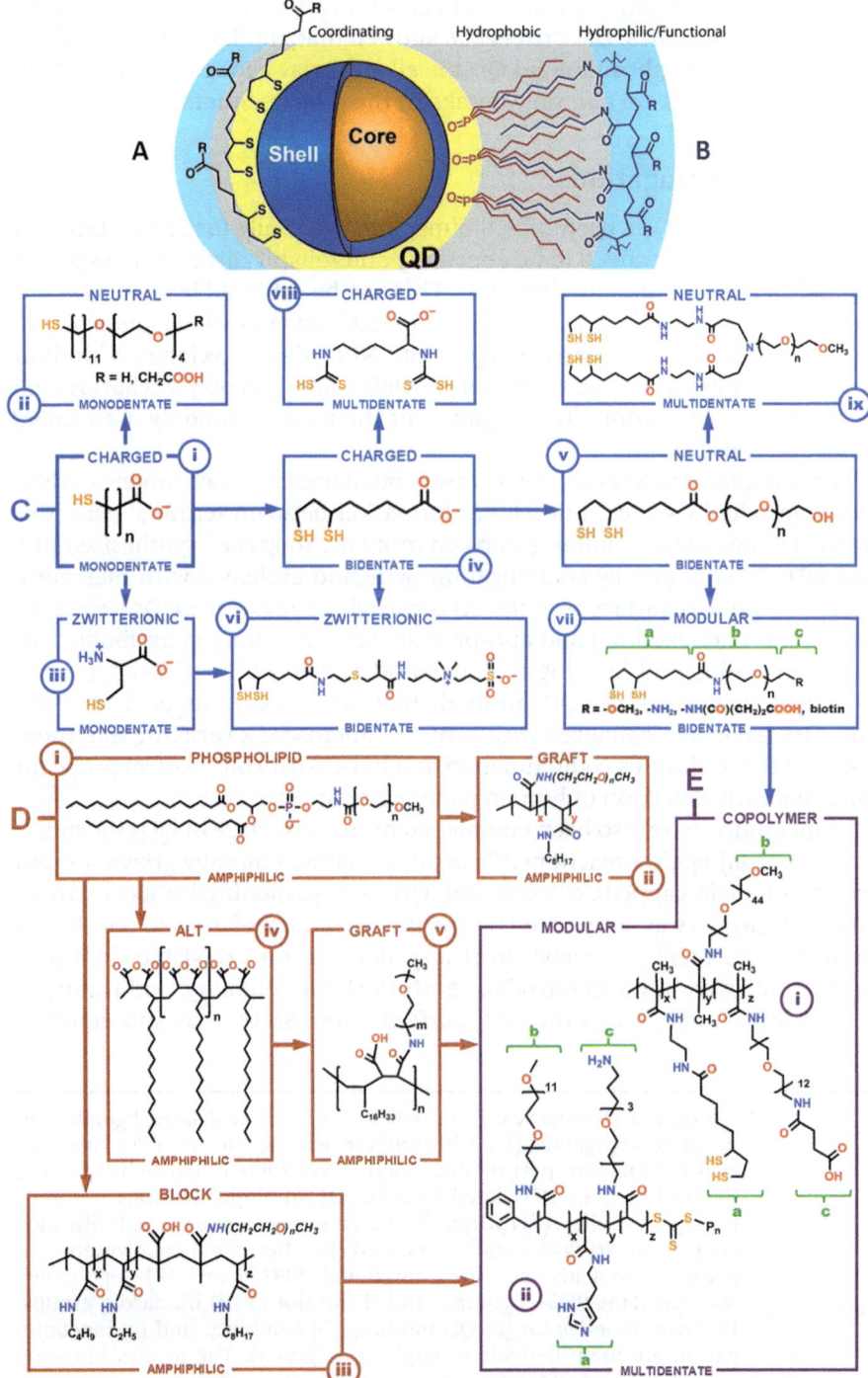

Figure 2.5 Surface chemistries on QDs. (A) Ligand binding at the QD surface. (B) Association of an amphipol (blue) with the native QD ligands (red).

provides highly versatile surface chemistry for conjugating multiple chemicals and biomolecules. Dubertret *et al.* showed that, unlike other general QD polymer coatings, phospholipid QD micelles display a long circulation half-life in the bloodstream and slow uptake by the reticuloendothelial system.[105]

2.3.2   Bioconjugation

In order to make QDs useful for bioimaging, especially for *in vivo* targeted imaging, ideally they need to be effectively and reliably directed to a specific biological site without any alteration. This can be achieved by attaching targeting molecules to the QD surface (Figure 2.6). So far, various kinds of biomolecules have been used to conjugate with NIR QDs to develop new kinds of bioprobes, such as peptides, proteins, nucleic acids, and polysaccharides, all of which originate from bioconjugation of the most commonly used CdSe/ZnS QDs (Figure 2.7).

For example, EDC/NHS chemistry is very often used to attach proteins to NIR QDs, typically by direct amide bond formation between terminal carboxyls on the QD ligands and amine groups on proteins. Yong *et al.* synthesized InP/ZnS QDs.[108] Followed by solubilization by ligand exchange with mercaptosuccinic acid, the surface-terminated carboxyl groups of the QDs were EDC coupled to anti-claudin 4 and anti-prostate stem cell antigen antibodies for specific staining and imaging of pancreatic cancer cell lines. Receptor-mediated endocytosis of the QD–antibody bioconjugates by targeted MiaPaCa and XPA3 cells was visualized with confocal microscopy, verifying that these covalent bioconjugates could function as a low-toxicity biomedical probe for potential early detection of human pancreatic cancer.

Thiol groups have also been engineered onto the surface of QDs for subsequent bioconjugation reactions.[109] An SiO_2 shell was initially grown around thioglycolic acid-capped CdTe core-only QDs using a modified Stöber method, and shell growth was accelerated by precipitating more SiO_2 on to the already formed shell. Mercaptopropyltrimethoxysilane and a PEGylated derivative of the same were used to introduce both PEG, for solubility, and thiols, as a chemical handle, on to the SiO_2 surface. Sulfo-SMCC was subsequently

---

(C) Ligand chemistries: (i) thioalkyl acids, (ii) PEGylated ligands, (iii) zwitterionic ligands, (iv) dihydrolipoic acid ligands, (v) PEGylated, (vi) zwitterionic, and (vii) modular derivatives thereof; (viii) charged and (ix) multidentate PEGylated ligands. (D) Amphipol coatings: (i) phospholipid micelles, (ii) hydrophilic polymer backbones grafted with alkyl chains, (iii) triblock copolymers, and (iv) alternating copolymers that hydrolyze to acids or (v) are grafted with PEG chains. (E) Copolymers with pendant PEG oligomers and (i) dithiol or (ii) imidazole groups. Discrete moieties for (a) QD binding, (b) solubility, and (c) bioconjugation are identified where applicable (green). The arrows illustrate a conceptual progression and not synthetic pathways or chronological development. (Reproduced with permission from W. R. Algar, K. Susumu, J. B. Delehanty and I. L. Medintz, *Anal. Chem.*, 2011, **83**, 8826. Copyright (2011) American Chemical Society.)[104]

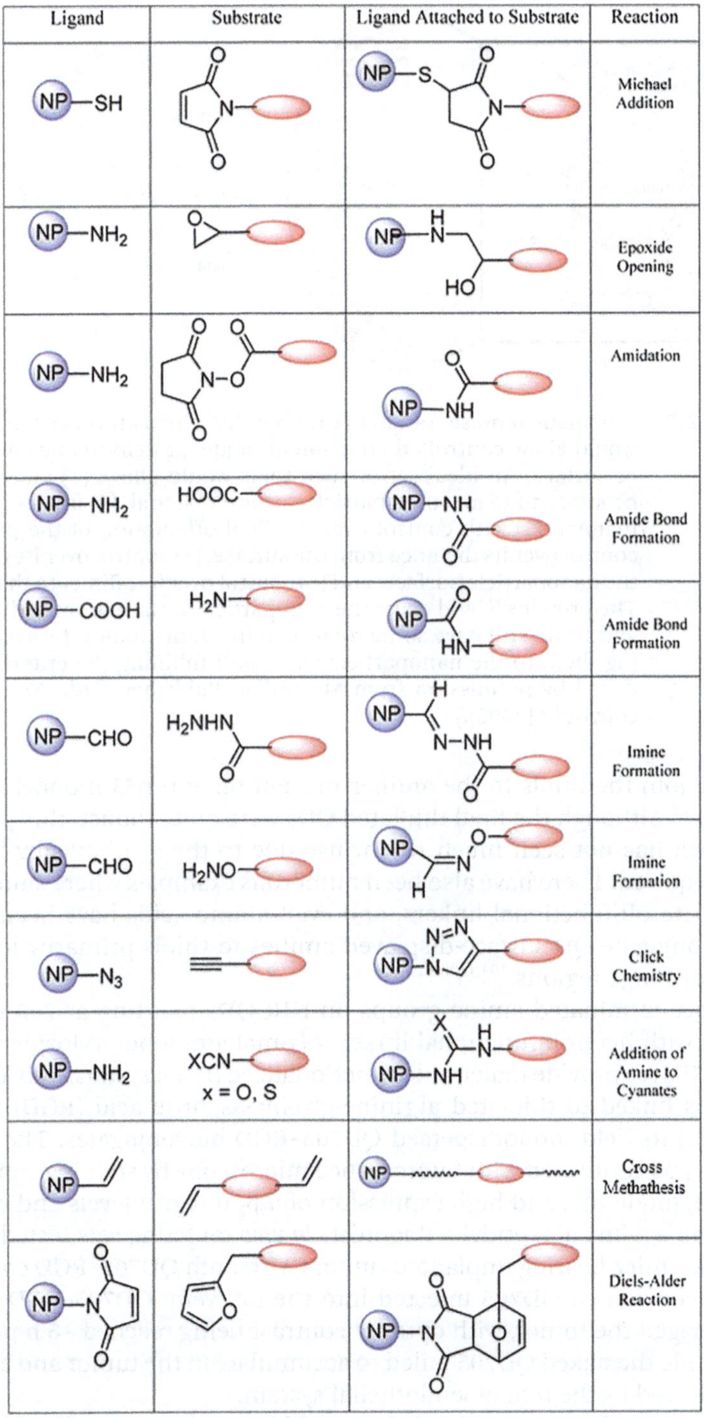
Ligand
Substrate
Ligand Attached to Substrate
Reaction
NP—SH
Michael Addition
NP—NH2
Epoxide Opening
NP—NH2
Amidation
NP—NH2
HOOC—
Amide Bond Formation
NP—COOH
H2N—
Amide Bond Formation
NP—CHO
H2NHN
Imine Formation
NP—CHO
H2NO—
Imine Formation
NP—N3
Click Chemistry
NP—NH2
XCN—
x = O, S
Addition of Amine to Cyanates
Cross Methathesis
Diels-Alder Reaction

Figure 2.6 Functionalization chemistry of nanoparticles. (Reproduced with permission from N. Erathodiyil and J. Y. Ying, *Acc. Chem. Res.*, 2011, **44**, 925. Copyright (2011) American Chemical Society.)[106]

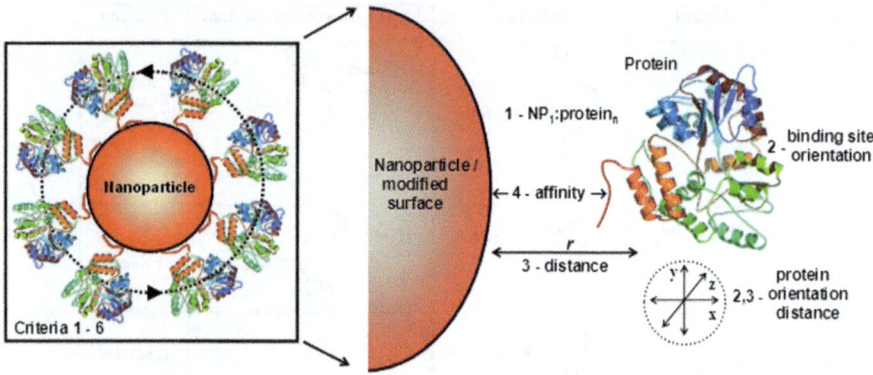

Figure 2.7   Schematic representation of the six criteria for a universal "toolset" that would allow controlled attachment of any protein to any nanoparticle or surface. An ideal set of such tools would allow: (1) any protein to be attached to any nanoparticle/surface material, (2) in a homogenous manner, (3) with control over the final orientation of the protein, (4) control over its distance from the surface, (5) control over its density on the nanoparticle/surface, and (6) control over its affinity to that surface. The proteins would cover the nanoparticle surface in three dimensions and could still have some rotational freedom around the axis connecting them to the nanoparticle while still fulfilling the criteria. (Reproduced by permission from Macmillan Publishers Ltd.: *Nat. Mater.*,[107] copyright (2006).)

used to join the thiols to the amines present on anti-p53 monoclonal antibodies.[109] Although the final thiolated QDs were quite robust, this particular approach has not seen much recent use due to the synthetically intensive steps required. There have also been numerous examples where sulfo-SMCC, other heterobifunctional linkers, and even amino acids have been used to cross-conjugate QD surface-displayed amines to thiols primarily located in antibody hinge regions.[104,110,111]

Surface-terminated amine groups on NIR QDs emitting at 705 nm were treated with heterobifunctional linker N-(γ-maleimidobutyryloxy)-succinimide (GMBS) to provide maleimide-functionalized nanocrystals. Subsequently, this was linked to thiolated arginine-glycine-aspartic acid (RGD) peptides (RGDSH) to yield monodispersed QD705–RGD bioconjugates. The binding affinity was confirmed with fluorescence microscopy by staining cancer cells with no, moderate, and high expression of $\alpha_v\beta_3$ integrin levels and challenging them against a peptidyl antagonist. *In vivo* targeting was tested in athymic nude mice bearing implanted tumors with both QD705–RGD conjugates and unconjugated QD705 injected into the tail vein. QD705–RGD successfully imaged the tumor, with optimal contrast being reached ~6 h postinjection, while the naked QD705 failed to accumulate in the tumor and appeared to be cleared by the reticuloendothelial system.

Cai *et al.* labeled viruses with NIR QDs with a QY of 60% based on bioorthogonal chemistry.[112] The method was used for the targeting of virus to cells.

2.3.3 Bioimaging Based on NIR QDs

2.3.3.1 Multicolor Bioimaging

Optical imaging is the one of the best methods for the simultaneous collection of multichannel signals. By changing the filters in the instruments, multichannel imaging can be achieved on the same sample with bioprobes emitting at different wavelengths. Instead of the conventional single-channel imaging, multicolor imaging has been used in simultaneous imaging of multiple physiological states or pathological states,[113-116] simultaneous evaluation of *in vivo* effects of drugs and drug metabolism kinetics,[117,118] and simultaneous non-invasive multicolor imaging of receptors and proteins on cancer cells,[119-123] providing information for cancer diagnosis and therapies. In 2007, five-channel imaging of five different lymph nodes was achieved by Kobayashi *et al.*[124] Nie *et al.* reported imaging based on the labeling of different antibody-labeled multicolor QDs on cancer cell surfaces, which is promising in the diagnosis of life-threatening cancers.[122,123]

2.3.3.2 Multimodality Bioimaging

Although bioprobes have eliminated tissue autofluorescence, the normal penetration depth of just a few millimeters still hinders imaging. Several groups have focused on the combination of NIR QD imaging and other imaging modes to get molecular functional and structure information for biosystems. For example, Park *et al.* reported on imaging based on bioprobes composed of NIR QDs and Fe_3O_4 nanoparticles.[125] Jin *et al.* reported on dual-mode *in vivo* imaging based on Gd^{3+}-functionalized NIR CdSeTe/CdS QDs.[126] Erogbogbo *et al.* prepared dual-mode bioprobes based on NIR Si QDs and Fe_3O_4, which had superior penetration and were suitable for measuring *in vivo* structural distribution.[127] Chen *et al.* reported on a series of bioprobes based on NIR QDs and PET QDs, which can be used to quantify markers for tumor blood vessels.[128,129] Au-doped NIR QDs were synthesized and used for NIR fluorescence imaging and SPECT imaging. Results showed that they have outstanding penetration and can be used to map *in vivo* tissue structure.[130]

2.3.3.3 Bioimaging Based on Resonance Energy Transfer

Many NIR QD bioprobes were proposed based on Förster (fluorescence) resonance energy transfer (FRET). Figure 2.8 shows the size of β-phycoerythrin (b-PE) in comparison to the QD and also highlights the large distances over which viable FRET was observed. Since both maximum overlap of the donor emission spectra and the acceptor absorbance and a relatively short distance between donor and acceptor are required, key factors for designing FRET-based probes are finding a suitable donor, acceptor, and spacer. FRET-based bioprobes usually result in a higher signal-to-noise ratio than

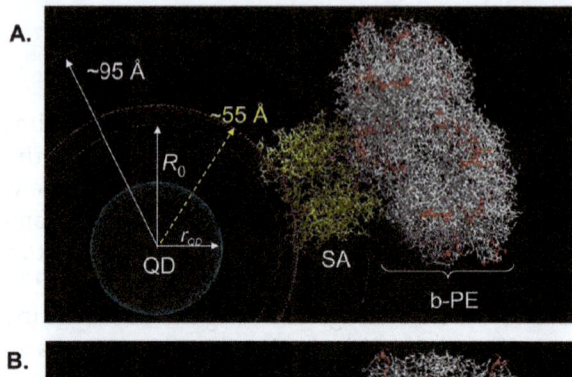

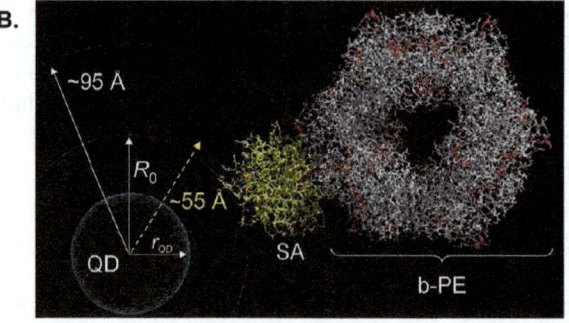

Figure 2.8   Models of the QD–b-PE conjugate structure/conformation. (A) b-PE is parallel to the QD surface, and (B) b-PE is fully extended away from the QD. The central QD with a radius of ~29 Å shown in blue is surrounded by a crimson shell of ~25 Å thickness representing the DHLA–PEG–biotin. The intermediary streptavidin (SA) protein is shown in yellow with biotin binding sites highlighted in purple. The b-PE ring structure is shown in white, with the multiple chromophores highlighted in red. The inner concentric white circle corresponds to the predicted 53 Å Förster distance (R_0) for the 540 nm QD–b-PE assembly. The second outer white circle is a visual distance marker set at ~95 Å from the QD center and represents the closest approach of the b-PE to the QD. (Reproduced with permission from I. L. Medintz, T. Pons, K. Susumu, K. Boeneman, A. M. Dennis, D. Farrell, J. R. Deschamps, J. S. Melinger, G. Bao and H. Mattoussi, *J. Phys. Chem. C,* 2009, **113**, 18552. Copyright (2009) American Chemical Society.)[134]

other bioprobes. Many metalloenzymes, such as matrix metalloproteinase, cathepsin, and caspase, can be detected with NIR QD-based FRET bioprobes. Current results show that many enzymes or proteins are highly expressed in cancer tissues.[131] The pH, temperature, and reducing agents in cancer tissue are all different from those in normal tissues.[132]

Although QDs can be used as outstanding fluorophores for molecular bioimaging, the requirement for external light sources hinders their application. For example, high-energy laser light can cause damage to irradiated areas. Bioluminescence resonance energy transfer (BRET) is a good way to solve the problem. In a BRET system, luciferase works as a donor by catalyzing the chemical reaction of an organic substrate to result in fluorescence.

Rao *et al.*[25,133] proposed the first BRET system based on QDs and luciferase for biosensing and bioimaging. In these systems, *Renilla* luciferase (Luc8) was chosen as the donor and conjugated to the acceptor QD surfaces. In the presence of coelenterazine, fluorescence at 480 nm can be produced by Luc8 and excite QDs. The advantage of the BRET system is the absence of an external laser, leading to low autofluorescence from the tissue and avoidance of tissue damage by laser illumination. However, the existence of a luciferase substrate is a prerequisite for BRET, which hinders its applications.[135,136]

During fluorescent bioimaging, external illumination often causes strong background autofluorescence and is ineffective for exciting fluorophores at depth within the tissue due to absorption and scattering. Seeking to overcome reliance on external illumination, Rao's group used EDC chemistry to attach a mutationally optimized *Renilla reniformis* Luc8 to QDs. This yielded self-illuminating QD–Luc8 bioconjugates that were driven *via* chemically induced BRET.[131,133]

2.3.3.4 Imaging of Tumors

Prepared doped NIR QDs with tunable wavelength and lifetime were wrapped into microbeads. Based on fluorescence lifetime and wavelength characteristics, a novel QD-based barcode was developed.[137] The fluorescence lifetime imaging mode (FLIM) based on QDs with small size (3.5 nm), long lifetime (1 μs), NIR emission (720 nm), and pH sensitivity was also used for *in vivo* bioimaging and biotracking.[138]

In vivo imaging of tumor fluorescence with NIR QDs has attracted much attention in tumor diagnosis and therapy during the past decade since it offers a novel technique for the precise detection of tumor location and margins. In the process of tumor growth, new blood vessels develop (angiogenesis) with an unstable structure and larger endothelial cells, different from those of normal tissues. These cells may provide channels allowing larger molecules (~400 nm) to penetrate. Since such large molecules cannot be drained by the lymph nodes, they accumulate in the tumor microenvironment. Based on this enhanced permeation and retention effect (EPR), passive targeting imaging methods have been proposed for tumor imaging. However, EPR is not applicable for tumor detection or microtumor metastasis.[139]

Cancer cells are in contact with the body's blood circulation, thus receptor molecules or specific markers on their surface are present in the bloodstream. *In vivo* imaging of cancer receptors offers potential opportunities for early diagnosis of cancer and research into the mechanism of metastasis. In 2004, Shuming Nie *et al.* reported on coating QDs with amphiphilic molecules and conjugation with antibody to human prostate specific antigen (PSA). After intravenous injection of QDs, results showed that the active target imaging resulted in a stronger fluorescence signal in the mice than that of passive targeting imaging. This is the first report on active target imaging, although the QDs used in the experiments emit in the visible range.[134] Subsequently, many groups labeled NIR QDs and used them in *in vivo* tumor imaging.[38,40,64,82,110,140–148] Cheng *et al.* proposed several requirements for NIR QDs in cancer imaging:

- Relatively strong NIR fluorescence.
- Good biocompatibility.
- Good stability and photostability in blood.
- Small size, to eliminate recognition by macrophages.
- Good renal clearance, to decrease toxicity.

By modifying InP/ZnS NIR QDs with RGD peptides, the *in vivo* targeted imaging of a tumor with high $\alpha_v\beta_3$ expression was achieved.[149]

2.3.3.5   Imaging of Lymph Nodes

Lymph nodes are among the most important organs of the human immune system. Their main functions are to filter lymph and participate in the body's immune response. When the body is infected by bacteria, viruses, or cancer cells, these can invade along the lymphatic vessels, causing local lymph node enlargement. If the lymphocytes in the lymph nodes cannot eliminate the infection or metastasis, it may spread along the drainage direction of the lymphatic vessels. All lymphatic drainage and most tumor metastasis passes through sentinel lymph nodes (SLNs).

Current methods for the detection of infected or cancerous lymph nodes have many disadvantages, such as the use of radioactive tracers and long hospitalization time. Fluorescent techniques based on NIR QDs are thus urgently needed for the imaging of SLNs.[24,63,82,150–154] In 2004, Frangioni *et al.* first proposed the imaging of lymph nodes at 1 cm depth in healthy mice with NIR CdTe/CdSe QDs.[150] InAs-based core–shell NIR QDs were also used in lymph node imaging in mice. However, since all these NIR QDs contain toxic elements such as Cd and As, they cannot be used in clinical studies[63] and new types of NIR QDs without toxic heavy metal elements are needed.[24,82,153,154] For example, CdTeSe/CdZnS QDs have more chance of detecting lymphatic inflammation than $CuInS_2$/ZnS QDs.[153] These NIR QDs will allow the detection of SLNs within a few minutes, which greatly simplifies the treatment program and is expected to be used in clinical trials.

2.3.3.6   Imaging of Blood Vessels

In general, after being injected intravenously, QDs will quickly be distributed around the body in the bloodstream, enabling blood vessels to be visualized on the basis of NIR QDs. Lim *et al.* first imaged NIR QDs in the NIR-I and NIR-II regions in the rat coronary artery system.[155] Subsequently, it was found that NIR QDs are more suitable for *in vivo* imaging of chicken embryos than QDs emitting in the visible region.[156] This is mainly due to the elimination of autofluorescence from organizations in the NIR region and the improvement of depth-of-field imaging.

After QDs enter the venous circulation, they may adhere to or interact with blood components or cells, so the distribution of QDs *in vivo* depends mainly on their size and surface properties.[157,158] NIR QDs (~5 nm) will gradually be eliminated from the body;[159] the larger the hydrodynamic diameter, the longer the retention time.[160–162] Some QDs, especially those with larger hydrodynamic diameters, will undergo endocytosis by macrophages in the reticuloendothelial system, thereby prolonging their stay in the body.

In addition, many QDs are functionalized with a peptide or antibody to achieve the vascular imaging of a tumor *in vivo*.[163–166] The RGD peptide interacts specifically with integrin on the tumor cell surface. Smith *et al.* functionalized NIR QDs with RGD (RGD-QDs) and directly observed tumor vascular imaging in mice with SKOV-3 tumors. The RGD-QDs specifically gathered near blood vessels in tumors, but this phenomenon was not observed when they used NIR QDs functionalized with arginine-alanine-aspartic acid (RAD-QDs) as a blank control.[166] In short, studying the distribution of QDs NIR *in vivo* and vascular imaging provides a basis for the evaluation of molecular target imaging.

2.3.3.7   In vivo *Cell Imaging*

In vivo imaging of cells labeled with QDs is of great interest for observing cell development and provides an intuitive basis for the regulation of cell growth, development, invasion, and metastasis of cancer. After injection of QD-labeled live cells, they can be observed in the blood circulation *in vivo*. The QDs must be labeled effectively *in vitro*, to ensure that live cells injected into living cells *in vivo* are not separated from the QDs, so that the fluorescence signal remains unchanged. In 2002, Dubertret *et al.* injected labeled QDs into the cells of *Xenopus* embryos. As time went on, the QD-labeled cells differentiated and the QDs translocated to the nucleus. Thus this method was suitable for the observation of embryonic development,[105] and could also be used to observe the distribution of QD-labeled tumor cells *in vivo*, and help us to understand the process and mechanism of tumor metastasis *in vivo*.[167–171]

In recent years, stem cell transplantation has been widely used in the treatment of nervous system diseases, blood diseases, and heart disease. Because stem cells have the potential for multiple differentiation, and can be widely used in regenerative medicine, the survival, differentiation, and migration of these living cells is important.[172] At present, QD labeling is internalized into cells mainly by cell phagocytosis, peptide-mediated transport, and receptor-mediated internalization.[172–178] Lei and others introduced QDs conjugated with Tat peptide into mesenchymal stem cells which were then injected into the tail of mice. The QDs were observed in liver, spleen, and lung.[174] Yukawa *et al.* investigated NIR QD labeling of octa-arginine peptide (R8) for adipose tissue-derived stem cells (ASCs) and its application for *in vivo* fluorescence imaging in mice with acute liver failure. The behavior and organ-specific accumulation of ASCs that were transplanted alone or in combination with heparin were imaged and analyzed using fluorescence imaging. The results showed that heparin was effective in increasing the accumulation of transplanted ASCs in the liver.[178]

2.3.3.8   Deep Tissue Imaging

Owing to the much lower photon scattering and tissue autofluorescence at longer wavelengths,[155,179–181] biological imaging in the NIR-II window (1000–1700 nm) benefits from deeper tissue penetration, higher signal-to-background ratio, and better image fidelity, allowing the imaging of sub-10 μm features at millimeter depths inside the highly scattering animal tissue.[182,183] However, one of the major limitations of NIR-II fluorescence for biological imaging remains the restricted choice of available NIR-II fluorophores that are biocompatible. Moreover, another issue with NIR-II fluorophores is their relatively low fluorescence QYs in comparison with their counterparts in the shorter-wavelength visible window (400–750 nm) and the traditional NIR-I window (750–900 nm).[184–187] The suboptimal QYs of NIR-II fluorophores limit their performance for *in vivo* NIR-II fluorescence imaging with sufficient temporal resolution.[188] Due to the insufficient fluorescence brightness, the difficulty of biofunctionalization, and the limited selection of current NIR-II fluorophores, there has always been an urgent need to develop new NIR-II fluorescent probes with higher fluorescence QY and good biocompatibility. QDs such as PbS,[185] PbSe,[52] and $CdHgTe$[189] have been reported with fluorescence in the NIR-II window. However, the highly toxic nature of heavy metal elements such as Pb, Cd, and Hg and the difficulty of surface functionalization of these QDs with biocompatible coatings have prohibited their *in vivo* applications.[190] Hong *et al.* demonstrated that Ag_2S QDs have good biocompatibility and facile surface functionalization with PEG and various targeting ligands. The bright NIR-II fluorescence emission, low toxicity, and small size of Ag_2S QDs have enabled the selective molecular imaging of cancer cells *in vitro* and tumor-specific imaging with high tumor-to-background ratio *in vivo*.

2.4   Conclusions

In summary, taking advantage of the superior optical characteristics of QDs, bioimaging with NIR QDs offers new opportunities for the discovery of biological molecular mechanisms, understanding cellular transportation pathways, and *in vivo* imaging. The illumination mechanism, preparation, surface chemistry, and bioconjugation methods for NIR QDs are very similar to those of the visible-range QDs such as CdSe. However, NIR QDs usually have disadvantages such as larger size, toxic ion content, and relatively low QY with stability/photostability not as good as visible-range QDs. Thus, before NIR QDs can be widely used, more interest needs to focus on crucial processes such as novel synthesis methods for NIR QDs with optimized fluorescence. Furthermore, milder surface chemistry and bioconjugation methods are required, allowing water-soluble NIR QDs to retain their structures and optical properties. Last but not least, applications of NIR QDs should focus on fields that take advantage of their special properties, such as multicolor imaging or deep tissue imaging.

Acknowledgements

The support of the National Basic Research Program of China (973 Program, 2011CB933600), the National Natural Science Foundation of China (21275111, 21535005), the 111 Project (111-2-10), the China Scholarship Council, and the Collaborative Innovation Center for Chemistry and Molecular Medicine is gratefully acknowledged.

References

1. S. Kawata, O. Nakamura, T. Kaneko, M. Hashimoto, K. Goto, N. I. Smith, T. Sugiura, I. Fujimasa and H. Matsumoto, in *Biological imaging and sensing*, ed. T. Furukawa, Springer, Berlin, 2004.
2. P. Suetens, *Fundamentals of medical imaging*, Cambridge University Press, New York, 2nd edn, 2009.
3. F. C. J. M. van Veggel, *Chem. Mater.*, 2014, **26**, 111.
4. L. T. Lu, L. D. Tung, I. Robinson, D. Ung, B. Tan, J. Long, A. I. Cooper, D. G. Fernig and N. T. K. Thanh, *J. Mater. Chem.*, 2008, **18**, 2453.
5. C. B. Murray, C. Kagan and M. Bawendi, *Annu. Rev. Mater. Res.*, 2000, **30**, 545.
6. A. P. Alivisatos, *Science*, 1996, **271**, 933.
7. C. Burda, X. B. Chen, R. Narayanan and M. A. El-Sayed, *Chem. Rev.*, 2005, **105**, 1025.
8. G. S. Hong, PhD Thesis, Stanford University, 2014.
9. M. F. Ashby, P. J. Ferreira and D. L. Schodek, *Nanomaterials, nanotechnologies and design: an introduction for engineers and architects*. Butterworth-Heinemann, Slovenia, 2012.
10. A. P. Alivisatos, *J. Phys. Chem. B*, 1996, **100**, 13226.
11. M. Smith and S. Nie, *Acc. Chem. Res.*, 2009, **43**, 190.
12. J. Yao, M. Yang and Y. Duan, *Chem. Rev.*, 2014, **114**, 6130.
13. C. M. Donega, *Chem. Soc. Rev.*, 2011, **40**, 1512.
14. Z. Deng, H. Yan and Y. Liu, *J. Am. Chem. Soc.*, 2009, **131**, 17744.
15. U. Resch-Genger, M. Grabolle, S. Cavaliere-Jaricot, R. Nitschke and T. Nann, *Nat. Methods*, 2008, **5**, 763.
16. J. K. Jaiswal, H. Mattoussi, J. M. Mauro and S. M. Simon, *Nat. Biotechnol.*, 2003, **21**, 47.
17. K. I. Hanaki, A. Momo, T. Oku, A. Komoto, S. Maenosono, Y. Yamaguchi and K. Yamamoto, *Biochem. Biophys. Res. Commun.*, 2003, **302**, 496.
18. A. Sukhanova, J. Devy, L. Venteo, H. Kaplan, M. Artemyev, V. Oleinikov, D. Klinov, M. Pluot, J. H. Cohen and I. Nabiev, *Anal. Biochem.*, 2004, **324**, 60.

19. X. Y. Wu, H. J. Liu, J. Q. Liu, K. N. Haley, J. A. Treadway, J. P. Larson, N. F. Ge, F. Peale and M. P. Bruchez, *Nat. Biotechnol.*, 2003, **21**, 41.
20. M. Smith and S. M. Nie, *Acc. Chem. Res.*, 2010, **43**, 190.
21. M. A. Malik, M. Afzaal and P. O'Brien, *Chem. Rev.*, 2010, **110**, 4417.
22. M. A. El-Sayed, *Acc. Chem. Res.*, 2004, **37**, 326.
23. D. Hu, P. Zhang, P. Gong, S. Lian, Y. Lu, D. Gao and L. Cai, *Nanoscale*, 2011, **3**, 4724.
24. E. Cassette, T. Pons, C. Bouet, M. Helle, L. Bezdetnaya, F. Marchal and B. Dubertret, *Chem. Mater.*, 2010, **22**, 6117.
25. N. Ma, A. F. Marshall and J. Rao, *J. Am. Chem. Soc.*, 2010, **132**, 6884.
26. J. B. Rivest and P. K. Jain, *Chem. Soc. Rev.*, 2013, **42**, 89.
27. D. Donega, *Chem. Soc. Rev.*, 2011, **40**, 1512.
28. S. V. Kershaw, A. S. Susha and A. L. Rogach, *Chem. Soc. Rev.*, 2013, **42**, 3033.
29. V. Lesnyak, N. Gaponik and A. Eychmüller, *Chem. Soc. Rev.*, 2013, **42**, 2905.
30. V. K. Lamer and R. H. Dinegar, *J. Am. Chem. Soc.*, 1950, **72**, 4847.
31. B. Murray, C. R. Kagan and M. G. Bawendi, *Annu. Rev. Mater. Sci.*, 2000, **30**, 545.
32. J. De Yoreo, *Nat. Mater.*, 2013, **12**, 284.
33. J. Baumgartner, A. Dey, P. H. H. Bomans, C. Le Coadou, P. Fratzl, N. A. J. M. Sommerdijk and D. Faivre, *Nat. Mater.*, 2013, **12**, 310.
34. J. Joo, J. M. Pietryga, J. A. McGuire, S. H. Jeon, D. J. Williams, H. L. Wang and V. I. Klimov, *J. Am. Chem. Soc.*, 2009, **131**, 10620.
35. D. Gao, P. Zhang, J. Jia, M. Li, Z. Sheng, D. Hu, P. Gong, Q. Wan and L. Cai, *RSC Adv.*, 2013, **3**, 21247.
36. N. Singh, S. Charan, K. Sanjiv, S. H. Huang, Y. C. Hsiao, C. W. Kuo, F. C. Chien, T. C. Lee and P. Chen, *Bioconjugate Chem.*, 2012, **23**, 421.
37. D. Gao, P. Zhang, Z. Sheng, D. Hu, P. Gong, C. Chen, Q. Wan, G. Gao and L. Cai, *Adv. Funct. Mater.*, 2014, **24**, 3897.
38. H. S. Choi, W. Liu, F. Liu, K. Nasr, P. Misra, M. G. Bawendi and J. V. Frangioni, *Nat. Nanotechnol.*, 2010, **5**, 42.
39. L. N. Chen, J. Wang, W. T. Li and H. Y. Han, *Chem. Commun.*, 2012, **48**, 4971.
40. K. T. Yong, I. Roy, W. C. Law and R. Hu, *Chem. Commun.*, 2010, **469**, 7136.
41. J. Jia, P. Zhang, D. Gao, Z. Sheng, D. Hu, P. Gong, C. Wu, J. Chen and L. Cai, *Chem. Commun.*, 2013, **49**, 4492.
42. S. Keuleyan, E. Lhuillier and P. Guyot-Sionnest, *J. Am. Chem. Soc.*, 2011, **133**, 16422.
43. N. Goswami, A. Giri, S. Kar, M. S. Bootharaju, R. John, P. L. Xavier, T. Pradeep and S. K. Pal, *Small*, 2012, **8**, 3175.
44. J. M. Pietryga, D. J. Werder, D. J. Williams, J. L. Casson, R. D. Schaller, V. I. Klimov and J. A. Hollingsworth, *J. Am. Chem. Soc.*, 2008, **130**, 4879.
45. H. Zhao, D. Wang, T. Zhang, M. Chaker and D. Ma, *Chem. Commun.*, 2010, **46**, 5301.

46. M. A. Hines and G. D. Scholes, *Adv. Mater.*, 2003, **15**, 1844.
47. I. Moreels, Y. Justo, B. De Geyter, K. Haustraete, J. C. Martins and Z. Hens, *ACS Nano*, 2011, **5**, 2004.
48. L. Cademartiri, E. Montanari, G. Calestani, A. Migliori, A. Guagliardi and G. A. Ozin, *J. Am. Chem. Soc.*, 2006, **128**, 10337.
49. R. B. Soriano, C. D. Malliakas, J. S. Wu and M. G. Kanatzidis, *J. Am. Chem. Soc.*, 2012, **134**, 3228.
50. K. E. Knowles, M. Malicki, R. Parameswaran, L. C. Cass and E. A. Weiss, *J. Am. Chem. Soc.*, 2013, **135**, 7264.
51. C. B. Murray, S. H. Sun, W. Gaschler, H. Doyle, T. A. Betley and C. R. Kagan, *IBM J. Res. Dev.*, 2001, **45**, 47.
52. B. L. Wehrenberg, C. J. Wang and P. Guyot-Sionnest, *J. Phys. Chem. B*, 2002, **106**, 10634.
53. H. Du, C. L. Chen, R. Krishnan, T. D. Krauss, J. M. Harbold, F. W. Wise, M. G. Thomas and J. Silcox, *Nano Lett.*, 2002, **2**, 1321.
54. C. E. Finlayson, A. Amezcua, P. J. A. Sazio, P. S. Walker, M. C. Grossel, R. J. Curry, D. C. Smith and J. J. Baumberg, *J. Mod. Opt.*, 2005, **52**, 955.
55. I. Moreels, B. Fritzinger, J. C. Martins and Z. Hens, *J. Am. Chem. Soc.*, 2008, **130**, 15081.
56. D. K. Britt, Y. Yoon, P. Ercius, T. D. Ewers and A. P. Alivisatos, *Chem. Mater.*, 2013, **25**, 2544.
57. R. Koole, G. Allan, C. Delerue, A. Meijerink, D. Vanmaekelbergh and A. J. Houtepen, *Small*, 2008, **4**, 127.
58. D. K. Smith, J. M. Luther, O. E. Semonin, A. J. Nozik and M. C. Beard, *ACS Nano*, 2011, **5**, 183.
59. Y. W. Liu, D. K. Ko, S. J. Oh, T. R. Gordon, D. N. Vicky, T. Paik, Y. J. Kang, X. C. Ye, L. H. Jin, C. R. Kagan, and C. B. Murray, *Chem. Mater.*, 2011, **23**, 4657.
60. R. B. Soriano, I. U. Arachchige, C. D. Malliakas, J. S. Wu and M. G. Kanatzidis, *J. Am. Chem. Soc.*, 2013, **135**, 768.
61. C. Chubilleau, B. Lenoir, S. Migot and A. Dauscher, *J. Colloid Interface Sci.*, 2011, **357**, 13.
62. A. Aharoni, T. Mokari, I. Popov and U. Banin, *J. Am. Chem. Soc.*, 2006, **128**, 257.
63. J. P. Zimmer, S. W. Kim, S. Ohnishi, E. Tanaka, J. V. Frangioni and M. G. Bawendi, *J. Am. Chem. Soc.*, 2006, **128**, 2526.
64. R. Xie, K. Chen, X. Chen and X. Peng, *Nano Res.*, 2008, **1**, 457.
65. R. G. Xie and X. G. Peng, *J. Am. Chem. Soc.*, 2009, **131**, 10645.
66. Y. Du, B. Xu, T. Fu, M. Cai, F. Li, Y. Zhang and Q. Wang, *J. Am. Chem. Soc.*, 2010, **132**, 1470.
67. M. Yarema, S. Pichler, M. Sytnyk, R. Seyrkammer, R. T. Lechner, G. Fritz-Popovski, D. Jarzab, K. Szendrei, R. Resel and O. Korovyanko, *ACS Nano*, 2011, **5**, 3758.
68. P. Jiang, Z. Q. Tian, C. N. Zhu, Z. L. Zhang and D. W. Pang, *Chem. Mater.*, 2011, **24**, 3.
69. Y. P. Gu, R. Cui, Z. L. Zhang, Z. X. Xie and D. W. Pang, *J. Am. Chem. Soc.*, 2011, **134**, 79.

70. P. Jiang, C. N. Zhu, Z. L. Zhang, Z. Q. Tian and D. W. Pang, *Biomaterials*, 2012, **33**, 5130.
71. I. Hocaoglu, M. N. Çizmeciyan, R. Erdem, C. Ozen, A. Kurt, A. Sennaroglu and H. Y. Acar, *J. Mater. Chem.*, 2012, **22**, 14674.
72. C. Wang, Y. Wang, L. Xu, D. Zhang, M. Liu, X. Li, H. Sun, Q. Lin and B. Yang, *Small*, 2012, **8**, 3137.
73. J. M. Wang, C. N. Zhu, D. L. Zhu, Z. Q. Tian, Y. Lin and D. W. Pang, *Chem. J. Chin. Univ.*, 2015, 1264.
74. B. Chen, H. Zhong and B. Zou, *Prog. Chem.*, 2011, **23**, 2276.
75. P. M. Allen and M. G. Bawendi, *J. Am. Chem. Soc.*, 2008, **130**, 9240.
76. L. Li, A. Pandey, D. J. Werder, B. P. Khanal, J. M. Pietryga and V. I. Klimov, *J. Am. Chem. Soc.*, 2011, **133**, 1176.
77. S. Liu, H. Zhang, Y. Qiao and X. Su, *RSC Adv.*, 2012, **2**, 819.
78. B. Chen, H. Zhong, W. Zhang, Z. a. Tan, Y. Li, C. Yu, T. Zhai, Y. Bando, S. Yang and B. Zou, *Adv. Funct. Mater.*, 2012, **22**, 2081.
79. H. Zhong, Z. Bai and B. Zou, *J. Phys. Chem. Lett.*, 2012, **3**, 3167.
80. H. Zhong, Z. Wang, E. Bovero, Z. Lu, F. C. J. M. van Veggel and G. D. Scholes, *J. Phys. Chem. C*, 2011, **115**, 12396.
81. L. Mangolini, E. Thimsen and U. Kortshagen, *Nano Lett.*, 2005, **5**, 655.
82. F. Erogbogbo, K. T. Yong, I. Roy, R. Hu, W. C. Law, W. Zhao, H. Ding, F. Wu, R. Kumar and M. T. Swihart, *ACS Nano*, 2010, **5**, 413.
83. Y. He, Y. Zhong, F. Peng, X. Wei, Y. Su, Y. Lu, S. Su, W. Gu, L. Liao and S. T. Lee, *J. Am. Chem. Soc.*, 2011, **133**, 14192.
84. J. H. Park, L. Gu, G. von Maltzahn, E. Ruoslahti, S. N. Bhatia and M. J. Sailor, *Nat. Mater.*, 2009, **8**, 331.
85. D. C. Lee, J. M. Pietryga, I. Robel, D. J. Werder, R. D. Schaller and V. I. Klimov, *J. Am. Chem. Soc.*, 2009, **131**, 3436.
86. D. A. Ruddy, J. C. Johnson, E. R. Smith and N. R. Neale, *ACS Nano*, 2010, **4**, 7459.
87. E. J. Henderson, M. Seino, D. P. Puzzo and G. A. Ozin, *ACS Nano*, 2010, **4**, 7683.
88. C. C. Fu, H. Y. Lee, K. Chen, T. S. Lim, H. Y. Wu, P. K. Lin, P. K. Wei, P. H. Tsao, H. C. Chang and W. Fann, *Proc. Natl. Acad. Sci. U. S. A.*, 2007, **104**, 727.
89. Y. P. Sun, B. Zhou, Y. Lin, W. Wang, K. S. Fernando, P. Pathak, M. J. Meziani, B. A. Harruff, X. Wang and H. Wang, *J. Am. Chem. Soc.*, 2006, **128**, 7756.
90. Z. Liu, S. M. Tabakman, Z. Chen and H. Dai, *Nat. Protoc.*, 2009, **4**, 1372.
91. V. N. Mochalin, O. Shenderova, D. Ho and Y. Gogotsi, *Nat. Nanotechnol.*, 2012, **7**, 11.
92. S. N. Baker and G. A. Baker, *Angew. Chem., Int. Ed.*, 2010, **49**, 6726.
93. H. Mattoussi, G. Palui and H. B. Na, *Adv. Drug Delivery*, 2012, **64**, 138.
94. P. Zrazhevskiy, M. Sena and X. Gao, *Chem. Soc. Rev.*, 2010, **39**, 4326.
95. W. W. Yu, J. C. Falkner, B. S. Shih and V. L. Colvin, *Chem. Mater.*, 2004, **16**, 3318.

96. B. R. Hyun, H. Y. Chen, D. A. Rey, F. W. Wise and C. A. Batt, *J. Phys. Chem. B*, 2007, **111**, 5726.

97. S. Hinds, S. Myrskog, L. Levina, G. Koleilat, J. Yang, S. O. Kelley and E. H. Sargent, *J. Am. Chem. Soc.*, 2007, **129**, 7218.

98. L. Etgar, E. Lifshitz and R. Tannenbaum, *J. Mater. Res.*, 2008, **23**, 899.

99. W. Lin, K. Fritz, G. Guerin, G. R. Bardajee, S. Hinds, V. Sukhovatkin, E. H. Sargent, G. D. Scholes and M. A. Winnik, *Langmuir*, 2008, **24**, 8215.

100. G. Zhao, D. F. Wang, T. Zhang, M. Chaker and D. L. Ma, *Chem. Commun.*, 2010, **46**, 5301.

101. H. G. Zhao, D. F. Wang, M. Chaker and D. L. Ma, *J. Phys. Chem. C*, 2011, **115**, 1620.

102. H. G. Zhao, M. Chaker and D. L. Ma, *Phys. Chem. Chem. Phys.*, 2010, **12**, 14754.

103. T. Zhang, B. Hu, L. F. Sun, R. Hovden, F. W. Wise, D. A. Muller and R. D. Robinson, *Nano Lett.*, 2011, **11**, 5356.

104. W. R. Algar, K. Susumu, J. B. Delehanty and I. L. Medintz, *Anal. Chem.*, 2011, **83**, 8826.

105. B. Dubertret, P. Skourides, D. J. Norris, V. Noireaux, A. H. Brivanlou and A. Libchaber, *Science*, 2002, **298**, 1759.

106. N. Erathodiyil and J. Y. Ying, *Acc. Chem. Res.*, 2011, **44**, 925.

107. I. Medintz, *Nat. Mater.*, 2006, **5**, 842.

108. T. Yong, H. Ding, I. Roy, W. C. Law, E. J. Bergey, A. Maitra and P. N. Prasad, *ACS Nano*, 2009, **3**, 502.

109. A. Wolcott, D. Gerion, M. Visconte, J. Sun, A. Schwartzberg, S. W. Chen and J. Z. Zhang, *J. Phys. Chem. B*, 2006, **110**, 5779.

110. P. Diagaradjane, J. M. Orenstein-Cardona, N. E. Colon-Casasnovas, A. Deorukhkar, S. Shentu, N. Kuno, D. L. Schwartz, J. G. Gelovani and S. Krishnan, *Clin. Cancer Res.*, 2008, **14**, 731.

111. J. Qian, K. T. Yong, I. Roy, T. Y. Ohulchanskyy, E. J. Bergey, H. H. Lee, K. M. Tramposch, S. He, A. Maitra and P. N. Prasad, *J. Phys. Chem. B*, 2007, **111**, 6969.

112. P. Zhang, S. Liu, D. Gao, D. Hu, P. Gong, Z. Sheng, J. Deng, Y. Ma and L. Cai, *J. Am. Chem. Soc.*, 2012, **134**, 8388.

113. H. Shroff, C. G. Galbraith, J. A. Galbraith, H. White, J. Gillette, S. Olenych, M. W. Davidson and E. Betzig, *Proc. Natl. Acad. Sci. U. S. A.*, 2007, **104**, 20308.

114. H. Kobayashi, Y. Koyama, T. Barrett, Y. Hama, C. A. Regino, I. S. Shin, B. S. Jang, N. Le, C. H. Paik and P. L. Choyke, *ACS Nano*, 2007, **1**, 258.

115. Y. Hama, Y. Koyama, Y. Urano, P. L. Choyke and H. Kobayashi, *J. Invest. Dermatol.*, 2007, **127**, 2351.

116. T. Egawa, K. Hanaoka, Y. Koide, S. Ujita, N. Takahashi, Y. Ikegaya, N. Matsuki, T. Terai, T. Ueno, T. Komatsu and T. Nagano, *J. Am. Chem. Soc.*, 2011, **133**, 14157.

117. K. Yamauchi, M. Yang, P. Jiang, M. Xu, N. Yamamoto, H. Tsuchiya, K. Tomita, A. R. Moossa, M. Bouvet and R. M. Hoffman, *Cancer Res.*, 2006, **66**, 4208.

118. M. Ogawa, N. Kosaka, P. L. Choyke and H. Kobayashi, *Cancer Res.*, 2009, **69**, 1268.
119. J. Guo, S. Wang, N. Dai, Y. N. Teo and E. T. Kool, *Proc. Natl. Acad. Sci. U. S. A.*, 2011, **108**, 3493.
120. S. T. Laughlin, J. M. Baskin, S. L. Amacher and C. R. Bertozzi, *Science*, 2008, **320**, 664.
121. J. Liu, S. K. Lau, V. A. Varma, R. A. Moffitt, M. Caldwell, T. Liu, A. N. Young, J. A. Petros, A. O. Osunkoya and T. Krogstad, *ACS Nano*, 2010, **4**, 2755.
122. V. Yezhelyev, A. Al-Hajj, C. Morris, A. I. Marcus, T. Liu, M. Lewis, C. Cohen, P. Zrazhevskiy, J. W. Simons, A. Rogatko, S. Nie, X. Gao and R. M. O'Regan, *Adv. Mater.*, 2007, **19**, 3146.
123. J. Liu, S. K. Lau, V. A. Varma, B. A. Kairdolf and S. Nie, *Anal. Chem.*, 2010, **82**, 6237.
124. H. Kobayashi, Y. Hama, Y. Koyama, T. Barrett, C. A. Regino, Y. Urano and P. L. Choyke, *Nano Lett.*, 2007, **7**, 1711.
125. J. H. Park, G. von Maltzahn, E. Ruoslahti, S. N. Bhatia and M. J. Sailor, *Angew. Chem., Int. Ed.*, 2008, **120**, 7394.
126. T. Jin, Y. Yoshioka, F. Fujii, Y. Komai, J. Seki and A. Seiyama, *Chem. Commun.*, 2008, 5764.
127. F. Erogbogbo, K. T. Yong, R. Hu, W. C. Law, H. Ding, C. W. Chang, P. N. Prasad and M. T. Swihart, *ACS Nano*, 2010, **4**, 5131.
128. W. Cai, K. Chen, Z. B. Li, S. S. Gambhir and X. Chen, *J. Nucl. Med.*, 2007, **48**, 1862.
129. K. Chen, Z. B. Li, H. Wang, W. Cai and X. Chen, *Eur. J. Nucl. Med. Mol. Imaging*, 2008, **35**, 2235.
130. C. Zhou, G. Hao, P. Thomas, J. Liu, M. Yu, S. Sun, O. K. Oz, X. Sun and J. Zheng, *Angew. Chem., Int. Ed.*, 2012, **51**, 10118.
131. A. Razgulin, N. Ma and J. Rao, *Chem. Soc. Rev.*, 2011, **40**, 4186.
132. M. S. Shim and Y. J. Kwon, *Adv. Drug Delivery Rev.*, 2012, **64**, 1046.
133. M. K. So, C. Xu, A. M. Loening, S. S. Gambhir and J. Rao, *Nat. Biotechnol.*, 2006, **24**, 339.
134. I. L. Medintz, T. Pons, K. Susumu, K. Boeneman, A. M. Dennis, D. Farrell, J. R. Deschamps, J. S. Melinger, G. Bao and H. Mattoussi, *J. Phys. Chem. C*, 2009, **113**, 18552.
135. T. Machleidt, C. C. Woodroofe, M. Schwinn, J. Mendez, M. B. Robers, K. Zimmerman, P. Otto, D. L. Daniels, T. A. Kirkland and K. V. Woods, *ACS Chem. Biol.*, 2015, **10**, 1797.
136. C. E. Badr and B. A. Tannous, *Trends Biotechnol.*, 2011, **29**, 624.
137. C. Chen, P. Zhang, G. Gao, D. Gao, Y. Yang, H. Liu, Y. Wang, P. Gong and L. Cai, *Adv. Mater.*, 2014, **26**, 6313.
138. C. Chen, P. Zhang, L. Zhang, D. Gao, G. Gao, Y. Yang, W. Li, P. Gong and L. Cai, *Chem. Commun.*, 2015, **51**, 11162.
139. M. Smith, H. Duan, A. M. Mohs and S. Nie, *Adv. Drug Delivery Rev.*, 2008, **60**, 1226.
140. K. T. Yong, I. Roy, W. C. Law and R. Hu, *Chem. Commun.*, 2010, **46**, 7136.

141. X. Gao, Y. Cui, R. M. Levenson, L. W. Chung and S. Nie, *Nat. Biotechnol.*, 2004, **22**, 969.
142. Z. Yan, G. Hong, Y. Zhang, G. Chen, F. Li, H. Dai and Q. Wang, *ACS Nano*, 2012, **6**, 3695.
143. J. Gao, K. Chen, R. Xie, J. Xie, Y. Yan, Z. Cheng, X. Peng and X. Chen, *Bioconjugate Chem.*, 2010, **21**, 604.
144. J. Gao, K. Chen, Z. Miao, G. Ren, X. Chen, S. S. Gambhir and Z. Cheng, *Biomaterials*, 2011, **32**, 2141.
145. D. Deng, Y. Chen, J. Cao, J. Tian, Z. Qian, S. Achilefu and Y. Gu, *Chem. Mater.*, 2012, **24**, 3029.
146. K. T. Yong, I. Roy, R. Hu, H. Ding, H. Cai, J. Zhu, X. Zhang, E. J. Bergey and P. N. Prasad, *Integr. Biol.*, 2010, **2**, 121.
147. H. Chen, B. Li, X. Ren, S. Li, Y. Ma, S. Cui and Y. Gu, *Biomaterials*, 2012, **33**, 8461.
148. G. Hong, J. T. Robinson, Y. Zhang, S. Diao, A. L. Antaris, Q. Wang and H. Dai, *Angew. Chem., Int. Ed.*, 2012, **51**, 9818.
149. J. Gao, K. Chen, R. Luong, D. M. Bouley, H. Mao, T. Qiao, S. S. Gambhir and Z. Cheng, *Nano Lett.*, 2012, **12**, 281.
150. S. Kim, Y. T. Lim, E. G. Soltesz, A. M. De Grand, J. Lee, A. Nakayama, J. A. Parker, T. Mihaljevic, R. G. Laurence, D. M. Dor, L. H. Cohn, M. G. Bawendi and J. V. Frangioni, *Nat. Biotechnol.*, 2004, **22**, 93.
151. B. Ballou, L. A. Ernst, S. Andreko, T. Harper, J. A. Fitzpatrick, A. S. Waggoner and M. P. Bruchez, *Bioconjugate Chem.*, 2007, **18**, 389.
152. E. Pic, T. Pons, L. Bezdetnaya, A. Leroux, F. Guillemin, B. Dubertret and F. Marchal, *Mol. Imaging. Biol.*, 2010, **12**, 394.
153. T. Pons, E. Pic, N. Lequeux, E. Cassette, L. Bezdetnaya, F. Guillemin, F. Marchal and B. Dubertret, *ACS Nano*, 2010, **4**, 2531.
154. M. Helle, E. Cassette, L. Bezdetnaya, T. Pons, A. Leroux, F. Plenat, F. Guillemin, B. Dubertret and F. Marchal, *PLoS One*, 2012, **7**, e44433.
155. Y. T. Lim, S. Kim, A. Nakayama, N. E. Stott, M. G. Bawendi and J. V. Frangioni, *Mol. Imaging*, 2003, **2**, 50.
156. J. D. Smith, G. W. Fisher, A. S. Waggoner and P. G. Campbell, *Microvasc. Res.*, 2007, **73**, 75.
157. H. S. Choi, W. Liu, P. Misra, E. Tanaka, J. P. Zimmer, B. Itty Ipe, M. G. Bawendi and J. V. Frangioni, *Nat. Biotechnol.*, 2007, **25**, 1165.
158. H. S. Han, J. D. Martin, J. Lee, D. K. Harris, D. Fukumura, R. K. Jain and M. Bawendi, *Angew. Chem., Int. Ed.*, 2013, **52**, 1414.
159. M. L. Schipper, Z. Cheng, S. W. Lee, L. A. Bentolila, G. Iyer, J. Rao, X. Chen, A. M. Wu, S. Weiss and S. S. Gambhir, *J. Nucl. Med.*, 2007, **8**, 1511.
160. H. S. Choi, B. I. Ipe, P. Misra, J. H. Lee, M. G. Bawendi and J. V. Frangioni, *Nano Lett.*, 2009, **9**, 2354.
161. M. L. Schipper, G. Iyer, A. L. Koh, Z. Cheng, Y. Ebenstein, A. Aharoni, S. Keren, L. A. Bentolila, J. Li, J. Rao, X. Chen, U. Banin, A. M. Wu, R. Sinclair, S. Weiss and S. S. Gambhir, *Small*, 2009, **5**, 126.
162. Y. Zhang, Y. Zhang, G. Hong, W. He, K. Zhou, K. Yang, F. Li, G. Chen, Z. Liu, H. Dai and Q. Wang, *Biomaterials*, 2013, **34**, 3639.

163. A. Jayagopal, P. K. Russ and F. R. Haselton, *Bioconjugate Chem.*, 2007, **18**, 1424.
164. W. Cai and X. Chen, *Nat. Protoc.*, 2008, **3**, 89.
165. R. Hu, K. T. Yong, I. Roy, H. Ding, W. C. Law, H. Cai, X. Zhang, L. A. Vathy, E. J. Bergey and P. N. Prasad, *Nanotechnology*, 2010, **21**, 145105.
166. B. R. Smith, Z. Cheng, A. De, A. L. Koh, R. Sinclair and S. S. Gambhir, *Nano Lett.*, 2008, **8**, 2599.
167. E. B. Voura, J. K. Jaiswal, H. Mattoussi and S. M. Simon, *Nat. Med.*, 2004, **10**, 993.
168. M. Stroh, J. P. Zimmer, D. G. Duda, T. S. Levchenko, K. S. Cohen, E. B. Brown, D. T. Scadden, V. P. Torchilin, M. G. Bawendi, D. Fukumura and R. K. Jain, *Nat. Med.*, 2005, **11**, 678.
169. E. B. Garon, L. Marcu, Q. Luong, O. Tcherniantchouk, G. M. Crooks and H. P. Koeffler, *Leuk. Res.*, 2007, **31**, 643.
170. C. Shi, Y. Zhu, Z. Xie, W. Qian, C. L. Hsieh, S. Nie, Y. Su, H. E. Zhau and L. W. Chung, *Urology*, 2009, **74**, 446.
171. H. H. Wang, C. A. J. Lin, C. H. Lee, Y. C. Lin, Y. M. Tseng, C. L. Hsieh, C. H. Chen, C. H. Tsai, C. T. Hsieh and J. L. Shen, *ACS Nano*, 2011, **5**, 4337.
172. O. Seleverstov, O. Zabirnyk, M. Zscharnack, L. Bulavina, M. Nowicki, J. M. Heinrich, M. Yezhelyev, F. Emmrich, R. O'Regan and A. Bader, *Nano Lett.*, 2006, **6**, 2826.
173. B. S. Shah, P. A. Clark, E. K. Moioli, M. A. Stroscio and J. J. Mao, *Nano Lett.*, 2007, **7**, 3071.
174. Y. Lei, H. Tang, L. Yao, R. Yu, M. Feng and B. Zou, *Bioconjugate Chem.*, 2007, **19**, 421.
175. Y. Higuchi, C. Wu, K. L. Chang, K. Irie, S. Kawakami, F. Yamashita and M. Hashida, *Biomaterials*, 2011, **32**, 6676.
176. S. Ranjbarvaziri, S. Kiani, A. Akhlaghi, A. Vosough, H. Baharvand and N. Aghdami, *Biomaterials*, 2011, **32**, 5195.
177. H. Yukawa, Y. Kagami, M. Watanabe, K. Oishi, Y. Miyamoto, Y. Okamoto, M. Tokeshi, N. Kaji, H. Noguchi, K. Ono, M. Sawada, Y. Baba, N. Hamajima and S. Hayashi, *Biomaterials*, 2010, **31**, 4094.
178. H. Yukawa, M. Watanabe, N. Kaji, Y. Okamoto, M. Tokeshi, Y. Miyamoto, H. Noguchi, Y. Baba and S. Hayashi, *Biomaterials*, 2012, **33**, 2177.
179. A. M. Smith, M. C. Mancini and S. M. Nie, *Nat. Nanotechnol.*, 2009, **4**, 710.
180. J. V. Frangioni, *Curr. Opin. Chem. Biol.*, 2003, **7**, 626.
181. A. N. Bashkatov, E. A. Genina, V. I. Kochubey and V. V. Tuchin, *J. Phys. D: Appl. Phys.*, 2005, **38**, 2543.
182. G. S. Hong, J. C. Lee, J. T. Robinson, U. Raaz, L. M. Xie, N. F. Huang, J. P. Cooke and H. J. Dai, *Nat. Med.*, 2012, **18**, 1841.
183. G. S. Hong, J. C. Lee, A. Jha, S. Diao, K. H. Nakayama, L. Hou, T. C. Doyle, J. T. Robinson, A. L. Antaris, H. Dai, J. P. Cooke and N. F. Huang, *Circ. Cardiovasc. Imaging*, 2014, **7**, 517.

184. M. J. O'Connell, S. M. Bachilo, C. B. Huffman, V. C. Moore, M. S. Strano, E. H. Haroz, K. L. Rialon, P. J. Boul, W. H. Noon, C. Kittrell, J. P. Ma, R. H. Hauge, R. B. Weisman and R. E. Smalley, *Science*, 2002, **297**, 593.
185. L. Bakueva, I. Gorelikov, S. Musikhin, X. S. Zhao, E. H. Sargent and E. Kumacheva, *Adv. Mater.*, 2004, **16**, 926.
186. J. Crochet, M. Clemens and T. Hertel, *J. Am. Chem. Soc.*, 2007, **129**, 8058.
187. L. J. Carlson, S. E. Maccagnano, M. Zheng, J. Silcox and T. D. Krauss, *Nano Lett.*, 2007, 7, 3698.
188. K. Welsher, Z. Liu, S. P. Sherlock, J. T. Robinson, Z. Chen, D. Daranciang and H. J. Dai, *Nat. Nanotechnol.*, 2009, **4**, 773.
189. M. Harrison, S. Kershaw, M. Burt, A. Eychmüller, H. Weller and A. Rogach, *Mater. Sci. Eng., B*, 2000, **69**, 355.
190. P. Zrazhevskiy, M. Sena and X. Gao, *Chem. Soc. Rev.*, 2010, **39**, 4326.

CHAPTER 3

Bioimaging Nanomaterials Based on Carbon Dots

ZHENHUI KANG*[a] AND YANG LIU[a]

[a]Jiangsu Key Laboratory for Carbon-Based Functional Materials & Devices, Institute of Functional Nano & Soft Materials (FUNSOM), Soochow University, 199 Ren'ai Road, Suzhou, 215123, Jiangsu, P. R. China
*E-mail: zhkang@suda.edu.cn

3.1   Synthesis Methods

3.1.1   Synthesis of CDs

Methods of carbon dot (CD) fabrication proposed during the last decade can be roughly classified into "top-down" and "bottom-up" approaches. The top-down route is implemented *via* either physical or chemical techniques, most often the latter. The mechanism of chemical type top-down routes can be described as defect-mediated fragmentation processes in which oxygen-containing functional groups create defects on graphite sheets and serve as chemically reactive sites, thus allowing graphite to be cleaved into smaller fragments. Typical top-down methods include electron beam lithography, laser ablation, acidic exfoliation, electrochemical oxidation, and microwave-assisted hydrothermal synthesis. Top-down routes for the preparation of CDs have the advantages of abundant raw materials, large-scale production, and simple operation.

RSC Nanoscience & Nanotechnology No. 40
Near Infrared Nanomaterials: Preparation, Bioimaging and Therapy Applications
Edited by Fan Zhang
© The Royal Society of Chemistry 2016
Published by the Royal Society of Chemistry, www.rsc.org

CDs can also be prepared through bottom-up routes, including solution chemistry, cyclodehydrogenation of polyphenylene precursors, carbonization of some special organic precursors, or the fragmentation of suitable precursors, *e.g.* C_{60}. Compared with the top-down routes, reports concerning the bottom-up routes are relatively scarce. Bottom-up methods offer opportunities to control the CDs with well-defined molecular size, shape, and thus properties. Nevertheless, these methods always involve complex synthetic procedures, and the special organic precursors may be difficult to obtain.

Whichever route is chosen, there are three problems facing CDs fabrication that need to be considered:

- carbonaceous aggregation during carbonization, which can be avoided by using electrochemical synthesis, confined pyrolysis, or solution chemistry methods
- size control and uniformity, which is important for uniform properties and mechanistic study, and can be optimized *via* post-treatment, such as gel electrophoresis, centrifugation, and dialysis
- surface properties that are critical for solubility and selected applications, which can be tuned during preparation or post-treatment.

For particular applications, it is important to control the sizes of CDs to obtain uniform properties. Many approaches have been proposed to obtain uniform CDs during preparation or post-treatment. In most reports, the as-synthesized CDs were purified *via* post-treatments such as filtration, dialysis, centrifugation, column chromatography, and gel electrophoresis.

Surface modification is a powerful method of tuning the surface properties of materials for selected applications. There are many approaches for functionalizing the surface of CDs through surface chemistry or interactions, such as covalent bonding, coordination, p–p interactions, and sol–gel technology. Doping is a widely used approach to tune the photoluminescence (PL) properties of CDs; and various doping methods with a range of elements such as N, S, and P have also been reported to tune the PL and related optical properties of CDs.

3.1.2 Synthesis of NDs

Many methods have been reported for production of nanodiamonds (NDs) such as laser ablation,[1-3] plasma-assisted chemical vapor deposition,[4] autoclave synthesis from supercritical fluids,[5] ion irradiation of graphite,[6] chlorination of carbides,[7] electron irradiation of carbon onions,[8] and ultrasound cavitation.[9] Smaller NDs can be prepared by detonation processes that yield aggregates of NDs with sizes of 4–5 nm embedded in a detonation soot composed of other carbon allotropes and impurities.[10,11] An explosive mixture having an overall negative oxygen balance provides a source of both carbon and energy for the conversion.[12,13] Because of their small size (2–10 nm) detonation NDs have also been referred to as ultradispersed,[11,13] nanocrystalline diamonds,[12] or single-digit NDs. Actually, detonation NDs form extremely tight core aggregates that withstand conventional disintegration technique (*e.g.* ultrasonication).[14] Ozone-purified 3–5 nm detonation NDs[15] form a stable acidic hydrosol that can be disaggregated by milling with ceramic microbeads (ZrO_2, *etc.*) or by microbead-assisted ultrasonic disintegration.[16] NDs used for bioimaging can also be prepared from commercial diamond powders resulting from high-energy ball milling of high-pressure high-temperature diamond microcrystals.[17] These materials have a nitrogen content suitable for the creation of nitrogen-vacancy (NV) color centers emitting in the far red. Green-emitting NDs, on the other hand, are prepared by pulverizing natural, nitrogen-rich diamonds.[18] NDs synthesized by detonation techniques[10,19] have also been proposed as PL contrast agents.

3.2   Structures and Properties

3.2.1   Components and Structure

3.2.1.1   *Components and Structure of CDs*

In general, the average size of CDs is <10 nm, and usually depends on the preparation method. Technically speaking, CDs are composed only of elemental C and H. However, due to their strong tendency for aggregation due to face-to-face attraction and the preparation methods such as oxidation-derived exfoliation, CDs reported so far are always partially oxidized, with hydroxyl, epoxy/ether, carbonyl, and carboxylic acid groups on the surfaces. Fourier transform infrared (FTIR) and X-ray photoelectron spectroscopy (XPS) spectra are commonly adopted to analyze their components. The crystalline nature of CDs can be investigated through X-ray diffraction (XRD) patterns, Raman spectroscopy, and high-resolution transmission electron microscopy (HRTEM) observations. Both (002) interlayer spacing and (100) in-plane lattice spacing exist in CDs, and the former has been widely studied. The interlayer spacing of CDs depends strongly on their degree of oxidation: that is, the attached hydroxyl, epoxy/ether, carbonyl, and carboxylic acid groups can increase the interlayer spacing of CDs. The Raman technique is also a powerful and non-destructive tool for the characterization of CDs. The G band is assigned to the E_{2g} vibrational modes of the aromatic domains, whereas the D band arises from the breathing modes of the graphitic domains. Traditionally, the intensity ratio of "disordered" D to crystalline G (I_D/I_G) is used to compare the structural order between crystalline and amorphous graphitic systems. The I_D/I_G values of CDs vary significantly depending on the preparation methods. The HRTEM images of CDs features two kinds of lattice fringes, namely (002) interlayer spacing and (110) in-plane lattice spacing. Similar to the XRD pattern, interlayer spacing centered at about 0.34 nm has been observed from CDs prepared by acidic oxidation from carbon black, microwave-assisted method, or the electrochemical cutting method. The in-plane lattice spacing, mostly centered at about 0.24 nm, has been observed from CDs synthesized *via* a microwave-hydrothermal protocol, amino-hydrothermal method, K intercalation, acidic oxidation from carbon fibers, and a photo-Fenton reaction. The exception is in-plane lattice spacing ~0.21 nm *via* hydrothermal cutting strategy and the glucose carbonization method. These two kinds of lattice fringes have also been simultaneously been observed in CDs synthesized by the electrochemical method from carbon nanotubes (CNTs) and acidic oxidation from natural graphite. Moreover, CDs are not always crystalline: amorphous CDs have also been obtained *via* the hydrothermal method and citric acid carbonization.

3.2.1.2 Components and Structure of NDs

The most important effect of reducing the size of diamonds to the nanoscale is the huge increase in the fraction of surface atoms compared to bulk ones.[1,20–22] Surface energies, indeed, are fundamental in the stabilization of NDs. In the case of hydrogen-terminated structures it was calculated that diamonds <3 nm in diameter are energetically favored over graphitic structures.[23] For sizes of ~3 nm, particles with bare reconstructed surfaces, arising from transition from an sp^3 to an sp^2 lattice,[24] become thermodynamically more stable than those with hydrogenated surfaces.[25] Transmission electron microscope (TEM) images of purified NDs confirm that they consists of polyhedral diamond cores of sp^3 carbon atoms, partially coated by a graphitic shell or amorphous carbon. From the optical point of view, NDs are strongly scattering objects because of their high refractive index but they do not absorb light and do not show any luminescence unless they contain structural crystal defects.[26] Interestingly, diamonds may hosts over 500 kinds of photon-absorbing defects (color centers) some of which emit as bright single-photon sources at room temperature.[27] FTIR can detect functional groups and adsorbed molecules on the surface[28–34] and can also detect changes in the surface chemistry of functionalized NDs.[35–38] Nitrogen defects in NDs manifest themselves as two broad bands in the region 1100–2500 cm^{-1}, which overlaps with the peaks of surface functional groups. The most characteristic features of NDs after oxidative purification include O–H stretch (3200–3600 cm^{-1}) and bend (1630–1640 cm^{-1}) with bands originating from both adsorbed species and from O–H groups covalently attached to the NDs surface; C=O stretch at 1700–1800 cm^{-1} where C=O can be part of ketone, aldehyde, carboxylic acid, ester, anhydride, cyclic ketone, lactone, or lactam; C–H stretch at 2850–3000 cm^{-1}, with bands originating from asymmetric and symmetric C–H stretch vibrations in –CH$_2$ and –CH$_3$ groups. Many NDs also have a very broad absorption feature at 1000–1500 cm^{-1} that is a combination of many overlapping peaks. These include O–H deformational and C–O–C stretch vibrations, epoxy C–O stretch, C–C stretch, amide C–N stretch, and C–N–H deformational vibrations; peaks related to nitrogen defects; and vibrations of other groups. Raman spectroscopy, on the other hand, provides information on the structure, composition, and homogeneity of the material, and also on the surface groups. For example, the peak at ~1640 cm^{-1} in the spectrum of detonation NDs was only recently associated with O–H vibrations.[31] The Raman spectra of NDs with a high sp^2 carbon content are dominated by the D and G bands of graphitic carbon and the diamond signal is weak or absent. With increasing sp^3 content, the intensity of the diamond peak increases, while the D band weakens. Because of the small Raman scattering cross-section of diamond and the shielding effect of graphitic and amorphous carbon around the diamond core, ultraviolet lasers with excitation energy close to the bandgap of diamond (~5.5 eV) are needed to amplify the Raman signal of NDs and suppress the D band of graphitic carbon that may overlap with a weak diamond peak. The NDs peak, is broadened and downshifted with respect to bulk diamond with a shoulder at about 1250 cm^{-1},[29,39–41] originating from smaller NDs or smaller coherent scattering domains separated by defects in larger NDs.

3.2.2 Properties

3.2.2.1 *Properties of CDs*

3.2.2.1.1 Absorption.   CDs typically show strong optical absorption in the ultraviolet (UV) region, with a tail extending out into the visible range. There may be some absorption shoulders attributed to the π–π* transition of the C=C bonds, the n–π* transition of C=O bonds, and/or others. Moreover, CDs prepared *via* different methods also showed different absorption behaviors and the absorption peak position was also dependent on the preparation method. Besides, the variation of oxygen content was reported to play an important role in deciding the absorption peak position of CDs.

3.2.2.1.2 Photoluminescence.   One of the most fascinating features of CDs is their PL. So far, variously sized CDs with different PL colors, ranging from the visible into the near infrared (NIR) region, have been prepared *via* various synthetic approaches. CDs prepared *via* different approaches can emit PL with different colors, including UV, blue, green, yellow, and red. Typically, the luminescence mechanism may derive from intrinsic state emission and defect state emission. However, the exact mechanism of PL for CDs remains unsettled. The luminescence has been tentatively suggested to arise from excitons of carbon, emissive traps, a quantum confinement effect, aromatic structures, oxygen-containing groups, free zigzag sites, and edge defects. Further systematic investigation is required before a mechanism for luminescence emission from CDs is widely accepted. Anyway, the PL of CDs may be attributed to either a combining effect or competition between intrinsic state emission and defect state emission. CDs prepared *via* various methods probably exhibit distinct PL mechanisms, leading to different dependences of their PL on size, excitation wavelength, pH, solvent, and concentration. The quantum yield (QY) of CDs varies with the fabrication method and the surface chemistry involved. For unpassivated CDs, the QYs ranged between 2% and 30%, which are observed in CDs prepared *via* stepwise solution chemistry and microwave-assisted acidic oxidation, respectively. The CDs commonly contain carboxylic and epoxide groups, which can act as non-radiative electron–hole recombination centers. Therefore, the removal of these oxygen-containing groups may improves the QY, either by reduction or surface passivation. Recently, a significantly enhanced QY of CDs (~72%) was reported by Sun *et al.*[42]

3.2.2.1.3 Photoinduced Electron Transfer Property.

It was found that the PL from a CD solution could be efficiently quenched in the presence of either electron acceptors such as 4-nitrotoluene and 2,4-dinitrotoluene or electron donors such as *N,N*-diethylaniline. The photoexcited CDs are excellent as both electron donors and electron acceptors. Efficient PL quenching in CDs by surface-doped metals through disrupting the excited-state redox processes also observed. Electron transfer in nanocomposites of CDs–GO (graphene oxide), CDs–CNTs, and CDs–TiO$_2$ nanoparticles (NPs) without linker molecules was also studied. Significant PL quenching was observed in the CDs–GO system, which was attributed to the ultrafast electron transfer from CDs to GO with a time constant of 400 fs. In comparison, addition of CNTs resulted in static quenching of fluorescence in CDs. No charge transfer was observed in either CDs–MWNT (multiwall nanotubes) or CDs–TiO$_2$ composites. These interesting photoinduced electron transfer properties of CDs as an electron donor/acceptor may offer new opportunities for light energy conversion, catalysis, and related applications, as well as mechanistic elucidation.

3.2.2.1.4 Proton Adsorption.

Because the surface of CDs contains many functional groups, such as –OH, –COOH, and –C=O, the uncontrolled hydrophilic groups of CDs may help to draw H$^+$ closer to the surface of CDs, which is good for enhancing the adsorption capacity. On account of the excellent water solubility of CDs, the adsorption experiments were conducted using a dialysis method. When the adsorption behavior of CDs was studied the results indicated that the adsorption of H$^+$ was extraordinary rapid in the first 4 min, then gradually increased as the contact time was prolonged. After 6 min of adsorption, the amount of H$^+$ remains constant, which suggests that 6 min is the equilibrium time in this adsorption experiment. The amount of H$^+$ adsorbed (based on HCl) was about 10–24 mg g^{-1} for CDs. The sorption equilibrium was further evaluated by two well-known models of Langmuir and Freundlich isotherms. The Langmuir isotherms for CDs have higher correlation coefficients than that of the Freundlich isotherm model, indicating that monolayer adsorption of H$^+$ takes place on the homogeneous surface of CD adsorbents. Moreover, the adsorption is favorable and not readily reversible. To evaluate adsorption kinetics, two common models were applied to the experimental data obtained for adsorption processes. Since calculated correlation coefficients are closer to unity for a pseudo-first-order kinetic model than for a pseudo-second-order kinetic model, the present sorption systems are considered to follow predominantly the first-order model.[43]

3.2.2.1.5 Toxicity.   The toxicity of CDs is a natural concern because of their potential for biological applications. The cytotoxicity of CDs has been evaluated by various research groups, revealing that CDs appear to possess low toxicity. However, Markovic *et al.* demonstrated that the defects and free radicals at the surface of CDs could result in the generation of singlet oxygen. An *in vitro* photodynamic cytotoxicity study showed that photoexcited CDs could cause programmed cell death *via* apoptosis and autophagy. Fortunately, this feature could be exploited in photodynamic therapy.[44] Wu *et al.* investigated the cytotoxicity of CDs in detail. The cytotoxicity of CDs was lower than that of GO sheets, which can be proven by the effects on cell viability, internal cellular reactive oxygen species levels, damage to mitochondrial membrane potential, and cell cycle. The toxicity of CDs did not dramatically increase with an increase in concentration. These results also demonstrated that the CDs are internalized primarily through caveolae-mediated endocytosis.[45]

3.2.2.2 Properties of NDs

NDs have many unique properties include superior hardness and Young's modulus, biocompatibility, optical properties and fluorescence, high thermal conductivity and electrical resistivity, chemical stability, and resistance to harsh environments. Only fluorescence and biocompatibility are discussed in this section; other useful properties of NDs can be found in recent reviews.[28]

3.2.2.2.1 Fluorescence.   The presence of NV centers in NDs leads to useful fluorescence properties. NV centers can be created by irradiating NDs with high-energy particles (electrons, protons, helium ions), followed by vacuum annealing treatment.[46,47] Irradiation forms vacancies in the diamond structure,[48] and during annealing these vacancies migrate and are trapped by the N atoms that are always present in diamond. Two types of NV centers are neutral (NV^0) and negatively charged (NV^-) vacancies, which have different emission spectra. The NV^- center is of particular interest because it has an $S = 1$ spin ground state that can be spin-polarized by optical pumping and manipulated using electron paramagnetic resonance. Moreover, it has a long spin coherence time. Fluorescent NV centers in isotopically clean diamond are of particular interest for quantum computing,[49] and NV centers in NDs are also being investigated for applications in high-resolution magnetic sensing,[50-52] fluorescence resonance energy transfer,[53] and biomedical imaging. Although the concentration of the fluorescent NV defects produced by electron irradiation does not, to a first approximation, depend on the size of the nanocrystals,[54] the fraction of NV^- defects produced by the same technique decreases with decreasing size of the NDs,[47] presumably because of electron traps located at the surface. Intermittent luminescence originating from NV centers has been observed in individual pristine 5 nm detonation NDs produced from a trinitrotoluene (TNT) and hexogen precursor.[52] Stable luminescence has also recently been detected from NV centers trapped in larger

(>20 nm) NDs produced from TNT and hexogen[55] or graphite and hexogen[56] precursors. These results suggest that various factors (such as the amount of N in the precursors and the cooling conditions) have a role in the N incorporation into the NDs core and in possible *in situ* formation of NV centers.[52,56] Fluorescent particles can also be produced by linking[57,58] or adsorbing[59] various fluorophores onto NDs. Fluorophore-conjugated NDs can travel through cellular compartments of varying pH without degradation of the surface-conjugated fluorophore or alteration of cell viability over extended periods of time.[60] Bright blue fluorescent NDs have been produced by covalent linking of octadecylamine to carboxylic groups on NDs surface.[35]

3.2.2.2.2 Cytotoxicity. After preliminary demonstration of the short-term low toxicity of NDs produced from synthetic diamond powders[61] by Chang's group, cytotoxicity of NDs[1] was investigated more in the detail considering parameters such as incubation time, nature of the cells, method of production, and functionalization of the diamond particles, as well as their shape and size.[62] Schrand *et al.* assessed the cytotoxicity of detonation NDs ranging in size from 2 to 10 nm, resulting from the synthesis or after termination with carboxylate or sulfonate groups. Assays of cell viability based on mitochondrial function (MTT) and luminescent ATP production[63] demonstrated that, independently of the functionalization, NDs were not toxic to a variety of cell types (including neuroblastoma and macrophages) and that they did not produce significant reactive oxygen species. Moreover, the authors confirmed that cells can grow on ND-coated substrates without morphological changes compared to control populations. Viability assays were performed after 24 h incubation of cells with NDs. Faklaris *et al.*[64] extended the cytotoxicity study to an incubation time of 48 h and to an ND concentration of 480 μg mL^{-1} in the case of 25 nm NDs prepared from Ib-type diamond powders. Further, MTT assays showed a decrease of cell viability of about 20% and 30% at a probe concentration of 53 and 480 μg mL^{-1} respectively, after 48 h incubation. Although such cytotoxicity levels are quite modest they were not completely in agreement with the previously reported extreme biocompatibility of NDs. This partial contradiction led Vaijayanthimala *et al.* to re-examine the issue of ND cytotoxicity.[65] These authors performed an MTS assay on HeLa cells at 48 h after incubation with different concentrations of NDs and reported the absence of cytotoxicity of NDs on HeLa cells at a concentration up to 200 μg mL^{-1} in the culture. Moreover, they investigated the possibility of extending the use of NDs in research applications with tissue stem cells and progenitor cells. They chose two cell models, 3T3-L1 pre-adipocytes and 489.2 osteoprogenitors, which can undergo adipogenesis and osteogenesis respectively when switched to an appropriate differentiation-inducing medium. At 14 days after induction of differentiation, ND-treated cells showed no significant difference with respect to untreated cultures. Weng *et al.* compared the toxicity on HeLa cells of type Ib 140 nm NDs after functionalization with either transferrin or amino groups.[66] Both transferrin and amine functionalization enforced the inhibition of cellular growth (~50% after 24 h), an effect that was not observed for carboxylated NDs. Moreover, transferrin-functionalized NDs showed a moderate photo-toxicity to HeLa cells when irradiated with a 532 nm continuous laser. The experiment was carried out under powerful light exposure and the green laser itself was reported to kill the untreated HeLa cells after 10 min exposure at 75 W cm^{-2}, while the same effect on transferrin ND-treated cells was observed using less than half the energy required for untreated cells. The actual mechanism of the process was demonstrated by Raman spectroscopy, which revealed a temperature increase of the ND cluster during irradiation, as expected in the case of a photothermal effect. Nevertheless, as expected for highly luminescent materials such as NDs, the heating effect was modest since absorbed energy is mostly released *via* radiative pathways. NDs also show almost no neuronal toxicity up to a concentration of 250 μg mL^{-1}. Nevertheless, time-lapse live cell imaging showed a reduction of neurite length due to the spatial hindrance of NDs on the advancing axonal growth cone. The authors concluded that although NDs indeed exhibited low neuronal toxicity, the interference with neuronal morphogenesis should be taken into consideration when applications involve actively growing neurites, as in the case of nerve regeneration.[67]

3.2.2.2.3 Biocompatibility and Fate in the Body. *In vitro* and *in vivo* studies have been conducted to examine characteristics as diverse as cell viability, gene programme activity, and *in vivo* mechanistic and physiological behavior.[28,68–74] NDs instilled within the trachea were reported to be of low pulmonary toxicity, with the amount of NDs in the alveolar region decreasing with time, and macrophages burdened with NDs were observed in the bronchi for 28 days after exposure.[72] Intravenously administered NDs complexes at high dosages did not change serum indicators of liver and systemic toxicity.[74] To evaluate the fate of NDs and their impact on stress response activity and worm reproduction, fluorescent ND aggregates with average hydrodynamic size of about 120 nm were fed and microinjected into the translucent *Caenorhabditis elegans* worm, and then tracked for several days.[73] Bare NDs typically remained in the worm lumen, whereas NDs coated with dextran or bovine serum albumin (BSA) were absorbed into the intestinal cells. NDs that were microinjected into worm gonads were transferred into the larvae and offspring, but this had no impact on the reproductive capabilities or survival of the worms. Further experiments involving DAF-16:GFP (DAF-16 is a gene group that regulates the stress and immune response of cells; GFP is green fluorescent protein) confirmed that fluorescent NDs are not toxic and do not induce stress in the worm model, thus providing support for their use in *in vivo* imaging. But given the number of surface modifications that are possible, it is important to be certain that functionalized NDs intended for biomedical applications remain safe. Therefore, we have recently compared the cytotoxicity and osteoblast proliferation and gene expression effects of carboxylated NDs, octadecylamine-modified NDs (NDs–ODA), and composites of poly(L-lactic acid) with NDs–ODA.[68] Although no harmful effects were found in any of these materials, toxicity and biocompatibility testing of new NDs-based materials should continue.

3.2.2.2.4   Internalization.   Weng *et al.* compared the kinetics of the uptake of NDs by HeLa cells after functionalization with either transferrin or amino groups.[66] During the cellular uptake, transferrin-bearing particles reached their saturation values with a half-time of about 0.8 h, about twice as fast as the amino-terminated NDs. The authors concluded that the receptor-mediated endocytosis was more effective than the endocytic process involving surface electrostatic interactions. The internalization pathways of 25 nm NDs was investigated by Faklaris *et al.* in HeLa cancer cells with endosomal marking and colocalization analyses.[64] A rather low degree of colocalization was observed between early endosomes and the ND emission. In particular, after 2 h of incubation colocalization was observed for about 21% of the particles with an error margin of 6%, analyzing 20 cells and 256 internalized NDs. They proposed two possible interpretations of this result: either the NDs do not enter the cells by endocytosis, or they are liberated early from the endosomes in the cytosol or in the lysosomes. In a later study, the same authors analyzed in detail the internalization mechanism of NDs with 46 nm hydrodynamic diameter in living HeLa cells.[75] Selectively blocking the different possible internalization pathways by using specific drugs, the authors concluded that NDs enter the cancer cells mainly by clathrin-mediated endocytosis. Intracellular localization of NDs was analyzed by immunofluorescence and TEM. Results confirmed that smallest particles appeared to be free in the cytosol. This study also revealed an effect of the NDs' size on their internalization by cancer cells. A high degree of colocalization between vesicles and the biggest nanoparticles or aggregates was in fact detected. Internalization of NDs is expected to takes place with different efficiency and mechanism in different target cells. Perevedentseva *et al.* investigated this issue by comparing the interaction of NDs with cancer and non-cancer cells.[76] In particular, internalization of 100 nm NDs by A549 human lung adenocarcinoma cell, Beas-2b non-tumorigenic human bronchial epithelial cell, and HFL-1 fibroblast-like human fetal lung cell were studied, comparing the dependence on dose and time of the uptake by confocal fluorescence imaging and Raman imaging methods. The main mechanism of internalization by cells was confirmed to be *via* clathrin-dependent endocytosis, for both cancer and non-cancer cells. Moreover, the uptake was quantitatively larger in the case of the cancer cells. Shape is another factor that regulates the interaction of NPs with cellular systems. According to Chu *et al.*, the effect of morphology of NDs is sufficiently relevant to determine independently their cellular fate.[77] This result was demonstrate using NDs with a physical size distribution of tens to hundreds of nanometers and an aspect ratio close to 1 but which exhibited at HRTEM irregular shapes and in most cases at least one or two sharp corners.[77] A combination of TEM and confocal microscopy, using fluorescence-based colocalization techniques, revealed that NDs having irregular shapes with sharp corners escaped from endosome to cytoplasm soon (a couple of hours on average) after their cellular uptake.[78] On the contrary, rounded NDs, obtained by direct chemical etching of the prickly ones, demonstrated stable endosomal residence and effective cellular excretion.

3.3 Bioimaging Based on CDs

3.3.1 Cellular Uptake and Fluorescence Imaging (Ref. 79)

CDs in most configurations are readily taken up by cells, enabling fluorescence imaging of cells with both one- and multiple-photon excitations, as demonstrated in early studies of CDs.[80–82] More recently, Zhang *et al.* prepared dots from polydopamine (PDA-FONs) for cell imaging. At 405 nm and 458 nm excitation, bright green and yellow fluorescence emissions, respectively, were observed in the cytoplasm, but not in the cell nucleus.[83] Yan *et al.* used CDs from cellulose and cyclodextrin *via* hydrothermal synthesis for cell imaging. Upon incubation with mouse melanoma cells (B16-F10) for 5 h, the CDs were taken up by the cells and meaningful blue, yellow, and red fluorescence emissions were observed corresponding to different excitation wavelengths, though the imaging resolution was not high enough for a more precise determination of the locations of the dots.[84] Zhang *et al.* reported the synthesis of aqueous compatible CDs in one-pot hydrothermal processing of ND.[85] The CDs were internalized into the NIH-3T3 cells upon incubation, with green and yellow fluorescence emissions at 405 nm and 458 nm excitations, respectively, in the cell cytoplasm.[85] Wei *et al.* prepared CDs from paper ash and demonstrated that these dots could be efficiently taken up by human L02 hepatic cells, with rather low cytotoxicity.[86] In some imaging experiments CDs were also found in the cell membranes. For example, Chen *et al.* prepared CDs by carbonizing sucrose with oleic acid for imaging 16HBE cells. Green fluorescence emissions were observed around the cell membrane, in addition to the cytoplasm, though only much weaker fluorescence was detected in the cell nucleus.[87] Zhang *et al.* used one-pot hydrothermal synthesis with BSA as the carbon source and 4,7,10-trioxa-1,13-tridecanediamine (TTDDA) as the surface passivation agent to prepare CDs for imaging colorectal carcinoma cells (SW1116).[88] After incubation for only 3 h, the CDs were internalized into the cells, exhibiting blue fluorescence emissions around the cell membrane and cytoplasm region under UV excitation.[88] Similarly, Ding *et al.* applied the hydrothermal synthesis to the preparation of TTDDA-passivated CDs.[89] These CDs were used to label HeLa cells, exhibiting low toxicity on the cells even at high concentrations of up to 5 mg mL^{-1}.[89] CDs from natural products or more biocompatible precursors were also evaluated in cell imaging experiments. For example, Sahu *et al.* used CDs from orange juice in the imaging of MG-63 cells.[90] Blue and green fluorescence emissions at 405 nm and 488 nm excitations, respectively, were observed in the cell cytoplasm, but no fluorescence was found in cell nuclei.[90] Sachdev *et al.* applied microwave-assisted pyrolysis to the one-step synthesis of PEG-4000 passivated CDs from chitosan.[91] The CDs were evaluated for their potential use in biolabeling in *S. aureus* and recombinant GFP-expressing *E. coli* as model systems.[91] Saxena *et al.* isolated water-soluble CDs from charred bread for the fluorescence imaging of human red blood cells.[92] The CDs were found to be relatively non-toxic to the human erythrocytes, and even exhibited higher fluorescence QYs the presence of the blood plasma, which was rationalized as being due to interactions of biomolecules in the blood plasma with the CDs.[92] Saxena *et al.* also synthesized CDs from wood wool for fluorescence imaging over the full life cycle of different species of mosquitoes (*Anopheles* sp., *Aedes* sp., and *Culex* sp. larvae). At higher concentrations the CDs could have growth restriction effects, slowing down the metabolism in the mosquito larvae, thus preventing their reaching the pupa stage and eventually leading to the death of the mosquitoes.[93]

Several reports have highlighted the use of doped CDs for cell imaging. Xu *et al.* used N-doped CDs for the fluorescence labeling of HeLa cells and HepG2 cells. Multicolor fluorescence emissions corresponding to different excitation wavelengths were observed in the cell cytoplasm.[94] Chandra *et al.* prepared S-doped CDs from TMA. These dots could be readily dispersed in water, with high photostability. In the imaging of *E. coli* cells, blue fluorescence emissions were observed at 405 nm excitation. The negatively charged O and S groups in the doped CDs could apparently bind with positively charged DNA–PEI complexes, resulting in bright fluorescence.[95] Wu *et al.* synthesized N-doped amphoteric CDs from natural silk, as *Bombyx mori* silk contains a high percentage of N. The doped CDs were amphoteric, depending on the solution pH, and stable in resisting relatively harsh chemical environments. The use of these dots as probes for the fluorescence imaging of A549 cells was evaluated.[96] A549 cells were also used in the study by Qu *et al.*, with CDs from dopamine as precursor in the hydrothermal synthesis.[97] Beyond cell imaging, the CDs were employed in the determination of Fe^{3+} ions in water samples and dopamine in human urine and serum samples, taking advantage of the distinctive catechol groups on the carbon NP surface.

3.3.2 Specific Targeting

Han *et al.* used polyethyleneimine-modified CDs to examine the fluorescence labeling of HeLa cells. The CDs conjugated with the CEA8 antibody could label HeLa cells upon incubation for only 90 min, and the labeling was visualized by the green fluorescence contour of the cell shape. In the control experiment without the CEA8 antibody, CDs were not found under UV excitation.[98] Bhunia *et al.* applied the CDs from the carbonization of carbohydrates to the fluorescence imaging of HeLa cells.[99] However, the dots had very low non-specific binding to cells, probably due to their small hydrodynamic sizes and low surface charge. When the dots were linked to TAT peptide or folate, their cellular uptake was significantly enhanced, with fluorescence emissions readily detected under a confocal microscope.[99] Lee *et al.* prepared CDs with the maleimide-terminated TTA1 aptamer, targeting tenascin C (Tnc) proteins, for the fluorescence imaging of cancer cells.[100] Since Tnc proteins are highly expressed in HeLa cells and C6 cells, but rarely expressed in CHO cells, the TTA1–CDs were found to be significantly selective to HeLa cells and C6 cells, with only minor uptake by CHO cells.[100]

There is another case for specific targeted imaging with CDs. As known, most CDs can label both the cell membrane and the cytoplasm without significantly reaching the nucleus, indicating that these functionalized CDs still cannot reach organelles selectively although they can easily enter into living cells. Kang and Kong synthesized CDs by refluxing polyethylene glycol with sodium hydroxide. This kind of CDs has also been shown to exhibit excellent good biocompatibility, strong visible emission, and stable fluorescence properties with different pH sensitivity, ionic strength, temperature, and time. In addition, the CDs can achieve selective staining for the cell nucleoli of HeLa cells and the effect is similar to some commercial available dyes (such as Hoechst). The surface of CDs is covered with hydrophilic oxygen functional groups, such as –OH and –COOH groups. The CDs also have really low cytoxicity. The QY of the present CDs, at an excitation wavelength of 400 nm, was estimated to be about 45% by calibrating against quinine sulfate as a reference. Moreover, the fluorescence intensity of CDs remained almost the same at various ionic strengths, even in the solution with high ionic strength (2 mol L^{-1} NaCl). The fluorescence intensity reached the strongest emission when the pH was ~7.0. The fluorescence images of CDs-stained cells show no obvious change even after illumination for 40 min. Although the CDs were stored for 3 months at room temperature, the CD-stained HeLa cells still exhibit strong fluorescence. Further, the time-dependent stability comparison of Hoechst-stained HeLa cells indicated that the fluorescence signals decreased sharply after 20 min irradiation. This demonstrates the CDs are particularly suitable for long-term cellular imaging, which is in good accord with the strong fluorescence and excellent photostability of CDs discussed above. Significantly, these CDs are specifically targeted to the nucleolus of HeLa cells. Owing to the large number of O functional groups on their surface, the CDs can be protonized in weak acidic conditions.[79] HeLa cells have optimal viability in weakly acidic conditions, but the chromatin inside the nucleus is weakly alkaline. So the CDs can easily interact with the intranuclear chromatin rather than the cell membrane, nuclear membrane, and organelle. Based on these properties, it is similar to the commercial available Hoechst stain. CDs can stain some highly decondensed DNAs a little more easily (Figure 3.1), which optimizes the fluorescence imaging for detection. Compared with CDs, most commercially available dyes are extremely expensive and commercial nucleoli staining kits must be stored between −20 and −80 °C whereas CDs can be stored at room temperature.

3.3.3 Fluorescence Imaging *In vivo*

Earlier studies showed that CDs were applicable to fluorescence imaging *in vivo*.[101-103] Recent efforts have confirmed or extended the application potential. For example, Wu *et al.* used CDs from honey for *in vivo* imaging experiments in a mouse model, with results showing high contrast enhancement

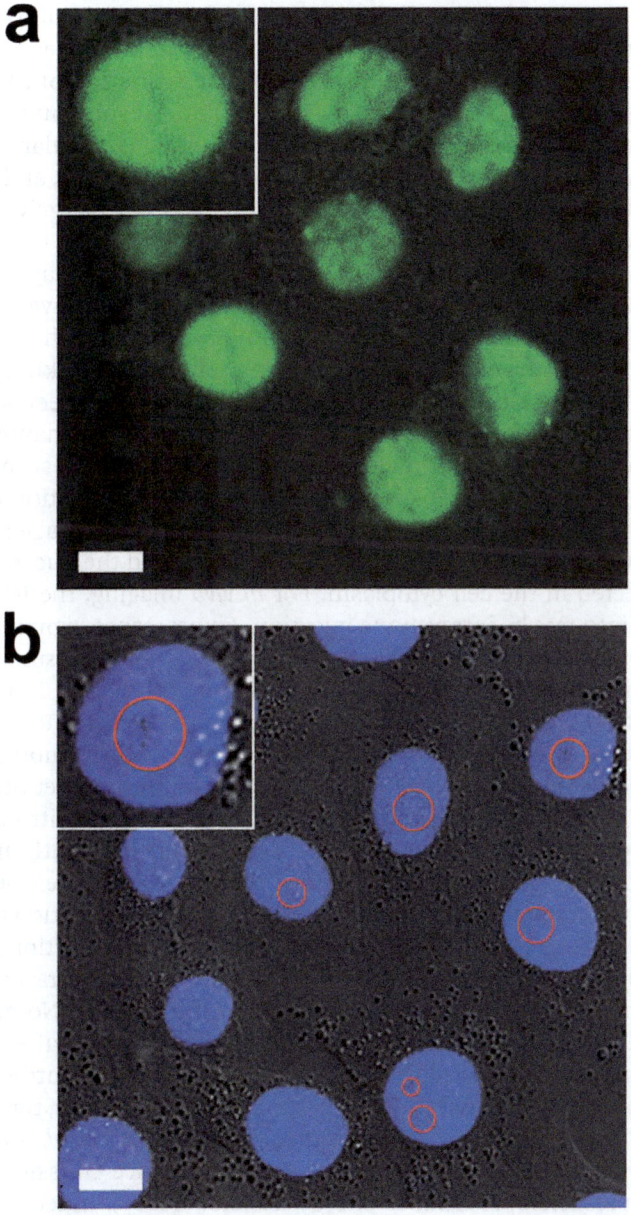

Figure 3.1 Comparison of HeLa cells stained with FCDs (a) and Hoechst stain (b). HeLa cells stained with Hoechst stain show void regions (marked by red circles). Scale bar = 10 μm.[100b] Reproduced from ref. 100b with permission from the Royal Society of Chemistry.

in the auxiliary lymph node.[104] As pointed out by the authors, the rapid lymphatic transport of these dots might be a valuable property, offering greater convenience and reduced procedural expense, as well as improved surgical advantage as the subject is positioned on the table for easier resection.[104] Li *et al.* investigated the cellular uptake and biodistribution of CDs for optical imaging.[105] The colocalization of CDs (blue fluorescence) and fluorescent markers (red fluorescence) was used to probe various cellular organelles, including lysosomes/endosomes, Golgi bodies, mitochondria, and endoplasmic reticulum. The results suggested that the CDs were largely trapped in lysosomes/endosomes, though their presence in Golgi bodies, mitochondria, and endoplasmic reticulum was also observed. For *in vivo* imaging, the CDs were introduced into mice by intravenous injection. Organs were harvested and sliced for fluorescence imaging at 500 nm emission with 405 nm excitation. Blue fluorescence was observed in heart, liver, spleen, kidneys, lungs, brain, and small intestine, with the brightest being in the spleen sample.[105]

Srivastava *et al.* fabricated iron oxide-doped carbogenic nanocomposite (IO–CNC) for multimodality (magnetic resonance (MR)/fluorescence) bioimaging. The IO–CNC was prepared by the thermal decomposition of organic precursors in the presence of small Fe_3O_4 nanoparticles (average size 6 nm). The IO–CNC could be taken up by RAW 264.7 cells, and the fluorescence was mainly detected in the cell cytoplasm. For *in vivo* imaging, the IO–CNC was introduced into rats by intravenous injection. Fluorescence signals due to the IO–CNC were observed in the spleen slide samples. The MRI results suggested enhanced signals in the brain blood vessels under both T1 and T2 models.[106]

Huang *et al.* extended the spectral range of CDs to red/NIR by attaching the fluorescent dye ZW800 to the dots.[107] These red/NIR fluorescent dots were tracked *ex vivo* and *in vivo* for their biodistribution, excretion, and passive tumor uptake by using both NIR fluorescence and positron emission tomography (PET) imaging techniques. The CDs were efficiently and rapidly excreted from the body after injection by different routes. Their blood clearance was quick, only 1 h after intravenous injection. The retention times were somewhat longer after subcutaneous and intramuscular injections. The CDs were slightly trapped in liver, spleen, and lungs at 1 h after intravenous injection. Very bright fluorescence was observed in the kidneys. No meaningful fluorescence was detected in the above-mentioned samples after 24 h. The urine accumulation followed the sequence of intravenous > intramuscular > subcutaneous injection routes. For the three injection models, tumor uptake of the CDs was also found.[107] In a related study, Huang *et al.* attached the fluorescence dye Ce6 to CDs to impart red fluorescence emissions through Förster (fluorescence) resonance energy transfer (FRET). Indeed, the conjugate of Ce6–CDs could be excited at 430 nm for red fluorescence emissions (668 nm). For imaging and photodynamic therapy, the conjugate could be taken up by cells, and under laser irradiation cell death was observed. After intravenous injection, accumulation of the conjugate in tumors was detected. Probably as a result, laser treatment of mice exposed to the Ce6–CD conjugate significantly inhibited tumor growth.[108]

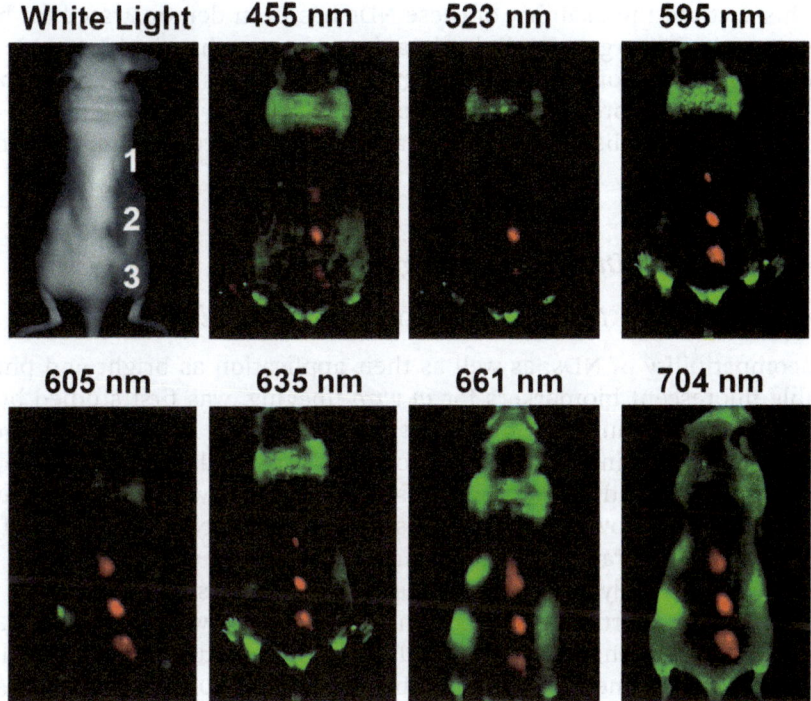

Figure 3.2 *In vivo* fluorescence images of a CD-injected mouse. The images were taken under various excitation wavelengths at 455, 523, 595, 605, 635, 661, and 704 nm. Red and green represent fluorescent signals of CDs and tissue autofluorescence, respectively.[103] Adapted from ref. 103 with permission from John Wiley & Sons. Copyright © 2012 Wiley-VCH Verlag GmbH & Co. KGaA, Weinheim.

Kang, Liu *et al.* demonstrated that CDs could serve as novel fluorescent probes for *in vivo* imaging in live mice by using a wide range of excitation wavelengths, with excellent signal-to-background separation under NIR excitation (Figure 3.2). They further studied the *in vivo* biodistribution of CDs by a radiolabeling method. After intravenous injection, the CDs exhibit high accumulation in the reticuloendothelial system as well as in the kidney, and they are gradually excreted *via* both renal and fecal pathways. Importantly, it was observed that CDs at a dose of 20 mg kg^{-1} appear to be safe to the treated animals over a period of 3 months as evidenced by the systematic time-course blood chemical analysis and complete blood panel and histological analyses.[103]

3.4 Bioimaging Based on NDs (Ref. 1)

A major difficulty in using NDs for bioimaging is to avoid their aggregation in the physiological environment. Nevertheless, internalization of NDs as small aggregates or as individual objects has been reported. As will be shown in this section, the usability of these NDs has been demonstrated for both untargeted and targeted cellular imaging. Also, in this field, examples of imaging based on organic fluorophores bound to NDs as probes have been reported. Since fluorescence is not produced directly by the NDs in these systems, they are not discussed here; detailed discussion can be found in recent review articles.

3.4.1 NDs for *In vitro* Bioimaging

3.4.1.1 *NDs for Non-Targeted* In vitro *Bioimaging*

Biocompatibility of NDs, as well as their application as bright and photostable fluorescent biomarkers for *in vitro* imaging, was first studied by Yu *et al.* using 293T human kidney cells.[61] The emission wavelength range of the produced ND in the 600–900 nm window is ideal for low-background *in vitro* imaging and *in vivo* imaging since it matches well with the optimal transparency window of biological tissues. The extreme photostability of the NDs was demonstrated by comparing their emission with that of 100 nm red fluorescent polystyrene carboxylated nanospheres containing 104 dye equivalents per particle under continuous excitation with an Hg lamp. No sign of photobleaching was observed for NDs even after 8 h excitation; by contrast, the polymeric beads photobleached within 0.5 h under the same excitation conditions. Interaction of NDs with living cells was investigated by incubating the NPs with 293T human kidney cells. Uptake of the NDs was confirmed by vertical cross-sectional images of the cells acquired by a confocal fluorescence microscope. According to the authors, observation of the bright red spots inside the reddish envelope was indicative of NDs translocating through the cell membrane, possibly by endocytosis, but also of NP aggregation. Considering the detrimental effect of aggregation on the performance of NDs for bioimaging, especially in view of single ND observation,[109] Yu *et al.* also proposed replacing the 100 nm probes with smaller diamond particles that they demonstrated to be detectable, as bright and photostable individually fluorescent biomarkers, in cancer cells.[110] For preparation of the probes, commercial type Ib NDs, with a nominal size of 35 nm, were purified by acidic oxidative treatment and exposed to proton-beam irradiation followed by thermal annealing to generate luminescent NV defect centers. The 35 nm diameter, COOH-terminated, NDs were incubated with HeLa cells and the uptake process was investigated by fluorescence microscopy. Translocation of the NDs through the cellular membrane was observed. Most of the uptaken NDs were seen to distribute in the cytoplasm where they were individually tracked,[110] presenting reasonable stability against aggregation. The authors demonstrated that fluorescence of a single 35 nm ND is significantly brighter and considerably more photostable than that of a single dye molecule such as Alexa Fluor 546 under the same excitation conditions. Long-term photostability and absence of fluorescence blinking are size-independent properties of NDs. Comparing these results with the ones relating to

100 nm NDs, they estimated that their method would create about 30 emitting defects in a hypothetical 10 nm ND.

Quite different results on the interaction of red-emitting NDs (~50 nm) with living HeLa cells were reported by Neugart *et al.*, who investigated the motional dynamics of NDs in aqueous dispersion and in the cellular system.[111] Confocal fluorescence images demonstrated that these NDs strongly interact with HeLa cells but also that most of them are simply adsorbed on the cell membrane. About 6% of the total particles (estimated from fluorescence imaging) were internalized and immobilized inside the cell in a short time, suggesting an endosomal pathway for uptake. Only rare cases of freely diffusing single NDs, showing a diffusion constant compatible with the 50 nm size of the primary particles, could be observed.

Analogously to red/NIR-emitting NDs, green-emitting ones were also used for *in vitro* imaging.[112] Green NDs were prepared by radiation damage of type Ia natural diamond nanocrystallites with a N concentration in the range of 900 ppm and size of 350, 140, 70, and 50 nm. For these NDs, the H3 centers emitted green light upon illumination by a blue laser. Application of these NDs as fluorescent cellular markers was demonstrated by confocal fluorescence microscopy, which revealed that the 70 nm particles were internalized by live HeLa cells.[112]

3.4.1.2 NDs for Targeted In vitro Bioimaging

NDs for targeted cellular imaging were synthesized by Weng *et al.*[113] by conjugating transferrin to the carboxylate surface groups of 100 nm diameter red-emitting diamond NPs. According to the authors these spectral features allowed them to eliminate the interference of cellular autofluorescence at 520–650 nm by detecting fluorescence at 660 nm. Transferrin receptors are overexpressed on HeLa cells and the transferrin-functionalized diamond nanoprobes were expected to be internalized *via* a receptor-mediated uptake mechanism. The process was investigated by confocal microscopy and, in order to demonstrate the nature and specificity of the uptake, two populations of HeLa cells were investigated, one of which had been presaturated with simple transferrin. Internalization of the NDs was observed only in the unsaturated cells, indicating that the transferrin receptors play a fundamental role in the transmembrane permeation of the probe. On the basis of these results, the authors concluded not only that internalization of the NDs was mediated by transferrin uptake, but also that transferrin activity was not altered after conjugation to the NDs. Moreover, the lack of quenching of the NDs color centers by the iron ions complexed into the protein was interpreted as proof of a large distance existing between the metal ions and the emissive centers, compatible with a localization of the defect deeply inside the crystal lattice rather than on the surface.

Detonation NDs have also been proposed for targeted fluorescent cellular bioimaging.[114] Mkandawire *et al.* prepared fluorescent cellular biomarkers by conjugating detonation NDs to antibodies and using fourth-generation dendrimers, cationic liposomes, and protamine sulfate as transfecting agents. The size of the starting NDs was about 5 nm in diameter and the detonation product was purified and functionalized by strong acidic oxidation to achieve the formation of surface carboxylate groups suitable for bioconjugation using conventional protocols.[115]

3.4.2   NDs for Long-Term *In vivo* Imaging

Due to their photochemical and chemical inertness, and in virtue of their emission in the NIR region, NDs can be detected during very long time lapses during *in vivo* imaging experiments. This makes it possible to investigate their long-term biodistribution and fate in the organs of living animals as well as to analyze their possible toxic effects. All the *in vivo* experiments carried out on rats and mice, discussed in this section, demonstrated the extreme chemical and photochemical stability of NDs in living organisms, as well as a unique long-term biocompatibility of these materials. Surprisingly, Lin *et al.* reported toxic effects of NDs incorporated into much simpler animal models.[116] Lin *et al.* also compared the interaction of commercial carboxylated NDs of different nominal sizes, 5 and 100 nm, with protozoans *Paramecium caudatum* and *Tetrahymena thermophila*. Fluorescence imaging experiments showed that the NDs could be excreted by the microorganisms, which partially preserved functioning such as division. Nevertheless, significant toxicity was observed following the growth rate of the microorganism as a function of the probe concentration and size. In particular, smaller NDs turned out to be more toxic. These results, which are in contrast with the well-documented high biocompatibility of NDs, may be in part influenced by the poor colloidal stability of the diamond particles used.[117]

3.4.2.1   *Long-Term* In vivo *Imaging in* C. elegans

Applicability of NDs to long-term *in vivo* optical imaging was demonstrated by Mohan *et al.*[118] who explored the long-term interaction of red-emitting NDs with the translucent model hermaphrodite organism *C. elegans* (Figure 3.3). These authors demonstrated the lack of toxicity of these materials after incorporation, by analyzing parameters such as lifespan, brood size, and reactive oxygen species (ROS) levels. Alteration of feeding behavior and reproduction were also considered. *C. elegans* is a suitable model for studying biological molecular processes since its genome has been completely sequenced. Moreover, this worm has a simple and well-defined anatomy arising from the organization of 959 cells into complex structures, which include intestine, muscle, hypodermis, gonads, and nerve systems. In Mohan *et al.*'s studies, biofunctionalized NPs, as well as bare carboxylated NDs, were incorporated into *C. elegans* either by feeding or by injection into the gonad. Effects of the surface functionalization of the NDs, as well as the administration method, on the physiological response of the animal model were investigated. In the case of oral administration, epifluorescence *in vivo* images

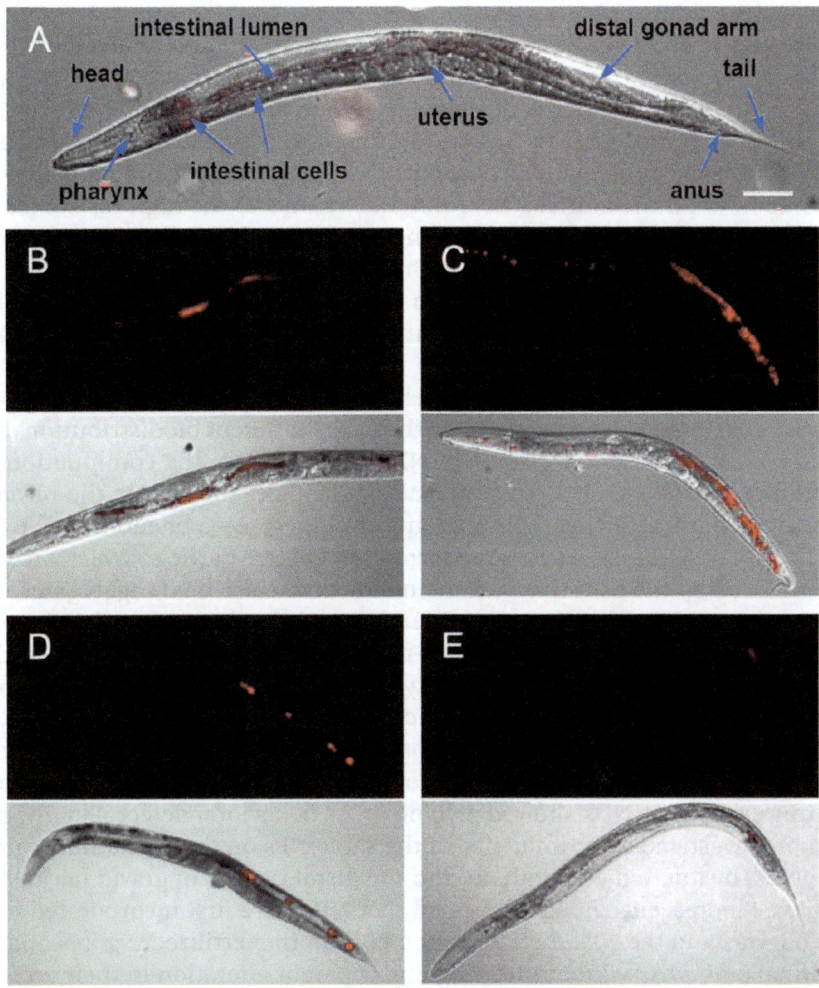

Figure 3.3 Epifluorescence and epifluorescence/DIC-merged images of wild-type *C. elegans*. (A) An untreated young adult. (B and C) Worms fed with bare NDs for 2 h (B) and 12 h (C). The NDs stayed inside the gut and were not excreted when the worms were deprived of food. Excretion of NDs occurred upon feeding with *E. coli*. (D and E) Worms were provided with *E. coli* for 20 min (D) and 40 min (E), after being fed with bare NDs for 2 h. Almost all, if not all, FNDs were excreted within 1 h. The upper panels in (B–E) show the epifluorescence images; (A) and the lower panels in (B–E) show the epifluorescence/DIC-merged images. Anterior is left and dorsal is up in all figures. Scale bar is 50 μm. Reprinted with permission from N. Mohan, C. S. Chen, H. H. Hsieh, Y. C. Wu and H. C. Chang, *Nano Lett.*, 2010, **10**, 3692. Copyright (2010) American Chemical Society.[118]

proved that both carboxylated and functionalized NPs were incorporated by the worms. Hence, the NDs were not regarded as repellent by the organisms during the feeding process.

The nature of the NP surface, on the other hand, strongly affected the bio-distribution and the fate of the probe in the model animal. Bare NDs accumulated in the lumen of the digestive system in an amount of about 1×10^6 particles per worm. After incorporation, these NPs were excreted within 1 h if the worms were fed with *E. coli*, but they persisted in the organism even after 12 hours if the animals were not fed. Thanks to the high resistance of the emitters to photobleaching and chemical degradation, even in the presence of robust digestive enzymes in the gut, the entire digestive tract of the worm was imaged for more than 48 h showing no translocation of the NDs to the intestinal cells. A completely different fate was observed for BSA- and dextran-coated NDs. These NPs were readily internalized by the intestinal cells where they remained located even 24 h after recovery of the worms on nutrient bacteria. According to the authors, the different biodistribution and fate observed for the two classes of NDs resulted from the combination of two different effects. First, bioconjugation stabilizes the NDs in the physiological medium, reducing their tendency to aggregate—a phenomenon that, in the case of bare NDs, causes the formation of large particles which are not internalized by the intestinal cells. Small functionalized ND aggregates, on the contrary, are readily taken up *via* endocytosis by the same cells. The second effect is the activation of receptor-mediated endocytosis pathways by the biomolecules. Interestingly, no toxicity upon internalization of the NDs into the intestinal cells was observed considering parameters such as the lifespan and number of progeny per worm. Moreover, the level of ROS of the animals treated with NDs perfectly matched those of a control, untreated, population. Worms exposed to NDs showed also no gross behavioral defect. Finally, the authors investigated the influence of the same NPs on the formation of new embryos, by microinjecting them into the distal gonads of gravid hermaphrodites. Fluorescence images revealed that NDs were first incorporated into the oocytes and the NPs then persisted both in the fertilized zygotes and in the final embryos, without inducing any apparent alteration in their growth and development.[118]

Kuo *et al.* extended the study of *C. elegans* to the intercellular transport of yolk lipoproteins in the animal, by using fluorescence lifetime imaging microscopy (FLIM).[119] FLIM allowed these authors to remove background fluorescence by time-gated acquisition, with an important decrease of the signal-to-noise ratio. The yolk lipoproteins in the nematode are similar to human serum low density lipoproteins, serving as an intercellular transporter of fat molecules and cholesterol. To study this fundamental transport process, NDs were first coated with yolk lipoprotein complexes, and then microinjected into the intestinal cells of the living organism. Real-time imaging revealed the secretion of the biofunctionalized NDs from the intestine to the pseudocoelomic space, followed by transport into oocytes and subsequent accumulation in the multicellular embryos derived from the oocytes.

These data confirmed the generally accepted mechanism for the energy supply process that involves yolk lipoproteins. Moreover, the results proved the value of NDs as biomolecular nanocarriers for intercellular transport, cell-specific targeting, and long-term imaging applications *in vivo*.

3.4.2.2 Long-Term In vivo *Imaging in Mice and Rats*

Vaijayanthimala *et al.* investigated the long-term stability and biocompatibility of red-emitting 100 nm NDs in rats and mice, as well as the potential use of these materials for sentinel lymph node mapping.[120] The carboxylate-terminated surface of the NDs was functionalized with BSA by simple electrostatic absorption. In a first experiment bare NDs were subcutaneously injected in a rat to evaluate the long-term stability of the fluorescence signal. Stable emission was detected over the surprisingly long period of 37 days after dosing. This demonstrated that NDs are robust *in vivo* and potentially useful as long-term imaging agents for living animals. In an additional study the same NPs were administrated *via* intraperitoneal injection. In stark contrast to what was observed after subcutaneous injection, the intensity of the far-red fluorescence in this case gradually decreased and disappeared in 6 min. According to the authors, the fading of the fluorescence signal resulted from the gradual dispersion of the NDs throughout the peritoneal cavity rather than from actual degradation or digestion. This conclusion was motivated by the detection of NDs in the organs and tissues explanted from the animals treated with NPs. Lastly, in a very long-term experiment, mice were observed over a 5 month period after intradermal injection of BSA-conjugated NDs. Combination of *in vivo* and *ex vivo* fluorescence imaging, as well as transmission electron microscopy, showed that ND particles were drained from the injection sites by macrophages and selectively accumulated in the axillary lymph nodes of the treated mice.[120] The physiological response of the animals to the ND administration was analyzed by measuring water and fodder consumption, body weight, and organ index. Observations revealed no significant alterations with respect to a control mouse population.

Wu *et al.* investigated the application of NDs to track how lung stem cells incorporate and regenerate themselves *in vivo* over time in a mouse model.[121] In their study, isolated pulmonary cells were labeled by incubation with the NDs, determined by flow cytometric analysis. Immediately after labeling, the mean fluorescence intensity of the cells was 45-fold greater than that of unlabeled control cells and it decreased ~50% every 48 h because of ND redistribution resulting from cell division. When injected into the tail veins of adult healthy normal mice, ND-labeled cells localized in the lungs, and not in other organs, demonstrating their maintained activity and the actual possibility of tracking them in living animals. Nevertheless, although this result was clear in flow cytometric analysis, direct imaging of the *ex vivo* tissues by confocal microscopy was complicated by strong background signal arising from the fluorescence of haematoxylin and eosin dyes used to stain the samples for morphological analysis. Time-gated imaging made it possible to distinguish the NDs emission clearly from the background. Distribution of the fluorescent cells in the pulmonary tissues was compared in the case of healthy and naphthalene-injured mice, showing different localization. In the healthy animals labeled cells were primarily located in the subepithelium of bronchiolar airways. In injured mice, time-gated fluorescence images[121] displayed engraftment of the transplanted ND-labeled cells to terminal bronchioles also in cluster form. Moreover, lung epithelia of the injured mice were restored more rapidly in the population of transplanted animals than in a control animal treated with a saline solution. These results allowed the authors to reach the conclusion that ND labeling permitted quantitative assessment of the distribution of transplanted cells in tissue.

Huang *et al.*[67] studied the compatibility of NDs in the nervous system of mice. These authors also injected NDs into the hippocampi of postweaned juvenile rats. After injection, the body weight and fodder and water consumption were assessed on a daily basis for 1 week. There were no significant differences in these parameters between ND- and saline-injected rats. The authors also performed a behavioral test to determine whether NDs had more subtle effects of on the hippocampus. The novel object recognition test (NORT)[122] is based on the natural preference of rats to explore novel objects more than familiar ones. It has been shown that drugs which can damage the hippocampus lower the discriminating index of NORT. Interestingly neither the procedure of intracranial injection nor the injection of NDs to the hippocampus altered the discriminating index of NORT. According to the authors these results suggested that NDs did not interfere with the general function of the hippocampus in live animals.

3.4.3 Background-Free *In vivo* Imaging by ND Fluorescence Modulation

The fluorescence intensity of NV defects in NDs depends on their spin states.[123,124] Magnetic sensitivity of the NDs emission has therefore been proposed as a powerful instrument to develop new *in vivo* imaging methods with reduced background and, as a consequence, high contrast. In a demonstrative application in mice, NDs injected at a depth of 0.4–0.5 mm could be visualized completely erasing the background signals.[123] Hegyi *et al.* demonstrated the use of NDs for molecular imaging by optically detected electron spin resonance as a proof of principle.[125] This method is suitable for 3D, background-free, optical tomography. The triplet ground state of NV defects in NDs is weakly coupled to the diamond lattice, and has long spin relaxation and spin coherence lifetimes, even at room temperature.[126] A prototype imaging system was also developed by using permanent magnets to create a magnetic field-free point. Only NDs localized near the field-free point are resonant with a microwave field at 2.869 GHz and respond to microwave modulation. Thus, when the microwave source is ON, a decrease of the fluorescence, by an amount proportional to the ND concentration at the field-free point, is measured with respect to the intensity detected when the modulation signal is OFF. By sweeping the field-free point across an organism and tracking the changes in fluorescence, a quantitative map of the ND concentration as a function of position can be obtained. To image a 2D slice or a 3D volume of the ND concentration within an organism, the authors scanned the field-free point across the sample in two or three dimensions. Images were then combined using a standard reconstruction algorithm, as in computed tomography (CT). The applicability of the imaging setup was demonstrated in the case of an ND phantom inserted in chicken breast. The magnetic sensitivity of ND emission allowed Sarkar *et al.* to develop wide-field *in vivo* background-free imaging methods based on magnetic modulation, reducing the signal-to-background ratio for *in vivo* imaging up to 100-fold.[127] Fluorescence modulation was achieved mechanically by alternately positioning and removing a permanent magnet close to the sample in a commercial imaging system, to induce a local field that oscillate as a square wave from 0 to 100 G with a 0.1 Hz frequency. Processing of the acquired images by lock-in detection of the fluorescence led to a strong background decrease in imaging NDs in the sentinel lymph nodes of a mouse after injection into the front pad of the animal.

3.5   Challenge and Perspectives

CDs/NDs have attracted much attention for their easy preparation, fascinating photophysical properties, excellent stability, low toxicity, and versatile surface chemistry. Engineering the surface functionality of CDs/NDs according to the multistep design and modification can generate CDs/NDs with high QY, good biocompatibility, and long-term stability, available for bioimaging in live cells, tissues, and animals. Recent research achievements have brought a change of focus from the synthesis of single-component probes to the design of hybrid nanostructures composed of multiple targeting, imaging, and therapeutic modules. However, currently available CD/ND probes are far from ideal, leaving plenty of room for the development of novel designs for the functional surface of CDs/NDs. It should be noted that the application of CDs/NDs for bioimaging is still at an early stage, with many important issues remained to be addressed, as follows.

- For biological applications size is very important since it is one of the parameters that control the interaction of probes with the biological targets, as well as their fate in the living organisms. In particular sub-10 nm probes are expected to be really cleared and to accumulate less in the body than larger particles do. Although NDs with sizes of ~5 nm can be prepared by detonation techniques, these materials aggregate strongly in aqueous environments. The use of single-digit NDs as individual NPs, and not as aggregates, in bioimaging is one of the challenging issue in future applications.

- Brightness is a very important feature for luminescent probes since it limits the detection sensitivity. Nevertheless, the few data available on the molar extinction coefficient of these emitters indicate values which are one order of magnitude lower than organic dyes. Presence of surface graphitic carbon has also been reported to cause partial PL quenching, reducing the NDs' brightness. Super-bright NDs can be prepared by creating a large number of emitting defects, but the effect on the PL of the interdefect interactions has not been completely explored. Moreover, the actual number of color centers produced in NDs in different activation conditions is still not well established. On the other hand, CDs with emission wavelengths ranging from UV to NIR have been synthesized *via* various bottom-up and top-down approaches, and the QYs of CDs reported so far are lower than conventional semiconductor quantum dots (QDs) or dyes. Notably, CDs with bright red/NIR emissions are considered as ideal nanoprobes for bioimaging. For NDs, the variety of the emission color is less broad, their PL being related to localized states with peculiar, not tunable, features. NV centers show emission in the far red suitable for *in vivo* detection. Nevertheless these centers absorb light in the green region, which is acceptable but not ideal for *in vivo* applications. Considering the wide variety of defect centers in NDs, the application of new color centers may be demonstrated in the future to be suitable for bioimaging, switching the excitation to the NIR spectral window.
- Targeted imaging of tumors based on CDs/NDs has rarely been demonstrated *in vivo*. The next generation of cancer diagnostic agents for small animals requires improvement in their accumulation efficiency in tumor tissues. Edge functionalization of CDs with antibodies or peptides for targeted imaging of specific cancers still needs to be further explored.
- Development of multifunctional nanoprobes to cover a broad spectrum of imaging modalities is always an interesting topic. It is still a major challenge to design CD/ND-based nanoprobes for simultaneous optical imaging, MRI, and CT monitoring. In these cases, the optical and radioactive properties of CDs seem promising for simultaneous imaging and imaging-guide therapy. On the other hand, by optimizing the multiphoton excitation and emission properties of CDs, their novel applications in biomedical areas could be exploited. The response of ND luminescence to magnetic fields has been demonstrated to be a unique, powerful feature for background elimination in bioimaging and it promises to become more and more important in the development of new high-sensitivity and high-contrast imaging systems.
- Cytotoxicity and gene toxicity studies of CDs with different components, sizes, shapes, and surface coatings need to be investigated. Biodistribution studies of CDs in various animal models are also needed. Continued research and development efforts in this emerging field are of great value. This will revolutionize the way in which future biomedical tests

and clinical diagnoses may be performed that could affect many aspects of our lives. A definitive assessment of the safety of the whole class of nanomaterials on human health and the environment is a hard task. NDs can be considered highly biocompatible materials both *in vitro* and *in vivo*, although some toxic effects have been observed, at high doses. The environmental safety of NDs has been poorly investigated, but actual environmental concern about these materials arises more from the procedures and the chemicals used for their synthesis than from their actual impact on ecosystems.

References

1. M. Montalti, A. Cantelli and G. Battistelli, *Chem. Soc. Rev.*, 2015, **44**, 4853.
2. G. W. Yang, J. B. Wang and Q. X. Liu, *J. Phys.: Condens. Matter*, 1998, **10**, 7923.
3. J. Xiao, G. Ouyang, P. Liu, C. X. Wang and G. W. Yang, *Nano Lett.*, 2014, **14**, 3645.
4. M. Frenklach, W. Howard, D. Huang, J. Yuan, K. E. Spear and R. Koba, *Appl. Phys. Lett.*, 1991, **59**, 546.
5. Y. G. Gogotsi, K. G. Nickel, D. Bahloul-Hourlier, T. Merle-Mejean, G. E. Khomenko and K. P. Skjerlie, *J. Mater. Chem.*, 1996, **6**, 595.
6. T. L. Daulton, M. A. Kirk, R. S. Lewis and L. E. Rehn, *Nucl. Instrum. Methods*, 2001, **175**, 12.
7. S. Welz, Y. Gogotsi and M. J. Mcnallan, *J. Appl. Phys.*, 2003, **93**, 4207.
8. F. Banhart and P. M. Ajayan, *Nature*, 1996, **382**, 433.
9. E. M. Galimov, A. M. Kudin, V. N. Skorobogatskii, V. G. Plotnichenko, O. L. Bondarev, B. G. Zarubin, V. V. Strazdovskii, A. S. Aronin and A. V. Fisenko, *Dokl. Phys.*, 2004, **49**, 150.
10. V. N. Mochalin, S. Olga, H. Dean and G. Yury, *Nat. Nanotechnol.*, 2012, **7**, 11.
11. V. V. Danilenko, *Phys. Solid State*, 2004, **46**, 595.
12. D. M. Gruen, O. A. Shenderova, and A. Y. Vul, *Synthesis, Properties and Applications of Ultrananocrystalline Diamond*, Springer, Netherlands, 2005, vol. 192, 31, pp. 373–382, ISBN:978-1-4020-3320-9, DOI: 10.1007/1-4020-3322-2.
13. V. Y. Dolmatov, *Russ. Chem. Rev.*, 2001, **70**, 687.
14. A. Krüger, F. Kataoka, M. Ozawa, T. Fujino, Y. Suzuki, A. E. Aleksenskiif, A. Y. Vul' and E. Ōsawa, *Carbon*, 2005, **43**, 1722.
15. O. Shenderova, A. Koscheev, N. Zaripov, I. Petrov, Y. Skryabin, P. Detkov, S. Turner and G. V. Tendeloo, *J. Phys. Chem. C*, 2011, **115**, 9827.
16. M. Ozawa, M. Inaguma, M. Takahashi, F. Kataoka and A. Krüger, *Adv. Mater.*, 2007, **19**, 1201.
17. J. P. Boudou, P. A. Curmi, F. Jelezko, J. Wrachtrup, P. Aubert, M. Sennour, G. Balasubramanian, R. Reuter, A. Thorel and E. Gaffet, *Nat. Nanotechnol.*, 2009, **20**, 10314.

18. M. H. Hsu, P. Yu, J. H. Huang, C. H. Chang, C. W. Wu, Y. C. Cheng and C. W. Chu, *Appl. Phys. Lett.*, 2011, **98**, 119904.
19. M. V. Korobov, D. S. Volkov, N. V. Avramenko, L. A. Belyaeva, P. I. Semenyuk and M. A. Proskurnin, *Nanoscale*, 2013, **5**, 1529.
20. V. N. Mochalin, S. Olga, H. Dean and G. Yury, *Nat. Nanotechnol.*, 2012, 7, 11.
21. Y. Y. Hui and H. Chang, *J. Chin. Chem. Soc.*, 2014, **61**, 67.
22. R. Kaur and I. Badea, *Int. J. Nanomed.*, 2013, **8**, 203.
23. P. Badziag, W. S. Verwoerd, W. P. Ellis and N. R. Greiner, *Nature*, 1990, **343**, 244.
24. Q. Xu and X. Zhao, *Phys. Rev. B*, 2012, **86**, 4093.
25. J. Y. Raty and G. Galli, *Nat. Mater.*, 2003, **2**, 792.
26. I. Aharonovich, S. Castelletto, D. A. Simpson, A. Stacey, J. McCallum, A. D. Greentree and S. Prawer, *Nano Lett.*, 2009, **9**, 3191.
27. A. M. Zaitsev, *Phys. Rev. B*, 2000, **61**, 12909.
28. V. N. Mochalin, S. Olga, H. Dean and G. Yury, *Nat. Nanotechnol.*, 2012, **7**, 11.
29. O. Sebastian, Y. Gleb, M. Vadym, S. O. Kucheyev and G. Yury, *J. Am. Chem. Soc.*, 2006, **128**, 11635.
30. O. A. Williams, J. Hees, C. Dieker, W. Jäger, L. Kirste and C. E. Nebel, *Acs. Nano*, 2010, **4**, 4824.
31. V. Mochalin, S. Osswald and Y. Gogotsi, *Chem. Mater.*, 2009, **21**, 273.
32. I. I. Kulakova, *Phys. Solid State*, 2004, **46**, 636.
33. T. Jiang, K. Xu and S. Ji, *J. Chem. Soc., Faraday Trans.*, 1996, **92**, 3401.
34. S. Ji, T. Jiang, K. Xu and S. Li, *Appl. Surf. Sci.*, 1998, **133**, 231.
35. V. N. Mochalin and G. Yury, *J. Am. Chem. Soc.*, 2009, **131**, 4594.
36. W. W. Zheng, Y. H. Hsieh, Y. C. Chiu, S. J. Cai, C. L. Cheng and C. P. Chen, *J. Mater. Chem.*, 2009, **19**, 8432.
37. V. N. Mochalin, I. Neitzel, B. Etzold, A. M. Peterson, G. R. Palmese and Y. Gogotsi, *Acs. Nano*, 2011, **5**, 7494.
38. V. V. Korolkov, I. I. Kulakova, B. N. Tarasevich and G. V. Lisichkin, *Diamond Relat. Mater.*, 2007, **16**, 2129.
39. O. O. Mykhaylyk, Y. M. Solonin, D. N. Batchelder and R. Brydson, *Appl. Phys.*, 2005, **97**, 074302.
40. E. D. Obraztsova, K. G. Korotushenko, S. M. Pimenov, V. G. Ralchenko, A. A. Smolina, V. I. Konova and E. N. Loubnin, *Nanostruct. Mater.*, 1995, **6**, 827.
41. G. N. Yushin, S. Osswald, V. I. Padalko, G. P. Bogatyreva and Y. Gogotsi, *Diamond Relat. Mater.*, 2005, **14**, 1721.
42. J. Sun, S. W. Yang, Z. Y. Wang, H. Shen, T. Xu, L. T. Sun, H. Li, W. W. Chen, X. Y. Jiang, G. Q. Ding, Z. H. Kang, X. M. Xie and M. H. Jiang, *Part. Part. Syst. Charact.*, 2015, **32**, 434.
43. (a) H. Li, J. Liu, S. J. Guo, Y. L. Zhang, H. Huang, Y. Liu and Z. H. Kang, *J. Mater. Chem. B*, 2015, **3**, 2378; (b) S. M. Henrichs and S. F. Sugai, *Geochim. Cosmochim. Acta*, 1993, **57**, 823; (c) E. C. Cho, L. Au, Q. Zhang and Y. N. Xia, *Small*, 2010, **6**, 517–522; (d) R. H. Liu, J. Liu, W. Q. Kong, H. Huang, X. Han, X. Zhang, Y. Liu and Z. H. Kang, *Dalton Trans.*, 2014, **43**, 10920.

44. (a) Y. Song, S. Zhu and B. Yang, *RSC Adv.*, 2014, **4**, 27184; (b) Z. M. Markovic, B. Z. Ristic, K. M. Arsikin, D. G. Klisic, L. M. Harhaji-Trajkovic, B. M. Todorovic-Markovic, D. P. Kepic, T. K. Kravic-Stevovic, S. P. Jovanovic, M. M. Milenkovic, D. D. Milivojevic, V. Z. Bumbasirevic, M. D. Dramicanin and V. S. Trajkovic, *Biomaterials*, 2012, **33**, 7084.

45. C. Y. Wu, C. Wang, T. Han, X. J. Zhou, S. W. Guo and J. Y. Zhang, *Adv. Healthcare Mater.*, 2013, **2**, 1613.

46. Y. R. Chang, H. Y. Lee, K. Chen, C. C. Chang, D. S. Tsai, C. C. Fu, T. S. Lim, Y. K. Tzeng, C. Y. Fang, C. C. Han, H. C. Chang and W. Fann, *Nat. Nanotechnol.*, 2008, **3**, 284.

47. L. Rondin, G. Dantelle, A. Slablab, F. Treussart, P. Bergonzo, S. Perruchas, T. Gacoin, M. Chaigneau, H.-C. Chang, V. Jacques and J.-F. Roch, *Phys. Rev. B*, 2010, **82**, 7174.

48. S. Brad, L. Istvan, G. Yury, H. V. David and K. Miklos, *Phys. Chem. Chem. Phys.*, 2010, **12**, 14017.

49. N. Philipp, B. Johannes, S. Matthias, R. Florian, F. Helmut, P. R. Hemmer, J. Wrachtrup and F. Jelezko, *Science*, 2010, **329**, 542.

50. J. R. Maze, P. L. Stanwix, J. S. Hodges, S. Hong, J. M. Taylor, P. Cappellaro, L. Jiang, M. V. Gurudev, M. V. Dutt, E. Togan, A. S. Zibrov, A. Yacoby, R. L. Walsworth and M. D. Lukin, *Nature*, 2008, **455**, 644.

51. G. Balasubramanian, I. Y. Chan, R. Kolesov, M. A. Hmoud, J. Tisler, C. Shin, C. Kim, A. Wojcik, P. R. Hemmer, A. Krueger, T. Hanke, A. Leitenstorfer, R. Bratschitsch, F. Jelezko and J. Wrachtrup, *Nature*, 2008, **455**, 648.

52. C. Bradac, T. Gaebel, N. Naidoo, M. J. Sellars, J. Twamley, L. J. Brown, A. S. Barnard, T. Plakhotnik, A. V. Zvyagin and J. R. Rabeau, *Nat. Nanotechnol.*, 2010, **5**, 345.

53. J. Tisler, R. Reuter, A. Lämmle, F. Jelezko, G. Balasubramanian, P. R. Hemmer, F. Reinhard and J. Wrachtrup, *ACS Nano*, 2011, **5**, 7893.

54. J. Tisler, G. Balasubramanian, B. Naydenov, R. Kolesov, B. Grotz, R. Reuter, J. P. Boudou, P. A. Curmi, M. Sennour, A. Thorel, M. Börsch, K. Aulenbacher, R. Erdmann, P. R. Hemmer, F. Jelezko and J. Wrachtrup, *ACS Nano*, 2009, **3**, 1959.

55. I. I. Vlasov, O. Shenderova, S. Turner, O. I. Lebedev, A. A. Basov, I. Sildos, M. Rähn, A. A. Shiryaev and G. V. Tendeloo, *Small*, 2010, **6**, 687.

56. O. A. Shenderova, I. I. Vlasov, S. Turner, G. V. Tendeloo, S. B. Orlinskii, A. A. Shiryaev, A. A. Khomich, S. N. Sulyanov, F. Jelezko and J. Wrachtrup, *J. Phys. Chem. C*, 2011, **115**, 14014.

57. A. M. Schrand, S. A. C. Hens and O. A. Shenderova, *Crit. Rev. Solid State*, 2009, **34**, 18.

58. S. C. Hens, G. Cunningham, T. Tyler, S. Moseenkovb, V. Kuznetsovb and O. Shenderova, *Diamond Relat. Mater.*, 2008, **17**, 1858.

59. L. C. H. Huang, *Langmuir*, 2004, **20**, 5879.

60. A. M. Schrand, J. B. Lin, S. C. Hens and S. M. Hussain, *Nanoscale*, 2011, **3**, 435.

61. S. J. Yu, M. W. Kang, H. C. Chang, K. M. Chen and Y. C. Yu, *J. Am. Chem. Soc.*, 2005, **127**, 17604.

62. Y. Zhu, J. Li, W. X. Li, Y. Zhang, X. F. Yang, N. Chen, Y. H. Sun, Y. Zhao, C. H. Fan and Q. Huang, *Theranostics*, 2012, **2**, 302.
63. A. M. Schrand, H. J. Huang, C. Carlson, J. J. Schlager, E. Ōsawa, S. M. Hussain and L. M. Dai, *J. Phys. Chem. B*, 2007, **111**, 2.
64. O. Faklaris, D. Garrot, V. Joshi, F. Druon, J. P. Boudou, T. Sauvage, P. Georges, P. A. Curmi and F. Treussart, *Small*, 2008, **4**, 2236.
65. V. Vaijayanthimala, Y. K. Tzeng, H. C. Chang and C. L. Li, *Nanotechnology*, 2009, **20**, 18436.
66. M. F. Weng, B. J. Chang, S. Y. Chiang, N. S. Wang and H. Niu, *Diamond Relat. Mater.*, 2012, **22**, 96.
67. Y. A. Huang, C. W. Kao, K. K. Liu, H. S. Huang, M. H. Chiang, C. R. Soo, H. C. Chang, T. W. Chiu, J. I. Chao and E. Hwang, *Sci. Rep.*, 2014, **4**, 6919.
68. Q. W. Zhang, V. N. Mochalin, I. Neitzel, I. Y. Knoke, J. J. Han, C. A. Klug, J. G. Zhou, P. I. Lelkes and Y. Gogotsi, *Biomaterials*, 2011, **32**, 87.
69. A. M. Schrand, J. Johnson, L. M. Dai, S. M. Hussain, J. J. Schlager, L. Zhu, Y. L. Hong and E. Ōsawa, *Nanostruct. Sci. Technol.*, 2008, 159.
70. A. M. Schrand, S. A. C. Hens and O. A. Shenderova, *Crit. Rev. Solid. State*, 2009, **34**, 18.
71. A. M. Schrand, H. J. Huang, C. Carlson, J. J. Schlager, E. Ōsawa, S. M. Hussain and L. M. Dai, *J. Phys. Chem. B*, 2007, **111**, 2.
72. Y. Yuan, X. Wang, G. Jia, J. H. Liu, T. C. Wang and X. Q. Gu, *Diamond Relat. Mater.*, 2010, **19**, 291.
73. N. Mohan, C. S. Chen, H. H. Hsieh, Y. C. Wu and H. C. Chang, *Nano Lett.*, 2010, **10**, 3692.
74. E. K. Chow, X. Q. Zhang, M. Chen, R. Lam, E. Robinson, H. J. Huang, D. Schaffer, E. Osawa, A. Goga and D. Ho, *Sci. Transl. Med.*, 2011, **3**, 772.
75. O. Faklaris, V. Joshi, T. Irinopoulou, P. Tauc, M. Sennour, H. Girard, C. Gesset, J. C. Arnault, A. Thorel, J. P. Boudou, P. A. Curmi and F. Treussart, *ACS Nano*, 2009, **3**, 3955.
76. E. Perevedentseva, S. F. Hong, K. J. Huang, I. T. Chiang, C. Y. Lee, Y. T. Tseng and C. L. Cheng, *J. Nano. Res.*, 2013, **15**, 8.
77. Z. Chu, S. Zhang, B. Zhang, C. Zhang, C. Y. Fang, I. Rehor, P. Cigler, H. C. Chang, G. Lin, R. Liu and Q. Li, *Sci. Rep.*, 2014, **4**, 4495.
78. O. A. Shenderova and D. M. Gruen, *Ultrananocrystalline Diamond: Synthesis, Properties and Applications*, William Andrew Publishing, U.S.A, 2006, vol. 4, pp. 134–147, ISBN: 0-8155-1524-1.
79. P. G. Luo, F. Yang, S. T. Yang, S. K. Sonkar, L. Yang, J. J. Broglie, Y. Liu and Y. P. Sun, *RSC Adv.*, 2014, **4**, 10791.
80. Y. P. Sun, B. Zhou, Y. Lin, W. Wang, K. A. S. Fernando, P. Pathak, M. J. Meziani, B. A. Harruff, X. Wang, H. F. Wang, P. G. Luo, H. Yang, M. E. Kose, B. Chen, L. M. Veca and S. Y. Xie, *J. Am. Chem. Soc.*, 2006, **128**, 7756.
81. L. Cao, X. Wang, M. J. Meziani, F. Lu, H. Wang, P. G. Luo, Y. Lin, B. A. Harruff, L. M. Veca, D. Murray, S. Y. Xie and Y. P. Sun, *J. Am. Chem. Soc.*, 2007, **129**, 11318.
82. Y. Wang, P. Anilkumar, L. Cao, J. H. Liu, P. G. Luo, K. N. Tackett, S. Sahu, P. Wang, X. Wang and Y. P. Sun, *Exp. Biol. Med.*, 2011, **236**, 1231.

83. X. Zhang, S. Wang, L. Xu, L. Feng, Y. Ji, L. Tao, S. Lia and Y. Wei, *Nanoscale*, 2012, **4**, 5581.
84. H. Yan, M. Tan, D. Zhang, F. Cheng, H. Wu, M. Fan, X. Ma and J. Wang, *Talanta*, 2013, **108**, 59.
85. X. Zhang, S. Wang, C. Zhu, M. Liu, Y. Ji, L. Feng, L. Tao and Y. Wei, *J. Colloid Interface Sci.*, 2013, **397**, 39.
86. J. Wei, J. Shen, X. Zhang, S. Guo, J. Pan, X. Hou, H. Zhang, L. Wang and B. Feng, *RSC Adv.*, 2013, **3**, 13119.
87. B. Chen, F. Li, S. Li, W. Weng, H. Guo, T. Guo, X. Zhang, Y. Chen, T. Huang, X. Hong, S. You, Y. Lin, K. Zeng and S. Chen, *Nanoscale*, 2013, **5**, 1967.
88. Z. Zhang, J. Hao, J. Zhang, B. Zhang and J. Tang, *RSC Adv.*, 2012, **2**, 8599.
89. H. Ding, L. W. Cheng, Y. Y. Ma, J. L. Kong and H. M. Xiong, *New J. Chem.*, 2013, **37**, 2515.
90. S. Swagatika, B. Birendra, T. K. Maiti and M. Sasmita, *Chem. Commun.*, 2012, **48**, 8835.
91. A. Sachdev, I. Matai, S. U. Kumar, B. Bhushan, P. Dubey and P. Gopinath, *RSC Adv.*, 2013, **3**, 16958.
92. M. Saxena and S. Sarkar, *Mater. Exp.*, 2013, **3**, 201.
93. M. Saxena, S. K. Sonkar and S. Sarkar, *Rsc. Adv.*, 2013, **3**, 22504.
94. Y. Xu, M. Wu, Y. Liu, Z. Feng, B. Yin, X. W. He and Y. K. Zhang, *Chemistry*, 2013, **19**, 2276.
95. S. Chandra, P. Patra, S. H. Pathan, S. Roy, S. Mitra, A. Layek, R. Bhar, P. Pramanik and A. Goswami, *J. Mater. Chem. B*, 2013, **1**, 2375.
96. Z. L. Wu, P. Zhang, M. X. Gao, C. F. Liu, W. Wang, F. Leng and C. Z. Huang, *J. Mater. Chem. B.*, 2013, **1**, 2868.
97. K. Qu, J. Wang, P. Ren and P. Qu, *Chemistry*, 2013, **19**, 7243.
98. B. Han, W. Wang, H. Wu and F. Fang, *Colloids Surf., B*, 2012, **100**, 209.
99. S. K. Bhunia, A. Saha, A. R. Maity, S. C. Ray and N. R. Jana, *Sci. Rep.*, 2013, **3**, 1473.
100. (a) C. H. Lee, R. Rajendran, M. S. Jeong, H. Y. Ko, J. Y. Joo, S. Cho, Y. W. Chang and S. Kim, *Chem. Commun.*, 2013, **49**, 6543; (b) W. Q. Kong, R. H. Liu, H. Li, J. Liu, H. Huang, Y. Liu and Z. H. Kang, *J. Mater. Chem. B*, 2014, **2**, 5077.
101. S. T. Yang, L. Cao, P. G. Luo, F. Lu, X. Wang, H. Wang, M. J. Meziani, Y. Liu, G. Qi and Y. P. Sun, *J. Am. Chem. Soc.*, 2009, **131**, 11308.
102. C. Li, S. T. Yang, X. Wang, P. G. Luo, J. H. Liu, S. Sahu, Y. Liu and Y. P. Sun, *Theranostics*, 2012, **2**, 295.
103. H. Tao, K. Yang, Z. Ma, J. Wan, Y. Zhang, Z. H. Kang and Z. Liu, *Small*, 2012, **8**, 281.
104. L. Wu, M. Luderer, X. Yang, C. Swain, H. Zhang, K. Nelson, A. J. Stacy, B. Shen, G. M. Lanza and D. Pan, *Theranostics*, 2013, **3**, 677.
105. N. Li, X. Liang, L. Wang, Z. H. Li, P. Li, Y. Zhu and J. Song, *J. Nanopart. Res.*, 2012, **14**, 1177.
106. S. Srivastava, R. Awasthi, D. Tripathi, M. K. Rai, V. Agarwal, N. S. Gajbhiye and R. K. Gupta, *Small*, 2012, **8**, 109.

107. X. Huang, F. Zhang, L. Zhu, K. Y. Choi, N. Guo, J. Guo, K. Tackett, P. Anil-kumar, G. Liu, Q. Quan, H. S. Choi, G. Niu, Y. P. Sun, S. Lee and X. Chen, *ACS Nano*, 2013, **7**, 5684.

108. P. Huang, J. Lin, X. Wang, Z. Wang, C. Zhang, Q. Wang, F. Chen, Z. Li, G. Shen, D. Cui and X. Chen, *Adv. Mater.*, 2012, **24**, 5104.

109. A. Gruber, *Science.*, 1997, **276**, 2012.

110. C. C. Fu, H. Y. Lee, K. Chen, T. S. Lim, H. Y. Wu, P. K. Lin, P. K. Wei, P. H. Tsao, H. C. Chang and W. Fann, *Proc. Natl. Acad. Sci. U. S. A.*, 2007, **104**, 727.

111. F. Neugart, A. Zappe, F. Jelezko, C. Tietz, J. P. Boudou, A. Krueger and J. Wrachtrup, *Nano Lett.*, 2007, **7**, 3588.

112. T. L. Wee, Y. W. Mau, C. Y. Fang and H. L. Hsu, *Diamand Relat. Mater.*, 2009, **18**, 567.

113. M. F. Weng, S. Y. Chiang, N. S. Wang and H. Niu, *Diamand Relat. Mater.*, 2009, **18**, 587.

114. Z. Wang, C. Xu and C. Liu, *J. Mater. Chem. C*, 2013, **1**, 6630.

115. M. Mkandawire, A. Pohl, T. Gubarevich, V. Lapina, D. Appelhans, G. Rödel, W. Pompe, J. Schreiber and J. Opitz, *J. Biophotonics*, 2009, **2**, 596.

116. Y. Lin, P. Elena, T. L. Wei, K. Wu and C. Cheng, *J. Biophotonics*, 2012, **5**, 838.

117. D. A. Simpson, A. J. Thompson, M. Kowarsky, N. F. Zeeshan, M. S. J. Barson, L. T. Hall, Y. Yan, S. Kaufmann, B. C. Johnson, T. Ohshima, F. Caruso, R. E. Scholten, R. B. Saint, M. J. Murray and L. C. L. Hollenberg, *Biomed. Opt. Express*, 2014, **5**, 1250.

118. N. Mohan, C. S. Chen, H. H. Hsieh, Y. C. Wu and H. C. Chang, *Nano Lett.*, 2010, **10**, 3692.

119. Y. Kuo, T. Y. Hsu, Y. C. Wu and H. C. Chang, *Biomaterials*, 2013, **34**, 8352.

120. V. Vaijayanthimala, P. Y. Cheng, S. H. Yeh, K. K. Liu, C. H. Hsiao, J. I. Chao and H. C. Chang, *Biomaterials*, 2012, **33**, 7794.

121. T. J. Wu, Y. K. Tzeng, W. W. Chang, C. A. Cheng, Y. Kuo, C. H. Chien, H. C. Chang and J. Yu, *Nat. Nanotechnol.*, 2013, **8**, 682.

122. A. Ennaceur and J. Delacour, *Behav. Brain Res.*, 1988, **31**, 47.

123. R. Igarashi, Y. Yoshinari, H. Yokota, T. Sugi, F. Sugihara, K. Ikeda, H. Sumiya, S. Tsujio, I. Mori, H. Tochio, Y. Harada and M. Shirakawa, *Nano Lett.*, 2012, **12**, 5726.

124. R. J. Epstein, F. M. Mendoza, Y. K. Kato and D. D. Awschalom, *Nat. Phys.*, 2005, **1**, 94.

125. A. Hegyi and E. Yablonovitch, *Nano Lett.*, 2013, **13**, 1173.

126. P. G. Baranov, A. A. Soltamova, D. O. Tolmachev, N. G. Romanov, R. A. Babunts, F. M. Shakhov, S. V. Kidalov, A. Y. Vul', G. V. Mamin, S. B. Orlinskii and N. I. Silkin, *Small*, 2011, **7**, 1533.

127. S. K. Sarkar, A. Bumb, X. Wu, K. A. Sochacki, P. Kellman, M. W. Brech-biel and K. C. Neuman, *Biomed. Opt. Express*, 2014, **5**, 1190.

CHAPTER 4

Near Infrared-Emitting Gold Nanoparticles for In vivo Tumor Imaging

JINBIN LIU[a] AND JIE ZHENG*[b]

[a]School of Chemistry and Chemistry Engineering, South China University of Technology, Guangzhou, 510640, P. R. China; [b]Department of Chemistry, The University of Texas at Dallas, Richardson, TX 75080, USA
*E-mail: jiezheng@utdallas.edu

4.1 Introduction

The interdisciplinary field of *in vivo* imaging focuses on non-invasive quantitation and visualization processes.[1,2] The critical points in this research area are reporters that are detectable by various imaging techniques. Besides radiotracers and magnetic probes,[3,4] fluorescent agents have played a major role in this field.[5] Fluorescence-based optical imaging is therefore becoming a promising non-invasive, real-time, and high-resolution approach for cancer diagnosis.[6-8] Among the non-invasive fluorescent imaging technologies, near infrared (NIR) fluorescence imaging with a wavelength window of 700–1000 nm has attracted increasing attention from the researchers due to the low absorption and autofluorescence from organisms, which can minimize the background interference, improve the tissue depth penetration, and increase the disease imaging sensitivity.[9-11] *In vivo* NIR imaging techniques

RSC Nanoscience & Nanotechnology No. 40
Near Infrared Nanomaterials: Preparation, Bioimaging and Therapy Applications
Edited by Fan Zhang
© The Royal Society of Chemistry 2016
Published by the Royal Society of Chemistry, www.rsc.org

have become indispensable low-cost and high-sensitivity preclinical tools for fundamental understandings of tumor angiogenesis, and they are to be translated into clinical trials to improve the reliability of surgery outcomes.[12] The rapid development of inorganic nanoparticles (NPs) with large surface areas, tunable material properties, and strong signal output will have significant impact on the future personalized oncology.[13–16]

NIR-emitting probes are predominantly used in research settings, although some of them have received approval from the United States Food and Drug Administration (FDA) for clinical use. Many NIR-emitting fluorophores are now commercially available that have high quantum yield (QY), water solubility, and good photostability. In general, NIR-emitting agents for bioimaging can be divided into two groups: small-molecule agents (molecular probes) and NP-based probes (nanoprobes). The majority of the recent developments in probe design focus on the nanoprobes.

To develop ideal nanoprobes for *in vivo* tumor imaging, the following features of nanoprobes should be considered:

- NIR emission, to minimize the tissue absorption and scattering effects.
- High stability at physiological pH (6.0–7.4), to maintain chemical stability and photostability and resist non-specific binding to serum proteins.
- Small size, to minimize recognition by macrophages and facilitate rapid distribution during circulation around the body.
- Renal clearance, to minimize the potential long-term toxicity and lower the background signal after performing the diagnostic task (*e.g.*, fluorescence imaging or drug delivery).[17,18]

Luminescent gold NPs (AuNPs) with the above advantages that can preferentially accumulate in the tumor site through enhanced permeability and retention (EPR) effect without requiring chemical conjugation of tumor-targeting ligands have been recently identified and have shown unique properties for cancer *in vivo* imaging.[19,20] Compared with the commonly used blue or green fluorescent noble metal NPs for biological labeling, water-soluble AuNPs with NIR emission have higher penetration capability so that they can be ideal for live animal imaging. In the past decade, significant progress has been made on the development of NIR-emitting AuNPs with molecular-like pharmacokinetics and tumor-targeting properties.[20–23] Although these newly developed NIR-emitting AuNPs have not yet been as widely used as conventional organic dyes in bioimaging, their biocompatibility, good water solubility, physiological stability, facile synthesis process, and diverse photophysical properties make them to be promising candidates.[24–26] In this review, we summarize the synthesis methods and *in vivo* behavior of NIR-emitting AuNPs, focusing on renal clearance, pharmacokinetics, and *in vivo* tumor imaging. We also discuss some critical challenges and our perspectives on the trends in this exciting research field.

4.2   Synthesis Strategies

The controllable synthesis of NIR-emitting AuNPs is of great importance to their *in vivo* imaging applications. In order to obtain high-quality NIR-emitting AuNPs, the reaction parameters such as temperature, time, and pH should be strictly optimized to control their size, gold oxidation state, surface properties, and optical properties.[27–29] Nowadays, significant efforts have been made to synthesize NIR-emitting AuNPs, and various approaches have been developed.[30–36] Typically, the surface ligand and valence state of gold have been reported to have an important influence on the optical properties of AuNPs.[37,38] In this section we summarize synthesis strategies for NIR-emitting AuNPs according to the two important factors: the surface ligand effect and the valence state effect.

4.2.1   Surface Ligand Effect

Stable NIR-emitting AuNPs were typically prepared through the reduction of Au^{3+} to Au^+ and Au^0 in the presence of reducing and capping agents. Due to strong interaction between Au and S, thiolated ligands are preferred capping agents for the synthesis of NIR-emitting AuNPs.[21,39–43] First, small thiol molecules have been reported to synthesize luminescent AuNPs *via* chemical reduction of the Au precursor, and fluorescence emission of AuNPs could be tuned from the visible to the NIR spectrum window.[27,29,44] For example, in 2005, Wang *et al.* reported the observation of NIR emissions ranging from 800 nm to 1.3 µm from AuNPs (size ~2 nm) capped with thiol molecules such as hexanethiolate, dodecanethiolate, and phenylethanethiol (Figure 4.1).[24]

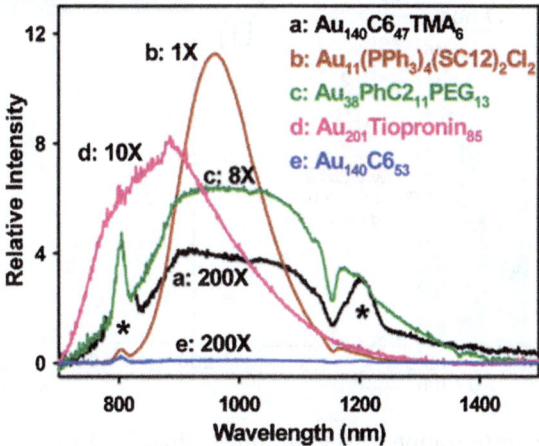

Figure 4.1   NIR-emitting AuNPs with different core sizes and surface ligands. C6, C12, PhC2, PEG, and PPh_3 represent hexanethiolate, dodecanethiolate, phenylethanethiol, poly(ethylene glycol) (MW 350) thiolate, and triphenylphosphine, respectively. (Reproduced with permission from G. L. Wang, T. Huang, R. W. Murray, L. Menard and R. G. Nuzzo, *J. Am. Chem. Soc.*, 2005, **127**, 812. Copyright (2005) American Chemical Society.)[24]

They proposed that size-independent emission originated from localized core surface states with size-independent energetics. They also found in their later work that the polarity of the thiolated ligands influences the QYs of NIR-emitting AuNPs, further demonstrating that emission fundamentally arises from the surface states of molecular AuNPs. In 2011, Tu *et al.* described the synthesis of NIR-emitting AuNPs using glutathione (a small tripeptide) and observed strong emission (maximum peak at 810 nm) from ~2.5 nm glutathione-coated AuNPs (GS–AuNPs) with a QY of 1% in water.[29] In 2012, by introduction of radiotracer [198]Au ions to the synthesis, we obtained NIR-emitting [[198]Au]AuNPs with a NIR emission maximum located at 810 nm (Figure 4.2),[45] which is identical to the NIR-emitting GS–AuNPs.[29] Moreover, the radiolabeling purity of [198]Au was assessed using radio high-performance liquid chromatography (radio-HPLC) with two detectors (UV and gamma counters), indicating that the radiochemical purity of the GS–[[198]Au]AuNPs was nearly 100%.

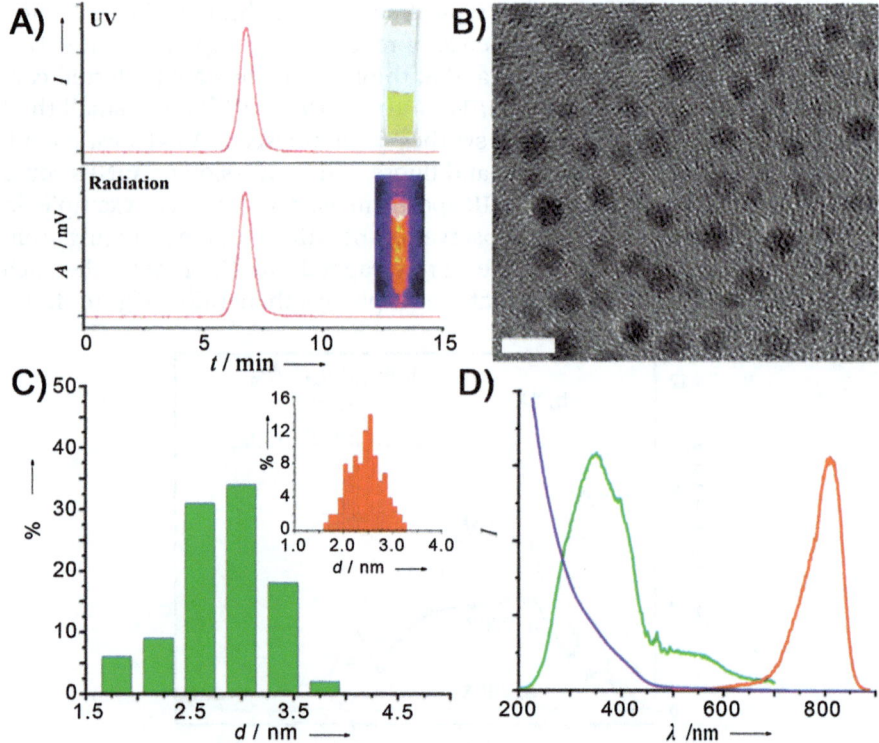

Figure 4.2   Characterization of NIR-emitting radioactive AuNPs (GS–[[198]Au]AuNPs). (A) Radio high-performance liquid chromatography (radio-HPLC) showing the radiochemical purity of GS–[[198]Au]AuNPs is 100%. Inset: bright-field image and radiation image of GS–[198Au]AuNPs. (B) Typical transmission electron microscopy (TEM; scale bar = 5 nm) image of the synthesized GS–[[198]Au]AuNPs. (C) Size distribution of GS–[[198]Au] AuNPs. (D) The absorption (blue), excitation (green), and emission (red) spectra of GS–[[198]Au]AuNPs in aqueous solution. (Reproduced from ref. 45 with permission from John Wiley and Sons. Copyright © 2012 Wiley-VCH Verlag GmbH & Co. KGaA, Weinheim.)

Ligand etching and exchange is another method of preparing NIR-emitting AuNPs using small thiol molecules.[46,47] For example, didodecyldimethylammonium bromide (DDAB)-stabilized AuNPs were introduced to aqueous phase upon ligand exchange with dihydrolipoic acid (DHLA) by Chang's group,[46] resulting in formation of bright red-emitting AuNPs that exhibited an emission maximum at 650–700 nm with an average core diameter of 1.56 ± 0.3 nm and QYs of 1–3%, reduced photobleaching compared to organic fluorophores, and very good colloidal stability. It was also found that the already synthesized AuNPs could be converted to other AuNPs through ligand exchange reaction.

In addition to small thiol compounds, polymers such as thiolated polyethylene glycol (PEG)[48-50] can be used as stable capping agents to prepare NIR-emitting AuNPs. In 2013, we created NIR-emitting AuNPs through a facile one-step synthesis by thermally reducing $HAuCl_4$ in the presence of thiolated PEG (PEG-SH) ligands with a molecular weight of 1 kDa in aqueous solution (Figure 4.3).[48] The core size of the PEG–AuNPs was measured

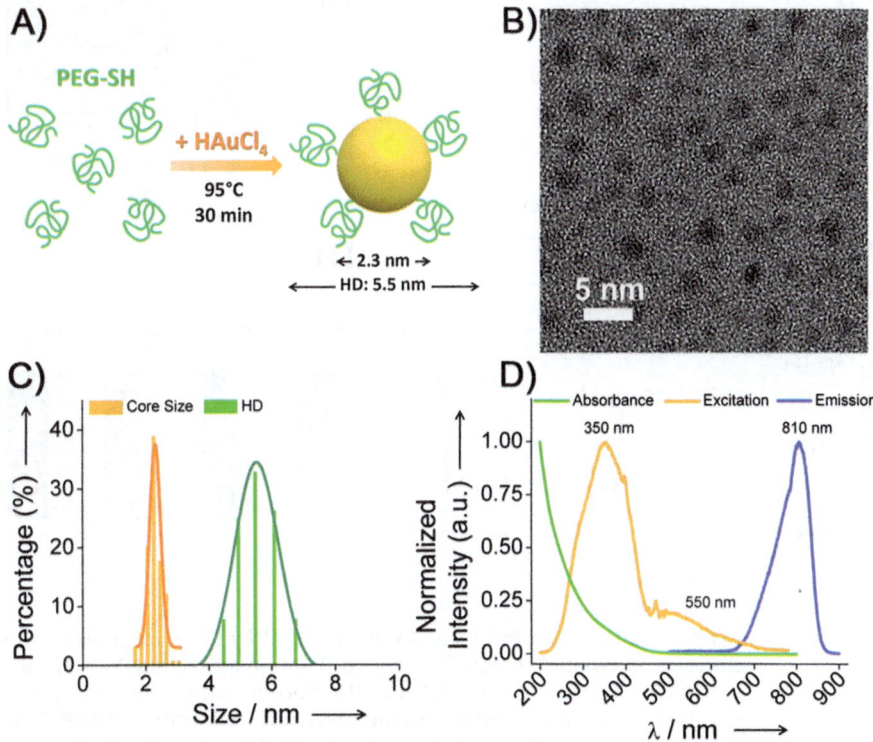

Figure 4.3 Characterization of NIR-emitting PEGylated AuNPs (PEG–AuNPs). (A) Scheme of the particle synthesis. (B) Typical TEM image of the synthesized PEG–AuNPs. (C) Core size measured by TEM, and hydrodynamic diameter (HD) in PBS measured by dynamic light scattering (DLS). (D) Absorption, excitation, and emission (λ_{ex} = 350 nm) spectra of the PEG–AuNPs in PBS at pH 7.4. (Reproduced from ref. 48 with permission from John Wiley and Sons. Copyright © 2013 Wiley-VCH Verlag GmbH & Co. KGaA, Weinheim.)

to be 2.3 nm, similar to that of the NIR-emitting GS–AuNPs (2.5 nm), and the PEG–AuNPs exhibited strong emission with a maximum at 810 nm, which is identical to the value of the NIR-emitting GS–AuNPs.[20,29] We also found that neither shorter (0.35 kDa) nor longer (5 kDa) PEG chains can generate highly luminescent or stable AuNPs, suggesting that the PEG chain length plays a crucial role in the synthesis of NIR-emitting AuNPs. Oh *et al.* also reported the synthesis of NIR-emitting AuNPs (size 1.5 nm) with emission maximum at 820 nm in water using PEG-dithiolane ligands terminating in either a carboxyl, amine, azide, or methoxy group (Figure 4.4).[50] The QYs were 4–8%, depending on the terminal functional groups. Furthermore, DNA and dye-conjugation reactions confirmed that the carboxyl, amine, and azide groups can be utilized on the NIR-emitting AuNPs for carbodiimide, succinimidyl ester, and Cu(I)-assisted cycloaddition chemistry, respectively.

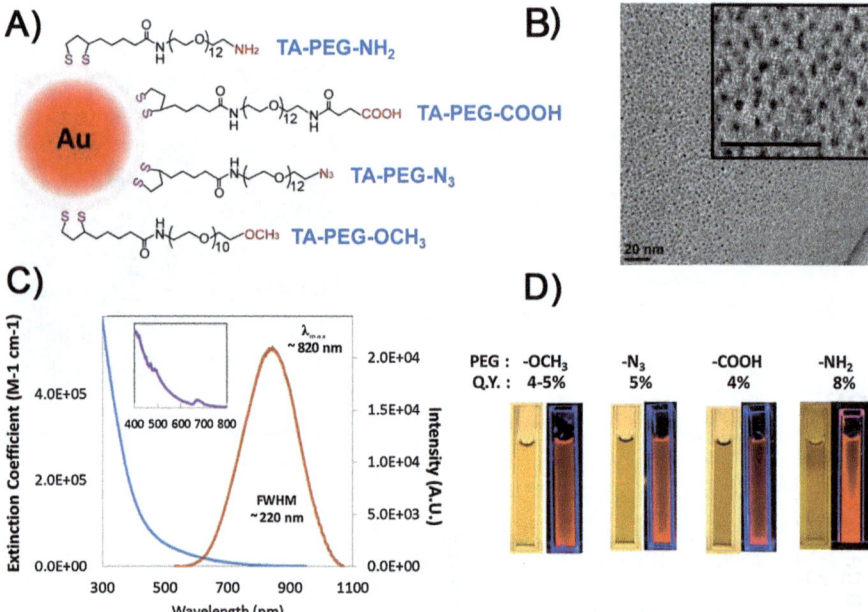

Figure 4.4   (A) Schematic diagram of AuNPs surrounded by the structures of the different PEGylated surface ligands. (B) TEM images of AuNP–NH$_2$ with a size distribution of 1.5 ± 0.3 nm. (C) Normalized UV-Vis absorption spectrum (blue line) and emission spectrum (red line) of AuNP–NH$_2$ excited at 450 nm. The inset shows the excitation spectrum (monitored at 820 nm) revealing some electronic structure (lowest energy absorption maximum at 670 nm). (D) Colloidal AuNPs displaying different functional groups with/without 365 nm UV excitation. (Reproduced from ref. 50 with permission from John Wiley and Sons. Copyright © 2013 Wiley-VCH Verlag GmbH & Co. KGaA, Weinheim.)

Some other biological molecules such as proteins and enzymes can also be used as surface templates for the preparation of NIR-emitting AuNPs. For example, Xie *et al.* reported a simple, one-pot and "green" synthetic route for preparation of AuNPs at physiological temperature (37 °C) with red emission ($\lambda_{em\,max}$ = 640 nm, QY ≈ 6%) using bovine serum albumin (BSA) as a bioligand and adjusting the reaction pH to 12.[51] Subsequently, Lu's group found that lysozyme[52] could also act as a good reducing and stabilizing agent for the preparation of highly fluorescent AuNPs with the same synthesis condition as BSA–AuNPs.[51] The average size was 1 nm and QY was 5.6% with the emission at 657 nm. Furthermore, pepsin,[53] horseradish peroxidase,[54] insulin,[55] and transferrin protein[56,57] have all been used as scaffolds for the preparation of NIR-emitting AuNPs. These functionalized AuNPs all exhibit excellent properties and have great potential for biological applications.

4.2.2   Valence State Effect

Fundamental understanding and control of photoluminescence from AuNPs have greatly expanded their applications in bioimaging. While luminescence from ultrasmall AuNPs is often attributed to a quantum size effect and surface ligands, the gold atom valence state also has a significant impact on photoluminescence from AuNPs.[28,37] For instance, dendrimer-encapsulated AuNPs will not emit fluorescence until their sizes become comparable to the electron Fermi wavelength (0.5 nm).[37] At this size, nanosecond emission was very sensitive to the number of gold atoms in the NPs and can be well explained by a free-electron model.[51,58] On the other hand, microsecond photoluminescence was also observed from 2 to 3 nm thiolated AuNPs,[28,29,43,59–62] where luminescence is dependent on S–Au charge transfer rather than particle size.[28] We were able to synthesize 1.7 nm orange-emitting AuNPs (OGS–AuNPs) with a maximum at 565 nm and 2.1 nm yellow-emitting AuNPs (YGS–AuNPs) with a maximum at 545 nm at different molar ratios of glutathione to $HAuCl_4$ (1:1, 2:1) by taking the advantages of the mild reducing ability of glutathione and the unique dissociation process of glutathione (GS)–Au(I) polymers.[28] QYs of OGS–AuNPs and YGS–AuNPs were measured to be 4.0 ± 0.4% and 4.3 ± 0.3% respectively, and element analysis indicates that the ratio of Au and GS in orange-emitting AuNPs is 22:10. Unlike these non-luminescent AuNPs, the luminescent GS–AuNPs contain a large number of Au(I) atoms measured by X-ray photoelectron spectroscopy (XPS). It was found that nearly 49% and 40% respectively of Au atoms in the OGS–AuNPs and YGS–AuNPs are in the Au(I) oxidation state. This is quite different from NIR-emitting AuNPs, where the Au(I) percentage was only 36.8%.[29,45] By comparing the Au(I) oxidation state percentages of OGS–AuNPs, YGS–AuNPs, and NIR-emitting AuNPs, it is obvious that the valence state is a critical factor in their emission maximum.

The luminescence lifetime studies unraveled a very intriguing phenomenon in the electronic structures of luminescent AuNPs. As for NIR-emitting AuNPs, only microsecond lifetimes were obtained regardless of the excitation wavelengths, which suggests that Au–S hybrid states must be involved in the emission process.[28,29,37] While strong surface Au(I)–S charge transfer is involved in both visible and NIR emission of GS–AuNPs, the exact origin of these emissions from AuNPs with the same glutathione surface ligand is still not fully understood. Since Au(I) atoms fail to donate free electrons to the particles, no surface plasmon absorption was observed from these visible or NIR-emitting GS–AuNPs. After strong reduction of the Au(I) with NaBH$_4$, the luminescence of the AuNPs vanished and very weak surface plasmon absorption of AuNPs started to emerge even though very little change in particle size was observed before or after further reduction.[28] These observations further confirmed that luminescence from these AuNPs is strongly dependent on the valence states of gold atoms in the particles.

4.3   Renal Clearance and Pharmacokinetics

The FDA requires all injected contrast agents to be completely cleared from the body in a reasonable time period.[63] In general, renal clearance is an obvious excretion pathway for contrast agents due to the rapid elimination and low cellular internalization/metabolism involved, thus minimizing the non-specific accumulation of the contrast agents in reticuloendothelial system (RES) organs (liver, spleen, *etc.*). Small molecular probes are rapidly eliminated from the body through the urinary system because they are smaller than the kidney filtration threshold (KFT, ~5.5 nm).[64–66] For instance, 2-deoxy-2-[^{18}F]-fluoro-D-glucose([^{18}F] FDG),[67] Gd–DTPA complex,[68] and iomeprol[69] all exhibit efficient renal clearance. On the contrary, nanoprobes with hydrodynamic diameter (HD) ranging from 10 to 200 nm are eventually accumulated and retained in the RES organs for a long time, which raises concerns about the long-term risk of adverse effects.[63] Therefore, renal clearance has been a major roadblock for the clinical translation of nanoprobes even though they often exhibit much stronger signal outputs and more biomedical functionalities than the common clinically used small molecular probes.[70–72] Therefore, developing inorganic NPs integrated with the merits of molecular probes in efficient renal clearance is highly desirable for catalyzing the shift of our current medical paradigm to "earlier detection and prevention".[8,73,74]

In addition, contrast agents used in clinics often exhibit the following pharmacokinetics: rapid diffusion (short distribution half-life, $t_{1/2\alpha}$), relatively long blood circulation time (long elimination half-life, $t_{1/2\beta}$). These specific pharmacokinetic features not only ensure the success of clinical imaging but also minimize the potential health hazards caused by the introduction of contrast agents into the body. We now summarize the types of renally clearable luminescent AuNPs developed recently and discuss their strengths in renal clearance and molecular-like pharmacokinetics.

4.3.1 Renal Clearance

The renal clearance of NPs is dependent on size, surface chemistry, and shape, which determines glomerular filtration in the kidneys (Figure 4.5).[75] Generally, a globular NP with a HD < 6 nm (KFT) can easily pass through the glomerular capillary wall. In addition, shape is also an important factor in kidney filtration. Large linear-structured NPs such as carbon nanotubes (CNTs) with a width smaller than the KFT can also be filtered through the kidney. For the ultrasmall AuNPs, surface chemistry also plays an important role in renal clearance. For example, nearly 50% of the injection dose (ID) of 1.4 nm AuNPs was detected in the liver and only about 9% ID can be excreted into urine 24 h after injection;[76] this may be due to the binding by serum protein during circulation around the body.

In 2007, Choi *et al.*[63] synthesized a series of QDs with defined HDs ranging from 4.36 to 8.65 nm. Their results showed that QDs with sizes <5.5 nm cleared from the body *via* the urine with an efficiency of >50% ID 4 h after

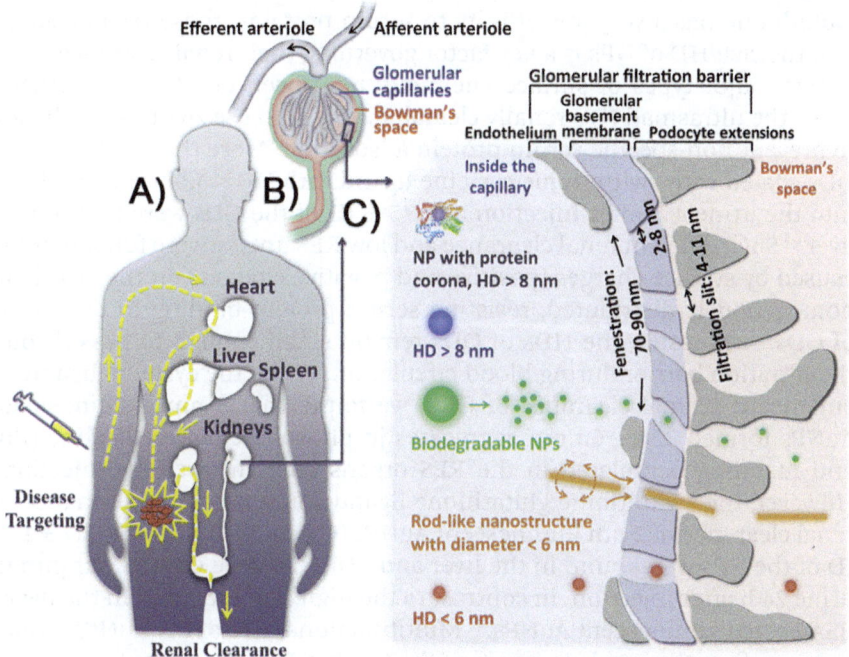

Figure 4.5 (A) Ideal renally clearable NPs for disease targeting in clinical practice: the NPs target the diseases and the untargeted ones are rapidly cleared out through the urinary system. (B) Schematic structure of the kidney corpuscle. (C) Glomerular filtration is a nanoscale phenomenon. The glomerular capillary wall is made up of three specialized layers: fenestrated endothelium, glomerular basement membrane, and podocyte extensions. (Reproduced from J. Liu *et al.*, Renal clearable inorganic nanoparticles: a new frontier of bionanotechnology, *Mater. Today*, 2013, **16**, 477–486. Copyright (2013) with permission from Elsevier.[75])

injection, while most of the 8.65 nm QDs accumulated in the liver. Since this landmark work on the renal clearance of QDs, more and more renally clearable NPs of various sizes and shapes have been synthesized.[77] For example, Li *et al.* reported a single-compartment, multifunctional ultrasmall copper sulfide nanodot (size 4.3 nm) coated with polyvinylpyrrolidone that can escape the RES and rapidly pass through the kidney filtration barrier.[78] Single-walled carbon nanotubes (SWCNTs) with length 100–1000 nm and diameter 0.8–1.2 nm were investigated and the results showed that the renal clearance efficiency was as high as 65% ID 20 min after injection.[79] Our recent investigations showed that glutathione can act as an effective surface ligand to minimize non-specific accumulation of luminescent AuNPs in the RES organs, and more than 50% ID of the GS–AuNPs is cleared from the circulation through the urinary system 48 h after injection.[19,20] To further understand how the particle size influences the renal clearance of GS–AuNPs, we investigated the renal clearance of 2, 6, and 13 nm (HD) AuNPs coated with glutathione and found that their renal clearance efficiencies were 50% ID, 4% ID, and 0.5% ID respectively, 24 h after injection (Figure 4.6).[19] Although glutathione has a very low affinity to serum proteins, these results suggest that the size/HD of NPs is a key factor governing their renal clearance.

Two major types of surface chemistry have been developed in order to make the ultrasmall NPs renally clearable. One is to use zwitterionic ligands to prevent non-specific serum protein adsorption. More than 50% ID of the QDs coated with zwitterionic cysteine ligand with HD < 5.5 nm were cleared into the urine 4 h after injection and <5% ID of the QDs were found in the liver.[63] Such efficient renal clearance and low RES uptake were fundamentally caused by surface charges (positive and negative charges) on the QDs being homogenously distributed, resisting serum protein binding to the surface of QDs. As a result, the HDs of QDs remain small enough to pass through the filtration barrier during blood circulation. While the cysteine ligand can enhance the renal clearance of QDs, we found that 3 nm cysteine-coated AuNPs formed 220 ± 60 nm aggregates in phosphate-buffered saline (PBS) and mainly accumulated in the RES organs after intravenous injection.[19] However, the zwitterionic glutathione ligand could be used to enhance the renal clearance of 2 nm luminescent AuNPs (GS–AuNPs).[19,20] Only 3.7 ± 1.9% ID of the NPs were found in the liver and >50% ID were cleared out into the urine 24 h after injection, in contrast to the high accumulation in the liver of BSA-coated luminescent AuNPs.[80] Multifunctionalities of the AuNPs allowed us to confirm the renal clearance of the GS–AuNPs with computed tomography (CT) for the bladder imaging due to the strong X-ray attenuation of gold atoms (Figure 4.7).[19] The further introduction of the radiotracer [198]Au in the synthesis of NIR-emitting [[198]Au]AuNPs provided an opportunity for *in vivo* single-photon emission CT (SPECT) imaging of both the kidneys and bladder (Figure 4.8A).[45] In addition, real-time fluorescence imaging can be conveniently used to image the bladder fluorescence signal kinetics in the initial stage (Figure 4.8B), which showed that the initial urinary excretion half-life of the AuNPs was only 6 min. Interestingly, the emission profile of the NPs

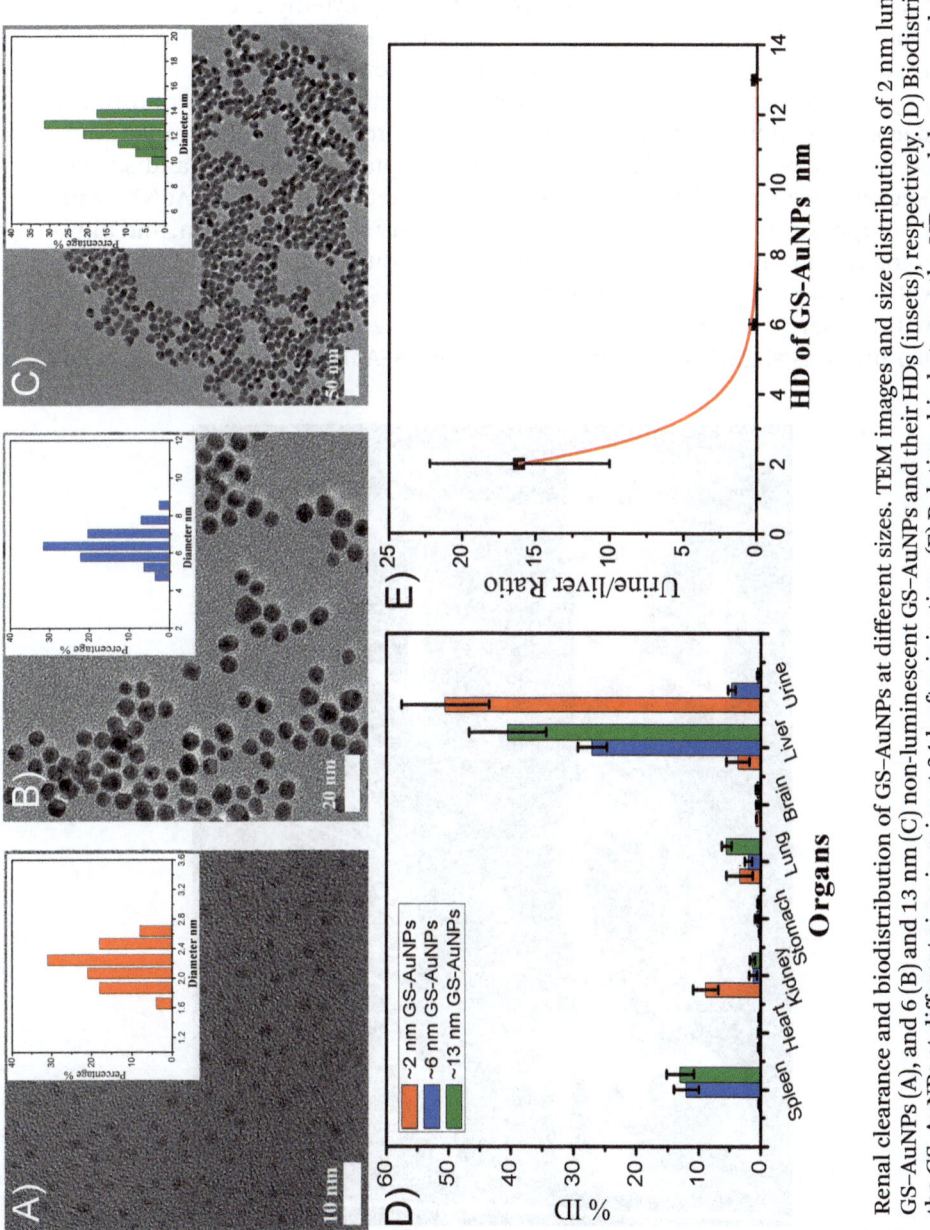

Figure 4.6 Renal clearance and biodistribution of GS-AuNPs at different sizes. TEM images and size distributions of 2 nm luminescent GS-AuNPs (A), and 6 (B) and 13 nm (C) non-luminescent GS-AuNPs and their HDs (insets), respectively. (D) Biodistribution of the GS-AuNPs at different sizes in mice at 24 h after injection. (E) Relationship between the HDs and the accumulated urine/liver ratio. (Reproduced from ref. 19 with permission from John Wiley and Sons. Copyright © 2011 Wiley-VCH Verlag GmbH & Co. KGaA, Weinheim.)

remained unchanged before the injection and after excretion in the urine, suggesting that the GS–AuNPs were fairly stable during blood circulation.

PEGylation is another surface chemistry strategy to make luminescent AuNPs renally clearable.[75] The strengths of poly(ethylene glycol) (PEG) ligands over conventional anionic or cationic ligands arise from the fact that the PEG moiety on the particle surface not only stabilizes NPs in the physiological environment but also creates steric hindrance to significantly decrease protein (opsonin) adsorption and improve the blood compatibility, so that the NPs can evade the RES with less uptake by the liver and spleen. Recently, we created renally clearable PEGylated NIR-emitting AuNPs with photophysical properties, core sizes, low affinity to serum protein, and physiological stability almost identical to our previous zwitterionic NIR-emitting GS–AuNPs.[20,48] The core size of the PEG–AuNPs was measured to be 2.3 ± 0.3 nm, very similar to that of the GS–AuNPs (2.5 nm), but their HD (5.5 ± 0.4 nm) was slightly larger than that of the GS–AuNPs (3.3 nm). In terms of

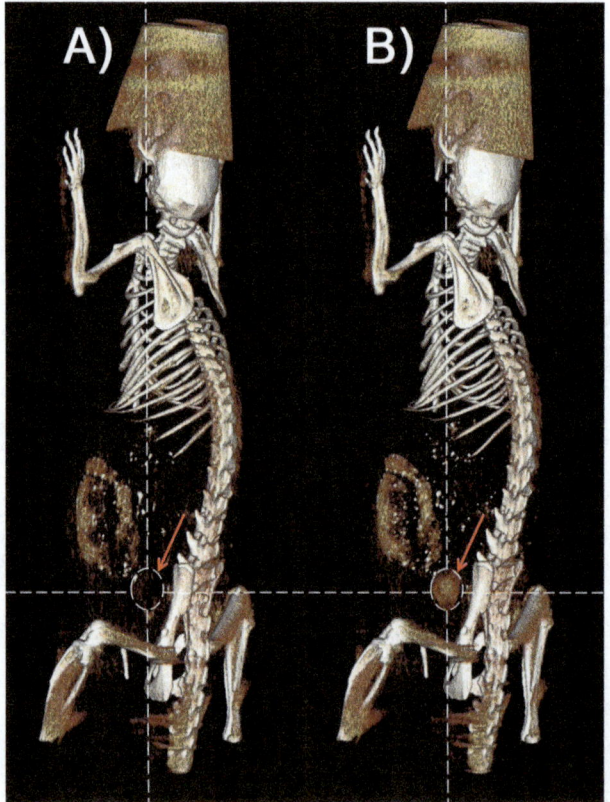

Figure 4.7   X-ray CT images of a live mouse (A) before and (B) 30 min after intravenous injection of GS–AuNPs. (Reproduced from ref. 19 with permission from John Wiley and Sons. Copyright © 2011 Wiley-VCH Verlag GmbH & Co. KGaA, Weinheim.)

liver uptake of the PEG–AuNPs, the maximum accumulation was 4.35% ID g^{-1} at 1 h after injection and the amount remained roughly constant (3.27–3.76% ID g^{-1}) during the 48 h body circulation, a result comparable to the liver accumulation of the GS–AuNPs. We found that a comparable amount of PEG–AuNPs and GS–AuNPs were detected in the urine 24 h after injection (~50%) (Figure 4.9),[48] but the PEG–AuNPs showed slow renal clearance in the initial

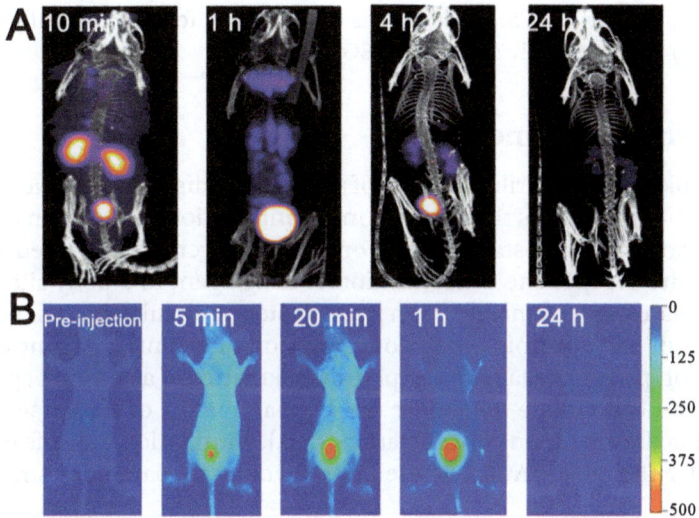

Figure 4.8 (A) Representative SPECT images (top row) of balb/c mice injected with GS–[^{198}Au]AuNPs. (B) *In vivo* fluorescence imaging of a live mouse after intravenous injection of GS–[^{198}Au]AuNPs. (Reproduced from ref. 45 with permission from John Wiley and Sons. Copyright © 2012 Wiley-VCH Verlag GmbH & Co. KGaA, Weinheim.)

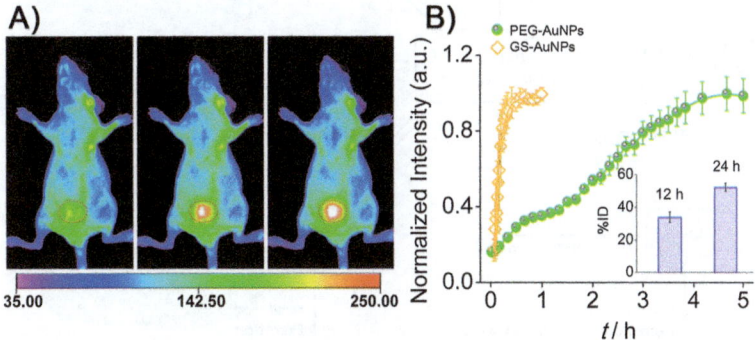

Figure 4.9 (A) Fluorescent imaging of the renal clearance kinetics of PEG–AuNPs in nude mice. (B) Renal clearance kinetics of PEG–AuNPs and GS–AuNPs (inset: PEG–AuNPs found in the urine measured by ICP–MS at 12 and 24 h after injection). (Reproduced from ref. 48 with permission from John Wiley and Sons. Copyright © 2013 Wiley-VCH Verlag GmbH & Co. KGaA, Weinheim.)

stage compared to the GS–AuNPs, thus demonstrating a slow diffusion in the body for the PEG–AuNPs. The results showed that the length of the PEG ligands was very important. If the PEG chain was too long, the HDs of PEGylated NPs exceeded the KFT and thus the NPs eventually accumulated in the RES. If the PEG ligands were too small, the physiological stability of the NPs was dramatically decreased and they significantly increase in size during circulation around the body, which also induces high non-specific accumulation in the RES. PEG lengths should therefore be carefully selected in order to create renally clearable PEGylated NPs.

4.3.2 Pharmacokinetics

Pharmacokinetics describes a series of absorption, distribution, metabolism, and excretion processes that happen in the interaction between an organism and an exogenous substance (drug or contrast agent) introduced into the body (Figure 4.10).[81] The concentration of drug/agent in a body site in most cases is related to its concentration in the systemic circulation. Therefore, we can monitor the dynamic blood concentration of a contrast agent until the elimination phase to evaluate the pharmacokinetics of a contrast agent, and then use a fitting mode to achieve the key parameters of pharmacokinetics such as maximum blood concentration (C_{max}), elimination half-life ($t_{1/2}$), and area under the curve (AUC). These pharmacokinetic parameters reflect the

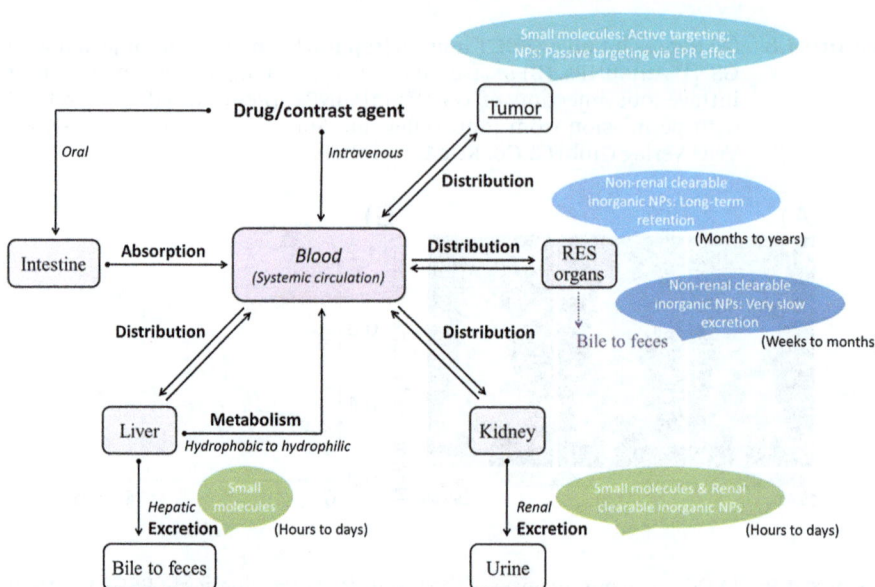

Figure 4.10   The pharmacokinetics of adsorption, distribution, metabolism, and excretion of a drug or contrast agent. (Reproduced with permission from M. X. Yu and J. Zheng, *ACS Nano*, 2015, **9**, 6655. Copyright (2015) American Chemical Society.)[81]

exposure of organs and tissues to the contrast agents, helping to analyze not only the toxicity but also the tumor-targeting efficiency of a contrast agent. Generally, a high blood concentration, long blood elimination half-life, and large AUC are favorable for enhancing the EPR effect of an NP, and thus improving its tumor-targeting efficiency.[48,75]

For small molecular probes, the typical pharmacokinetics of is a two-compartment model with a short distribution half-life $(t_{1/2\alpha})$ and a short elimination half-life $(t_{1/2\beta})$ due to the rapid renal clearance. For example, iomeprol (a commercially available CT contrast agent) exhibits a $t_{1/2\alpha}$ of 16.2 min and a $t_{1/2\beta}$ of 2.34 h after intravenous injection.[69] Among conventional nanoprobes, most NPs exhibited a one-compartment pharmacokinetic model after intravenous administration with a typical half-life of 4–7 h,[81] but some PEGylated NPs showed two-compartment model pharmacokinetics. For instance, PEGylated graphene showed a $t_{1/2\alpha}$ of 0.39 h and $t_{1/2\beta}$ of 6.97 h after intravenous injection.[82] NIR-emitting zwitterionic GS–AuNPs and PEGylated AuNPs followed a classical two-compartment pharmacokinetics model.[20,48,81] Results showed that GS–AuNPs exhibited a short $t_{1/2\alpha}$ of 5.4 ± 1.2 min and a relatively long $t_{1/2\beta}$ of 8.5 ± 2.1 h. Their short $t_{1/2\alpha}$ indicated that GS–AuNPs could rapidly distribute into tissues during the body circulation, very similar to molecular probes. Moreover, the relatively long $t_{1/2\beta}$ was more like a conventional nanoprobe, which allowed the NPs to circulate in the blood at a relatively high concentration. The $t_{1/2\alpha}$ of renally clearable PEG–AuNP (56.1 ± 1.2 min) was 10 times longer than that of GS–AuNP, but the $t_{1/2\beta}$ of PEG–AuNPs and GS–AuNPs were comparable, resulting in a larger AUC for PEG–AuNP, which would significantly affect their *in vivo* tumor-targeting behavior.

4.4 *In vivo* Tumor Imaging

The EPR effect is a great strength of nanoprobes in cancer diagnosis and therapy.[83–85] Selective targeting of tumor tissues with nanoprobes can be achieved by taking advantage of the unique tumor structure (hypervasculature, defective vascular architecture, and impaired lymphatic drainage).[86] Nanoprobes can accumulate in tumor sites at much higher concentrations and for longer times than small molecules through the EPR effect.[87,88] As a result, nanoprobes have great potential to address some critical challenges in early cancer diagnosis and therapy.[89] This unique strength of nanoprobes in tumor targeting is fundamentally due to the fact that they can escape rapid kidney filtration and accumulate in a tumor after prolonged circulation at high concentrations in the blood plasma. However, once nanoprobes become renally clearable, many fundamental questions such as whether the renally clearable nanoprobes can still target tumors through the EPR effect need to be answered.[20] The renally clearable NIR-emitting AuNPs possess an attractive series of features such as ultrasmall size and molecular-like pharmacokinetics, which have been recently demonstrated to enable non-invasive tumor imaging.[20,48] Here, we summarize the differences in tumor-targeting behavior between the renally clearable NIR-emitting AuNPs, conventional non-renally clearable nanoprobes, and small molecules.

The small-molecule-based contrast agents have been widely used for clinical passive tumor diagnosis based on the differences in vascular structure or metabolism between tumor and normal tissues.[90,91] Most small molecular probes have the pharmacokinetic merits in of a short $t_{1/2\alpha}$, a relatively long $t_{1/2\beta}$, and efficient renal clearance, which ensure that tumors can be imaged in a desired time window. More importantly, the unbound probes can be cleared out rapidly with low toxicity. However, small molecular probes are often rapidly distributed in the whole body and target the tumor randomly, which cause low imaging contrast index and low tumor-targeting efficiency. For example, the passive tumor targeting efficiencies of the renally clearable small molecules IRDye 800CW and ^{111}In complexes were only 0.22% ID g^{-1} (at 12 h after injection) and 2.0% ID g^{-1} (at 24 h after injection),[20,92] respectively. The rapid renal clearance of small molecules significantly lowers their tumor-targeting efficiency. However, as for the conventional non-renally clearable nanoprobes, the EPR effect is the most widely used strategy for delivering the nanoprobe to tumor sites in either passive or active tumor targeting.[93,94] However, unlike the small molecular probes, nanoprobes often markedly accumulate in RES organs (liver, spleen, *etc.*), resulting in low targeting specificity (defined as the amount of probes in tumor *versus* that in liver) and potential long-term toxicity.[95,96] For example, gold nanocages (AuNCs) coated with 5 kDa PEG exhibited a tumor accumulation of 15.3 ± 2.9% ID g^{-1} at 24 h after injection, but nearly 65% ID g^{-1} and ~30% ID g^{-1} of the AuNCs were found in the liver and spleen, respectively.[97] Such severe long-term accumulation in RES potentially induces health hazards, hampering further clinical translation.

Renally clearable NIR-emitting AuNPs behave like small molecules in renal clearance and two-compartment pharmacokinetics due to their small size and a surface chemistry that enables them to resist absorption by serum proteins. However, whether they can still behave sufficiently like conventional NPs to show the EPR effect in passive tumor targeting is unknown, but important to their future applications in tumor diagnosis and therapy.[98] We conducted a head-to-head comparison between a renally clearable inorganic NP, GS–AuNPs, and an organic dye (IRDye 800CW) in passive tumor targeting under the same conditions (Figure 4.11). Both of the NPs and the dye molecules showed rapid distribution in the mouse body and tumor targeting ability (Figure 4.11A). However, the amount of GS–AuNPs that was retained in the tumor site ten was times higher than that of the dye molecules at 12 h after injection, demonstrating that tumor retained time of GS–AuNPs is much longer than that of the dye molecules, and the renally clearable GS–AuNPs do show the EPR effect (Figure 4.11B). More interestingly, the clearance of GS–AuNPs from normal tissues is more than three times faster than that of the dye molecules (Figure 4.11C). As a result, the contrast index (CI) value of GS–AuNPs reached a threshold for substantial tumor detection (CI = 2.5) earlier than the small dye molecules (3 h *versus* 8 h) (Figure 4.11D). It took only

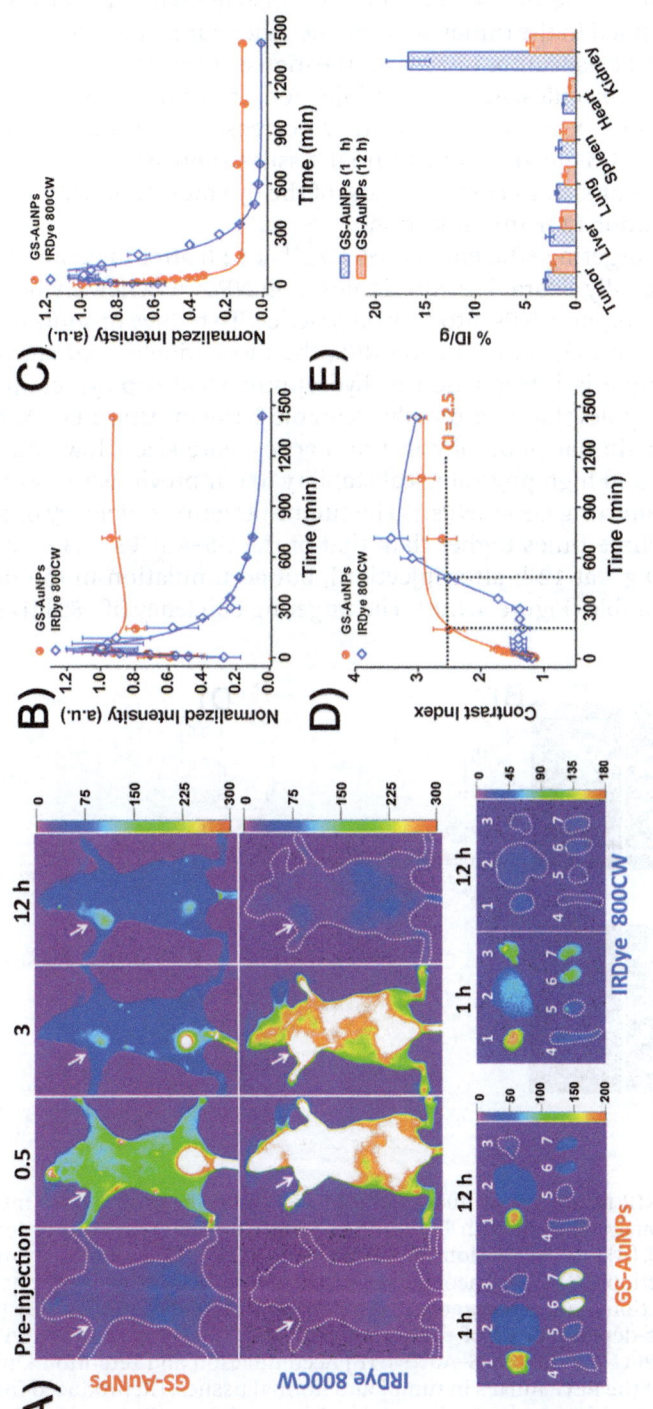

Figure 4.11 (A) Representative *in vivo* NIR fluorescence images of MCF-7 tumor-bearing mice intravenously injected with GS–AuNPs and IRDye 800CW. The tumor areas are indicated by arrows. Labels: 1, tumor; 2, liver; 3, lung; 4, spleen; 5, heart; 6, kidney (left); 7, kidney (right). (B) Time-dependent contrast index (CI) tumor after probe injection. Accumulation and retention kinetics of the two probes in tumor tissue and (C) normal 0–24 h after injection. (D) CI of GS–AuNPs and IRDye 800CW at different after injection time points. (E) Ratios of the GS–AuNP concentration in tumor to that in liver at 1 and 12 h after injection. (Reproduced with permission from J. B. Liu, M. X. Yu, C. Zhou, S. Y. Yang, X. H. Ning and J. Zheng, *J. Am. Chem. Soc.*, 2013, 135, 4978. Copyright (2013) American Chemical Society.)[20]

3 h after injection for the mouse intravenously injected with GS–AuNPs to become readily defined in the tumor area. On the other hand, it would take an additional 5 h to get clear tumor images for the mouse intravenously injected with the small dye molecules. At 12 h after injection, the tumor area was even more distinguishable because of the further decrease in the fluorescence background of the GS–AuNPs from normal tissues, and well-maintained signal from the tumor site. In contrast, most of the dye molecules were eliminated from the tumor 12 h after injection.

The low tumor-targeting efficiency (2.3% ID g^{-1} at 12 h after injection, Figure 4.11E) of the renally clearable zwitterionic GS–AuNPs was a new challenge for translation of inorganic NPs into clinical practice. To reduce opsonization by blood proteins and clearance by the RES, the most widely used surface decoration technique is introduction of hydrophilic stealth polymers (*e.g.* PEG).[99] We recently developed a renally clearable NIR-emitting PEG–AuNP (Figure 4.12) with similar photophysical properties, core sizes, low affinity to serum protein, and high physiological stability to our previously reported zwitterionic NIR-emitting GS–AuNPs.[48] The tumor-targeting efficiency of the PEG–AuNPs was three times higher than that of the GS–AuNPs (8.3 ± 0.9% ID g^{-1} *vs.* 2.3% ID g^{-1} at 12 h after injection), but accumulation in the RES organs was comparable (Figure 4.12B). The targeting efficiency of ~8% ID g^{-1}

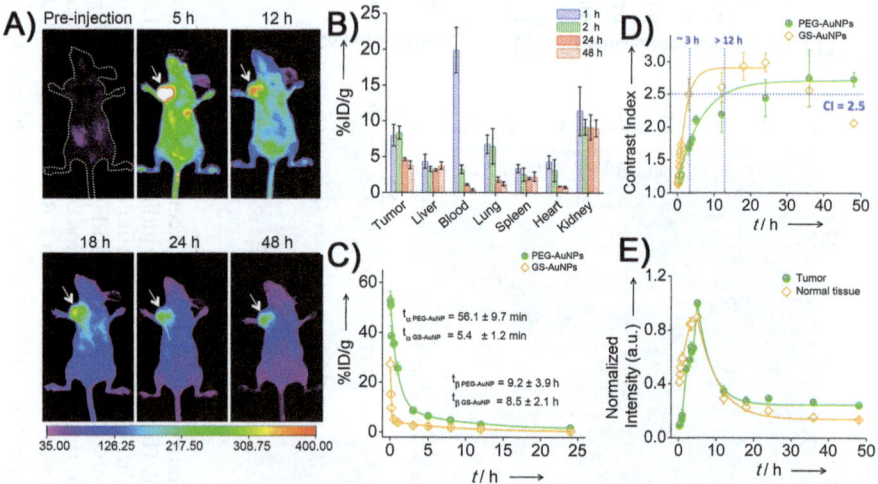

Figure 4.12 (A) NIR fluorescence images of a MCF-7 tumor-bearing mouse intravenously injected with PEG–AuNPs (arrow indicates the tumor location). (B) Biodistributions of the PEG–AuNPs at 1, 12, 24, and 48 h after injection. (C) Pharmacokinetics of the two probes after intravenous injection (data presented as the mean ± standard deviation, $n = 3$). (D) Time-dependent CI of the tumor area after intravenous injection of the PEG–AuNPs or GS–AuNPs. (E) Accumulation and retention kinetics of the PEG–AuNPs in tumor and normal tissues. (Reproduced from ref. 48 with permission from John Wiley and Sons. Copyright © 2013 Wiley-VCH Verlag GmbH & Co. KGaA, Weinheim.)

was ten times higher than that of renally clearable silica NPs $(0.9\% \text{ ID g}^{-1})$,[100] and also comparable to those of reported non-renally clearable NPs with strong EPR effect, such as PEGylated 20 nm AuNPs $(6.63\% \text{ ID g}^{-1})$,[101] and six-armed PEGylated Ag_2S QDs $(7-15\% \text{ ID g}^{-1})$.[102] The high tumor-targeting efficiency of these PEG–AuNPs is attributed to the strong EPR effect enhanced by extended distribution half-life and increased AUC of the NPs (Figure 4.12C). The difference in tumor and liver uptake behaviors between PEG–AuNPs and GS–AuNPs, suggested that PEGylation of renally clearable AuNPs could further enhance tumor-targeting specificity without increasing the RES uptake.

However, the limitation of PEGylation in tumor imaging is that it takes much longer for PEG–AuNPs to reach the CI threshold value (CI = 2.5) than for zwitterionic GS–AuNPs (Figure 4.12D). It took >12 h for the tumor area to reach the CI threshold (CI = 2.5) after intravenous injection of PEG–AuNPs, much longer than the time needed for GS–AuNPs (~3 h), even if the tumor-targeting efficiency of PEG–AuNPs is three times higher than that of the GS–AuNPs. This is because of the slow diffusion and clearance of PEG–AuNPs in the normal tissues (Figure 4.12E). These differences in the tumor targeting of luminescent AuNPs indicate that PEGylated AuNPs have more potential in cancer therapy due to the high targeting efficiency and long tumor retention, whereas zwitterionic GS–AuNPs are potentially more appropriate in cancer diagnosis owing to their short detection time and rapid clearance from normal tissues.

4.5 Conclusion and Outlook

In this chapter, we briefly summarized the recent advances in the synthesis and *in vivo* behaviors (including the renal clearance, pharmacokinetics, and tumor imaging) of NIR-emitting AuNPs. The NIR-emitting AuNPs have been developed as a new class of nanofluorophore with excellent optical properties and have been considered as attractive nanoprobes for *in vivo* tumor imaging. Significant progress has been achieved in the field of synthesis of NIR-emitting AuNPs after a decade of effort, and the synthesis of NIR-emitting AuNPs is no longer a great difficulty. However, several challenges remain and need to be addressed.

At present, some NIR-emitting AuNPs still exhibit strong non-specific accumulation in RES organs after administration due to their surface chemistries, although their core sizes are small (<2.5 nm). Bioimaging applications require carefully tuning of the surface chemistries of the NIR-emitting AuNPs to improve their *in vivo* behaviors. Intravenously injected NIR-emitting AuNPs should be effectively cleared out of the body and should not accumulate in healthy organs, to minimize long-term toxicity. While the emergence of two types of surface chemistries, zwitterionization and PEGylation, have partially addressed the issue of non-specific RES accumulation, the surface chemistries of the NIR-emitting NPs should be studied more thoroughly and applied to more NPs.[75] In addition, the emergence of renally clearable NIR-emitting AuNPs provides unprecedented opportunities to address long-term challenges such as high nanotoxicity and low specificity in the clinical translation of inorganic NPs. While fluorescent imaging is a major tool used for the study of the NIR-emitting AuNPs due to its low cost and high temporal and spatial resolution, the applications of NIR-emitting AuNPs are not limited to optical imaging. By introducing radioactive isotopes to the synthesis, we can develop multimode imaging systems including many clinically available imaging techniques, such as positron emission tomography (PET) and SPECT, so that deep-tissue imaging with NIR-emitting AuNPs can be achieved.

While there is still a long way to go before NIR-emitting AuNPs are as widely used in clinical practice as the small molecular contrast agents, with the rapid development of renally clearable NIR-emitting AuNPs in the past few years, we believe that these NPs will find wide applications in clinical diagnosis and therapy in the future.

Acknowledgement

The support from NIH R01DK103363, CPRIT (RP120588, RP140544) and the start-up fund from the University of Texas at Dallas (J.Z.)

References

1. R. Weissleder and M. J. Pittet, *Nature*, 2008, **452**, 580.
2. J. V. Frangioni, *J. Clin. Oncol.*, 2008, **26**, 4012.
3. W. J. Rieter, J. S. Kim, K. M. L. Taylor, H. An, W. Lin, T. Tarrant and W. Lin, *Angew. Chem., Int. Ed.*, 2007, **119**, 3754.
4. W. J. Rieter, J. S. Kim, K. M. L. Taylor, H. An, W. Lin, T. Tarrant and W. Lin, *Angew. Chem., Int. Ed.*, 2007, **46**, 3680.
5. S. Wolfbeis, *Chem. Soc. Rev.*, 2015, **44**, 4743.
6. L. Fass, *Mol. Oncol.*, 2008, **2**, 115.
7. M. Ferrari, *Nat. Rev. Cancer*, 2005, **5**, 161.
8. H. Xing, N. Y. Wong, Y. Xiang and Y. Lu, *Curr. Opin. Chem. Biol.*, 2012, **16**, 429.
9. S. A. Hilderbrand and R. Weissleder, *Curr. Opin. Chem. Biol.*, 2010, **14**, 71.
10. M. A. Pysz, S. S. Gambhir and J. K. Willmann, *Clin. Radiol.*, 2010, **65**, 500.
11. J. V. Frangioni, *Curr. Opin. Chem. Biol.*, 2003, **7**, 626.
12. S. L. Owens, *Br. J. Ophthalmol.*, 1996, **80**, 263.
13. S. R. Banerjee, M. Pullambhatla, Y. Byun, S. Nimmagadda, C. A. Foss, G. Green, J. J. Fox, S. E. Lupold, R. C. Mease and M. G. Pomper, *Angew. Chem., Int. Ed.*, 2011, **123**, 9333.
14. J. L. West and N. J. Halas, *Annu. Rev. Biomed. Eng.*, 2003, **5**, 285.
15. X. H. Huang, I. H. El-Sayed, W. Qian and M. A. El-Sayed, *J. Am. Chem. Soc.*, 2006, **128**, 2115.

16. V. P. Zharov, K. E. Mercer, E. N. Galitovskaya and M. S. Smeltzer, *Biophys. J.*, 2006, **90**, 619.
17. J. H. Gao, K. Chen, R. Luong, D. M. Bouley, H. Mao, T. C. Qiao, S. S. Gambhir and Z. Cheng, *Nano Lett.*, 2012, **12**, 281.
18. Q. J. He, Z. W. Zhang, F. Gao, Y. P. Li and J. L. Shi, *Small*, 2011, **7**, 271.
19. C. Zhou, M. Long, Y. P. Qin, X. K. Sun and J. Zheng, *Angew. Chem., Int. Ed.*, 2011, **123**, 3226.
20. J. B. Liu, M. X. Yu, C. Zhou, S. Y. Yang, X. H. Ning and J. Zheng, *J. Am. Chem. Soc.*, 2013, **135**, 4978.
21. M. X. Yu, C. Zhou, J. B. Liu, J. D. Hankins and J. Zheng, *J. Am. Chem. Soc.*, 2011, **133**, 11014.
22. H. S. Choi, W. H. Liu, F. B. Liu, K. Nasr, P. Misra, M. G. Bawendi and J. V. Frangioni, *Nat. Nanotechnol.*, 2010, **5**, 42.
23. Z. T. Luo, K. Y. Zheng and J. P. Xie, *Chem. Commun.*, 2014, **50**, 5143.
24. G. L. Wang, T. Huang, R. W. Murray, L. Menard and R. G. Nuzzo, *J. Am. Chem. Soc.*, 2005, **127**, 812.
25. M. Bakr, V. Amendola, C. M. Aikens, W. Wenseleers, R. Li, L. Dal Negro, G. C. Schatz and F. Stellacci, *Angew. Chem., Int. Ed.*, 2009, **121**, 6035.
26. A. Retnakumari, S. Setua, D. Menon, P. Ravindran, H. Muhammed, T. Pradeep, S. Nair and M. Koyakutty, *Nanotechnology*, 2010, **21**, 55103.
27. Z. T. Luo, X. Yuan, Y. Yu, Q. B. Zhang, D. T. Leong, J. Y. Lee and J. P. Xie, *J. Am. Chem. Soc.*, 2012, **134**, 16662.
28. C. Zhou, C. Sun, M. X. Yu, Y. P. Qin, J. G. Wang, M. Kim and J. Zheng, *J. Phys. Chem. C*, 2010, **114**, 7727.
29. X. J. Tu, W. B. Chen and X. Q. Guo, *Nanotechnology*, 2011, **22**, 095701.
30. L. Y. Chen, C. W. Wang, Z. Q. Yuan and H. T. Chang, *Anal. Chem.*, 2015, **87**, 216.
31. A. Mathew and T. Pradeep, *Part. Part. Syst. Charact.*, 2014, **31**, 1017.
32. Y. Lu and W. Chen, *Chem. Soc. Rev.*, 2012, **41**, 3594.
33. S. Choi, R. M. Dickson and J. Yu, *Chem. Soc. Rev.*, 2012, **41**, 1867.
34. T. Udayabhaskararao and T. Pradeep, *J. Phys. Chem. Lett.*, 2013, **4**, 1553.
35. J. P. Wilcoxon and B. L. Abrams, *Chem. Soc. Rev.*, 2006, **35**, 1162.
36. L. Shang, S. Dong and G. U. Nienhaus, *Nano Today*, 2011, **6**, 401.
37. J. Zheng, C. Zhou, M. X. Yu and J. B. Liu, *Nanoscale*, 2012, **4**, 4073.
38. C. Zhou, S. Y. Yang, J. B. Liu, M. X. Yu and J. Zheng, *Exp. Biol. Med.*, 2013, **238**, 1199.
39. Y. Negishi, K. Nobusada and T. Tsukuda, *J. Am. Chem. Soc.*, 2005, **127**, 5261.
40. D. E. Jiang, *Nanoscale*, 2013, **5**, 7149.
41. A. S. Castillo, C. Noguez and I. L. Garzon, *J. Am. Chem. Soc.*, 2010, **132**, 1504.
42. O. Varnavski, G. Ramakrishna, J. Kim, D. Lee and T. Goodson, *J. Am. Chem. Soc.*, 2010, **132**, 16.
43. L. Shang, N. Azadfar, F. Stockmar, W. Send, V. Trouillet, M. Bruns, D. Gerthsen and G. U. Nienhaus, *Small*, 2011, **7**, 2614.

44. Y. Yu, J. G. Li, T. K. Chen, Y. N. Tan and J. P. Xie, *J. Phys. Chem. C*, 2015, **119**, 10910.
45. C. Zhou, G. Hao, P. Thomas, J. Liu, M. Yu, S. Sun, O. K. Öz, X. Sun and J. Zheng, *Angew. Chem., Int. Ed.*, 2012, **51**, 10118.
46. C. A. J. Lin, T. Y. Yang, C. H. Lee, S. H. Huang, R. A. Sperling, M. Zanella, J. K. Li, J. L. Shen, H. H. Wang, H. I. Yeh, W. J. Parak and W. H. Chang, *ACS Nano*, 2009, **3**, 395.
47. H. Duan and S. Nie, *J. Am. Chem. Soc.*, 2007, **129**, 2412.
48. J. B. Liu, M. X. Yu, X. H. Ning, C. Zhou, S. Y. Yang and J. Zheng, *Angew. Chem., Int. Ed.*, 2013, **52**, 12572.
49. F. Aldeek, M. A. H. Muhammed, G. Palui, N. Q. Zhan and H. Mattoussi, *ACS Nano*, 2013, **7**, 2509.
50. O. H. Eunkeu, F. K. Fatemi, M. Currie, B. J. Delehanty, P. Thomas, F. Alexandra, L. F. Sandrine, G. Ramasis, S. Kimihiro, A. L. Huston and L. M. Igor, *Part. Part. Syst. Charact.*, 2013, **30**, 453.
51. J. Xie, Y. Zheng and J. Y. Ying, *J. Am. Chem. Soc.*, 2009, **131**, 888.
52. H. Wei, Z. Wang, L. Yang, S. Tian, C. Hou and Y. Lu, *Analyst*, 2010, **135**, 1406.
53. H. Kawasaki, K. Hamaguchi, I. Osaka and R. Arakawa, *Adv. Funct. Mater.*, 2011, **21**, 3508.
54. F. Wen, Y. Dong, L. Feng, S. Wang, S. Zhang and X. Zhang, *Anal. Chem.*, 2011, **83**, 1193.
55. C. L. Liu, H. T. Wu, Y. H. Hsiao, C. W. Lai, C. W. Shih, Y. K. Peng, K. C. Tang, H. W. Chang, Y. C. Chien, J. K. Hsiao, J. T. Cheng and P. T. Chou, *Angew. Chem., Int. Ed.*, 2011, **50**, 7056.
56. P. L. Xavier, K. Chaudhari, P. K. Verma, S. K. Pal and T. Pradeep, *Nanoscale*, 2010, **2**, 2769.
57. C. Sun, H. Yang, Y. Yuan, X. Tian, L. Wang, Y. Guo, L. Xu, J. Lei, N. Gao, G. J. Anderson, X. J. Liang, C. Chen, Y. Zhao and G. Nie, *J. Am. Chem. Soc.*, 2011, **133**, 8617.
58. K. Clemenger, *Phys. Rev. B*, 1985, **32**, 1359.
59. Z. Tang, D. A. Robinson, N. Bokossa, B. Xu, S. Wang and G. Wang, *J. Am. Chem. Soc.*, 2011, **133**, 16037.
60. T. Zhou, Y. Huang, W. Li, Z. Cai, F. Luo, C. J. Yang and X. Chen, *Nanoscale*, 2012, **4**, 5312.
61. T. K. Mandal, M. S. Fleming and D. R. Walt, *Nano Lett.*, 2002, 2, 3.
62. Y. Cui, Y. Wang, R. Liu, Z. Sun, Y. Wei, Y. Zhao and X. Gao, *ACS Nano*, 2011, **5**, 8684.
63. H. S. Choi, W. H. Liu, P. Misra, E. Tanaka, J. P. Zimmer, B. I. Ipe, M. G. Bawendi and J. V. Frangioni, *Nat. Biotechnol.*, 2007, **25**, 1165.
64. V. Lorusso, P. Taroni, S. Alviino and A. Spinazzi, *Invest. Radiol.*, 2001, **36**, 309.
65. J. C. Miller and J. H. Thrall, *J. Am. Coll. Radiol.*, 2004, **1**, 4.
66. K. W. Y. Chan and W. T. Wong, *Coord. Chem. Rev.*, 2007, **251**, 2428.
67. R. S. Liu, T. K. Chou, C. H. Chang, C. Y. Wu, C. W. Chang, T. J. Chang, S. J. Wang, W. J. Lin and H. E. Wang, *Nucl. Med. Biol.*, 2009, **36**, 305.

68. H. Kobayashi, N. Sato, A. Hiraga, T. Saga, Y. Nakamoto, H. Ueda, J. Konishi, K. Togashi and M. W. Brechbiel, *Magn. Reson. Med.*, 2001, **45**, 454.

69. V. Lorusso, F. Luzzani, F. Bertani, P. Tirone and C. D. Haën, *Eur. J. Radiol.*, 1994, **18**, 13.

70. P. Zrazhevskiy, M. Sena and X. H. Gao, *Chem. Soc. Rev.*, 2010, **39**, 4326.

71. Y. D. Liu, X. G. Han, L. He and Y. D. Yin, *Angew. Chem., Int. Ed.*, 2012, **51**, 6373.

72. J. H. Lee, K. H. Park and S. H. Kang, *Quant. Imaging Med. Surg.*, 2012, **12**, 266.

73. C. M. Cobley, J. Y. Chen, E. C. Cho, L. V. Wang and Y. N. Xia, *Chem. Soc. Rev.*, 2011, **40**, 44.

74. D. Ling, W. Park, Y. Park, N. Lee, F. Li, C. Song, S. G. Yang, S. H. Choi, K. Na and T. Hyeon, *Angew. Chem., Int. Ed.*, 2011, **50**, 11360.

75. J. B. Liu, M. X. Yu, C. Zhou and J. Zheng, *Mater. Today*, 2013, **16**, 477.

76. M. S. Behnke, W. G. Kreyling, J. Lipka, S. Fertsch, A. Wenk, S. Takenaka, G. Schmid and W. Brandau, *Small*, 2008, **4**, 2108.

77. J. H. Park, L. Gu, G. V. Maltzahn, E. Ruoslahti, S. N. Bhatia and M. J. Sailor, *Nat. Mater.*, 2009, **8**, 331.

78. M. Zhou, J. J. Li, S. Liang, A. K. Sood, D. Liang and C. Li, *ACS Nano*, 2015, **7**, 7085.

79. A. Ruggiero, C. H. Villa, E. Bander, D. A. Rey, M. Bergkvist, C. A. Batt, K. M. Todorova, W. M. Deen, D. A. Scheinberg and M. R. McDevitt, *Proc. Natl. Acad. Sci. U. S. A.*, 2010, **107**, 12369.

80. X. Wu, X. X. He, K. M. Wang, C. Xie, B. Zhou and Z. H. Qing, *Nanoscale*, 2010, **2**, 2244.

81. M. X. Yu and J. Zheng, *ACS Nano*, 2015, **9**, 6655.

82. K. Yang, J. Wan, S. Zhang, Y. Zhang, S. T. Lee and Z. Liu, *ACS Nano*, 2011, **5**, 516.

83. S. K. Hobbs, *Proc. Natl. Acad. Sci. U. S. A.*, 1998, **98**, 4607.

84. A. K. Iyer, G. Khaled, J. Fang and H. Maeda, *Drug Discovery Today*, 2006, **11**, 812.

85. Y. Matsumura and H. Maeda, *Cancer Res.*, 1986, **46**, 6387.

86. J. V. Burstin, S. Eser and J. Mages, *Int. J. Cancer*, 2008, **123**, 2138.

87. H. Maeda, *Bioconjugate Chem.*, 2010, **21**, 797.

88. R. Toy, L. Bauer, C. Hoimes, K. Ghaghada and E. Karathanasis, *Adv. Drug Delivery Rev.*, 2014, **76**, 79.

89. E. Mahon, A. Salvati, F. B. Bombelli, I. Lynch and K. A. Dawson, *J. Controlled Release*, 2012, **161**, 164.

90. F. Danhier, O. Feron and V. Preat, *J. Controlled Release*, 2010, **148**, 135.

91. J. R. Johnson, N. Fu, E. Arunkumar, W. M. Leevy, S. T. Gammon, D. W. Pawnica and B. D. Smith, *Angew. Chem., Int. Ed.*, 2007, **46**, 5528.

92. K. R. Bhushan, E. Tanaka and J. V. Frangioni, *Angew. Chem., Int. Ed.*, 2007, **46**, 7969.

93. H. Kareem, K. Sandström, R. Elia, L. Gedda, M. Anniko, H. Lundqvist and M. Nestor, *Tumor Biol.*, 2010, **31**, 79.

94. R. G. Pleijhuis, G. C. Langhout, W. Helfrich, G. Themelis, A. Saran-topoulos and L. M. Crane, *Eur. J. Surg. Oncol.*, 2011, **37**, 32.
95. J. Pauli, T. Vag, R. Haag, M. Spieles, M. Wenzel and W. A. Kaiser, *Eur. J. Med. Chem.*, 2009, **44**, 3496.
96. W. C. Zamboni, *Clin. Cancer Res.*, 2005, **11**, 8230.
97. R. H. Yang, *Environ. Health Perspect.*, 2007, **115**, 1339.
98. Y. Wang, Y. Liu, H. Luehmann, X. Xia, D. Wan, C. Cutler and Y. Xia, *Nano Lett.*, 2013, **13**, 581.
99. S. D. Perrault, C. Walkey, T. Jennings, H. C. Fischer and W. C. Chan, *Nano Lett.*, 2009, **9**, 1909.
100. M. Benezra, O. Penate-Medina, P. B. Zanzonico, D. Schaer, H. Ow, A. Burns, E. DeStanchina, V. Longo, E. Herz and S. Iyer, *J. Clin. Invest.*, 2011, **121**, 2768.
101. G. Zhang, Z. Yang, W. Lu, R. Zhang, Q. Huang, M. Tian, L. Li, D. Liang and C. Li, *Biomaterials*, 2009, **30**, 1928.
102. G. S. Hong, J. T. Robinson, Y. J. Zhang, S. Diao, A. L. Antaris, Q. B. Wang and H. J. Dai, *Angew. Chem., Int. Ed.*, 2012, **51**, 9818.

CHAPTER 5

Bioimaging Nanomaterials Based on Near Infrared Organic Dyes

ANDONG SHAO[a], XUMENG WU[a], AND WEIHONG ZHU*[a]

[a]Institute of Fine Chemicals, East China University of Science & Technology, Shanghai, 200237, P. R. China
*E-mail: whzhu@ecust.edu.cn

5.1 Introduction

Fluorescent sensors based on organic chromophores have been widely exploited as a useful tool to sense biologically important species *in vitro* and *in vivo* because of their simplicity, delicate sensitivity, and high spatial resolution of fluorescence. They allow on-site and real-time detection in an uncomplicated and inexpensive manner, along with qualitative and quantitative information. A typical fluorescent sensor consists of a receptor (*recognition site*) bridged to a fluorescent chromophore (*signal source*). When the receptor senses the analyte species, the resulting signal can be monitored in the form of fluorescence turn-off (quenching), turn-on (enhancement), or shift in the fluorescence maxima due to either photon-induced electron transfer (PET), intramolecular charge transfer (ICT), or Förster (fluorescence) resonance energy transfer (FRET) processes. As a signaling unit, fluorescent dyes have several desirable characteristics: straightforward synthesis,

RSC Nanoscience & Nanotechnology No. 40
Near Infrared Nanomaterials: Preparation, Bioimaging and Therapy Applications
Edited by Fan Zhang
© The Royal Society of Chemistry 2016
Published by the Royal Society of Chemistry, www.rsc.org

convenient wavelength modulation, and easy signal transduction. To date, long wavelength or near infrared (NIR) fluorescent dyes have emerged as promising candidates, which enable deep photon penetration in tissue, minimize photodamage to biological samples, and produce low background autofluorescence from biomolecules present in living systems. In this chapter, we focus on major NIR organic fluorescent chromophores, NIR dye-encapsulated nanoparticles for improving chromophore stability and brightness, and long wavelength aggregation-induced emission (AIE) nanoparticles for morphology control and tumor-targeting specificity. The photophysical and photochemical properties of conventional NIR organic dyes used for potential bioimaging applications are discussed, especially for improving *in vivo* performance of NIR imaging agents through encapsulation of NIR fluorescent probes in nanoparticles.

5.2   Major NIR Organic Fluorescent Chromophores

NIR fluorophores are generally considered as substances that emit fluorescence in the NIR region (650–900 nm). Particularly, the fluorescence quantum yield (QY) of NIR fluorophores is always lower than that of short wavelength emission ones. Over the past few decades, enormous progress has been made in the field of NIR fluorescent dyes. There are several major NIR organic fluorescent chromophores, such as bay-substituted perylene or naphthalene bisimides, cyanine dyes, BODIPYs, DPPs, and porphyrins.

5.2.1   Bay-Substituted Perylene or Naphthalene Bisimides

Perylene or naphthalene bisimide derivatives possess excellent thermal, photochemical, and photophysical stabilities with high extinction coefficients. They have been explored for potential applications in molecular electronic and optical devices, such as light-emitting diodes, photovoltaic cells, organic field-effect transistors (OFETs), dye lasers, and molecular wires. More recently, they have been used as a building block in supramolecular or artificial photosynthetic systems.[1,2] Generally, there are two ways to modify perylene bisimides with improved solubility (Figure 5.1): (1) introducing solubilizing substituents at the imide nitrogen; (2) incorporating halogen-, cyano-, nitro-, alkylamino-, alkoxy-substitute groups at the bay-area *via* the replacement of chlorine or bromine.[3,4]

Generally, an imide substituent has a negligible influence on the absorption and emission properties of perylene bisimides due to the nodes of the HOMO and LUMO orbitals at the imide nitrogens. There are always three peaks in the wide range of 400–550 nm, corresponding to the $0 \rightarrow 2$, $1 \rightarrow 2$, and $0 \rightarrow 1$ transitions. Notably, the emission and absorption spectra are well mirrored, giving strong evidence of the Franck–Condon principle. However, the chemical modification at bay positions (parent ring) usually has a pronounced effect on the absorption and emission properties of perylene bisimides. Especially, the absorption and fluorescence peaks of perylene bisimide 7 (Figure 5.1) containing two electron donors of piperidinyl groups

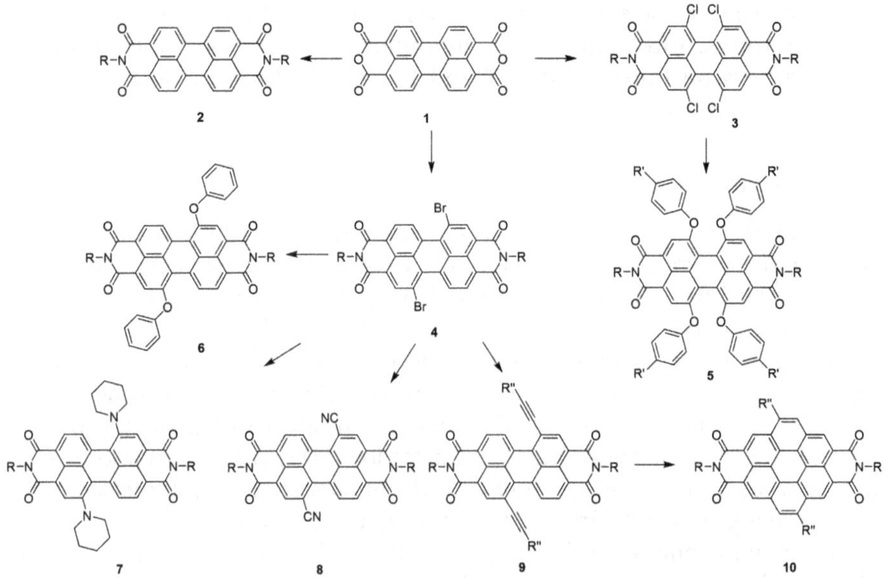

Figure 5.1 Two methods to modify perylene bisimides with increasing solubility.

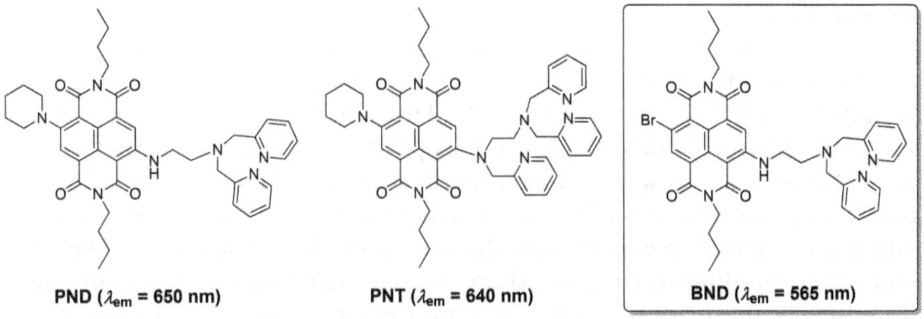

PND (λ_{em} = 650 nm)   PNT (λ_{em} = 640 nm)   BND (λ_{em} = 565 nm)

Figure 5.2 Target sensors PND and PNT for Zn^{2+} based on *N*-core substituted naphthalene bisimides as NIR chromophore.

are shifted dramatically. Actually, the emission of 7 bathochromically shifts to the infrared region with a peak at 759 nm, which really falls in the NIR fluorescence region.[5]

As illustrated in Figure 5.2, two NIR fluorescent turn-on sensors (PND and PNT) were developed for Zn^{2+} based on the chromophore of naphthalene bisimide. Notably, compared with bromo-substituted BND (λ_{em} = 565 nm), incorporation of a strong electron donor (a piperidine unit) to the naphthalene bisimide core by nucleophilic substitution red-shifted the emission wavelength by 85 nm. In this regard, extending the push–pull electronic system for PND can move the emission band centered at 650 nm, thus successfully tuning the determining wavelength to fall in the desirable NIR region. Among the available ionophore groups, two or three pyridine units were incorporated as two water-compatible and membrane-permeable chelators for comparing the selectivity of Zn^{2+} over other metal ions under physiological conditions. Interestingly, two ionophores exhibit completely different ligand effects on fluorescent PET and ICT channels due to their different coordination behaviors.[6]

5.2.2 Cyanine Dyes

The cyanine dyes belong to a class of synthetic polymethine dyes with long wavelength absorption and emission. The word "cyanin" means a shade of blue-green, derived from the Greek "kyanos". As typical dyes still used in industry, cyanines have become more attractive in biomedical fluorescence bioimaging (labeling, analysis) since their NIR absorption and emission spectra exhibit special advantages for deeply penetration into tissues, being poorly absorbed by biomolecules. A well-known cyanine dye, indocyanine green (ICG), was approved by the United States Food and Drug Administration (FDA) in 1958 as a fluorescence indicator for human use, *e.g.* in *in vivo* NIR fluorescence imaging tissue perfusion, protease activity, hydroxyapatite, and cancer detection.

The general chemical formula of cyanine dyes has a unique class of charged chromophores with an odd number of sp^2 carbon atoms forming a π-conjugated bridge between electron-donating and/or electron-accepting groups, controlling the photophysical and structural properties of the dye (Figure 5.3).[7]

Among cyanine dyes, tricarbocyanine with a rigid chlorocyclohexenyl or chlorocyclopentenyl ring in the methine chain can increase both the photostability and the fluorescence QY. The chlorocyclohexenyl-substituted methine linker provides an ideal nucleophilic site for further modification with alcohols, amines, and thiols. As illustrated in Figure 5.4, the turn-on NIR fluorescent probe CS790AM consists of a cyanine dye as a fluorophore and a Cu^+-selective receptor. Actually, the probe does not induce a spectral shift upon binding to Cu^+, but there is a distinct 15-fold enhancement in fluorescence intensity at 790 nm in response to Cu^+ as a consequence of suppression of the PET quenching mechanism. Generally, the longer excitation of tricarbocyanine makes fluorescence modulation with the PET channel difficult. Notably, incorporation of benzene moieties with a much higher HOMO energy level by nucleophilic substitution of the chlorine atom at the methine chain is an effective modulation strategy for the PET-based fluorescence quenching of tricarbocyanines.[8]

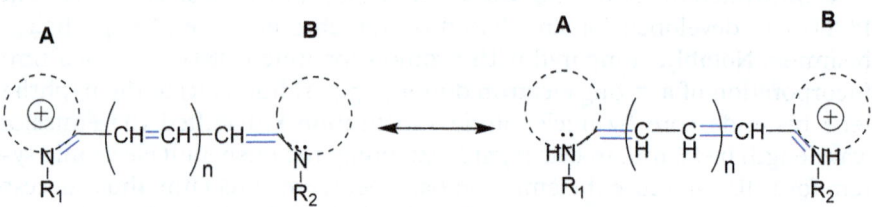

Figure 5.3　Different resonance structures of cyanine dyes.

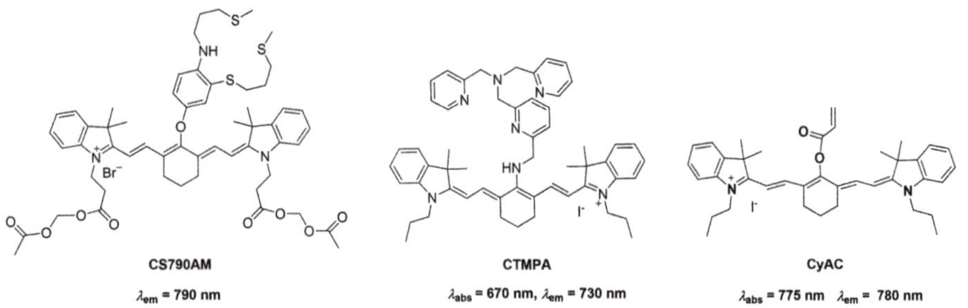

Figure 5.4 Structures of NIR fluorescent probes based on tricarbocyanine.

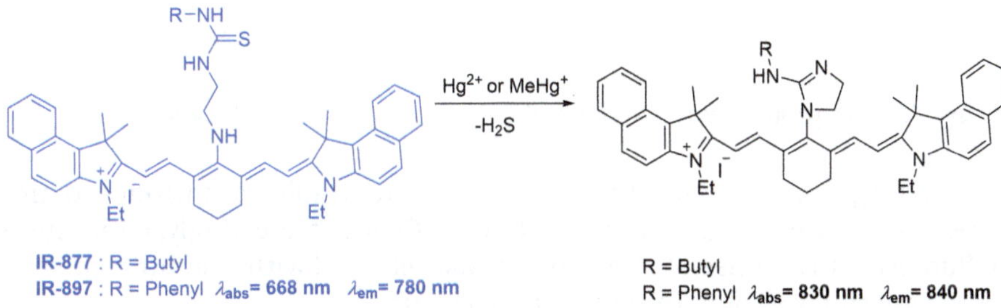

Figure 5.5 Molecular structure of NIR cyanine probes (IR-877 and IR-897) and their interaction mode with Hg^{2+} or $MeHg^+$.

Tricarbocyanines always exhibit intrinsically small Stokes shifts, which severely limits them as a scaffold for functional NIR fluorescent probes. When electron-donating amino substituents, are introduced, the resulting tricarbocyanines have a much larger Stokes shift than the parent cyanines. It is indicative that the lower the electron density on the amine, the longer the wavelength of the absorption maximum. As shown in Figure 5.5, amino-substituted tricarbocyanine dyes with different thiourea substituents were developed, in which the central bridging secondary amine is converted to the imidazoline tertiary amine by mercury-promoted desulfurization. As a consequence, the electron-donating ability of this nitrogen is attenuated and sharply reduced, which results in a large red shift in the absorption band for IR-877 (from 664 to 840 nm) and IR-897 (from 668 to 830 nm). This allows detection of color change by the naked eye. With regard to fluorescence, the IR-897 probe shows an emission band at 780 nm but mercury ion-induced turn-on fluorescence is observed at 840 nm.[9]

Modulation of the pull–push conjugated π-electron system in cyanine dyes is demonstrated as an efficient approach to tuning the fluorescent emission profile. Specifically, the removable trigger moiety on the backbone of the polymethine π-electron system was implemented for the design of ratiometric cyanine-based sensors. As illustrated in Figure 5.6, hydroxycyanine CyAE displayed remarkable blue shifts in absorption (~175 nm) and emission spectra (~115 nm) with reversible tautomeric forms of CyAE and CyAK or CyAD.

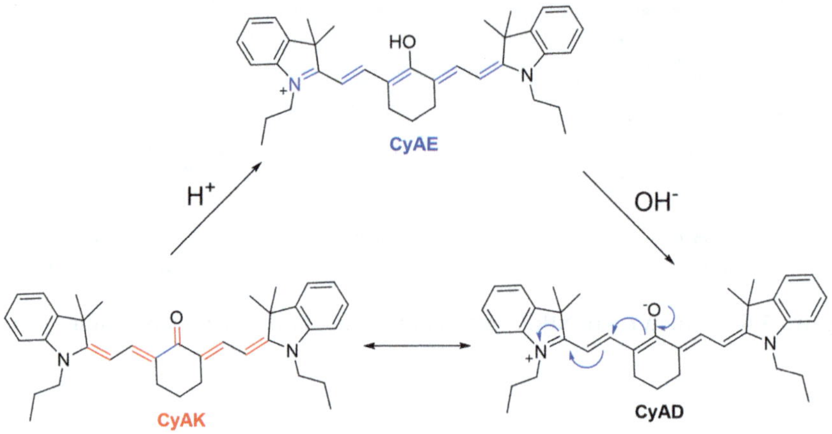

Figure 5.6   Deprotonation of CyAE and its resonance forms CyAK and CyAD.

An acrylate group as the specific trigger moiety for thiols was introduced into CyAK to generate the ratiometric NIR probe CyAC. Interestingly, a ratiometric fluorescent response of CyAC to Cys was achieved with a significant shift in the emission wavelength from 780 to 625 nm.[10]

5.2.3   BODIPY

Accumulated evidence has identified 4,4-difluoro-4-bora-3*a*,4*a*-diaza-*s*-indacene derivatives (hereafter abbreviated as BODIPY dyes) as a promising building block for the creation of molecular probes due to their excellent features, including high fluorescence QYs, high photo- and chemo-stability, and easy modification of the parent BODIPY structure. Generally, BODIPYs show absorption and emission within the visible region of 500–600 nm. A rich and sometimes surprising chemistry has been devoted to generating red and NIR BODIPYs as fluorescent probes. One general route to construct such NIR dyes is to fuse aryl moieties into parent BODIPY core, resulting in an extension of the π-system while retaining the rigidity and planarity of parent BODIPYs.[11] The so-called *Keio fluors* (Figure 5.7) have excellent and useful optical properties: vivid colors and sharp emission in the visible–NIR region (583–738 nm), high QYs (Φ 0.56–0.98), high extinction coefficients (185 000–288 000 M^{-1} cm^{-1}), and good photostabilities.

An alternative approach to generate NIR BODIPYs is to extend the π-conjugation by introduction of methylene units at the 3- and 5-positions *via* Knoevenagel condensation. The aromatic BODIPY core endows methyl groups at the 3- or 5- position with acidic characteristics, enabling Knoevenagel condensation with aromatic aldehydes to producing large π-systems with NIR fluorescence features. Utilizing this special strategy, attachment of versatile sensing units to the BODIPY core can result in promising probes for different targets.

Zhao *et al.* created an approach to red-shift the emission wavelength based on the excited state proton transfer (ESPT) process (Figure 5.8). In PCys-B, two Cys sensing sites (acrylate ester and aldehyde) were installed *ortho* to each

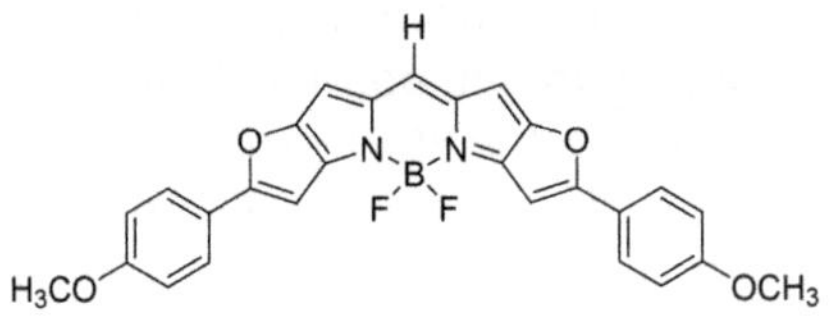

KFL-3
(λ_{ab} = 673 nm, λ_{em} = 683 nm)

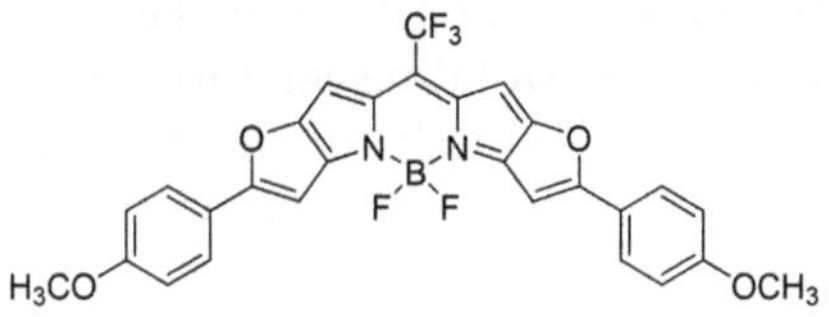

KFL-4
(λ_{ab} = 723 nm, λ_{em} = 738 nm)

Figure 5.7 Structures of NIR BODIPYs with aryl moieties fused to the parent BODIPY core.

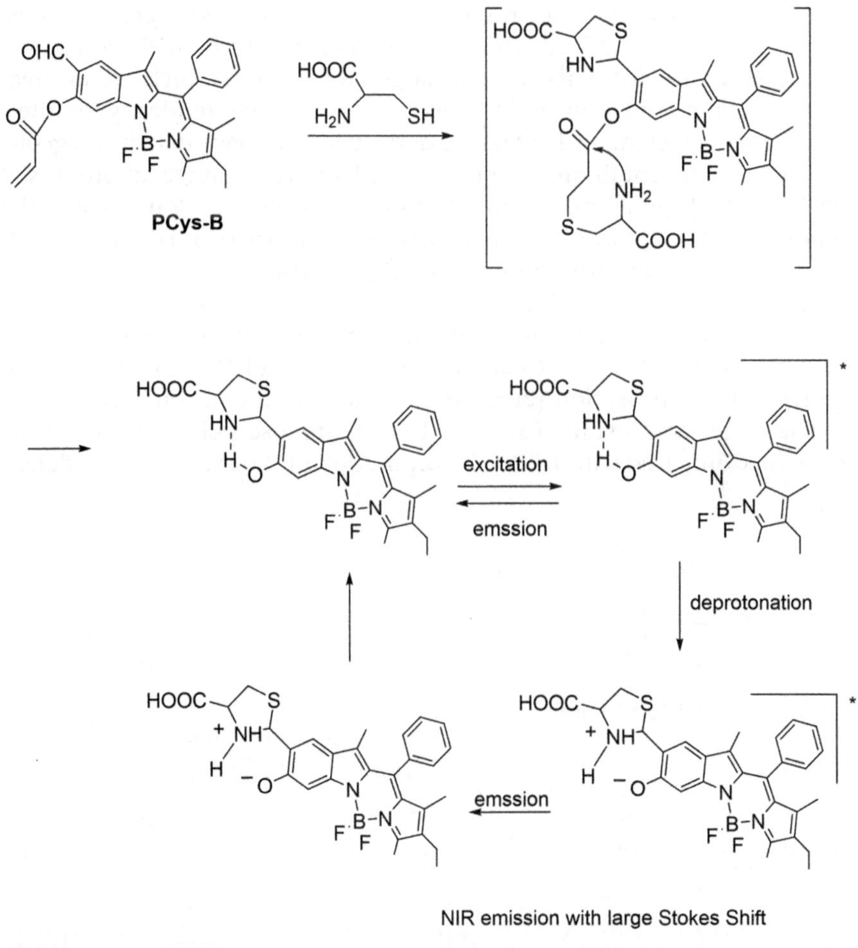

NIR emission with large Stokes Shift

Figure 5.8 NIR fluorescence of PCys-B triggered with Cys based on the ESPT process.

other into indole-based BODIPY.[12] The Cys triggers the release of phenolic hydroxy-based BODIPY together with the formation of thiazolidine. The amino group in thiazolidine can form an intramolecular hydrogen bond with phenolic hydroxy through a six-membered ring. This enables the ESPT process upon photo-excitation, thus resulting in a large red-shifted emission at 650 nm.

5.2.4   DPP

1,4-Diketo-3,6-diphenylpyrrole[3,4-c]pyrrole (DPP) and its derivatives have been extensively used in many fields, such as sensitizing solar cells, OFETs, OLEDs, fluorescent probes, and two-photon absorption sensors. DPP derivatives exhibit exceptional thermal and photochemical stability, large extinction coefficients, and high fluorescence QYs in both solution and solid films.

DPP is a typical kind of electron-deficient fluorophore with green fluorescence. When functionalized with electron-donor groups, DPP derivatives can exhibit red to NIR emission and large Stokes shift, which are of great benefit to their applications in bioimaging. However, the tendency for intermolecular π–π stacking of DPP-based derivatives can have disadvantageous effects on the efficient fluorescent emission. To solve the aggregation-caused quenching (ACQ) problem, a rational strategy is to attach AIE units onto the parent DPP chromophore, especially developing a variety of DPP-based red/NIR fluorescent probes with optimized AIE properties.[13]

By introducing propeller-like starburst triphenylamine as a strong donor to increase the emission wavelength and hinder the ACQ effect (Figure 5.9), two DPP-based red/NIR compounds (DPP-1 and DPP-2) exhibit both large two-photo absorption (2PA) cross-sections and excellent AIE properties. The large Stokes shifts ($\Delta\lambda \geq 3571$ cm^{-1}) are also helpful for bioimaging *in vivo*. Compared with DPP-1, incorporation of the extended π-deficient

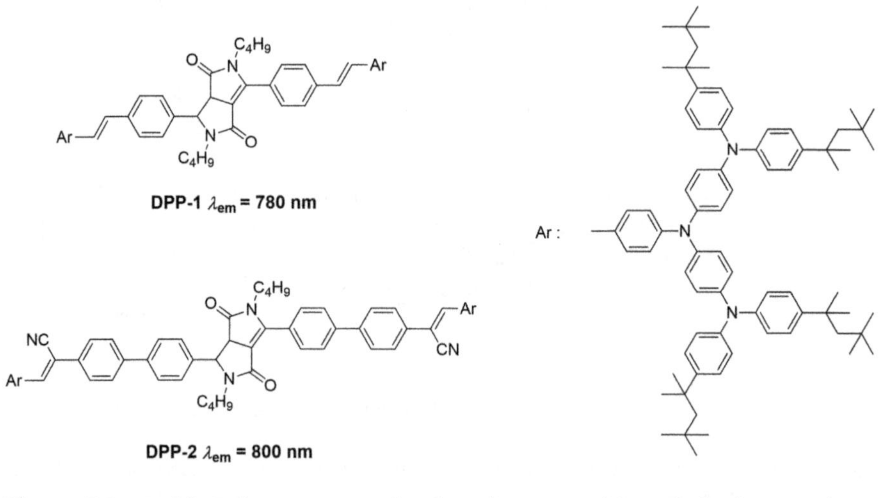

DPP-1 λ_{em} = 780 nm

DPP-2 λ_{em} = 800 nm

Ar :

Figure 5.9   Red/NIR fluorescent probes based on DPP with optimized 2PA and AIE properties.

phenylacrylonitrile assures DPP-2 with better planarity, longer emission wavelength (λ_{em} = 800 nm) and much higher QY of 0.11 in the solid state. After encapsulation into nanoparticles by DSPE–PEG–Mal, DPP-2 possesses good dispersibility and biocompatibility, which has been successfully exploited for cell imaging and two-photon imaging of the blood vasculature.[14]

Figure 5.10 shows two other DPP-based glycosyl probes (DDPM and DPPG) for the sensitive and selective detection of sugar–lectin interactions in the NIR region using the AIE mechanism. Here, methoxytriphenylamine groups are introduced as the electron-donating groups to guaranteeing the DPP derivatives with AIE activity, NIR emission, and large Stokes shift. When the amount of water (f_w) is further increased to >50%, the hydrophobic DPP-based compounds are prone to further aggregate into larger particles and quickly precipitate, resulting in strong fluorescence at approximately 650 nm. On the introduction of glycosides, the compounds possess the ability to selectively and sensitively detect plant lectins including concanavalin A (Con A, mannose-selective), lentil lectin (LcA, mannose-selective), and peanut agglutinin (PNA, galactose-selective). SEM analyses suggest that the resulting fluorescence signal probably originates from AIE in the divalent glycodyes upon complexation with the polymeric lectins.[15]

5.2.5 Porphyrin and Porphyrin Analogs

Porphyrins and their derivatives widely occur in natural and artificial opto-electronic systems. A porphyrin molecule contains four pyrrole units inter-connected *via meso*-methine bridges (=CH–) at their α-positions (Figure 5.11), which contains an 18-π conjugation framework. Porphyrins always exhibit intense absorption in the visible region and NIR fluorescence with emission

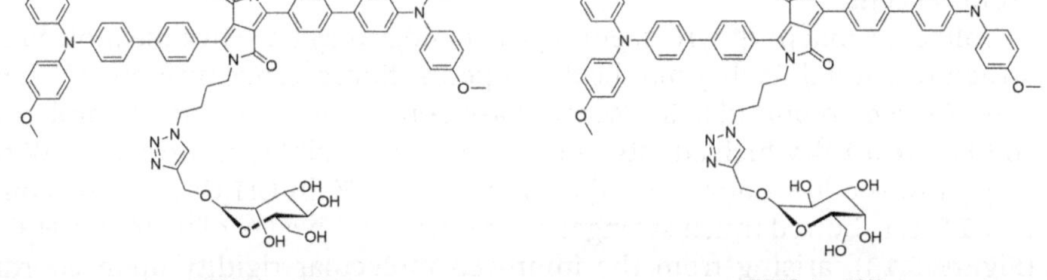

DPPM λ_{em} = 655 nm   DPPG λ_{em} = 643 nm

Figure 5.10 Two DPP-based glycodyes for detection of lectins using the AIE mechanism.

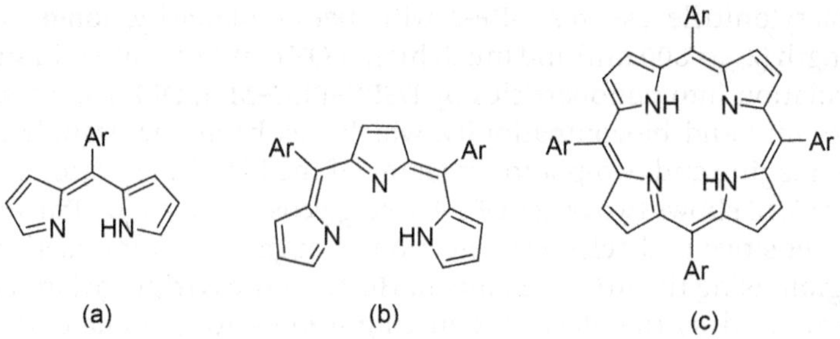

Figure 5.11 Molecular structures of dipyrrin (a), tripyrrin (b), and porphyrin (c).

wavelengths beyond 650 nm, highly desirable for developing NIR fluorescent sensors.[16]

In fact, linear di- and tripyrrins (Figure 5.11) have been developed by Xie *et al.* as fluorescence turn-on Zn^{2+} sensors based on the chelation-enhanced fluorescence (CHEF).[17] Actually, the di- and tripyrrin molecules are conformationally flexible, and their rigidity could be dramatically improved upon coordination with Zn^{2+}, thus decreasing the non-radiative energy loss of the excitation states. Unlike linear conjugated oligopyrrins, macrocyclic frameworks of porphyrins are rather rigid, and metal coordination cannot significantly improve the rigidity. In this regard, the fluorescence cannot be dramatically enhanced upon coordination with metal ions like Zn^{2+}. Therefore, macrocycle cannot be used simultaneously as the binding moiety as well as the reporting moiety for the design of fluorescence turn-on Zn^{2+} probes.

Fortunately, the porphyrin *meso* positions can be readily functionalized. As shown in Figure 5.12, by introducing a DPA unit as the Zn^{2+} binding site and three sulfonatophenyl groups to improve the water solubility, Lippard *et al.* developed PDP and its manganese complex PDP–Mn for Zn^{2+} sensing in living HEK-293 cells. Upon addition of Zn^{2+} to a buffered solution of PDP, a vivid fluorescence "turn-on" can be observed at 648 and 715 nm based on a PET mechanism.[18] More than 10-fold fluorescence enhancement was observed. Here probe PDP is selective for Zn^{2+} with slight interference from Cd^{2+} and Hg^{2+}, working well in a large pH range (4.5–10.1). PDP–Mn can be used as an magnetic resonance imaging (MRI) contrast agent for imaging Zn^{2+} in living HEK-293 cells.

Unlike normal porphyrins, porphyrin analogs may have flexible molecular structures, and thus they may be developed as fluorescence turn-on Zn^{2+} sensors. Xie *et al.* reported a Zn^{2+} sensor based on a unique expanded porphyrin (EP, Figure 5.13), which adopts a non-planar and flexible conformation. With respect to the low fluorescent QY, as low as 0.16% in CH_2Cl_2, its zinc complex EP–Zn showed much stronger fluorescence with a QY of 9.9% in CH_2Cl_2 (Figure 5.13), arising from the improved molecular rigidity upon coordination. Similarly, addition of Zn^{2+} to the methanol solution of EP resulted in dramatic fluorescence enhancement by 31-fold, with an emission wavelength of 736 nm in the NIR region.[19]

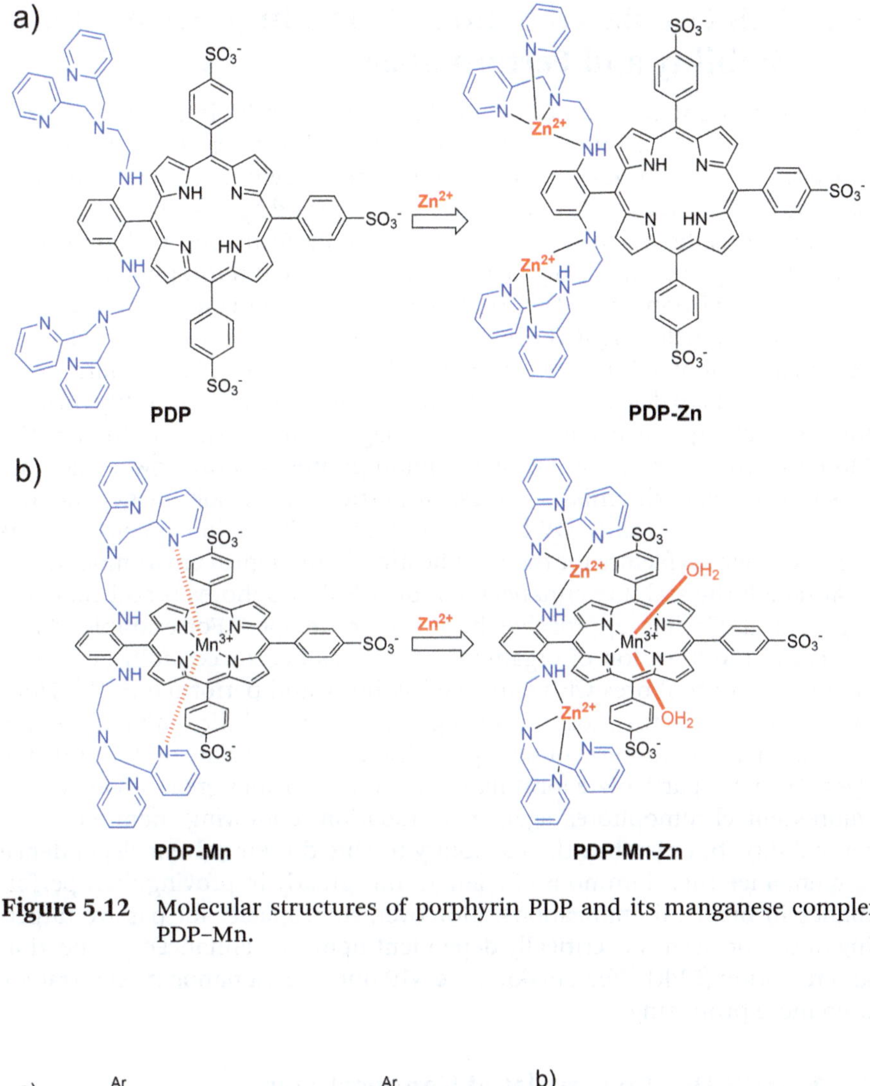

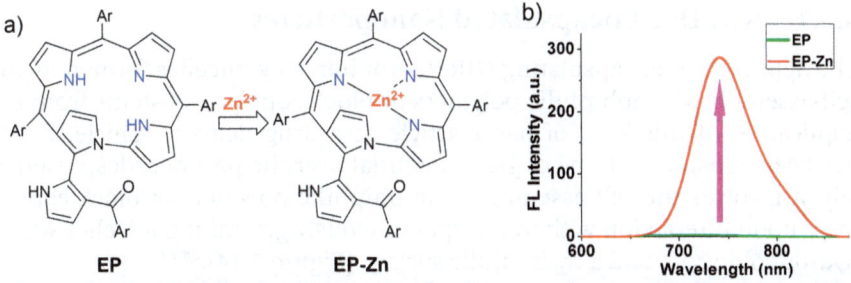

a)
PDP
PDP-Zn
b)
PDP-Mn
PDP-Mn-Zn

Figure 5.12 Molecular structures of porphyrin PDP and its manganese complex PDP–Mn.

a)
EP
EP-Zn
b)
EP
EP-Zn
FL Intensity (a.u.)
Wavelength (nm)

Figure 5.13 (a) Proposed sensing mechanism of expanded porphyrin (EP) for detecting Zn^{2+} (Ar = C_6F_5); (b) fluorescence spectra of EP and EP–Zn (10 μM) in CH_2Cl_2.

5.3 NIR Dye-Based Nanoparticles: Improvement of Stability and Performance

The practical application of organic fluorophores in bioimaging is severely limited by several factors, such as biocompatibility, *in vivo* toxicity, narrow emission, brightness, response time, and bleaching. Given the need to observe the NIR fluorescent marker for a long period against the background of intrinsic cellular emission, and excellent performance in bioimaging, photostability is always a key factor in the use of NIR fluorophores in a real biological environment.[20] Unfortunately, most of the commonly used NIR fluorophores, including ICG which is approved by the FDA for clinical application, inevitably suffer from comparatively poor stability. The exploration of novel methods to develop NIR fluorophores with enhanced stability is therefore of great significance for biological imaging, diagnosis, and therapy.[21,22] Moreover, the fluorescence QY of NIR fluorophores is always lower than that of short wavelength emission ones. In particular, the solvent dependence and the low fluorescence QY of most NIR fluorophores in aqueous systems are also negative factors for their application in biological environment.[23]

Although the stability enhancement of NIR fluorophores is still challenging, several efficient approaches have been developed. For example, direct structural modification of organic fluorophores has successfully generated some NIR fluorophores with enhanced stability and performance.[24,25] However, undoubtedly the most promising strategy for stability enhancement is to construct NIR fluorescent nanoparticles that encapsulate or dope organic dyes. The nanoparticulate formulation not only provides great protection for fluorescent chromophores against degradation, endowing them with better stability, but also has the capability to shield their solvent dependence and enhance their luminous efficiency, thus greatly improving their performance in practical applications. Moreover, the capacity for passive targeting of tumor tissues is critically dependent upon the enhanced permeation and retention (EPR) effect, making the NIR fluorescent nanoparticle strategy even more promising.[26]

5.3.1 NIR Dye-Encapsulated Nanoparticles

The approach of encapsulating NIR fluorophores by micelles formed by the self-assembly of amphiphilic polymers or block copolymers stems from the application of this kind of nanoparticle as a drug delivery vehicle, which has been considered to be of great potential over the past decades.[27] Generally, it involves the self-assembly of amphiphilic polymers or block copolymers upon interaction with water, spontaneously generating micelles with a hydrophobic core and a hydrophilic surface (Figure 5.14).[28,29]

The main reason why this strategy is so attractive is that it helps to resolve the problems of low stability and performance of free NIR fluorophores in aqueous systems, with its excellent ability to endow hydrophobic or instable drugs with high hydrophilicity (due to the outer hydrophilic block) and

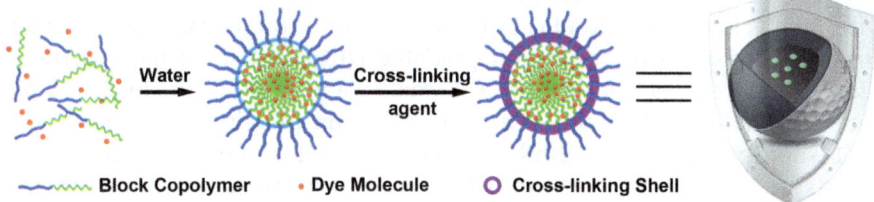

Figure 5.14   Schematic diagram of the formation of dye-encapsulated fluorescent nanoparticles based on diblock copolymer, and further cross-linking by a cross-linking agent. Adapted from ref. 23 with permission from the Royal Society of Chemistry.

stability (due to the protective nanoparticle environment). It is obviously helpful for their performance in the biological environment.[30] In 2004, a preliminary attempt to construct NIR fluorescent nanoparticles by the polymeric micelle strategy was made by Saxena *et al.*, utilizing poly(lactic-*co*-glycolic acid) (PLGA) to form nanomicelles in which the FDA-approved ICG as an NIR emitter was encapsulated. The resulting NIR fluorescent nanoparticles presented an distinct improvement of photo-, thermal, and aqueous stability with respect to free ICG. This was attributed to the isolation of encapsulated ICG molecules from the surrounding environment by the polymeric envelope.[31] Inspired by this result, more and more successful applications of the micelle system in fabricating NIR fluorescent nanoparticles sprang up, utilizing various diblock copolymers such as poly(styrene-*alt*-maleic anhydride)-*b*-poly(sytrene) (PSMA-*b*-PS) and poly(ethylene glycol)-*b*-poly(D,L-lactide) (PEG-*b*-PLA). These further prove the ability of NIR dye-encapsulated nanoparticles to stabilize both the organic dye molecules and their NIR fluorescence.[32,33] Even more interestingly, Tang *et al.* found that, using NIR fluorescent sensor as cargo, the NIR fluorescent nanoparticles could retain or even further enhance the response sensitivity and selectivity, indicative of more potential applications of this approach as NIR fluorescent sensors.[33] Also, the NIR fluorescent brightness was enhanced by the fluorophore-encapsulated nanomaterials.[34]

Liposomes, widely used as drug delivery vehicles, bestow favorable biocompatibility and easy surface modification, which could also be exploited as the hydrophobic core in a NIR fluorescent nanoparticle system.[35] Qian *et al.* utilized self-assembly of the lipid-containing amphiphilic substrate DSPE-mPEG$_{5000}$, in which the lipid part acts as the hydrophobic block and the PEG part is the hydrophilic block, to construct optically and chemically stable nanoparticles with the typical NIR fluorophore IR-820 encapsulated in the hydrophobic core. Considering their very low cytotoxicity, these NIR fluorescent nanoparticles could be utilized as an efficient optical nanoprobe in sentinel lymph node (SLN) mapping and whole-body functional *in vivo* bioimaging. After being intravenously injected into the ear and brain of mice, the vascular architecture could be revealed vividly with the assistance of a confocal microscope with red excitation.[36] A similar DSPE-PEG substrate is equipped with folic acid (FA) in the PEG block as the tumor-targeting ligand for constructing non-cytotoxic and biodegradable NIR fluorescent nanoparticles. The NIR fluorescent cyanine dye ICG was encapsulated in the hydrophobic core with the help of PLGA. In *in vitro* and *in vivo* imaging the resulting FA–PLGA–lipid nanoparticles exhibited better optical properties than the free ICG as well as highly selective tumor localization (resulting from the targeting ligand FA) and prolonged circulation time *in vivo* (Figure 5.15). This example highlights that the easy surface modification is a unique advantage of the NIR fluorophore-encapsulated nanoparticles for applications in tumor diagnosis and targeted imaging.[37]

Other than traditional block copolymers, polysaccharides such as hyaluronic acid (HA) and chitosan could also be used as substrates in the

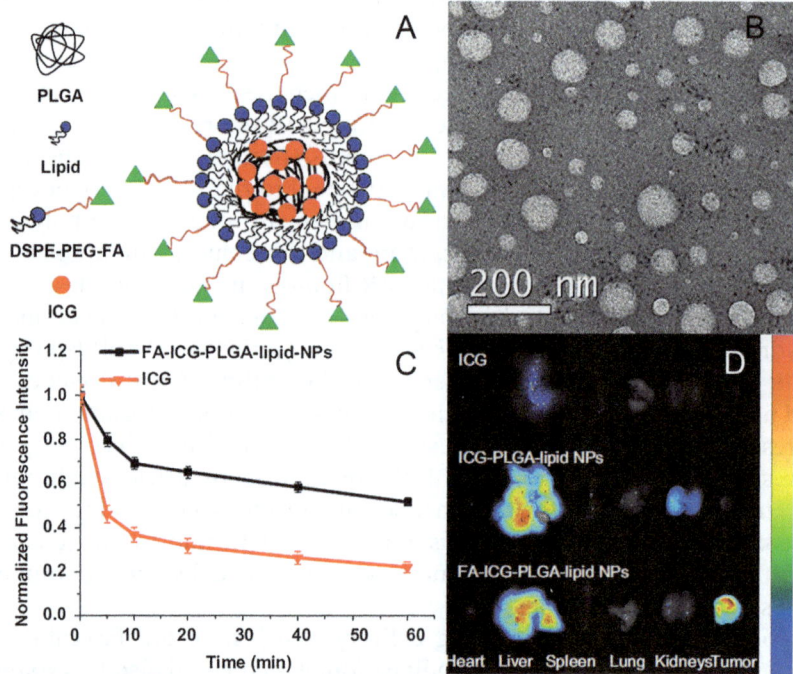

Figure 5.15   (A) Schematic illustration of the structure of FA–ICG–PLGA–lipid nanoparticles. (B) TEM image of FA–ICG–PLGA–lipid nanoparticles; the nanoparticles were negatively stained with 2% phosphotungstic acid. (C) Photobleaching experiment of FA–ICG–PLGA–lipid nanoparticles and free ICG, under 1 h continuous illumination from a 150 W xenon lamp. (D) Fluorescence images of organs and tumors in MCF-7 tumor-bearing mice after 24 h post-injection of free ICG, ICG–PLGA–lipid nanoparticles, and FA–ICG–PLGA–lipid nanoparticles. Adapted from C. Zheng *et al.*, Indocyanine green-loaded biodegradable tumor-targeting nanoprobes for *in vitro* and *in vivo* imaging, *Biomaterials*, 2012, **33**, Copyright (2012) with permission from Elsevier.[37]

fabrication of NIR dye-encapsulated nanoparticles due to their excellent biocompatibility and diverse biological functions as targeting ligands and signaling molecules. For example, Chung *et al.* first reported an ICG-encapsulated HA nanogel and its practical application in bioimaging, in which ICG molecules were conjugated with HA *via* an amide linker. They demonstrated that the resulting NIR fluorescent nanogels displayed enhanced photostability and accumulation in the body, facilitating longer-term visualization of targeted tissue *in vivo* than free ICG (Figure 5.16). Interestingly, the NIR fluorescence intensity of the HA–ICG nanogels is distinctly self-quenched during the formation process, and could be significantly enhanced by hyaluronidase (HAdase) which specifically degrades HA. Therefore, the NIR

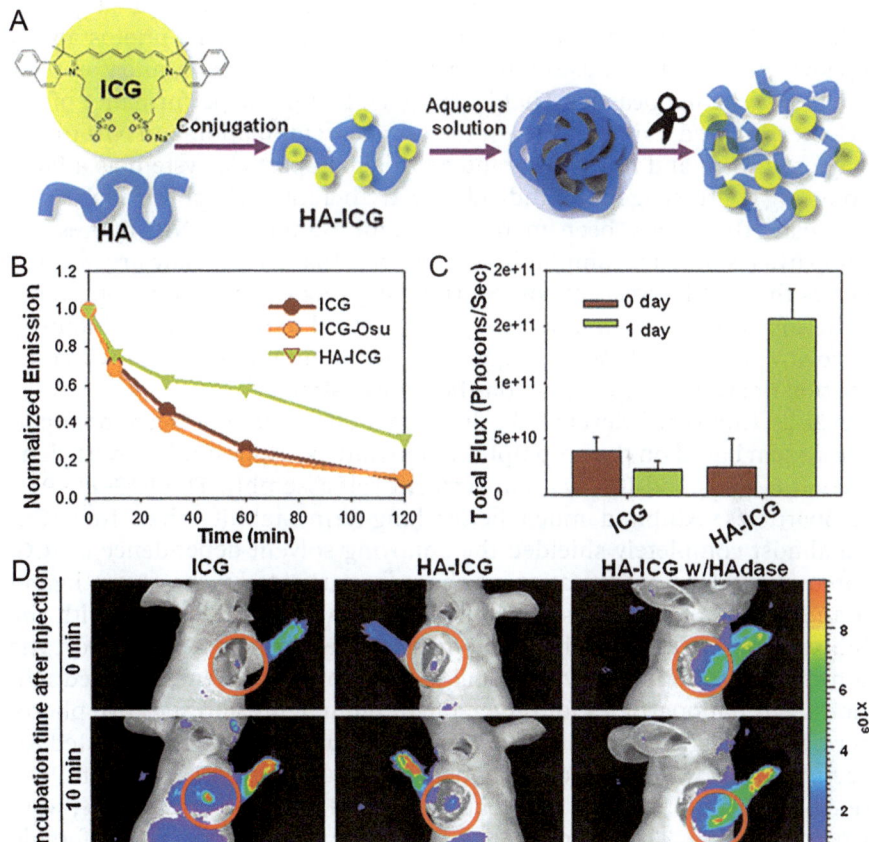

Figure 5.16 (A) Schematic illustration for the self-quenching and restoration of ICG fluorescence in HA–ICG nanogels. (B) Stability of ICG fluorescence signal in HA–ICG nanogels after exterior light exposure for a predetermined time. (C) ICG fluorescence intensities after transdermal injection. (D) Images of sentinel lymph nodes after ICG probes were intrademally injected into the forepaw pad of mice. Adapted from ref. 38 with permission from the Royal Society of Chemistry.

fluorescent nanogels could successfully be used in the targeted visualization of HAdase-overexpressing cells and tissues, such as some tumors and metastatic lymph nodes, *in vitro* and *in vivo*. Chung *et al.* also anticipated that the hydrophobic core of the nanogels could be a potential vehicle for drug delivery.[38] In another case, hydrophobic groups such as pyrene and octadecylamine were used for modifying HA *via* an amide linker to further facilitate the formation of HA-based and ICG-encapsulated NIR fluorescent nanoparticles. The resulting nanoparticles had low toxicity in cell culture and produced stronger contrast enhancement compared to free ICG. In this way, the potential application of image-guided surgery in the treatment of operable tumors could be realized.[39] Tan *et al.* utilized FA-modified chitosan to form crystalline ultra-small NIR fluorescent nanoparticles with FDA-approved ICG *via* self-assembly for *in vivo* targeted tumor imaging. Unsurprisingly, the FA-based NIR fluorescent nanoparticles exhibited excellent photostability and low cytotoxicity because of the usage of the biocompatible and non-toxic chitosan as the nanocarrier. Thanks to the FA group, the performance of the ICG-encapsulated nanoparticles in the tumor accumulation and NIR tumor imaging *in vitro* and *in vivo* guaranteed this nanoparticle system as a fairly promising contrast agent for individualized (theranostic) applications.[40]

To date, there have been many successful examples of NIR fluorescent nanoparticles based on amphiphilic substrates that encapsulate organic NIR dyes as the NIR fluorescent source. However, single-function nanoparticles as the contrast agents cannot satisfy the practical requirements of bioimaging to any great extent. Accordingly, NIR fluorescent nanoparticles with dual function or even multiple functions become desirable.[41]

In 2011, Chen *et al.* developed a dual-functional NIR fluorescent nanoparticle system based on the amphiphilic substrate phospholipid–polyethylene glycol (PL–PEG) and ICG *via* non-covalent self-assembly. The ICG–PL–PEG nanoparticles exhibited much better long-term stability than free ICG, and almost completely shielded the annoying solvent dependence of ICG. Moreover, surface modification of the nanoparticles by the target ligand FA and integrin $\alpha_v\beta_3$ monoclonal antibody (mAb) endowed them with the ability to target cancer cells *via* the ligand-receptor mediated endocytosis pathway. More importantly, the dual application of ICG-encapsulated NIR fluorescent nanoparticles for targeted NIR bioimaging and selective photothermal cell destruction both *in vitro* and *in vivo* has been demonstrated for the first time.[42] NIR fluorescent nanoparticles based on amphiphilic mPEG-*b*-PAsp, encapsulating ICG or Cypate as the NIR fluorescence source, were also constructed. The nanomicelles facilitated cellular uptake, preferable accumulation and long-term retention in tumor tissues with rapid elimination in normal tissues resulting from the EPR effect. Impressively, the NIR fluorescent micelles could trigger significant photothermal damage on cancer cells *via* the destabilization of organelles, achieving successful tumor necrosis and regression upon photoirradiation, which could engender extraordinary NIR fluorescence imaging contrast during a long-term imaging window.[43] Enhanced photostability of the loaded dye molecules with excellent photothermal properties could also be achieved by NIR fluorescent nanoparticles based on another block copolymers such as PEOz-PCL.[44] Based on heparin as a non-cytotoxic, biodegradable, and water-soluble natural polysaccharide, Cai *et al.* reported the FA-modified NIR fluorescent nanoparticles HF-IR-780, which is anticipated to be a safe theranostic agent for imaging-guided cancer treatment because of the good particle stability, tumor-targeting ability, and excellent performance in both NIR fluorescence bioimaging and photothermal therapy *in vitro* and *in vivo* for folate-overexpressing tumor tissues.[45] Cai *et al.* also attempted to simultaneously encapsulate ICG and the anticancer drug doxorubicin (DOX) by nanoparticles formed by self-assembly of a lecithin-based substrate DSPE-PEG with the help of PLGA. While retaining the ability to enhance the stability of ICG and producing higher localized temperature than free ICG, the nanoparticles could deliver DOX to tumor regions for combined NIR fluorescence imaging, photothermal therapy, and chemotherapy. The development of dual-functional or multifunctional NIR fluorescent nanoparticles is now focusing on a variety of potential applications in the NIR imaging-guided therapy.[46]

With NIR dye-encapsulated nanoparticles, there is still an inevitable drawback that the micelles might undergo spontaneous dissociation at concentrations below their critical micelle concentration (CMC), releasing the encapsulated dye molecules as a potentially toxic substance.[31] Locking the micellar structure with a cross-linking agent has been considered as an attractive strategy for constructing core–shell NIR fluorescent nanoparticles with enhanced stability of both the content and micelles.[47,48] By using 3-mercaptopropyltrimethoxy silane (MPTMS) as the cross-linking agent and polystyrene-*b*-polyacrylic acid (PS-*b*-PAA) as the block copolymer, Zhu *et al.* developed a kind of highly stable NIR fluorescent nanoparticle. In this strategy, the NIR cyanine dye modified with a long hydrophobic group was encapsulated within the hydrophobic core of micelles, then locked by a silica shell formed by MPTMS. Under this dual protection, the encapsulated cyanine dye exhibits almost no solvent dependence, 450-fold higher NIR fluorescence QY, and 50-fold longer half-life period than free dye (95-fold longer than ICG). Moreover, destructive substances such as reactive oxygen species are also excluded from the nanoparticles, guaranteeing much better chemical stability for the encapsulated dye molecules. Interestingly, these NIR fluorescent nanoparticles display excellent performance in tumor imaging because of the high stability and EPR effect.[49] In addition, calcium phosphate, which is considered as the most biocompatible inorganic material, could also be exploited for cross-linking the nanomicelles.[50]

Generally speaking, the development of NIR fluorophore-encapsulated nanomaterials has great significance in the field of NIR bioimaging, biomedicine, and NIR imaging-guided therapy. Meanwhile, advances in essential nanoparticles (such as amphiphilic substrates), encapsulated NIR fluorophores, and multifunctional nanoparticles are still making rapid progress.

5.3.2 NIR Dye-Doped Nanoparticles

As well as the NIR fluorescent nanoparticles based on organic substrates discussed above, NIR dye-doped silica nanoparticles are also being successfully used to overcome typical limitations of conventional organic dyes such as poor photostability, low QY, and aggregation under physiological conditions, thus realizing very effective tools for diagnostic and theranostic applications. The stability of silica nanoparticles themselves in the real biological environment and easy surface functionalization by targeting groups or polymers adds to their potential for bioimaging and diagnosis. In fact, silica nanoparticles have been approved by the FDA for human cancer clinical trials.[51,52]

As is well known, the most widely used methods for synthesizing silica nanoparticles are the Stöber method and the reverse microemulsion method (Figure 5.17). The Stöber method, also known as the sol–gel method, was developed by Stöber *et al.* in the late 1960s. Generally, a silica precursor such as tetraethoxysilane (TEOS) is utilized in controlled hydrolysis and condensation in alcohol and water with ammonia as catalyst to construct size-controllable, monodisperse, spherical nanoparticles.[53] This method can also be used in the synthesis of core–shell nanoparticles based on a pre-synthesized core.[54] Otherwise, mesoporous silica particles (MSNs) could be prepared by modifying the conditions of the traditional Stöber method in the presence of high concentrations of surfactant molecules.[55] The reverse microemulsion method was developed by Arriagada and Osseo-Asare in the early 1990s, using a silica precursor such as TEOS for ammonia-catalyzed polymerization in a reverse-phase or water-in-oil microemulsion. The reverse microemulsion method is considered as an ideal method for constructing highly monodisperse and perfectly spherical particles with sizes ranging from 20 to 100 nm.[56]

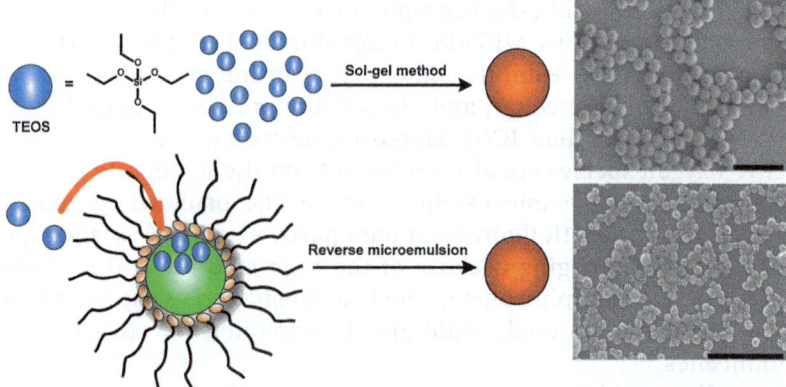

Figure 5.17 Schematic synthesis of silica nanoparticles by the Stöber method (top) and *via* reverse-phase microemulsion (bottom). The scale bars represent 1000 nm and 500 nm, respectively. Adapted from ref. 51 with permission from the Royal Society of Chemistry.

Generally, NIR fluorophores doped in the silica nanoparticles present brighter fluorescence and much better photostability than the free dyes, contributing to better performance and longer tracking time in bioanalysis and disease diagnosis.[57] Bringley *et al.* utilized triethoxysilane-modified cyanine dyes to form NIR dye-doped silica nanoparticles as versatile and unique platform for *in vivo* diagnostics. The NIR fluorescent silica nanoparticles could have extinction coefficients as high as 100×10^6 L mol^{-1} cm^{-1} in the NIR region with a high fluorescence QY of 8–10%. The unique advantage of NIR fluorescent silica nanoparticles is that they can be easily prepared in a range of sizes (10–100 nm); this is significant for nanoparticle-based diagnostics and therapeutics, making the silica nanoparticle system an even more promising platform.[58] Prodi *et al.* also used triethoxysilane-modified cyanine dyes to construct NIR fluorescent core–shell silica nanoparticles with PEG surface modification as efficient probes for *in vivo* mapping of regional lymph nodes in mice.[59] Moreover, two perylene diimide (PDI) dyes were modified by Farinha *et al.* with alkoxysilane groups to form visible and NIR fluorescent silica nanoparticles. The NIR fluorescent nanoparticles with diameters of 30–300 nm which homogeneously doped the dye molecules were further surface-modified with an arginine-glycine-aspartic acid (RGD) tumor-targeting oligopeptide for bioimaging. The brighter NIR fluorescence and higher photostability make the NIR-emitting nanoparticles a promising contrast agent in cells expressing high levels of fluorescent proteins which usually emit in the visible region.[60] Mesoporous silica nanoparticles (MSNs) were also considered a promising platform for fabricating an ultrabright NIR fluorescent contrast agent. Sokolov *et al.* encapsulated NIR fluorescent cyanine dye LS277 physically inside the nanochannels of MSNs with size 28 ± 3 nm and maximum excitation/emission 804 and 815 nm. In aqueous solution, the NIR fluorescent nanoparticles exhibit fluorescent brightness of each particle that was equivalent to 2070 ± 40 (or 650 ± 50) times the brightness of free dye molecules. Furthermore, the fluorescence QY of the dye was also increased by approximately five-fold. More importantly, the NIR fluorescent nanoparticles presented faster internalization, brighter NIR fluorescence, and a much longer imaging period in 4T1luc breast tumor cells, indicating much better potential of fluorescent nanoparticles than free organic dyes in bioimaging applications.[61]

Understandably, the development of NIR fluorescent silica nanoparticles is also directed toward multifunctionalization. For example, Lo *et al.* reported a system of trifunctional MSNs for NIR fluorescence imaging by doping ATTO 647N as the NIR fluorophore acting as a photosensitizer for photodynamic therapy, minimizing collateral damage to normal cells by surface modification with RGD-modified PEG. *In vitro* evaluation of the NIR MSNs as a theranostic platform demonstrated excellent targeting of NIR imaging and minimal collateral damage, as well as a highly potent therapeutic effect.[62] By chelating paramagnetic Gd to silica nanoparticles doped with ICG, Moudgil *et al.* developed NIR fluorescent and magnetic silica nanoparticles for NIR and MRI. The results from *in vitro* and *in vivo* studies illustrated that the photostability of ICG was enhanced by the silica nanoparticles (>1 week) and excellent NIR and MRI capabilities of the nanoparticles. Also, this type of nanoparticles has the potential to be used as non-invasive photo-initiated image-guided therapy vehicles *via* combined utilization of the NIR imaging ability and the photodynamic effect.[63] Another NIR fluorescent and magnetic silica nanoparticle was reported by Kang *et al.*, utilizing magnetic $CoFe_2O_4$ as the core and doped with NIR cyanine as the fluorophore. The enhancement ability of photostability and brightness of doped dye molecules was proved again by the resulting silica nanoparticles. More interestingly, after the radio-labeling of the nanoparticles with the radioisotope ^{68}Ga, a triple-modality nanoprobe was developed and successfully used to visualize the SLNs of mice *via* PET/MRI/NIRFI.[64] The possibility of co-encapsulating a functional substance with NIR dyes and easy surface modification of the silica nanoparticles gives them excellent potential as a platform for constructing multi-functional NIR fluorescent nanoprobes.

Calcium phosphate (CP), which is "generally regarded as safe" by the FDA, has been established as a fairly promising platform for bioimaging.[65,66] Adair *et al.* encapsulated ICG by biodegradable CP as a NIR fluorophore to construct a NIR fluorescent nanoparticle for sensitive diagnostic imaging. The ICG molecules inside the CP nanoparticles exhibit solvent-independent maximum fluorescence peak and higher emission intensity than the free fluorophore. The encapsulation of CP nanoparticles also sequesters the ICG molecules from the outside environment, endowing inside ICG a 4.7-fold longer fluorescence half-life at clinical imaging excitation power ranges and twofold increase in QY in phosphate buffered saline (PBS). Moreover, the CP nanoparticles with surface modification by PEG display prolonged circulation times *in vivo* with passive tumor accumulation, suggesting the excellent potential of CP nanoparticles for diagnostic applications.[67] Doping of CP nanoparticles with the NIR fluorophore ICG and Gd^{3+} as NIR and MR contrast, and surface tagging with ^{99m}Tc, led to a trimodal contrast agent for combined NIR, MRI, and nuclear imaging, thus greatly facilitating the merging of high spatial resolution and sensitivity of optical signals with clear anatomical and functional detail through MRI and detailed physiological information from nuclear imaging.[68]

5.4   NIR AIE Nanomaterials for Bioimaging

We have elaborated a variety of common NIR organic fluorescent chromophores and used different dye-encapsulation methods to improve chromophore stability and brightness. Nevertheless, these traditional NIR organic dyes used in biological diagnosis and therapy still have several limitations due to their inherent molecular structures. One of the most notorious phenomena is ACQ, which means that the fluorescence can be easily observed in dilute solution, but quenched in high concentrations or aggregated states.[69] An example of the ACQ effect is shown in Figure 5.18A. Although *N,N*-dicyclohexyl-1,7-dibromo-3,4,9,10-perylenetetracarboxylic diimide (DDPD)

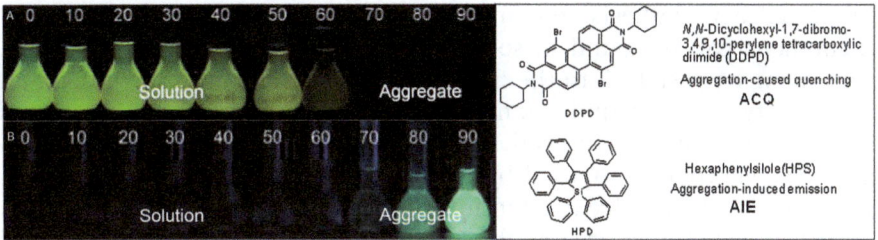

Figure 5.18 Fluorescent photographs of (A) DDPD (10 μM) and (B) HPS (10 μM) in THF–water mixtures with different water fraction (f_w) upon UV illumination. Adapted from ref. 70 with permission from the Royal Society of Chemistry.

exhibits strong green fluorescence in dilute tetrahydrofuran (THF) solution, its emission is gradually quenched when water content in the THF/water mixture increases, which significantly limits the sensitivity to bioanalytes from dye molecules in an aqueous biological environment. Great efforts have been made to solve the ACQ problem by chemical/physical methods or engineering processes,[70] but the results have been inefficient and have always led to other difficulties such as complicated synthesis or expensive instrumentation. Increased recognition of the aggregation environment formed by dye molecules and further utilizing the aggregate itself as a potential pattern for biomedical application is therefore highly desirable.

In 2001, Tang *et. al.* first discovered an unusual phenomenon of the propeller-like compound hexaphenylsilole (HPS, as shown in Figure 5.18B), which exhibits highly bright fluorescence when aggregated, and weak fluorescence when dissolved in solution.[71] Since the light emission derived from the aggregating process, they called this *aggregation-induced emission* (AIE). These AIE chromophores can exhibit bright luminescence in aggregate or high concentration, making them beneficial for improving the sensitivity of biosensors and completely solving the fluorescence quenching problem when bioimaging *in situ* or *in vivo*. In this section, we discuss construction strategies for extending the AIE emission from blue to the NIR region and the luminescence principles of AIE compounds, then mapping the nature of AIE-active organic molecules to finely control their excellent optical and morphological properties for NIR bioimaging *in vivo*.

5.4.1 Main Luminescent Principles of AIE Luminogens

As soon as the concept of AIE was introduced, it became a hot topic, giving rise to enthusiastic efforts to develop novel AIE building blocks,[72] especially by exploring the unique AIE-active characteristics in various applications such as photovoltaic devices, biosensors, and bioimaging. Tetraphenylethene (TPE),[73] one of the most shining "star molecules" reported by Tang *et al.*, serves as the AIE core to construct diverse functional TPE derivatives due to its easy synthesis and simple modulation. It is surprising that AIE phenomena can be found in some widely differing molecular structures, thus giving an insight into the luminescence mechanism, which offers an opportunity to deeply understand the relationship between AIE structures and properties, and rationally design novel AIE-active building blocks. Generally, the viewpoint proposed by Tang *et al.* in their recent review is that restriction of intramolecular motion (RIM), including molecular rotation and vibration in the aggregated state, is the mechanism responsible for most AIE effects.[72] As shown in Figure 5.19A, in the system of propeller-like structures such as TPE or HPS, several phenyl pendants rotate against the ethene or silole core on the single-bond axes in dilute solution, exhausting the excited energy as a non-radiative channel. In the aggregated or solid state, the rotation is restricted and the fluorescence turns on. However, for some other AIE derivatives such as THAB (1, Figure 5.19A) without any rotators in its molecular structure, the AIE properties derive from the restriction of intramolecular vibration in the aggregate because the two flexible parts can dissipate the excited energy in dilute solution in the way a butterfly vibrates its wings. In addition to the RIM mechanism as the main cause of AIE effects, sometimes the fluorescence yielded from the aggregation or solid state needs to be explained by the simultaneous assistance of other luminescence turn-on mechanisms such as *J*-aggregation, ESIPT, and TICT.[74-76] Recently, Tian and Chou *et al.* reported another relatively rare bent-to-planar motion system based on *N,N'*-disubstituted-dihydrophenazine derivatives,[77] which exhibit anomalous photophysical properties with wide tailoring of emission from red to deep blue or even to white light (Figure 5.19B). Different from the twisted-to-planar system, the authors ascribed the multifluorescence in *N,N'*-disubstituted-dihydrophenazine analogs to a sequential three-state electronic/structural relaxation process in combination, with an initial charge-transfer state, an intermediate state, and a final planarization state.

5.4.2 NIR Organic AIE Nanomaterials for Bioimaging

The high brightness of AIE derivatives in aggregate is highly desirable for developing biosensors and bioimaging agents in the living system. Indeed, many excellent research results have been reported by utilizing AIE building blocks to construct NIR AIE-active organic nanoparticles together with multiple functional properties for biological diagnosis and therapy.[78] Here we summarize the recent work on NIR AIE-active nanomaterials for bioimaging *in vivo*.

5.4.2.1 *Encapsulating an AIE Unit to Develop NIR Organic Nanomaterials*

The strategies of encapsulating organic dyes into nanoparticles have been successful in improving the hydrophilicity of luminogens and minimizing photobleaching in the living system. However, traditional dyes tend

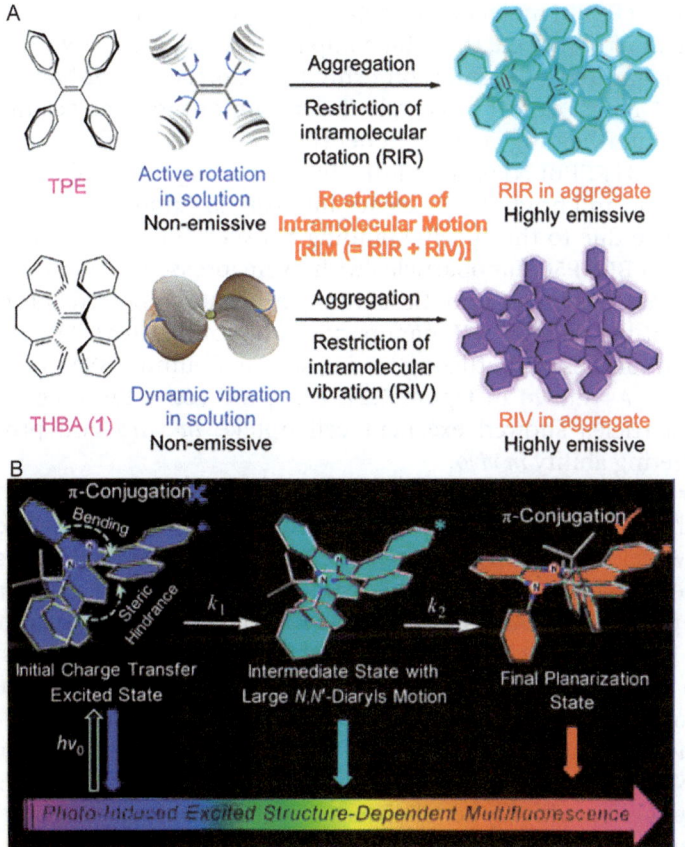

Figure 5.19 (A) RIM mechanism taking insight into the AIE effects: propeller-like luminogens such as TPE show AIE-active emission in aggregate due to the restricted intramolecular rotation (RIR) while shell-like THBA (1) serves as AIE luminogen due to the restriction of intramolecular vibration in aggregate state. Adapted from ref. 72 with permission from John Wiley and Sons. Copyright © 2014 Wiley-VCH Verlag GmbH & Co. KGaA, Weinheim. (B) Scheme of bent-to-planar motion system of *N,N'*-disubstituted-dihydrophenazine derivatives FIPAC by photo-induced excited structure-dependent multifluorescence emission. Adapted with permission from Z. Y. Zhang, Y.-S. Wu, K.-C. Tang, C.-L. Chen, J.-W. Ho, J. H. Su, H. Tian and P.-T. Chou, *J. Am. Chem. Soc.*, 2015, **137**, 8509–8520.[77] Copyright (2015) American Chemical Society.

to aggregate in high encapsulating concentration, which causes ACQ and decreases sensitivity in biosensing or bioimaging. Therefore, Tang *et al.* attempted to incorporate an AIE-active TPE unit into classical organic luminogen 3,4,9,10-tetracarboxylic perylene bisimide (PBI), and converted the ACQ-type PBI into an AIE-active molecule, BTPEPBI.[13,79] Furthermore, they employed amphiphilic copolymers DSPE-PEG$_{2000}$ and DSPE-PEG$_{5000}$-folate to encapsulate BTPEPBI molecules into nanoparticles for beneficial bioimaging *in vitro* and *in vivo*. Indeed, these biocompatible nanoparticles with uniform particle size and red–NIR emission exhibited fluorescence-enhanced imaging in MCF-7 breast cancer cells and efficient tumor-targeting accumulation in an H_{22} tumor-bearing mouse model (Figure 5.20). In addition, in comparison with BTPEPBI-NP0, BTPEPBI-NP50 had a higher uptake rate in the MCF-7 cell line and bright fluorescence intensity in tumor tissue after injection into mice due to the specific binding between folate units on the surface of BTPEPBI-NP50 nanoparticles with overexpressed folate receptors in cancer cell membrane or tumor tissue. Tang *et al.* later reported another type of biocompatible red–NIR AIE nanoparticles by combining a TPE unit with DCM chromophores and utilizing bovine serum albumin (BSA) as the polymer matrix.[80] As shown in Figure 5.21, these AIE-active luminogen-loaded BSA nanoparticles showed excellent cell uptake *in vitro* and prominent tumor-targeting ability *in vivo*.

Another efficient approach to achieve NIR AIE-active fluorescence signals is to use the Förster resonance energy transfer (FRET) strategy to overcome the short wavelength emission of the AIE luminogen as well as to make the synthesis easier. Recently, Liu *et al.* successfully obtained NIR nanoparticles by coencapsulation of AIE luminogen TPETPAFN with the traditional NIR fluorogen NIR775 using DSPE-PEG$_{2000}$ as the polymer matrix.[81] The spectral studies in Figure 5.22C indicate that the emission of TPETPAFN and the absorption of NIR775 overlap in the 550–800 nm range, leading to 47-fold fluorescence enhancement of NIR775 dye through FRET upon excitation of TPETPAFN at 510 nm. These biocompatible fluorescent nanoparticles with large Stokes shift are highly beneficial for bioimaging *in vivo* due to the bright luminescence derived from FRET.

In comparison with one-photon fluorescence bioimaging, there are several merits of two-photon fluorescence imaging such as deep penetration, low tissue autofluorescence perturbation, and less photobleaching for *in vivo* diagnosis and therapy. As early as 2007, Prasad *et al.* combined the advantages of two-photon properties and AIE-active characteristics to successfully achieve high luminogen-loading organic silica nanoparticles for cellular uptake *in vitro*.[82] Furthermore, Ng and Liu *et al.* reported ultrabright organic AIE dots by encapsulation of AIE luminogen BTPEBT with polymer matrix DSPE-PEG$_{2000}$ for real-time two-photon intravital vasculature imaging.[83] The AIE dots with large two-photon action cross-section value (6.3 × 10^4 GM) showed excellent TPFI ability in *in vivo* blood vasculature imaging after injection into mice. Similar AIE two-photon studies were reported by Hua *et al.,* further confirming the advantages of two-photon fluorescence with clear bioimaging of intravital blood vasculature up to 80 μm depth.[14] Moreover, He and Tang *et al.* even reported unique AIE dots TTF exhibiting three-/four-photon excited luminescence simultaneously with third/fifth harmonic generation in the solid state.[84] These aggregation-induced TTF-doped nanoparticles were also well used in multimodal imaging *in vitro* and *in vivo*.

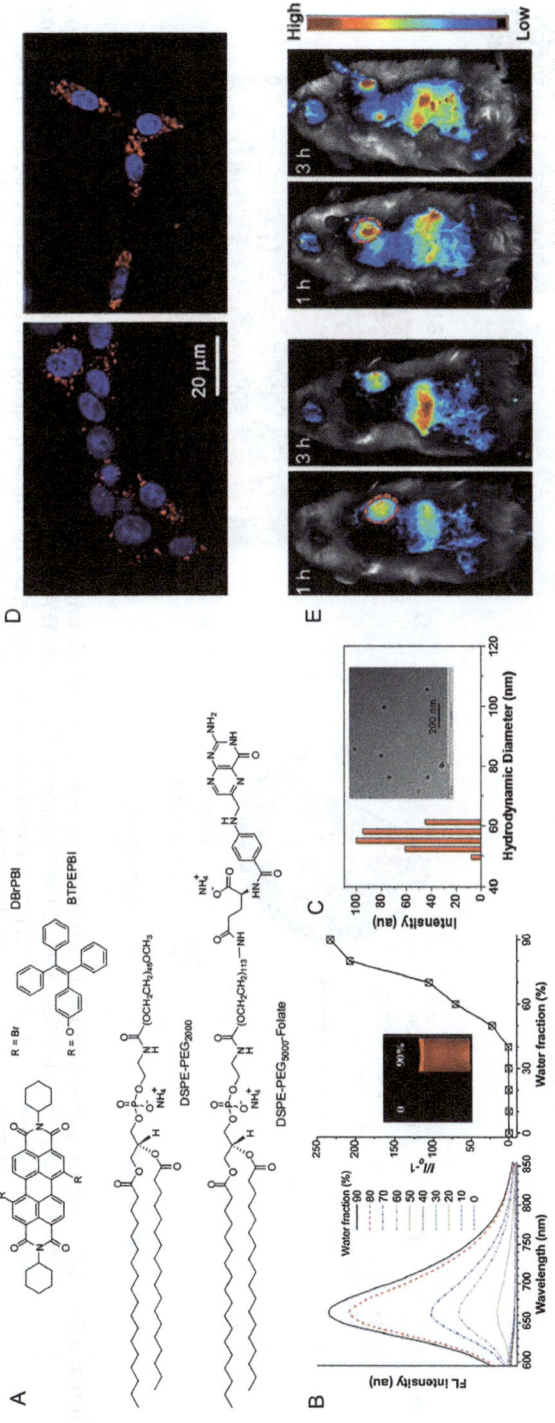

Figure 5.20 (A) Structures of AIE-active TPE unit modified traditional PBI dye and the polymer matrix DSPE-PEG$_{2000}$ and DSPE-PEG$_{5000}$-folate. (B) Photoluminescence properties of BTPEPBI in THF–water mixtures with different water fraction (f_w) excited at 538 nm. (C) Particle size of the BTPEPBI-encapsulated organic nanoparticles. (D) *In vitro* cell imaging stained by BTPEPBI–NP0 (left) and BTPEPBI–NP50 (right) for 2 h at 37 °C; (E) *in vivo* FL imaging of H22 tumor-bearing (red circle) mice after intravenous injection of BTPEPBI–NP0 (left) and BTPEPBI–NP50 (right) respectively. Adapted from ref. 79 with permission from the Royal Society of Chemistry.

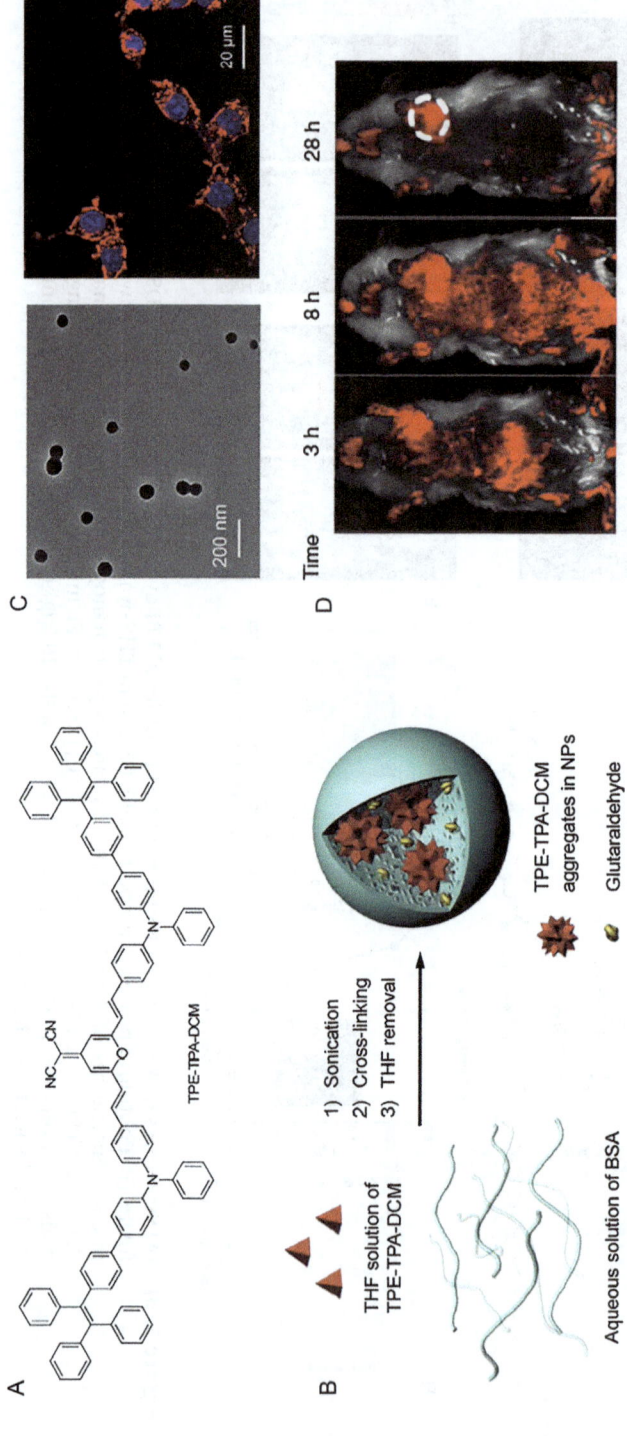

Figure 5.21 (A) Structures of AIE-active TPE unit incorporated into traditional DCM dye. (B) Schematic illustration of the fabrication of TPE–TPA–DCM-loaded BSA NPs. (C) TEM images of the TPE–TPA–DCM encapsulated organic nanoparticles and *in vitro* cell imaging stained for 2 h at 37 °C; (D) *in vivo* fluorescent imaging of H$_{22}$ tumor-bearing (white circle) mice after intravenous injection of BSA NPs loaded with TPE–TPA–DCM. Adapted from ref. 80 with permission from John Wiley and Sons. Copyright © 2012 Wiley-VCH Verlag GmbH & Co. KGaA, Weinheim.

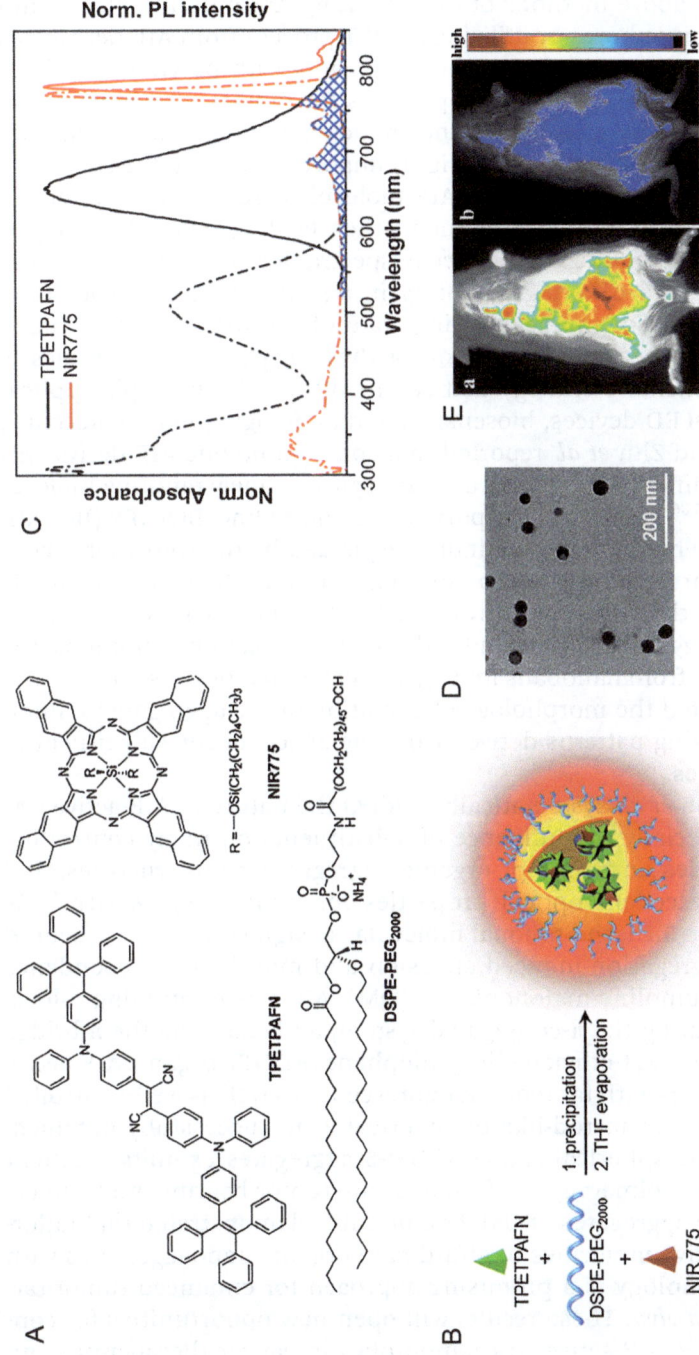

Figure 5.22   (A) Structures of AIE-active TPETPAFN, traditional dye NIR775, and the polymer matrix DSPE-PEG$_{2000}$. (B) Schematic illustration of the coencapsulation of TPETPAFN and traditional dye NIR775. (C) Normalized absorbance and fluorescence properties of TPETPAFN nanoparticles in water and NIR775 in THF. (D) TEM images of the coencapsulated organic nanoparticles. (E) *In vivo* FL imaging of mice after intravenous injection of coencapsulated organic nanoparticles excited at 523 nm (a) and 704 nm (b) respectively. Adapted from ref. 81 with permission from the Royal Society of Chemistry.

5.4.2.2   AIE Morphology Control and Tumor-Targeting Bioimaging

In addition the above methods of incorporating AIE building blocks into traditional chromophores or utilizing a FRET process from AIE derivatives to NIR dyes, then encapsulating these AIE dots with polymer matrix to obtain NIR AIE-active organic nanoparticles for biological diagnosis and therapy *in situ* and *in vivo*, mapping the unknown nature of the AIE lumino-gen in the aggregation state is also highly important. In fact, the structural and conformational differences of AIE molecules are responsible for the different aggregated microenvironment, thus leading to multiple aggre-gating characteristics such as AIE-active spectral features, shape-tailoring self-aggregates, and differing solubility in aqueous solution. Therefore, further exploration of the aggregating state of the AIE luminogen itself is greatly needed because integration of these aggregating properties is a novel comprehensive strategy that can be utilized for multiple applica-tions such as OLED devices, biosensing, and bioimaging *in vivo*. Recently, Pandey *et al.* and Zhu *et al.* reported that some quinolone AIE derivatives can formulate different aggregate morphologies by finely tailoring molecu-lar structures.[85,86] Pandey *et al.* reported three quinoline–BODIPY (BQ) AIE derivatives, in which the steric hindrance yielded by the thioether served as the trigger for a restricted intramolecular rotation (RIR) mechanism. By simply altering the substituent from –H to –CH$_3$ or –OCH$_3$ in molecular structure, the aggregate shape under the same fabrication conditions was totally changed: from nanoballs for BQ$_1$ and BQ$_2$ to nanofibrils for BQ$_3$. The authors attributed the morphological alteration of the aggregate to differ-ent crystal packing patterns derived from the effects of the substituent in the BQ structures.

Moreover, Zhu *et al.* systematically studied the nature of AIE-active QM molecules, especially the influence of substituents on finely controlling the morphologies and sizes of organic aggregated nanostructures, and therefore the excellent optical properties for bioimaging *in vivo*.[87] As shown in Figure 5.23, the rational molecular design strategy was focused for the NIR aggregation-induced emission and morphology dependence based on a quinoline-malononitrile (QM) AIE-active building block. Through elongating the π-conjugated system, and changing the π-bridge from thiophene to 3,4-ethylenedioxythiophene (EDOT), organic AIE-active QM nanomaterials with far red/NIR fluorescence as well as well-controlled aggregate shapes from rod-like to spherical were successfully obtained. Impressively, the spherical shape of QM-5 aggregates exhibits excellent tumor-targeted bioimaging performance in tumor-bearing nude mice, but the rod-like aggregates of QM-2 do not. It is demonstrated that tailor-ing NIR AIE-active molecules to afford bare organic nanoaggregates with desirable morphology is a promising approach for enhanced tumor-tar-geted efficacy *in vivo*. These results will open new opportunities for con-structing organic AIE-active NIR nanoprobes in cancer therapeutics and diagnostics.

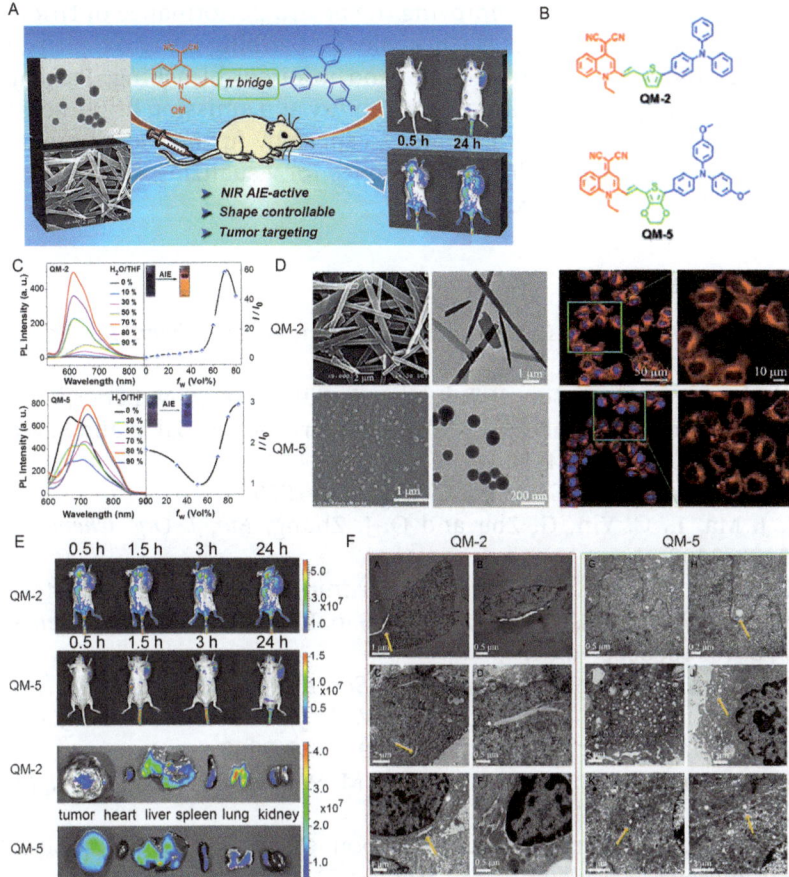

Figure 5.23   (A and B) Schemes and structures of NIR AIE molecule QM-2 and QM-5 derivatives. (C) Photoluminescence properties of QM-2, QM-5 in THF–water mixtures with different water fraction (f_w). (D) SEM and TEM imaging of QM-2 and QM-5 aggregates used for *in vitro* cell imaging. (E) *In vivo* non-invasive imaging of tumor-bearing mice after intravenous injection of QM-2 and QM-5 at different time periods (0.5, 1.5, 3, and 24 h). (F) TEM imaging of HeLa cells and tissues exposed to the QM nanoprobes. Adapted from ref. 87 with permission from John Wiley and Sons. Copyright © 2015 Wiley-VCH Verlag GmbH & Co. KGaA, Weinheim.

5.5   Conclusion

NIR organic dye-based nanomaterials have practical applications in the fields of biological and biomedical imaging. Indeed, NIR organic chromophores serve as a promising platform for construction of bioimaging sensors. Due to their intense fluorescence QYs and long emission characteristics, there has been great interest in the use of NIR organic chromophore-based probes for a range of applications. There is still much room to explore conventional NIR organic dyes, especially for improving the *in vivo* performance of NIR imaging agents in several respects, such as biocompatibility, *in vivo* toxicity, long emission, brightness, response time, and minimal bleaching. Particularly important are improvements in photostability and emission brightness with the dye-encapsulation and AIE strategy, especially further developments in photostability and morphology dependence for visualizing disease therapy, tumors, and biological processes.

References

1. P. D. Frischmann, V. Kunz and F. Würthner, *Angew. Chem., Int. Ed.*, 2015, **54**, 7285–7289.
2. S. Bhosale, A. L. Sisson, P. Talukdar, A. Fürstenberg, N. Banerji, E. Vauthey, G. Bollot, J. Mareda, C. Röger, F. Würthner, N. Sakai and S. Matile, *Science*, 2006, **313**, 84–86.
3. F. Würthner, *Chem. Commun.*, 2004, 1564–1579.
4. J. J. Ma, L. C. Yin, G. Zou and Q. J. Zhang, *Eur. J. Org. Chem.*, 2015, 3296–3302.
5. L. Q. Fan, Y. P. Xu and H. Tian, *Tetrahedron Lett.*, 2005, **46**, 4443–4447.
6. X. Y. Lu, W. H. Zhu, Y. S. Xie, X. Li, Y. Gao, F. Y. Li and H. Tian, *Chem.-Eur. J.*, 2010, **16**, 8355–8364.
7. Z. Q. Guo, S. Park and J. Yoon, *Chem. Soc. Rev.*, 2014, **43**, 16–29.
8. T. Hirayama, G. C. Van de Bittner, L. W. Gray, S. Lutsenko and C. J. Chang, *Proc. Natl. Acad. Sci. U. S. A.*, 2012, **109**, 2228–2233.
9. Z. Q. Guo, W. H. Zhu, M. M. Zhu, X. M. Wu and H. Tian, *Chem.-Eur. J.*, 2010, **16**, 14424–14432.
10. Z. Q. Guo, S. W. Nam, S. Park and J. Yoon, *Chem. Sci.*, 2012, **3**, 2760–2765.
11. K. Umezawa, Y. Nakamura, H. Makino, D. Citterio and K. Suzuki, *J. Am. Chem. Soc.*, 2008, **130**, 1550–1551.
12. C. Zhao, X. Li and F. Wang, *Chem.-Asian J.*, 2014, **9**, 1777–1781.
13. K. Li, D. Ding, Q. L. Zhao, J. Z. Sun, B. Z. Tang and B. Liu, *Sci. China: Chem.*, 2013, **56**, 1228–1233.
14. Y. T. Gao, G. X. Feng, T. Jiang, C. Goh, L. G. Ng, B. Liu, B. Liu, L. Yang, J. L. Huan and H. Tian, *Adv. Funct. Mater.*, 2015, **25**, 2857–2866.
15. Y. D. Hang, X.-P. He, L. Yang and J. L. Hua, *Biosens. Bioelectron.*, 2015, **65**, 420–426.
16. Y. B. Ding, Y. Y. Tang, W. H. Zhu and Y. S. Xie, *Chem. Soc. Rev.*, 2015, **44**, 1101–1112.
17. Y. B. Ding, Y. S. Xie, X. Li, J. P. Hill, W. B. Zhang and W. H. Zhu, *Chem. Commun.*, 2011, **47**, 5431–5433.
18. X. A. Zhang, K. S. Lovejoy, A. Jasano and S. J. Lippard, *Proc. Natl. Acad. Sci. U. S. A.*, 2007, **104**, 10780–10785.
19. Y. S. Xie, P. C. Wei, X. Li, T. Hong, K. Zhang and H. Furuta, *J. Am. Chem. Soc.*, 2013, **135**, 19119–19122.
20. S. L. Luo, E. L. Zhang, Y. P. Su, T. M. Cheng and C. M. Shi, *Biomaterials*, 2011, **32**, 7127–7138.

21. Y. Noh, H. S. Park, M. H. Sung and Y. T. Lim, *Biomaterials*, 2011, **32**, 6551–6557.
22. V. J. Pansare, S. Hejazi, W. J. Faenza and R. K. Prud'homme, *Chem. Mater.*, 2012, **24**, 812–827.
23. X. M. Wu and W. H. Zhu, *Chem. Soc. Rev.*, 2015, **44**, 4179–4184.
24. R. B. Altman, Q. S. Zheng, Z. Zhou, D. S. Terry, J. D. Warren and S. C. Blanchard, *Nat. Method*, 2012, **9**, 428–429.
25. A. Samanta, M. Vendrell, R. Das and Y. T. Chang, *Chem. Commun.*, 2010, **46**, 7406–7408.
26. S. Chang, X. M. Wu, Y. S. Li, D. C. Niu, Z. Ma, W. R. Zhao, J. L. Gu, W. J. Dong, F. Ding, W. H. Zhu and J. L. Shi, *Adv. Healthcare Mater.*, 2012, **1**, 475–479.
27. B. Jeong, Y. H. Bae, D. S. Lee and S. W. Kim, *Nature*, 1997, **388**, 860–862.
28. D. C. Niu, Y. S. Li, Z. Ma, H. Diao, J. L. Gu, H. R. Chen, W. R. Zhao, M. L. Ruan, Y. L. Zhang and J. L. Shi, *Adv. Funct. Mater.*, 2010, **20**, 773–780.
29. M. Akbulut, P. Ginart, M. E. Gindy, C. Theriault, K. H. Chin, W. Soboyejo and R. K. Prud'homme, *Adv. Funct. Mater.*, 2009, **19**, 718–725.
30. A. Rösler, G. W. M. Vandermeulen and H. A. Klok, *Adv. Drug Delivery Rev.*, 2001, **53**, 95–108.
31. V. Sexena, M. Sadoqi and J. Shao, *J. Photochem. Photobiol., B*, 2004, **74**, 29–38.
32. V. B. Rodriguez, S. M. Henry, A. S. Hoffman, P. S. Stayton, X. D. Li and S. H. Pun, *J. Biomed. Opt.*, 2008, **13**, 014025.
33. J. Tian, H. Chen, L. Zhuo, Y. Xie, N. Li and B. Tang, *Chem.–Eur. J.*, 2011, **17**, 6626–6634.
34. G. Sun, M. Y. Berezin, J. Fan, H. Lee, J. Ma, K. Zhang, K. L. Wooley and S. Achilefu, *Nanoscale*, 2010, **2**, 548–558.
35. T. Isabelle, G. Mathieu, D. Anabela, G. Laurent, D. Nadia, J. Véronique, N. Emmanuelle, B. Jérôme and V. Françoise, *J. Biomed. Opt.*, 2009, **14**, 054005.
36. L. Chu, S. Wang, K. Li, W. Xi, X. Zhao and J. Qian, *Biomed. Opt. Express*, 2014, **5**, 4076–4088.
37. C. Zheng, M. Zheng, P. Gong, D. Jia, P. Zhang, B. Shi, Z. Sheng, Y. Ma and L. Cai, *Biomaterials*, 2012, **33**, 5603–5609.
38. H. Mok, H. Jeong, S. Kim and B. H. Chung, *Chem. Commun.*, 2012, **48**, 8628–8630.
39. T. K. Hill, A. Abdulahad, S. S. Kelkar, F. C. Marini, T. E. Long, J. M. Provenzale and A. M. Mohs, *Bioconjugate Chem.*, 2015, **26**, 294–303.
40. H. Wu, H. Zhao, X. Song, S. Li, X. Ma and M. Tan, *J. Mat. Chem. B*, 2014, **2**, 5302–5308.
41. Z. Sheng, D. Hu, M. Xue, M. He, P. Gong and L. Cai, *Nano-Micro Lett.*, 2013, **5**, 145–150.
42. X. Zheng, D. Xing, F. Zhou, B. Wu and W. R. Chen, *Mol. Pharmaceutics*, 2011, **8**, 447–456.
43. H. Yang, H. Mao, Z. Wan, A. Zhu, M. Guo, Y. Li, X. Li, J. Wan, X. Yang, X. Shuai and H. Chen, *Biomaterials*, 2013, **34**, 9124–9133.

44. L. Yan and L. Qiu, *Nanomedicine*, 2015, **10**, 361–373.

45. C. Yue, P. Liu, M. Zheng, P. Zhao, Y. Wang, Y. Ma and L. Cai, *Biomaterials*, 2013, **34**, 6853–6861.

46. M. Zheng, C. Yue, Y. Ma, P. Gong, P. Zhao, C. Zheng, Z. Sheng, P. Zhang, Z. Wang and L. Cai, *ACS Nano*, 2013, 7, 2056–2067.

47. Q. Huo, J. Liu, Q. Wang, Y. Jiang, T. N. Lambert and E. Fang, *J. Am. Chem. Soc.*, 2006, **128**, 6447–6453.

48. A. Rösler, G. W. M. Vandermeulen and H. Klok, *Adv. Drug Delivery Rev.*, 2001, **53**, 95–108.

49. X. M. Wu, S. Chang, X. R. Sun, Z. Q. Guo, Y. S. Li, J. B. Tang, Y. Q. Shen, J. L. Shi, H. Tian and W. H. Zhu, *Chem. Sci.*, 2013, **4**, 1221–1228.

50. B. P. Bastakoti, Y. C. Hsu, S. H. Liao, K. C.-W. Wu, M. Inoue, S. Yusa, K. Nakashia and Y. Yamauchi, *Chem.–Asian J.*, 2013, **8**, 1301–1305.

51. J. L. Vivero-Escoto, R. C. Huxford-Phillips and W. Lin, *Chem. Soc. Rev.*, 2012, **41**, 2673–2685.

52. M. Montalti, L. Prodi, E. Rampazzo and N. Zaccheroni, *Chem. Soc. Rev.*, 2014, **43**, 4243–4268.

53. W. Stöber, A. Fink and E. Bohn, *J. Colloid Interface Sci.*, 1968, **26**, 62–69.

54. Y. Piao, A. Burns, J. Kim, U. Wiesner and T. Hyeon, *Adv. Funct. Mater.*, 2008, **18**, 3745–3758.

55. S. Wu, Y. Hung and C. Y. Mou, *Chem. Commun.*, 2011, **47**, 9972–9985.

56. F. J. Arriagada and K. Osseo-Asare, *Colloid Surf.*, 1992, **69**, 105–115.

57. E. Mahon, D. R. Hristov and K. A. Dawson, *Chem. Commun.*, 2012, **48**, 7970–7972.

58. J. F. Bringley, T. L. Penner, R. Wang, J. F. Harder, W. J. Harrison and L. Buonemani, *J. Colloid Interface Sci.*, 2008, **320**, 132–139.

59. M. Helle, E. Rampazzo, M. Monchanin, F. Marchal, F. Guillemin, S. Bonacchi, F. Salis, L. Prodi and L. Bezdetnaya, *ACS Nano*, 2013, 7, 8645–8657.

60. T. Ribeiro, S. Raja, A. S. Rodrigues, F. Fernandes, C. Baleizão and J. P. S. Farinha, *Dyes Pigm.*, 2014, **110**, 227–234.

61. S. Palantavida, R. Tang, G. P. Sudlow, W. J. Akers, S. Achilefu and I. Sokolov, *J. Mater. Chem. B*, 2014, **2**, 3107–3114.

62. S. H. Cheng, C. H. Lee, M. C. Chen, J. S. Souris, F. G. Tseng, C. S. Yang, C. Y. Mou, C. T. Chen and L. W. Lo, *J. Mater. Chem.*, 2010, **20**, 6149–6157.

63. P. Sharma, N. E. Bengtsson, G. A. Walter, H. B. Sohn, G. Zhou, N. Iwakuma, H. Zeng, S. R. Grobmyer, E. W. Scott and B. M. Moudgil, *Small*, 2012, **8**, 2856–2868.

64. J. S. Kim, Y. H. Kim, J. H. Kim, K. W. Kang, E. L. Tae, H. Youn, D. Kim, S. Kim, J. T. Kwon, M. H. Cho, Y. S. Lee, J. M. Jeong, J. K. Chung and D. S. Lee, *Nanomedicine*, 2012, 7, 219–229.

65. T. T. Morgan, H. S. Muddana, E. İ. Altinoğlu, S. M. Rouse, A. Tabaković, T. Tabouillot, T. J. Russin, S. S. Shanmugavelandy, P. J. Butler, P. C. Eklund, J. K. Yun, M. Kester and J. H. Adair, *Nano Lett.*, 2008, **8**, 4108–4115.

66. H. S. Muddana, T. T. Morgan, J. H. Adair and P. J. Butler, *Nano Lett.*, 2009, **9**, 1559–1566.

67. E. İ. Altinoğlu, T. J. Russin, J. M. Kaiser, B. M. Barth, P. C. Eklund, M. Kester and J. H. Adair, *ACS Nano*, 2008, **2**, 2075–2084.

68. A. Ashokan, G. S. Gowd, V. H. Somasundaram, A. Bhupathi, R. Peetham-baran, A. K. K. Unni, S. Palaniswamy, S. V. Nair and M. Koyakutty, *Biomaterials*, 2013, **34**, 7143–7157.

69. J. B. Birks, *Photophysics of Aromatic Molecules*, Wiley, London, 1970.

70. Y. N. Hong, J. W. Y. Lam and B. Z. Tang, *Chem. Soc. Rev.*, 2011, **40**, 5361–5388.

71. J. D. Luo, Z. L. Xie, J. W. Y. Lam, L. Cheng, H. Y. Chen, C. F. Qiu, H. S. Kwok, X. W. Zhan, Y. Q. Liu, D. B. Zhu and B. Z. Tang, *Chem. Commun.*, 2001, 1740–1741.

72. J. Mei, Y. N. Hong, J. W. Y. Lam, A. J. Qin, Y. H. Tang and B. Z. Tang, *Adv. Mater.*, 2014, **26**, 5429–5479.

73. R. T. K. Kwok, C. W. T. Leung, J. W. Y. Lam and B. Z. Tang, *Chem. Soc. Rev.*, 2015, **44**, 4228–4238.

74. C. X. Shi, Z. Q. Guo, Y. L. Yan, S. Q. Zhu, Y. S. Xie, Y. S. Zhao, W. H. Zhu and H. Tian, *ACS Appl. Mater. Interfaces*, 2013, **5**, 192–198.

75. Z. Zhao, J. W. Y. Lam and B. Z. Tang, *J. Mater. Chem.*, 2012, **22**, 23726–23740.

76. Z. Chi, X. Zhang, B. Xu, X. Zhou, C. Ma, Y. Zhang, S. Liu and J. Xu, *Chem. Sov. Rev.*, 2012, **41**, 3878–3896.

77. Z. Y. Zhang, Y.-S. Wu, K.-C. Tang, C.-L. Chen, J.-W. Ho, J. H. Su, H. Tian and P.-T. Chou, *J. Am. Chem. Soc.*, 2015, **137**, 8509–8520.

78. D. Ding, K. Li, B. Liu and B. Z. Tang, *Acc. Chem. Res.*, 2013, **46**, 2441–2453.

79. Q. L. Zhao, K. Li, S. J. Chen, A. J. Qin, D. Ding, S. Zhang, Y. Liu, B. Liu, J. Z. Sun and B. Z. Tang, *J. Mater. Chem.*, 2012, **22**, 15128–15135.

80. W. Qin, D. Ding, J. Z. Liu, W. Z. Yuan, Y. Hu, B. Liu and B. Z. Tang, *Adv. Funct. Mater.*, 2012, **22**, 771–779.

81. J. L. Geng, Z. S. Zhu, W. Qin, L. Ma, Y. Hu, G. G. Gurzadyan, B. Z. Tang and B. Liu, *Nanoscale*, 2014, **6**, 939–945.

82. B. S. Kim, H. E. Pudavar, A. Bonoiu and P. N. Prasad, *Adv. Mater.*, 2007, **19**, 3791–3795.

83. D. Ding, C. Goh, G. X. Feng, Z. J. Zhao, J. Liu, R. R. Liu, N. Tomczak, J. L. Geng, B. Z. Tang, L. G. Ng and B. Liu, *Adv. Mater.*, 2013, **25**, 6083–6088.

84. J. Qian, Z. F. Zhu, A. J. Qin, W. Qin, L. L. Chu, F. H. Cai, H. Q. Zhang, Q. Wu, R. R. Hu, B. Z. Tang and S. L. He, *Adv. Mater.*, 2015, **27**, 2332–2339.

85. R. S. Singh, R. K. Gupta, R. P. Paitandi, M. Dubey, G. Sharma, B. Koch and D. S. Pandey, *Chem. Commun.*, 2015, **51**, 9125–9128.

86. A. D. Shao, Z. Q. Guo, S. J. Zhu, S. Q. Zhu, P. Shi, H. Tian and W. H. Zhu, *Chem. Sci.*, 2014, **5**, 1383–1389.

87. A. D. Shao, Y. S. Xie, S. J. Zhu, Z. Q. Guo, S. Q. Zhu, J. Guo, P. Shi, T. D. James, H. Tian and W. H. Zhu, *Angew. Chem., Int. Ed.*, 2015, **54**, 7275–7280.

CHAPTER 6

Quantum Dots for Bioimaging-Related Bioanalysis

NAN MA[a]

[a]College of Chemistry, Chemical Engineering, and Materials Science,
Soochow University, 199 Ren'ai Road, Suzhou, 215123, P. R. China
*E-mail: nan.ma@suda.edu.cn

6.1 Introduction

Quantum dots (QDs), also known as semiconductor nanocrystals, exhibit unique size-dependent optical and electronic properties originating from quantum size effect.[1] QDs have proven to be powerful tools for bioimaging and bioanalysis because of their high photostability, broad absorption range, large extinction coefficient, and narrow and tunable emission wavelengths.[2-4] QDs have been widely used for biomolecular sensing both *in vitro* and *in vivo*, which is successfully applied to biomarker identification and disease diagnostics. The specificity and sensitivity of QD-based bioanalysis are determined by a variety of factors, including the size and emission wavelength of QDs and their colloidal stability, as well as their surface biofunctionalization. Simultaneous excitability make QDs versatile tools for multiplex biomarker detection. Direct biomolecule binding and Förster (fluorescence) resonance energy transfer (FRET)-based sensing are the two most frequently used strategies for biomolecule detection. Recently several novel methods have been developed for QD-based bioanalysis, which are also discussed in this

RSC Nanoscience & Nanotechnology No. 40
Near Infrared Nanomaterials: Preparation, Bioimaging and Therapy Applications
Edited by Fan Zhang
© The Royal Society of Chemistry 2016
Published by the Royal Society of Chemistry, www.rsc.org

chapter. In order to make QDs more compatible with *in vivo* studies, extensive research has been focused on developing non-toxic near infrared (NIR) QDs for *in vivo* imaging-based bioanalysis.

6.2   QD probe Chemistry and Synthetic Routes

QD emission wavelengths could be rationally tuned by changing the QD sizes and compositions. Group II–VI (CdS, CdSe, CdTe, *etc.*), group III–V (InP, InAs, *etc.*), and group IV–VI (PbS, PbSe, *etc.*) QDs are the three types most frequently used for biological applications. Doped QDs and alloyed QDs have also been synthesized with large Stokes shifts and finely tuned optical properties. Cd-based QDs usually emit visible light with high quantum yields (QYs) that are suitable for *in vitro* bioanalysis. However, these QDs raise toxicity concerns since they contain heavy metal ions. Overcoating QDs with non-toxic shells (inorganic semiconductor or organic molecule materials) helps to reduce QD toxicity. In-based and Ag-based QDs (*e.g.* InAs, Ag_2S) are less toxic and emit light in the near infrared (NIR) range, which offers better tissue penetration capability than visible light for *in vivo* whole-body imaging. The construction of QD probes usually proceeds in three steps: (1) colloidal synthesis of QDs; (2) surface modification and solubilization; (3) biofunctionalization. To construct a QD probe with optimal performance for biosensing, several considerations need to be taken into account including the overall size of the probe, its colloidal stability of the probe, and the number of ligands per probe. In general, small compact QD probes are preferred for imaging-based bioanalysis rather than larger bulky probes.

6.2.1   Conventional Synthetic Routes

Bottom-up colloidal synthesis is the most classical method of producing produce highly dispersed QDs for bioapplications. The synthesis may be conducted in either organic solvents or aqueous solution. Conventionally, QDs are produced through pyrolysis of organometallic precursors in organic solvents under high temperature and an inert environment.[5–8] The size and emission wavelength of the QDs can be rationally tuned by changing reaction conditions (temperature, precursor concentrations, time, *etc.*); see Figures 6.1 and 6.2. QDs synthesized in organic solvents usually possess good crystallinity and monodispersity, high QYs, and narrow emission peaks. However, these QDs are capped with hydrophobic molecules and need to be further solubilized in aqueous solution for bioapplications. The hydrophobic ligands can be displaced with thiol-containing hydrophilic molecules to solubilize the QDs; alternatively, the hydrophobic QDs can be encapsulated within amphipathic polymers or a water-soluble silica shell.[3,9–11] Subsequently, the QDs need to be further functionalized with biomolecules for biorecognition. Biofunctionalization can been achieved by covalent bioconjugation chemistry, biotin–avidin interaction, electrostatic interaction, and

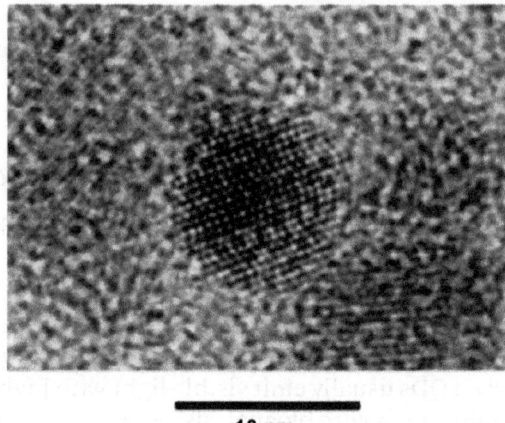

10 nm

Figure 6.1   High-resolution transmission electron microscopy (TEM) image of an
8 nm CdSe QD. Reproduced with permission from C. B. Murray, D. J.
Noms and M. G. Bawendi, *J. Am. Chem. Soc.*, 1993, **113**, 8706. Copyright
(1993) American Chemical Society.[5]

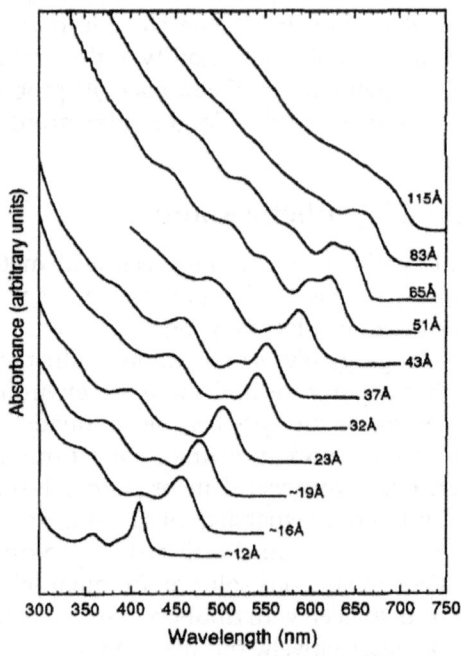

Figure 6.2   Absorption spectra of CdSe QDs dispersed in hexane and ranging in
size from 1.2 to 11.5 nm. Reproduced with permission from C. B. Murray, D. J. Noms and M. G. Bawendi, *J. Am. Chem. Soc.*, 1993, **113**, 8706.
Copyright (1993) American Chemical Society.[5]

other methods.[3,12,13] Instead of the organometallic synthetic route, some types of QDs can be directly synthesized in aqueous solution. This method has attracted much attention recently because of the straightforward synthetic procedure and mild reaction conditions. Also, the as-prepared QDs can be directly conjugated with biomolecules without further solubilization. Small thiol-containing molecules are frequently used as ligands for aqueous synthesis of QDs.[14-16] A variety of nanomaterials including carbon dots (CDs), QDs based on Ag, Pb, or In, and related core–shell, alloy, and doped QDs have been successfully synthesized in aqueous solution.[11,17-22] Similar to the organometallic route, the size and emission wavelength of QDs synthesized in aqueous solution can be tuned by altering reaction conditions.

6.2.2 Biomolecule-Templated Synthesis

The naturally occurring biomineralization process has inspired people to produce inorganic materials using biomolecules as synthetic templates. A variety of biomolecules including nucleic acids, peptides, proteins, and polysaccharides could be used to template the synthesis of QDs. Nucleic acid molecules could serve as an excellent template to promote nucleation, modulate QD growth, and finally passivate on the QD surface because of the strong metal-chelating properties of nucleic acid functional groups.[23-32] Biofunctionalized QDs could be produced in one step using chimeric DNA molecules as templates (Figure 6.3). The chimeric DNA contains a phosphorothioate domain for QD growth and passivation and a phosphate domain for biorecognition and biomolecular targeting.[28,31] Single DNA molecule functionalized QDs could be readily produced using this method. These nucleic acid-capped QDs exhibit robust biocompatibility and colloidal stability, making them well suited for cell imaging studies.[27-30] Peptide molecules containing cysteine residues (*e.g.* glutathione) have been widely used for the synthesis of QDs with excellent size-dependent optical properties and good biocompatibility.[33,34] The thiol group of the peptide is critical for QD passivation. Protein-directed QD synthesis needs to be conducted under very mild conditions to avoid protein denaturing, which somewhat compromises the quality of the QDs.[33,35-40] Recently a general strategy to synthesize highly luminescent QDs using protein molecule as ligands has been demonstrated.[33] This synthesis can be completed in 1 s and the protein molecules capped on QDs remain their activities for biotargeting or catalysis. Moreover, the synthesis of QDs in living organisms has also been demonstrated.[41-45]

6.2.3 Other Methods

There are a lot of other methods for QDs synthesis such as microwave-assisted synthesis,[46] cation-exchange-based synthesis,[47,48] seeded synthesis,[49] polymer-templated synthesis,[50] and a top-down approach.[51] These methods could serve as alternatives to the conventional methods of preparing QDs with the desired properties, especially for those that are difficult to prepare

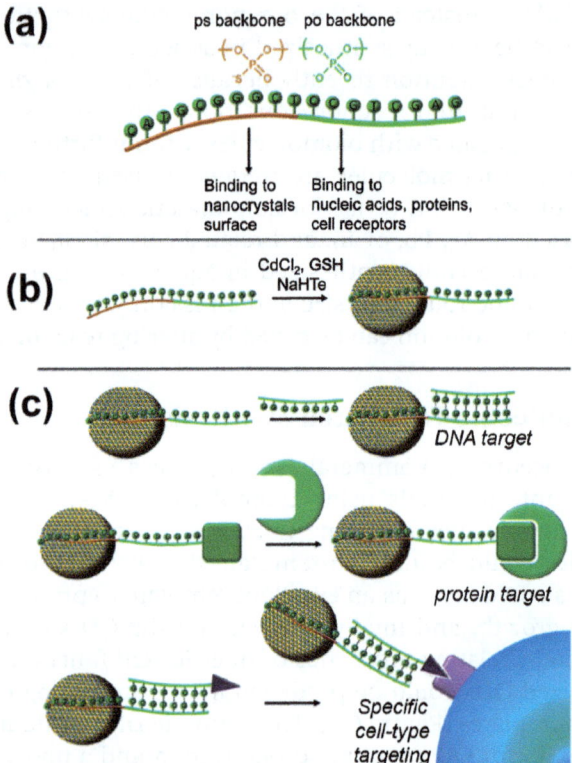

Figure 6.3   DNA-programmed and functionalized CdTe nanocrystals: (a) design of chimeric oligonucleotides with ligand (ps) and recognition (po) domains and schematic representation of one-pot biofunctionalized nanocrystal synthesis; (b) synthesis of CdTe nanocrystals using chimeric DNA molecules as templates; (c) binding to biological targets. Reproduced with permission from N. Ma, G. Tikhomirov and S. O. Kelley, *Acc. Chem. Res.*, 2010, **43**, 173. Copyright (2010) American Chemical Society.[25]

directly *via* the traditional routes. For example, microwave-assisted synthesis could provide higher temperature and pressure than normal synthetic methods, which could help to shorten the reaction time and increase the QD qualities.

6.3   New Types of QDs for Bioimaging and Bioanalysis

6.3.1   NIR QDs

NIR light possess better tissue penetration capability than visible light (Figure 6.4).[52] NIR QDs are superior contrast agents for optical imaging of living animals including tumor imaging, blood vessel imaging, and lymph

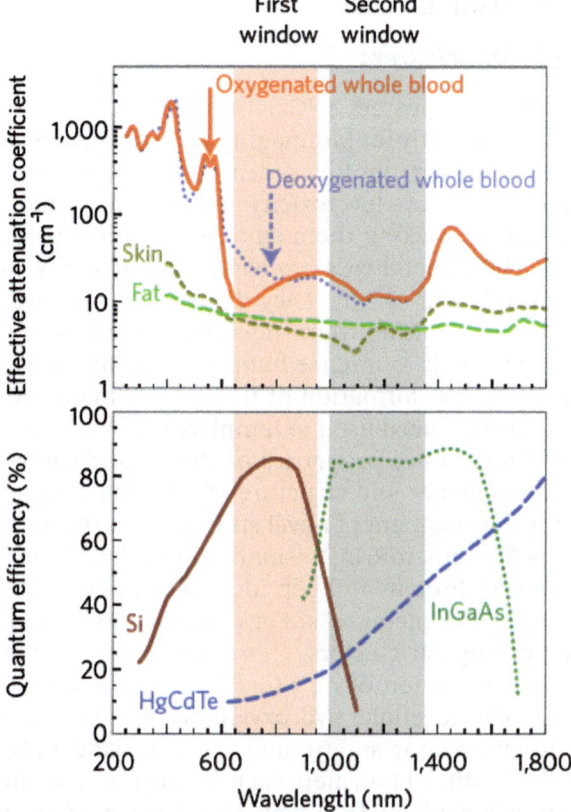

Figure 6.4   Optical windows in biological tissues. Top: effective attenuation coefficient *versus* wavelength; bottom: sensitivity curves for typical cameras based on Si, InGaAs, or HgCdTe sensors. Reproduced by permission from Macmillan Publishers Ltd.: *Nat. Nanotechnol.*[52] Copyright 2009.

node mapping, which serve as robust platforms for bioimaging-based analysis. The first generation of NIR QDs are composed of Cd, Hg, or Pb.[22,53-58] To minimize the risk of *in vivo* toxicity, these QDs are usually coated with protective layers, such as a silica coating or a QD shell based on non-toxic elements, to prevent the release of the heavy metal ions into the biological milieu. The second generation of QDs includes In-based and Ag-based QDs, which are less toxic and more popular for animal imaging studies. In particular, NIR QDs with emission wavelength in the second biological window (1000–1400 nm) are demonstrated to provide higher imaging resolution and depth than those emitting in the first biological window (700–900 nm).[59-61] The In- and Ag-based QDs can be synthesized either in organic solvent or in aqueous solution.[17,19,62-70] NIR emission can also be obtained from alloyed QDs,[56,57,71] type II QDs,[72] or core–shell QDs with lattice strain tuning.[21,73]

6.3.2 Non-Traditional "QDs"

6.3.2.1 Metal Nanoclusters

Fluorescent metal nanoclusters (NCs) (Au/Ag/Cu clusters) have gained increasing attention recently for bioimaging and biosensing. Some of these metal NCs possess strong photoluminescence and robust photostability.[74,75] Importantly, they have low toxicity and are much smaller than other types of nanoparticles, making them superior candidates for bioimaging applications.[76-87] These metal NCs are composed of several to a hundred atoms with an overall size <2 nm. They display discrete and size-tunable electronic transitions resulting from the strong quantum confinement of free electrons, which leads to intense fluorescence. Biomolecules are powerful tools to mediate the formation of fluorescent metal NCs. For example, DNA molecules have been used to template the synthesis of Ag and Cu NCs.[79,88-102] The fluorescence properties of the Ag and Cu NCs are highly dependent on the sequence and structure of DNA molecules,[92,94,96,97,101,103] which enables the development of novel strategies for the detection of DNA single mutations,[90-92] microRNA,[98] small molecules,[82,88,89,99] and metal ions[100,102] (Figure 6.5). Protein and peptide molecules have also been used to synthesize fluorescent metal NCs. For example, BSA could mediate the growth of red-emitting Au clusters,[104] and cysteine-containing peptides could promote the formation of Ag NCs.[105] Fluorescent Au NCs could be used for detection of intracellular hydroxyl radicals,[77,84] temperatures,[85] and glutathione.[82] *In vivo* imaging and tumor targeting using Au NCs have been achieved.[78,80,81,86,87] Artificial polymers such as poly(amidoamine) (PAMAM) and poly(methacrylic acid) (PMAA) have also been used to synthesize Au

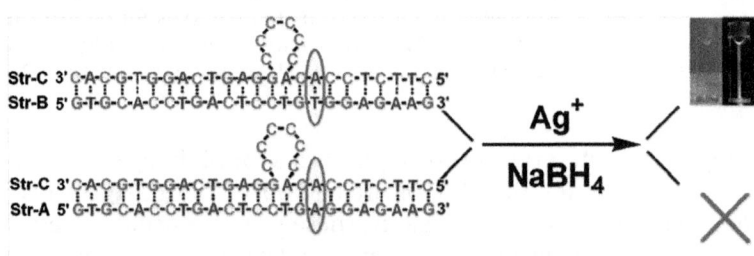

Str-A ▬ Target segment of the Homo sapiens hemoglobin beta chain (HBB) gene
Str-B ▬ Target segment of the variation of HBB gene
Str-C ▬ The designed cytosine-loop inserted probe strand
◯ ▬ The Sickle-Cell Anemia Gene Mutation Point

Figure 6.5   Identification of the sickle cell anemia gene mutation using two different DNA duplexes with inserted cytosine loops working as synthetic scaffolds to generate fluorescent silver clusters. Reproduced with permission from W. Guo, J. Yuan, Q. Dong and E. Wang, *J. Am. Chem. Soc.*, 2010, **132**, 932. Copyright (2010) American Chemical Society.[92]

and Ag NCs.[74,78,80,81,86,87,106] Small thiol ligands were able to synthesize Au and Ag NCs with ultrafine size.[105,107]

6.3.2.2 Carbon Dots (CDs)

Luminescent CDs are promising probes for bioimaging and bioanalysis because of their low toxicity, chemical inertness, and versatile surface chemistry.[108–110] They are usually produced by laser ablation of C in the presence of water and passivated with polymers.[111] The emission color of these CDs could be tuned when excited at different wavelengths. CDs have been used in a range of bioapplications including cancer cell targeting,[112] intracellular pH sensing,[113,114] and Cu^{2+} sensing.[114] The biodistribution, clearance, and tumor uptake of CDs have been systematically explored by *in vivo* animal imaging.[108] Photosensitizer-conjugated CDs have been used for simultaneous fluorescence imaging and photodynamic therapy.[115]

6.3.2.3 Graphene Quantum Dots (GQDs)

Graphene is a two-dimensional crystal composed of a one-atom thick monolayer of sp^2 C atoms arranged in a honeycomb lattice. Graphene quantum dots (GQDs) are usually obtained from graphite using the Hummers method[116] followed by cutting at high temperature.[117,118] The as-prepared GQDs contain carboxyl groups, leading to good water solubility,[119] and exhibit blue or green photoluminescence. The size and shape of GQDs have a crucial impact on their photoluminescence properties.[120] These GQDs can be easily biofunctionalized for sensing and imaging applications. For example, insulin-conjugated GQDs have been used for real-time tracking of dynamics of insulin receptors in cells;[121] polydopamine and hyaluronic acid-functionalized GQDs could be used for cell and animal imaging;[122,123] and aggregation-induced photoluminescence quenching of GQDs could be used for protein kinase sensing.[124] Carboxylated GQDs exhibit no acute toxicity either *in vitro* or *in vivo*.[125]

6.3.2.4 Silicon QDs

Silicon QDs possess strong photoluminescence, high photostability, low toxicity, and tunable emission wavelengths that are suitable for bioimaging applications.[126] Several methods have been established to synthesize luminescent Si QDs, including etching of porous Si, laser ablation, solution-phase methods, low-pressure plasma synthesis, thermal processing of silsesquioxane, and a two-step method based on laser pyrolysis of silane followed by acid etching.[127–129] This last method produced Si QDs with high QY and tunable emission wavelengths from 450 nm to 800 nm. Si QDs could be conjugated with biomolecules for specific cell targeting and labeling.[126–128] *In vivo* targeted cancer imaging,[130] sentinel lymph node mapping,[130] and multichannel imaging with biocompatible Si QDs[130] have also been demonstrated.

6.3.3 Self-Illuminating QDs

One potential limitation of QDs for optical imaging is that excitation light is required to produce an imaging signal. However, the excitation light usually possess low tissue penetration capability and could cause tissue autofluorescence to compromise the signal-to-background ratio. To overcome this barrier, a special type of self-illuminating QDs was developed that does not require any light excitation to illuminate light. The first kind of self-illuminating QDs is based on bioluminescence resonance energy transfer (BRET)[131,132] (Figure 6.6). In this construct luciferase is conjugated to the QDs and the QD emission can be induced by the addition of luciferin. These luciferase–QD conjugates were successfully applied to *in vivo* imaging with a reduced imaging background.[131,132] Additionally, these BRET-based QDs were

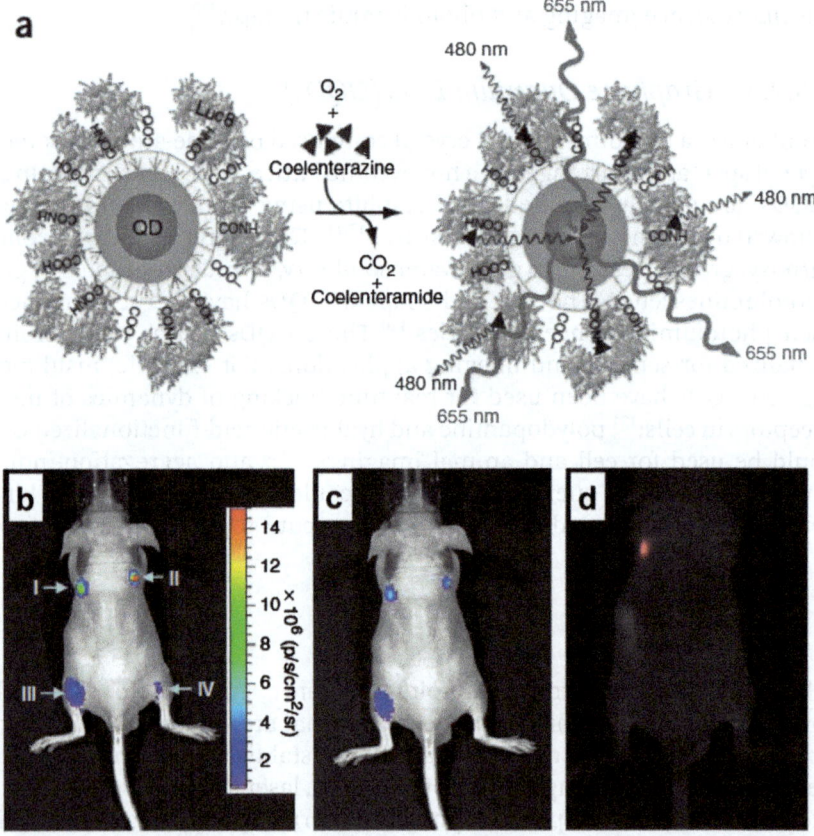

Figure 6.6 (a) Design of self-illuminating QDs with luciferase–QD conjugates. (b–d) Bioluminescence and fluorescence imaging of luciferase–QD conjugates injected subcutaneously (I and II) and intramuscularly (III and IV). (b) Open without filters; (c) with 575–650 nm filter; (d) fluorescence imaging with excitation filter 503–555 nm. Reproduced by permission from Macmillan Publishers Ltd.: *Nat. Biotechnol.*[132] Copyright (2006).

also used for detection of protease activities.[133] Similar to BRET, the chemiluminescence resonance energy transfer (CLRET)-based QDs[134] and Cerenkov resonance energy transfer (CRET)-based QDs[135–137] have also been reported. CRET-based QDs have been constructed by conjugating Cu^{64} to QDs and used for *in vivo* imaging studies.[138]

6.4 QDs for Bioimaging and Bioanalysis

6.4.1 Biolabeling and Bioimaging

6.4.1.1 Biofunctionalization

Biofunctionalization is a critical step to allow QDs to specifically recognize and bind with the biomolecular targets. Numerous methods have been developed to conjugate biomolecules to the surface of QDs. In general, bioconjugation methods can be divided into covalent conjugation and non-covalent interaction. Carboxyl groups, amine groups, and thiol groups are frequently used for biomolecule crosslinking.[3,139,140] In some cases a crosslinker is required to join two chemical groups together. Azide click chemistry has also been applied for nanoparticle bioconjugation.[141] Non-covalent interactions such as electrostatic interaction, metal chelation, and biotin–avidin binding have been used for QD bioconjugation. Electrostatic interaction could be used to attach negatively charged nucleic acids or protein molecules to positively charged QDs.[13,142,143] Alternatively, protein or peptides containing strong metal ion-chelating moieties could attach directly to the QD surface. For example, a polyhistidine tag fused to protein molecules could serve as a linker to conjugate proteins to the QD surface.[144] Also, as mentioned earlier, protein and DNA molecules could be directly used as ligands to synthesize biofunctionalized QDs where the synthesis and conjugation are combined into one step.[28,33] Biotin–avidin conjugation is another solution for attaching biomolecules to QDs.[12,145,146]

6.4.1.2 Cell Imaging

Biofunctionalized QDs have been extensively applied to cell surface receptor labeling,[145,152–158] intracellular biomolecule tracking and sensing,[12,147,159–168] organelle targeting,[169,170] *etc.* Because QDs have very strong photoluminescence and high photostability, they could offer much more intense and persistent signals than organic fluorophores, enabling QD-based single-particle tracking within cells.[12,147] Besides, QDs have been frequently used for multicolor imaging, *via* which multiple targets could be simultaneously detected and profiled.[146,148–151] A variety of biomolecule ligands including proteins (antibodies), peptides, aptamers, and small biomolecules have been used to functionalize QDs for specific labeling of overexpressed receptors on the cell surface. For example, prostate-specific membrane antigen, Her2, and EGFR that are overexpressed on cancer cells could be labeled with biofunctionalized

QDs for cancer diagnostics.[141,145,171] The arginine-glycine-aspartate (RGD) peptide which binds to cell surface integrin is also a popular ligand for cancer cell targeting.[172-174] Nucleic acid aptamers identified through *in vitro* selection possess very high affinity with their targets and are another type of important biomolecule ligands for cell imaging and targeting.[175] Compared to antibodies, aptamers are usually more stable, and easier to synthesize and manipulate. RNA and DNA aptamers have been used to target different cell types including prostate cancer cells, leukemia cells, and lymphoma cells.[28,176,177] Small molecules such as folate are another good choice for cell targeting. Folic acid receptors overexpressed on a variety of cancer cells could be imaged with folic acid-capped QDs.[178-180] The use of these small-molecule ligands could help to minimize the overall hydrodynamic size of the bioconjugated QDs, which is a critical parameter for *in vivo* imaging applications.

While cell surface labeling with QDs can be readily achieved, specific intracellular labeling with QDs in live cells is much more challenging. QDs are relatively big and could not directly penetrate cell membranes like small organic fluorophores. Instead, the QDs are internalized into the cells *via* endocytosis that is either non-specific or specific (*e.g.* receptor-mediated endocytosis) and are finally trapped into endosomes and lysosomes.[3,146,181] These compartments present barriers for the interaction of QDs with the intracellular biomolecules. Early attempts for QDs cytosolic delivery were achieved by conjugating QDs with cell penetrating peptides (CPP) or using liposome-based transfection agents. However, significant aggregation and endosomal sequestration of QDs were observed.[181-184] So far a few strategies have been developed to overcome this barrier and achieve cytosolic delivery of QDs. PEI-based endosomal disrupting polymers were capped on the QD surface to enable QDs to escape from endosomes and lysosomes. These polymers absorbed protons in the acidic compartments, building up an osmotic pressure across the membrane that could disrupt the membrane and allow the QDs to enter the cytosol.[185,186] This route is usually accompanied by pronounced cytotoxicity. Another method for cytosolic delivery of QDs is based on biodegradable polymeric nanospheres.[187] Recently, a new type of heterobivalent QD probe was developed to bypass the endocytosis delivery route and translocate into cytoplasm (Figure 6.7).[31] This probe features a nucleolin-targeting ligand (AS1411 aptamer) that can promote macropinocytosis of the probe; the internalized QD probe could easily enter the cytoplasm for FRET-based mRNA imaging. Microinjection is an efficient way to deliver QDs into the cell cytosol,[9,150,183] although it is quite tedious and has low throughput. Alternatively, the cells could be fixed before imaging studies. This method is more suitable for quantitative identification of specific intracellular biomolecules. Recent progress in QD-based cell imaging includes multiplex protein tracking, monitoring of intracellular protein interaction dynamics, stem cell labeling and imaging, and detection of gene expression.[12,147,159-168,175] As bright and robust fluorophores, QDs are especially useful for single molecule imaging.[188,189] This opens up the possibility of monitoring complex biological processes in real time.[163,190]

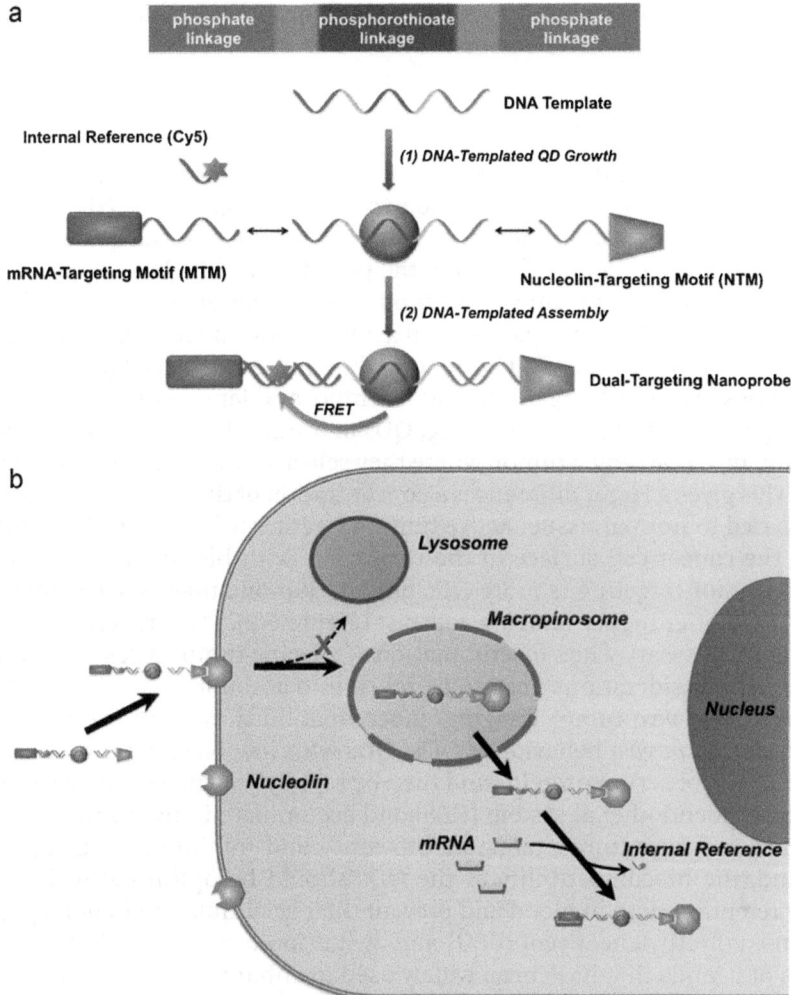

Figure 6.7 Schematic illustration of (a) the construction of a DNA-templated heterobivalent QD nanoprobe; (b) the dual-targeting strategy and intracellular delivery route of the heterobivalent QD nanoprobe. Reproduced from ref. 31 with permission from John Wiley and Sons. Copyright © 2014 Wiley-VCH Verlag GmbH & Co. KGaA, Weinheim.

Cytotoxicity is a major concern for QD-based imaging studies. The cytotoxicity mainly originates from the release of heavy metal ions (*e.g.* Cd^{2+}) from the QD surface. Enhancing the colloidal stability using multidentate capping molecules or growing a non-toxic inorganic shell on the surface of the QD core could prevent the release of heavy metal ions.[191,192] In addition, QDs are known to generate free radicals that can cause oxidative stress in cells.[193,194] QD capping ligands are another reason for toxicity.[195] Highly inert and biocompatible molecules are preferred to minimize potential cytotoxicity.[196,197]

The best solution to avoid the risk of toxicity for *in vivo* applications is to use QDs based on non-toxic elements.[198–201]

6.4.1.3   Animal Imaging

NIR QDs have found wide use for *in vivo* imaging including tumor targeting, lymph node mapping, and blood vessel imaging.[149,202,203] NIR light has better tissue penetration capability than visible light and can be detected in deep tissue (up to ~1 cm).[202] During the past decade QDs has become a very popular tool for *in vivo* cancer detection using xenografted tumor models in small animals.[149,174] Both passive and active tumor targeting can be applied to *in vivo* tumor imaging.[149] The endothelium (innermost cell layer) of tumor capillaries is less well organized and has more and larger pores than normal healthy vessels. In passive targeting, QDs of a suitable size stay in the blood stream until reaching a tumor, where they selectively leak into the tumor tissue. This gives a larger difference in concentration of the QDs in tumor tissue compared to normal tissue. Active tumor targeting refers to specific binding with the cancer cell surface in the tumor site with biofunctionalized QDs. Active tumor targeting is more efficient and durable than passive targeting and could offer higher imaging contrast (Figure 6.8).[174] Moreover, active targeting is necessary when discrimination of specific tumor types is required.

Several considerations need to be taken into account in the design of QD probes for *in vivo* tumor imaging. Biocompatibility is a critical parameter affecting the *in vivo* behavior of QDs. QDs with low biocompatibility could adsorb a lot of serum proteins and then be rapidly recognized and cleared by the reticuloendothelial system (RES) and accumulate in liver and spleen.[204] To ensure a high tumor targeting efficiency and minimum imaging background, the biocompatibility of the QDs should be optimized to increase their retention time in blood and prevent their accumulation in non-specific organs. Polyethylene glycol (PEG) and zwitterionic molecules[205–208] are two types of ligands that have been widely used to enhance the biocompatibility of QDs. PEG molecules are long neutral polymers that are highly resistant to non-specific protein absorption. These molecules could be conjugated to the surface of QDs to prevent RES clearance. Zwitterionic molecules can also avoid non-specific absorption; they contain an equal number of positive and negative charges and the overall charge is zero.[145,208] An additional advantage of zwitterionic molecules is that they are very small and therefore do not significantly increase the hydrodynamic size of the QDs after modification.[209] The size of QDs is found to be a determining factor for QD biodistribution. It is known that accumulation of QDs in normal tissues and organs can cause potential long-term toxicity and may hamper their clinical use. Therefore, efficient clearance of QDs from the body in an appropriate time period is highly preferred. Renal clearance is the most efficient way to clear nanoparticles from the body. The efficiency of renal clearance is highly dependent on the hydrodynamic size of the nanoparticles. A hydrodynamic diameter (HD) <5.5 nm resulted in rapid and efficient urinary excretion and elimination of

QDs from the body.[209] However, such a small HD is not easy to achieve for bio-functionalized QDs. The best way to construct QDs with both small HD and targeting capability is to use small zwitterionic molecules as capping ligands and small biorecognition molecules (*e.g.* folic acid) for biotargeting.[174,210] In addition, multidentate polymeric capping ligands could also offer compact

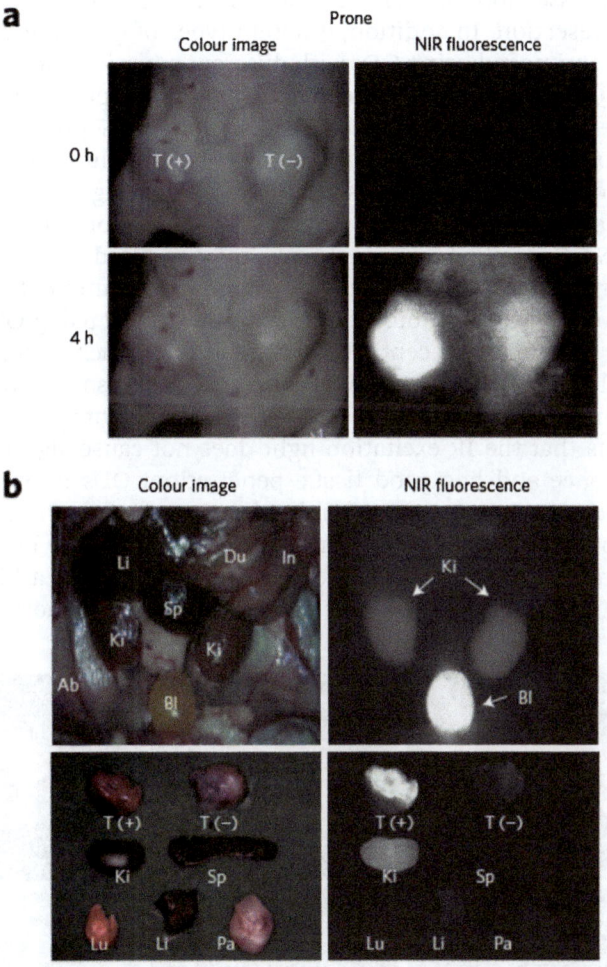

Figure 6.8 *In vivo* fluorescence imaging of human prostate cancer xenograft tumors. Mice were injected intravenously with GPI-functionalized QDs and observed for 4 h. (a) The prostate-specific membrane antigen (PSMA)-positive LNCaP tumor and PSMA-negative PC-3 tumor are indicated. Shown are representative images for animals in the prone position. (b) *In situ* (top row) and resected (bottom row) organs from a imaged at 4 h post-injection with color video and NIR fluorescence. (Ki, kidneys; Du, duodenum; Sp, spleen; In, intestines; Lu, lungs; Li, liver; Pa, pancreas; Ab, abdominal wall; Bl, bladder.) Reproduced by permission from Macmillan Publishers Ltd.: *Nat. Nanotechnol.*[174] Copyright (2009).

QDs with minimized HD sizes and robust colloidal stabilities.[211–213] Recently, simultaneous highly efficient QD-based tumor targeting and renal clearance have been achieved.[174,197] Another important *in vivo* application of QDs is lymph node imaging. NIR QDs permit sentinel lymph nodes (SLN) 1 cm deep to be imaged easily in real time.[202] This provides the surgeon with direct visual guidance throughout the entire SLN mapping procedure, minimizing incision and dissection inaccuracies and permitting real-time confirmation of complete resection. In addition, different types of lymph nodes could be simultaneously imaged using QDs with different emission wavelengths.[148]

Recently QDs with emission in the second biological optical window (1000–1400 nm) were successfully applied to *in vivo* imaging (Figure 6.9).[59–61,200,214] Compared to the first biological window (700–900 nm), fluorescence imaging in the second biological window is more desirable owing to reduced photon scattering, deeper tissue penetration, and lower autofluorescence. For example, NIR Ag_2S QDs applied to *in vivo* tumor imaging could provide deep inner organ registration, dynamic tumor contrast, and fast tumor detection.[199,200]

Two-photon and three-photon excitation microscopy using QDs as imaging contrast agents have recently been achieved. For each two-photon excitation, two low-energy photons of the IR light are absorbed to excite the QDs to generate a high-energy photon. The main advantage of two-photon microscopy is that the IR excitation light does not cause significant tissue autofluorescence and has good tissue penetration. QDs have a very high two-photon absorption cross-section that is necessary for efficient two-photon excitation. QD-based two-photon excitation microscopy has been used for *in vivo* imaging such as blood vessel imaging,[203] where a higher imaging resolution is obtained for QDs than organic fluorophores. Similarly,

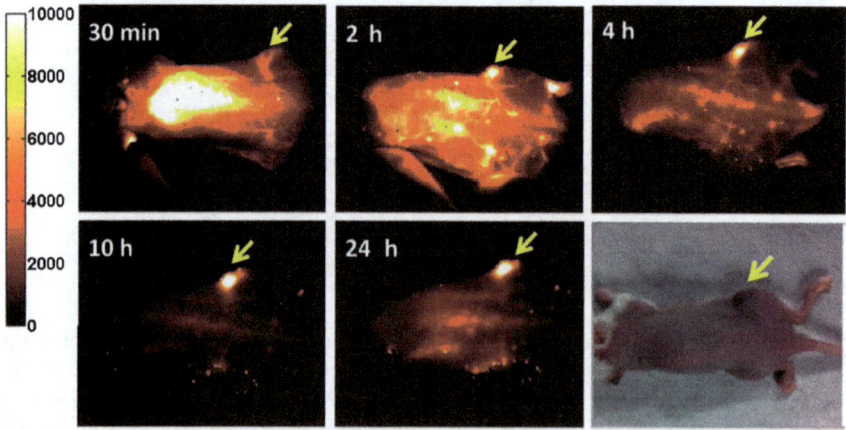

Figure 6.9   NIR-II fluorescence imaging of a xenograft tumor with high uptake of 6PEG–Ag_2S QDs. Time course of NIR-II fluorescence images and white-light optical image of the mouse injected with 6PEG–Ag_2S QDs. Reproduced from ref. 200 with permission from John Wiley and Sons. Copyright © 2012 Wiley-VCH Verlag GmbH & Co. KGaA, Weinheim.

QD-based three-photon excitation microscopy has been used for *in vivo* tumor imaging.[215]

The recently developed self-illuminating QDs are superior probes for *in vivo* animal imaging. Both BRET and CRET-based self-illuminating QDs have been successfully applied to *in vivo* animal imaging.[132,136–138] Because these QD probes do not require light excitation, little imaging background was obtained and thus the imaging sensitivity could be significantly improved.

6.4.2 QDs for Biosensing

QDs possess several unique features which make them versatile tools for the detection of various molecules ranging from metal ions and small molecules to large proteins and nucleic acids. Also, QDs could be used to monitor pH, temperature, and oxygen. Both *in vitro* and *in vivo* detection using QD-based biosensors have been accomplished.

6.4.2.1 Protein Detection

QDs have been extensively used as fluoro-tags for protein detection. The main principle for QD-based protein detection is similar to traditional immuno-assays.[216–224] Recently, several new protein detection strategies have been developed including DNA-based protein detection,[225–227] a doped QD-based multidimensional sensing platform for protein discrimination,[228] dual-aptamer-based protein detection,[229] AN electrochemiluminescent immuno-sensor,[216,230] and A ratiometric QD protein sensor.[231]

6.4.2.2 Protease Detection

Protease is an important type of biomarker for many diseases. The upregulated expression level and activity of protease in several diseases can be detected with QD-based biosensors for disease diagnostics. A FRET-based QD biosensor is the most popular tool to detect protease activities (Figure 6.10).[144,232] In this construct a FRET acceptor is attached to the QDs through a peptide linker which is the substrate of protease. Upon treatment with specific protease, the peptide linker is cleaved and the FRET signal is turned off. Several FRET acceptors have been used to construct the FRET pairing with QDs, including fluorescent proteins,[233] organic fluorophores,[234–239] gold nanoparticles,[240] and quencher dyes.[241,242] Multiplexed detection of different proteases has also been realized using multicolored QD-based biosensors.[243] Similar to FRET, BRET has also been used to detect protease activities, where bioluminescent proteins were used as BRET donors and QDs as BRET acceptors.[133] Recently new label-free strategies have been developed for highly sensitive detection of protease activities. For example, protease-triggered QD growth[244] and the inner filter effect of Au nanoparticles[245] were used to design protease-responsive biosensors to detect protease activities.

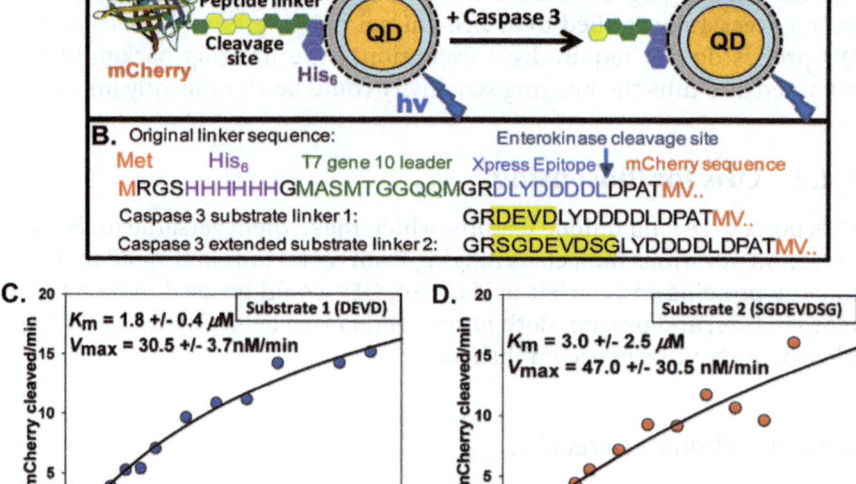

Figure 6.10 (a) Schematic of the QD–fluorescent protein sensor mCherry with an N-terminal linker expressing the caspase 3 cleavage site and a His6 sequence were self-assembled to the surface of CdSe–ZnS DHLA QDs, resulting in FRET quenching of the QD and sensitized emission from the mCherry acceptor (mCherry PDB structure 2H5Q). Caspase 3 cleaves the linker, reducing the FRET efficiency. (b) Linker sequences. (c and d) Proteolytic velocity *vs.* substrate concentration. Changes in FRET efficiency were converted to enzymatic velocity. Reproduced with permission from I. L. Medintz, A. R. Clapp, H. Mattoussi, E. R. Goldman, B. Fisher and J. M. Mauro, *Nat. Mater.*, 2003, **2**, 630. Copyright (2003) American Chemical Society.[233]

6.4.2.3   *DNA and RNA Detection*

QDs have been widely applied to high-sensitivity DNA and RNA detection. Because of their strong photoluminescence and multicolor properties, QDs could be used for multiplexed detection of different DNA targets simultaneously.[246] A FRET-based strategy is most popular for DNA detection. For example, the target of interest could serve as a bridge to bring the QD-labeled probe and a fluorophore-labeled probe together and turn on the FRET signal.[247] Alternatively, QD-tethered molecular beacons could be used to construct a FRET-based biosensor for DNA detection.[248,249] In some cases enzymes are used to amplify the target to improve the detection sensitivity.[250–252] In additional to FRET, the detection could be performed by bridging the QD-labeled DNA probe and magnetic nanoparticle-labeled DNA probe with the DNA target through hybridization. In this case the QDs are isolated by magnetic force and high-detection sensitivity could be attained.[253–255] A new label-free strategy has also been developed for high-sensitivity DNA detection based on QD doping.[256] Biocomputing-based nucleic acid detection has been reported, which could be potentially applied to intelligent diagnostics (Figure 6.11).[257] In this strategy different-colored QDs are assembled into one probe on a DNA template to construct different types of logic gates. Interaction of the probe with a complementary nucleic acid target would trigger a strand displacement reaction to release specific QDs, resulting in FRET signal changes.

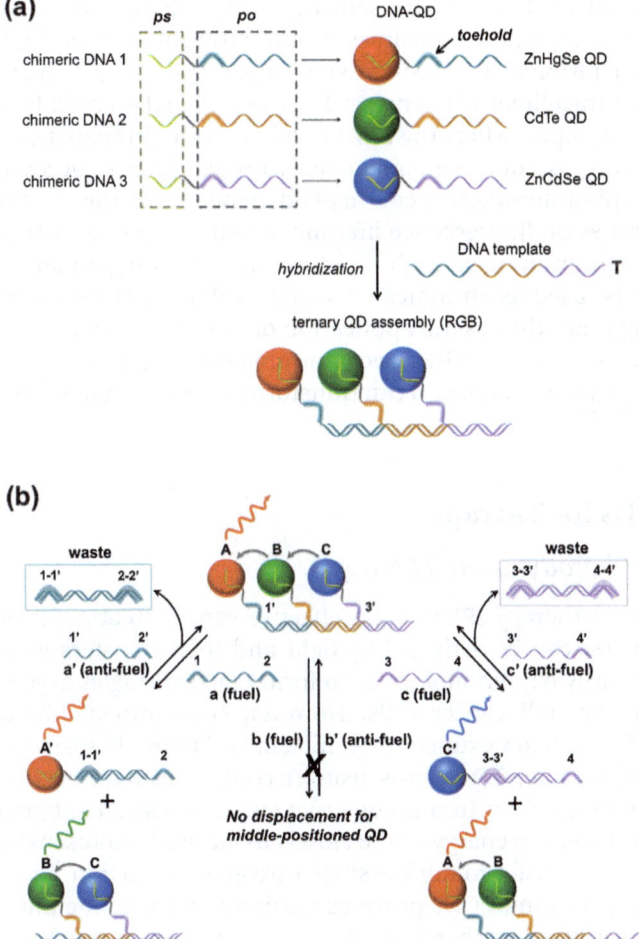

Figure 6.11   Illustration of (a) construction of ternary QD complex; (b) disassembly and reassembly of ternary QD complex through strand displacement reactions. Reproduced from ref. 257 with permission from John Wiley and Sons. Copyright © 2014 Wiley-VCH Verlag GmbH & Co. KGaA, Weinheim.

6.4.2.4   Small Molecule Detection

Small molecules such as dopamine,[258,259] ascorbic acid,[260,261] amino acids,[262] nucleotides,[263–265] NADH,[266] and glucose[267,268] have been successfully detected with QD-based biosensors. FRET-based fluorescence microscopy and electrochemistry are the most commonly used methods for small biomolecule detection using QD-based biosensors.[257,258,263,267,269,270]

6.4.3   Temperature/pH/Oxygen Sensing

The optical properties of QDs exhibit temperature-dependent behavior that could be used for temperature sensing.[271] High temperature sensitivity is realized *via* temperature-sensitive ligand molecule-induced QD transformation[272] or ratiometric sensing using doped QDs.[273] QDs have also been applied to intracellular pH sensing. For example, QDs could be conjugated with a FRET acceptor where the FRET efficiency is highly dependent on pH.[274] Alternatively, QDs could be conjugated with an electron acceptor that can quench QD photoluminescence in a pH-dependent manner.[275] A more direct approach relies on fluorescence lifetime imaging microscopy (FLIM).[276] The average photoluminescence (PL) lifetime of QDs is dependent on pH and thus could be used as an indicator for intracellular pH measurement. Oxygen sensing is another useful application of QDs. This is usually achieved by coupling the QDs with a FRET acceptor molecule (*e.g.* porphyrin, coenzyme Q) that reacts with oxygen and can transform to a state that would alter FRET efficiency.[277–279]

6.4.4   QDs for Therapy

6.4.4.1   Photodynamic Therapy

Photodynamic therapy (PDT) is an effective cancer treatment method. The photosensitizer can be activated by light and then transfers its triplet state energy to nearby oxygen molecules to form reactive singlet oxygen ($^1O_2$) species, which can kill cancer cells. However, conventional photosensitizers usually suffer from low extinction coefficient and poor photostability. To overcome this challenge, QD–photosensitizer conjugates have been constructed for cancer therapy.[280–283] In this construct QD serves as an antenna to absorb light and transfer the energy to the closely associated photosensitizer *via* resonance energy transfer to initiate singlet oxygen production.[281,284] The broad absorption spectrum of QDs provides flexibility to excite the photosensitizer within a wide wavelength range. Recently a ternary molecular beacon has been developed for simultaneous mRNA detection and cancer therapy (Figure 6.12).[285] This probe contains a red-emitting QD, a photosensitizer (Ce6), and a quencher molecule (BHQ3). The probe could be specifically activated within cancer cells to generate fluorescence as well as singlet oxygen for cancer theranostics. It could help to eliminate the side effects associated with PDT. Also,

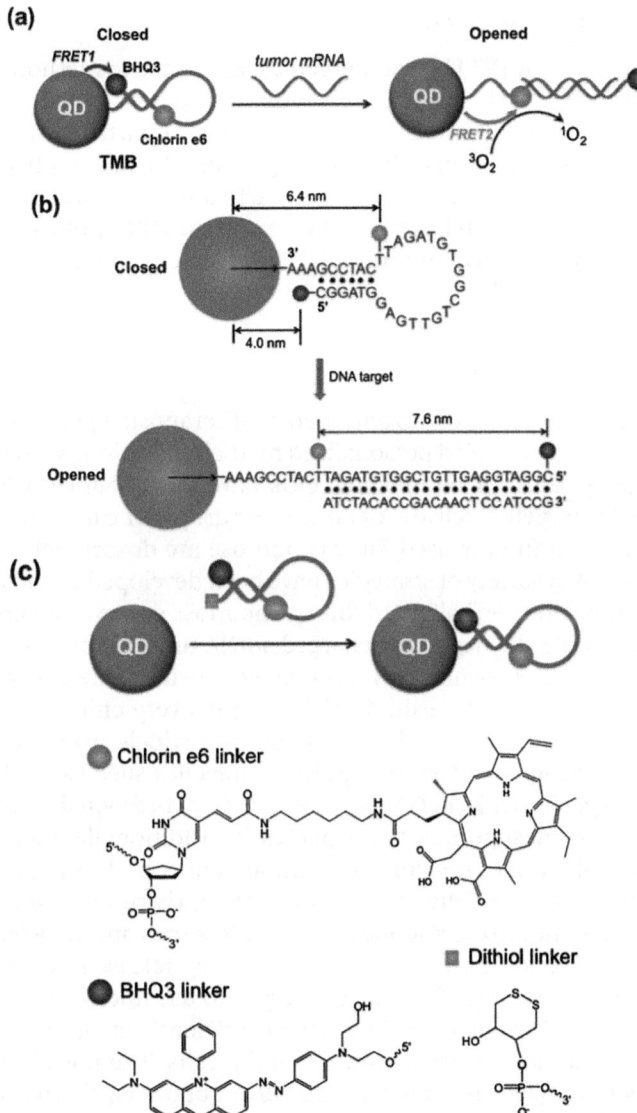

Figure 6.12 Schematic illustration of the ternary molecular beacon (TMB): (a) FRET modes of TMB in the closed and opened forms; (b) hairpin DNA sequence and the separation distance of each FRET pair; (c) chemical structures of Ce6, BHQ3, and dithiol linker. Reproduced from ref. 285 with permission from the Royal Society of Chemistry.

it could provide a stronger and more persistent fluorescence signal than Ce6 and an elevated level of singlet oxygen to effectively suppress tumor growth. Because of the large two-photon absorption cross-section of QDs, QD–photosensitizer conjugates have been successfully applied to two-photon PDT,[286] which cannot be adequately realized by a photosensitizer alone.

6.4.4.2 *Photothermal Therapy*

Photothermal therapy (PTT) has emerged as an attractive method for cancer treatment. It converts electromagnetic radiation to heat to kill cancer cells. Although a variety of nanomaterials such as Au nanoparticles and graphene have been used as PTT agents, the use of QDs for PTT has not been demonstrated until very recently. It is shown that CdTe and CdSe QDs could convert light energy into heat upon laser irradiation and the PTT application of these QDs was demonstrated in a tumor model with efficient suppression of tumor growth.[287-290]

6.4.4.3 *Drug Delivery*

Simultaneous cancer diagnostics and therapy (theranostics) has emerged as a promising technology toward personalized medicine.[291] Along with their superior optical properties for bioimaging, QDs have been coupled with various drug carriers for targeted delivery of anticancer drugs for cancer theranostics. The drugs most commonly used for this purpose are doxorubicin (DOX), cisplatin, and siRNA. A variety of strategies have been developed to construct QD–drug nanocomposites. For siRNA delivery, the most straightforward method is directly absorbing the negatively charged siRNA onto the positively charged QD surface (*e.g.* amine-terminated) through electrostatic interactions.[186,292,293] These QD–siRNA complexes exhibit high drug delivery efficiency and excellent gene silencing effect. Another common and widely used strategy is to encapsulate the drugs and QDs into polymer micelles such as biodegradable poly(lactic-*co*-glycolic acid) (PLGA) polymers,[294-296] pH-responsive amphiphilic copolymers,[297] or mesoporous nanospheres.[298] Additionally, biorecognition molecules are tethered to the surface of these nanocomplexes to achieve targeted imaging and drug delivery.[299-301] In general, these nanocomplexes are internalized *via* endocytosis and localize in endosomes and lysosomes where the low pH in these compartments can trigger drug release into cytosol.[302,303] Another interesting strategy for drug delivery is to use nucleic acid molecules as drug carriers for DOX delivery. DOX can be efficiently incorporated into the nucleic acid duplex and be transported into the cells. The nucleic acid-based drug carriers have been demonstrated for RNA aptamers,[304] DNA origami,[305] and self-assembled DNA–nanoparticle hybrid nanostructures.[306]

6.5 Outlook

More and better QD probes have been designed and constructed during the past several years, and they are getting closer and closer to clinical applications in the near future. The impact of QDs on immunoassay, diagnostics, and molecular pathology can already be seen. We believe that more robust and reproducible conjugation methods for QD probe production are key for their commercialization and widespread use. Additionally, the discovery of non-traditional QDs will pave the way for the *in vivo* clinical use of these nanomaterials.

References

1. A. P. Alivisatos, *Science*, 1996, **271**, 933.
2. X. Michalet, F. F. Pinaud, L. A. Bentolila, J. M. Tsay, S. Doose, J. J. Li, G. Sundaresan, A. M. Wu, S. S. Gambhir and S. Weiss, *Science*, 2005, **307**, 538.
3. W. Chan and S. Nie, *Science*, 1998, **281**, 2016.
4. M. Bruchez Jr., M. Moronne, P. Gin, S. Weiss and A. P. Alivisatos, *Science*, 1998, **281**, 2013.
5. C. B. Murray, D. J. Noms and M. G. Bawendi, *J. Am. Chem. Soc.*, 1993, **113**, 8706.
6. W. W. Yu, J. C. Falkner, B. S. Shih and V. L. Colvin, *Chem. Mater.*, 2004, **16**, 3318.
7. Z. A. Peng and X. Peng, *J. Am. Chem. Soc.*, 2001, **123**, 183.
8. A. A. Guzelian, J. E. B. Katari, A. V. Kadavanich, U. Banin, K. Hamad, E. Juban, A. P. Alivisatos, R. H. Wolters, C. C. Arnold and J. R. Heath, *J. Phys. Chem.*, 1996, **100**, 7212.
9. B. Dubertret, P. Skourides, D. J. Norris, V. Noireaux, A. H. Brivanlou and A. Libchaber, *Science*, 2002, **298**, 1759.
10. D. Gerion, F. Pinaud, S. C. Williams, W. J. Parak, D. Zanche, S. Weiss and A. P. Alivisatos, *J. Phys. Chem. B*, 2001, **105**, 8861.
11. J. Cao, B. Xue, H. Li, D. Deng and Y. Gu, *J. Colloid Interface Sci.*, 2010, **348**, 369.
12. M. Dahan, S. Levi, C. Luccardini, P. Rostaing, B. Riveau and A. Triller, *Science*, 2003, **302**, 442.
13. H. Mattoussi, J. M. Mauro, E. R. Goldman, G. P. Anderson, V. C. Sundar, F. V. Mikulec and M. G. Bawendi, *J. Am. Chem. Soc.*, 2000, **122**, 12142.
14. N. Gaponik, D. V. Talapin, A. L. Rogach, K. Hoppe, E. V. Shevchenko, A. Kornowski, A. Eychmuller and H. Weller, *J. Phys. Chem. B*, 2002, **106**, 7177.
15. W. C. Law, K. T. Yong, I. Roy, H. Ding, R. Hu, W. Zhao and P. N. Prasad, *Small*, 2009, **5**, 1302.
16. H. Qian, C. Dong, J. Weng and J. Ren, *Small*, 2006, **2**, 747.
17. W.-W. Xiong, G.-H. Yang, X.-C. Wu and J.-J. Zhu, *ACS Appl. Mater. Interfaces*, 2013, **5**, 8210.
18. Y. Wang, C. Ye, L. Wu and Y. Hu, *J. Pharm. Biomed. Anal.*, 2010, **53**, 235.
19. M. D. Regulacio, K. Y. Win, S. L. Lo, S. Y. Zhang, X. Zhang, S. Wang, M. Y. Han and Y. Zheng, *Nanoscale*, 2013, **5**, 2322.
20. L. Tan, A. Wan and H. Li, *ACS Appl. Mater. Interfaces*, 2013, **5**, 11163.
21. Z. Deng, A. Samanta, J. Nangreave, H. Yan and Y. Liu, *J. Am. Chem. Soc.*, 2012, **134**, 17424.
22. D. Deng, J. Xia, J. Cao, L. Qu, J. Tian, Z. Qian, Y. Gu and Z. Gu, *J. Colloid Interface Sci.*, 2012, **367**, 234.
23. S. Hinds, B. J. Taft, L. Levina, V. Sukhovatkin, C. J. Dooley, M. D. Roy, D. D. MacNeil, E. H. Sargent and S. O. Kelley, *J. Am. Chem. Soc.*, 2006, **128**, 64.
24. N. Ma, J. Yang, K. M. Stewart and S. O. Kelley, *Langmuir*, 2007, **23**, 12783.

25. N. Ma, G. Tikhomirov and S. O. Kelley, *Acc. Chem. Res.*, 2010, **43**, 173.
26. N. Ma, C. J. Dooley and S. O. Kelley, *J. Am. Chem. Soc.*, 2006, **128**, 12598.
27. C. Zhang, X. Ji, Y. Zhang, G. Zhou, X. Ke, H. Wang, P. Tinnefeld and Z. He, *Anal. Chem.*, 2013, **85**, 5843.
28. N. Ma, E. H. Sargent and S. O. Kelley, *Nat. Nanotechnol.*, 2009, **4**, 121.
29. L. Gao and N. Ma, *ACS Nano*, 2012, **6**, 689.
30. J. Farlow, D. Seo, K. E. Broaders, M. J. Taylor, Z. J. Gartner and Y. W. Jun, *Nat. Methods*, 2013, **10**, 1203.
31. W. Wei, X. He and N. Ma, *Angew. Chem., Int. Ed.*, 2014, **53**, 5573.
32. G. Tikhomirov, S. Hoogland, P. E. Lee, A. Fischer, E. H. Sargent and S. O. Kelley, *Nat. Nanotechnol.*, 2011, **6**, 485.
33. X. He, L. Gao and N. Ma, *Sci. Rep.*, 2013, **3**, 2825.
34. H. He, M. Feng, J. Hu, C. Chen, J. Wang, X. Wang, H. Xu and J. R. Lu, *ACS Appl. Mater. Interfaces*, 2012, **4**, 6362.
35. P. Wu, T. Zhao, Y. Tian, L. Wu and X. Hou, *Chem.–Eur. J.*, 2013, **19**, 7473.
36. W. Zhou, D. T. Schwartz and F. Baneyx, *J. Am. Chem. Soc.*, 2010, **132**, 4731.
37. Q. Wang, F. Ye, T. Fang, W. Niu, P. Liu, X. Min and X. Li, *J. Colloid Interface Sci.*, 2011, **355**, 9.
38. N. Ma, A. F. Marshall and J. Rao, *J. Am. Chem. Soc.*, 2010, **132**, 6884.
39. N. Goswami, A. Giri, S. Kar, M. S. Bootharaju, R. John, P. L. Xavier, T. Pradeep and S. K. Pal, *Small*, 2012, **8**, 3175.
40. W. Zhou and F. Baneyx, *ACS Nano*, 2011, **5**, 8013.
41. C. Mi, Y. Wang, J. Zhang, H. Huang, L. Xu, S. Wang, X. Fang, J. Fang, C. Mao and S. Xu, *J. Biotechnol.*, 2011, **153**, 125.
42. R. Cui, H.-H. Liu, H.-Y. Xie, Z.-L. Zhang, Y.-R. Yang, D.-W. Pang, Z.-X. Xie, B.-B. Chen, B. Hu and P. Shen, *Adv. Funct. Mater.*, 2009, **19**, 2359.
43. Y. Li, R. Cui, P. Zhang, B.-B. Chen, Z.-Q. Tian, L. Li, B. Hu, D.-W. Pang and Z.-X. Xie, *ACS Nano*, 2013, **7**, 2240.
44. L. Tan, A. Wan and H. Li, *ACS Appl. Mater. Interfaces*, 2014, **6**, 18.
45. S. R. Sturzenbaum, M. Hockner, A. Panneerselvam, J. Levitt, J. S. Bouillard, S. Taniguchi, L. A. Dailey, R. Ahmad Khanbeigi, E. V. Rosca, M. Thanou, K. Suhling, A. V. Zayats and M. Green, *Nat. Nanotechnol.*, 2013, **8**, 57.
46. H.-J. Zhan, P.-J. Zhou, Z.-Y. He and Y. Tian, *Eur. J. Inorg. Chem.*, 2012, **2012**, 2487.
47. D. H. Son, S. M. Hughes, Y. Yin and A. P. Alivisatos, *Science*, 2004, **306**, 1009.
48. C. Chen, X. He, L. Gao and N. Ma, *ACS Appl. Mater. Interfaces*, 2013, **5**, 1149.
49. A. M. Jawaid, S. Chattopadhyay, D. J. Wink, L. E. Page and P. T. Snee, *ACS Nano*, 2013, **7**, 3190.
50. X. Pang, L. Zhao, W. Han, X. Xin and Z. Lin, *Nat. Nanotechnol.*, 2013, **8**, 426.
51. J. Yang, T. Ling, W. T. Wu, H. Liu, M. R. Gao, C. Ling, L. Li and X. W. Du, *Nat. Commun.*, 2013, **4**, 1695.

52. A. M. Smith, M. C. Mancini and S. Nie, *Nat. Nanotechnol.*, 2009, **4**, 710–711.
53. S. Miao, S. G. Hickey, C. Waurisch, V. Lesnyak, T. Otto, B. Rellinghaus and A. Eychmuller, *ACS Nano*, 2012, **6**, 7059.
54. W. S. Ojo, S. Xu, F. Delpech, C. Nayral and B. Chaudret, *Angew. Chem., Int. Ed.*, 2012, **51**, 738.
55. D. K. Harris, P. M. Allen, H. S. Han, B. J. Walker, J. Lee and M. G. Bawendi, *J. Am. Chem. Soc.*, 2011, **133**, 4676.
56. A. M. Smith and S. Nie, *J. Am. Chem. Soc.*, 2011, **133**, 24.
57. S. Gupta, O. Zhovtiuk, A. Vaneski, Y.-C. Lin, W.-C. Chou, S. V. Kershaw and A. L. Rogach, *Part. Part. Syst. Charact.*, 2013, **30**, 346.
58. S. Keuleyan, E. Lhuillier and P. Guyot-Sionnest, *J. Am. Chem. Soc.*, 2011, **133**, 16422.
59. C.-N. Zhu, P. Jiang, Z.-L. Zhang, D.-L. Zhu, Z.-Q. Tian and D.-W. Pang, *ACS Appl. Mater. Interfaces*, 2013, **5**, 1186.
60. Y. Du, B. Xu, T. Fu, M. Cai, F. Li, Y. Zhang and Q. Wang, *J. Am. Chem. Soc.*, 2010, **132**, 1470.
61. P. Jiang, Z.-Q. Tian, C.-N. Zhu, Z.-L. Zhang and D.-W. Pang, *Chem. Mater.*, 2012, **24**, 3.
62. W. Zhang and X. Zhong, *Inorg. Chem.*, 2011, **50**, 4065.
63. W. Zhang, Q. Lou, W. Ji, J. Zhao and X. Zhong, *Chem. Mater.*, 2014, **26**, 1204.
64. J. Zhang, R. Xie and W. Yang, *Chem. Mater.*, 2011, **23**, 3357.
65. K. Yu, P. Ng, J. Ouyang, M. B. Zaman, A. Abulrob, T. N. Baral, D. Fatehi, Z. J. Jakubek, D. Kingston, X. Wu, X. Liu, C. Hebert, D. M. Leek and D. M. Whitfield, *ACS Appl. Mater. Interfaces*, 2013, **5**, 2870.
66. R. Xie and X. Peng, *J. Am. Chem. Soc.*, 2009, **131**, 10645.
67. B. Mao, C.-H. Chuang, F. Lu, L. Sang, J. Zhu and C. Burda, *J. Phys. Chem. C*, 2013, **117**, 648.
68. M. F. Foda, L. Huang, F. Shao and H. Y. Han, *ACS Appl. Mater. Interfaces*, 2014, **6**, 2011.
69. E. Cassette, T. Pons, C. Bouet, M. Helle, L. Bezdetnaya, F. Marchal and B. Dubertret, *Chem. Mater.*, 2010, **22**, 6117.
70. P. M. Allen and M. G. Bawendi, *J. Am. Chem. Soc.*, 2008, **130**, 9240.
71. F. Yang, P. Yang and L. Zhang, *Luminescence*, 2013, **28**, 836.
72. S. Kim, B. Fisher, H.-J. Eisler and M. G. Bawendi, *J. Am. Chem. Soc.*, 2013, **129**, 11466.
73. A. M. Smith, A. M. Mohs and S. Nie, *Nat. Nanotechnol.*, 2009, **4**, 56.
74. J. Zheng, P. R. Nicovich and R. M. Dickson, *Annu. Rev. Phys. Chem.*, 2007, **58**, 409.
75. X. Yuan, Z. Luo, Q. Zhang, X. Zhang, Y. Zheng, J. Y. Lee and J. Xie, *ACS Nano*, 2011, **5**, 8800.
76. V. Venkatesh, A. Shukla, S. Sivakumar and S. Verma, *ACS Appl. Mater. Interfaces*, 2014, **6**, 2185.
77. M. Zhuang, C. Ding, A. Zhu and Y. Tian, *Anal. Chem.*, 2014, **86**, 1829.
78. Y. Wang, J. Chen and J. Irudayaraj, *ACS Nano*, 2011, **5**, 9718.

79. J. Yu, S. Choi and R. M. Dickson, *Angew. Chem., Int. Ed.*, 2009, **48**, 318.

80. Y. Zhang, M. Wang, Y.-g. Zheng, H. Tan, B. Y.-w. Hsu, Z.-c. Yang, S. Y. Wong, A. Y.-c. Chang, M. Choolani, X. Li and J. Wang, *Chem. Mater.*, 2013, **25**, 2976.

81. J. Wang, G. Zhang, Q. Li, H. Jiang, C. Liu, C. Amatore and X. Wang, *Sci. Rep.*, 2013, **3**, 1157.

82. D. Tian, Z. Qian, Y. Xia and C. Zhu, *Langmuir*, 2012, **28**, 3945.

83. Y. Tao, E. Ju, Z. Li, J. Ren and X. Qu, *Adv. Funct. Mater.*, 2014, **24**, 1004.

84. E. S. Shibu, S. Sugino, K. Ono, H. Saito, A. Nishioka, Y. Yamamura, M. Sawada, Y. Nosaka and V. Biju, *Angew. Chem., Int. Ed.*, 2013, **52**, 10559.

85. L. Shang, F. Stockmar, N. Azadfar and G. U. Nienhaus, *Angew. Chem., Int. Ed.*, 2013, **52**, 11154.

86. J. Liu, M. Yu, C. Zhou, S. Yang, X. Ning and J. Zheng, *J. Am. Chem. Soc.*, 2013, **135**, 4978.

87. C.-A. J. Lin, W.-K. Chuang, Z.-Y. Huang, S.-T. Kang, C.-Y. Chang, C.-T. Chen, J.-L. Li, J. K. Li, H.-H. Wang, F.-C. Kung, J.-L. Shen, W.-H. Chan, C.-K. Yeh, H.-I. Yeh, W.-F. T. Lai and W. H. Chang, *ACS Nano*, 2012, **6**, 5111.

88. Z. Zhou, Y. Du and S. Dong, *Anal. Chem.*, 2011, **83**, 5122.

89. L. Zhang, J. Zhao, M. Duan, H. Zhang, J. Jiang and R. Yu, *Anal. Chem.*, 2013, **85**, 3797.

90. H. C. Yeh, J. Sharma, M. Shih Ie, D. M. Vu, J. S. Martinez and J. H. Werner, *J. Am. Chem. Soc.*, 2012, **134**, 11550.

91. X. Jia, J. Li, L. Han, J. Ren, X. Yang and E. Wang, *ACS Nano*, 2012, **6**, 3313.

92. W. Guo, J. Yuan, Q. Dong and E. Wang, *J. Am. Chem. Soc.*, 2010, **132**, 932.

93. T. Vosch, Y. Antoku, J. C. Hsiang, C. I. Richards, J. I. Gonzalez and R. M. Dickson, *Proc. Natl. Acad. Sci. U. S. A.*, 2007, **104**, 12616.

94. Y. Teng, X. Yang, L. Han and E. Wang, *Chem.–Eur. J.*, 2014, **20**, 1111.

95. S. A. Patel, C. I. Richards, J.-C. Hsiang and R. M. Dickson, *J. Am. Chem. Soc.*, 2008, **130**, 11602.

96. A. Rotaru, S. Dutta, E. Jentzsch, K. Gothelf and A. Mokhir, *Angew. Chem., Int. Ed.*, 2010, **49**, 5665.

97. Z. Qing, X. He, D. He, K. Wang, F. Xu, T. Qing and X. Yang, *Angew. Chem., Int. Ed.*, 2013, **52**, 9719.

98. P. Shah, A. Rørvig-Lund, S. B. Chaabane, P. W. Thulstrup, H. G. Kjaergaard, E. Fron, J. Hofkens, S. W. Yang and T. Vosch, *ACS Nano*, 2012, **6**, 8803.

99. X. Liu, F. Wang, A. Niazov-Elkan, W. Guo and I. Willner, *Nano Lett.*, 2013, **13**, 309.

100. J. Liu, J. Chen, Z. Fang and L. Zeng, *Analyst*, 2012, **137**, 5502.

101. C. I. Richards, S. Choi, J.-C. Hsiang, Y. Antoku, A. B. T. Vosch, Y.-L. Tzeng and R. M. Dickson, *J. Am. Chem. Soc.*, 2008, **130**, 5038.

102. J. Chen, J. Liu, Z. Fang and L. Zeng, *Chem. Commun.*, 2012, **48**, 1057.

103. H. C. Yeh, J. Sharma, J. J. Han, J. S. Martinez and J. H. Werner, *Nano Lett.*, 2010, **10**, 3106.

104. J. Xie, Y. Zheng and J. Y. Ying, *J. Am. Chem. Soc.*, 2009, **131**, 888.

105. Y. Cui, Y. Wang, R. Liu, Z. Sun, Y. Wei, Y. Zhao and X. Gao, *ACS Nano*, 2011, **5**, 8684.
106. J. Zheng and R. M. Dickson, *J. Am. Chem. Soc.*, 2002, **124**, 13982.
107. M. A. H. Muhammed, F. Aldeek, G. Palui, L. Trapiella-Alfonso and H. Mattoussi, *ACS Nano*, 2012, **6**, 8950.
108. X. Huang, F. Zhang, L. Zhu, K. Y. Choi, N. Guo, J. Guo, K. Tackett, P. Anilkumar, G. Liu, Q. Quan, H. S. Choi, G. Niu, Y.-P. Sun, S. Lee and X. Chen, *ACS Nano*, 2013, 7, 5684.
109. L. Cao, X. Wang, M. J. Meziani, F. Lu, H. Wang, P. G. Luo, Y. Lin, B. A. Harruff, L. M. Veca, D. Murray, S.-Y. Xie and Y.-P. Sun, *J. Am. Chem. Soc.*, 2007, **129**, 11318.
110. S.-T. Yang, L. Cao, P. G. Luo, F. Lu, X. Wang, H. Wang, M. J. Meziani, Y. Liu, G. Qi and Y.-P. Sun, *J. Am. Chem. Soc.*, 2009, **131**, 11308.
111. Y.-P. Sun, B. Zhou, Y. Lin, W. Wang, K. A. S. Fernando, P. Pathak, M. J. Meziani, B. A. Harruff, X. Wang, H. Wang, P. G. Luo, H. Yang, M. E. Kose, B. Chen, L. M. Veca and S.-Y. Xie, *J. Am. Chem. Soc.*, 2006, **128**, 7756.
112. Q. Li, T. Y. Ohulchanskyy, R. Liu, K. Koynov, D. Wu, A. Best, R. Kumar, A. Bonoiu and P. N. Prasad, *J. Phys. Chem. C*, 2010, **14**, 12062.
113. B. Kong, A. Zhu, C. Ding, X. Zhao, B. Li and Y. Tian, *Adv. Mater.*, 2012, **24**, 5844.
114. A. Zhu, Q. Qu, X. Shao, B. Kong and Y. Tian, *Angew. Chem., Int. Ed.*, 2012, **51**, 7185.
115. P. Huang, J. Lin, X. Wang, Z. Wang, C. Zhang, M. He, K. Wang, F. Chen, Z. Li, G. Shen, D. Cui and X. Chen, *Adv. Mater.*, 2012, **24**, 5104.
116. W. S. Hummers Jr. and R. E. Offeman, *J. Am. Chem. Soc.*, 1958, **80**, 1339.
117. F. Jiang, D. Chen, R. Li, Y. Wang, G. Zhang, S. Li, J. Zheng, N. Huang, Y. Gu, C. Wang and C. Shu, *Nanoscale*, 2013, **5**, 1137.
118. J. Peng, W. Gao, B. K. Gupta, Z. Liu, R. Romero-Aburto, L. Ge, L. Song, L. B. Alemany, X. Zhan, G. Gao, S. A. Vithayathil, B. A. Kaipparettu, A. A. Marti, T. Hayashi, J. J. Zhu and P. M. Ajayan, *Nano Lett.*, 2012, **12**, 844.
119. F. Liu, M. H. Jang, H. D. Ha, J. H. Kim, Y. H. Cho and T. S. Seo, *Adv. Mater.*, 2013, **25**, 3657.
120. S. Kim, S. W. Hwang, M.-K. Kim, D. Y. Shin, D. H. Shin, C. O. Kim, S. B. Yang, J. H. Park, E. Hwang, S.-H. Choi, G. Ko, S. Sim, C. Sone, H. J. Choi, S. Bae and B. H. Hong, *ACS Nano*, 2012, **6**, 8203.
121. X. T. Zheng, A. Than, A. Ananthanaraya, D.-H. Kim and P. Chen, *ACS Nano*, 2013, 77, 6276.
122. M. Nurunnabi, Z. Khatun, M. Nafiujjaman, D. G. Lee and Y. K. Lee, *ACS Appl. Mater. Interfaces*, 2013, **5**, 8246.
123. N. Abdullah Al, J. E. Lee, I. In, H. Lee, K. D. Lee, J. H. Jeong and S. Y. Park, *Mol. Pharm.*, 2013, **10**, 3736.
124. Y. Wang, L. Zhang, R. P. Liang, J. M. Bai and J. D. Qiu, *Anal. Chem.*, 2013, **85**, 9148.
125. M. Nurunnabi, Z. Khatun, K. M. Huh, S. Y. Park, D. Y. Lee, K. J. Cho and Y.-k. Lee, *ACS Nano*, 2013, 7, 6858.

126. Y. He, Y. Su, X. Yang, Z. Kang, T. Xu, R. Zhang, C. Fan and S.-T. Lee, *J. Am. Chem. Soc.*, 2009, **131**, 4434.

127. Y. Zhong, F. Peng, F. Bao, S. Wang, X. Ji, L. Yang, Y. Su, S.-T. Lee and Y. He, *J. Am. Chem. Soc.*, 2013, **135**, 8350.

128. Y. Zhong, F. Peng, X. Wei, Y. Zhou, J. Wang, X. Jiang, Y. Su, S. Su, S.-T. Lee and Y. He, *Angew. Chem., Int. Ed.*, 2012, **51**, 8485.

129. Z. Kang, Y. Liu, C. H. A. Tsang, D. D. D. Ma, X. Fan, N.-B. Wong and S.-T. Lee, *Adv. Mater.*, 2009, **21**, 661.

130. F. Erogbogbo, K.-T. Yong, I. Roy, R. Hu, W.-C. Law, W. Zhao, H. Ding, F. Wu, R. Kumar, M. T. Swihart and P. N. Prasad, *ACS Nano*, 2011, **5**, 413.

131. Y. Zhang, M. K. So, A. M. Loening, H. Yao, S. S. Gambhir and J. Rao, *Angew. Chem., Int. Ed.*, 2006, **45**, 4936.

132. M. K. So, C. Xu, A. M. Loening, S. S. Gambhir and J. Rao, *Nat. Biotechnol.*, 2006, **24**, 339.

133. H. Yao, Y. Zhang, F. Xiao, Z. Xia and J. Rao, *Angew. Chem., Int. Ed.*, 2007, **46**, 4346.

134. X. Huang, L. Li, H. Qian, C. Dong and J. Ren, *Angew. Chem., Int. Ed.*, 2006, **45**, 5140.

135. N. Kotagiri, D. M. Niedzwiedzki, K. Ohara and S. Achilefu, *Angew. Chem., Int. Ed.*, 2013, **52**, 7756.

136. H. Liu, X. Zhang, B. Xing, P. Han, S. S. Gambhir and Z. Cheng, *Small*, 2010, **6**, 1087.

137. R. S. Dothager, R. J. Goiffon, E. Jackson, S. Harpstrite and D. Piwnica-Worms, *PloS One*, 2010, **5**, e13300.

138. X. Sun, X. Huang, J. Guo, W. Zhu, Y. Ding, G. Niu, A. Wang, D. O. Kiesewetter, Z. L. Wang, S. Sun and X. Chen, *J. Am. Chem. Soc.*, 2014, **136**, 1706.

139. S. Wang, N. Mamedova, N. A. Kotov, W. Chen and J. Studer, *Nano Lett.*, 2002, **2**, 817.

140. A. Wolcott, D. Gerion, M. Visconte, J. Sun, A. Schwartzberg, S. Chen and J. Z. Zhang, *J. Phys. Chem. B*, 2006, **110**, 5779.

141. C. Schieber, A. Bestetti, J. P. Lim, A. D. Ryan, T.-L. Nguyen, R. Eldridge, A. R. White, P. A. Gleeson, P. S. Donnelly, S. J. Williams and P. Mulvaney, *Angew. Chem., Int. Ed.*, 2012, **51**, 10523.

142. G. Jiang, A. S. Susha, A. A. Lutich, F. D. Stefani, J. Feldmann and A. L. Rogach, *ACS Nano*, 2009, **3**, 4127.

143. A. R. Clapp, I. L. Medintz, J. M. Mauro, B. R. Fisher, M. G. Bawendi and H. Mattoussi, *J. Am. Chem. Soc.*, 2004, **126**, 301.

144. I. L. Medintz, A. R. Clapp, F. M. Brunel, T. Tiefenbrunn, H. T. Uyeda, E. L. Chang, J. R. Deschamps, P. E. Dawson and H. Mattoussi, *Nat. Mater.*, 2006, **5**, 581.

145. X. Wu, H. Liu, J. Liu, K. N. Haley, J. A. Treadway, J. P. Larson, N. Ge, F. Peale and M. P. Bruchez, *Nat. Biotechnol.*, 2003, **21**, 41.

146. J. K. Jaiswal, H. Mattoussi, J. M. Mauro and S. M. Simon, *Nat. Biotechnol.*, 2003, **21**, 47.

147. E. Zamir, P. H. Lommerse, A. Kinkhabwala, H. E. Grecco and P. I. Bastiaens, *Nat. Methods*, 2010, **7**, 295.
148. H. Kobayashi, Y. Hama, Y. Koyama, T. Barrett, C. A. S. Regino, Y. Urano and P. L. Choyke, *Nano Lett.*, 2007, **7**, 1711.
149. X. Gao, Y. Cui, R. M. Levenson, L. W. Chung and S. Nie, *Nat. Biotechnol.*, 2004, **22**, 969.
150. J. B. Delehanty, C. E. Bradburne, K. Susumu, K. Boeneman, B. C. Mei, D. Farrell, J. B. Blanco-Canosa, P. E. Dawson, H. Mattoussi and I. L. Medintz, *J. Am. Chem. Soc.*, 2011, **133**, 10482.
151. Y.-P. Ho, M. C. Kung, S. Yang and T.-H. Wang, *Nano Lett.*, 2005, **5**, 1693.
152. P. Zrazhevskiy and X. Gao, *Nat. Commun.*, 2013, **4**, 1619.
153. D. Ag, R. Bongartz, L. E. Dogan, M. Seleci, J. G. Walter, D. O. Demirkol, F. Stahl, S. Ozcelik, S. Timur and T. Scheper, *Colloids Surf., B*, 2014, **114**, 96.
154. R. Jeyadevi, T. Sivasudha, A. Rameshkumar, D. A. Ananth, G. S. Aseervatham, K. Kumaresan, L. D. Kumar, S. Jagadeeswari and R. Renganathan, *Colloids Surf., B*, 2013, **112**, 255.
155. A. A. P. Mansur, J. B. Saliba and H. S. Mansur, *Colloids Surf., B*, 2013, **111**, 60.
156. Y. Yu, L. Xu, J. Chen, H. Gao, S. Wang, J. Fang and S. Xu, *Colloids Surf., B*, 2012, **95**, 247.
157. A. Zajac, D. Song, W. Qian and T. Zhukov, *Colloids Surf., B*, 2007, **58**, 309.
158. Y. Zhang, A. Haage, E. M. Whitley, I. C. Schneider and A. R. Clapp, *Colloids Surf., B*, 2012, **94**, 27.
159. S. Barua and K. Rege, *Small*, 2009, **5**, 370.
160. S. Lee, K. J. Jung, H. S. Jung and S. Chang, *PloS One*, 2012, **7**, e38045.
161. B. Biermann, S. Sokoll, J. Klueva, M. Missler, J. S. Wiegert, J. B. Sibarita and M. Heine, *Nat. Commun.*, 2014, **5**, 3024.
162. Y. Choi, H. P. Kim, S. M. Hong, J. Y. Ryu, S. J. Han and R. Song, *Small*, 2009, **5**, 2085.
163. H. Y. Yeh, M. V. Yates, A. Mulchandani and W. Chen, *Chem. Commun.*, 2010, **46**, 3914.
164. M. L. Chen, Y. J. He, X. W. Chen and J. H. Wang, *Bioconjugate Chem.*, 2013, **24**, 387.
165. C. You, S. Wilmes, O. Beutel, S. Lochte, Y. Podoplelowa, F. Roder, C. Richter, T. Seine, D. Schaible, G. Uze, S. Clarke, F. Pinaud, M. Dahan and J. Piehler, *Angew. Chem., Int. Ed.*, 2010, **49**, 4108.
166. V. K. Sreenivasan, E. J. Kim, A. K. Goodchild, M. Connor and A. V. Zvyagin, *Nanomedicine*, 2012, **7**, 1551.
167. M. Z. Zhang, Y. Yu, R. N. Yu, M. Wan, R. Y. Zhang and Y. D. Zhao, *Small*, 2013, **9**, 4183.
168. M. P. Clausen and B. C. Lagerholm, *Nano Lett.*, 2013, **13**, 2332.
169. K. Yum, N. Wang and M. F. Yu, *Small*, 2010, **6**, 2109.
170. J. Xu, T. Teslaa, T. H. Wu, P. Y. Chiou, M. A. Teitell and S. Weiss, *Nano Lett.*, 2012, **12**, 5669.

171. L. Feng, H. Y. Long, R. K. Liu, D. N. Sun, C. Liu, L. L. Long, Y. Li, S. Chen and B. Xiao, *Cell. Mol. Neurobiol.*, 2013, **33**, 759.
172. K. Narayanan, S. K. Yen, Q. Dou, P. Padmanabhan, T. Sudhaharan, S. Ahmed, J. Y. Ying and S. T. Selvan, *Sci. Rep.*, 2013, **3**, 2184.
173. Y. Chen, M. Molnar, L. Li, P. Friberg, L.-M. Gan, H. Brismar and Y. Fu, *PloS One*, 2013, **8**, e83805.
174. H. S. Choi, W. Liu, F. Liu, K. Nasr, P. Misra, M. G. Bawendi and J. V. Frangioni, *Nat. Nanotechnol.*, 2010, **5**, 42.
175. L. Q. Chen, S. J. Xiao, P. P. Hu, L. Peng, J. Ma, L. F. Luo, Y. F. Li and C. Z. Huang, *Anal. Chem.*, 2012, **84**, 3099.
176. T. C. Chu, F. Shieh, L. A. Lavery, M. Levy, R. Richards-Kortum, B. A. Korgel and A. D. Ellington, *Biosens. Bioelectron.*, 2006, **21**, 1859.
177. O. C. Farokhzad, S. Jon, A. Khademhosseini, T.-N. T. Tran, D. A. LaVan and R. Langer, *Cancer Res.*, 2004, **64**, 7668.
178. Y. Zhang, J. M. Liu and X. P. Yan, *Anal. Chem.*, 2013, **85**, 228.
179. D. J. Bharali, D. W. Lucey, H. Jayakumar, H. E. Pudavar and P. N. Prasad, *J. Am. Chem. Soc.*, 2005, **127**, 11364.
180. E.-Q. Song, Z.-L. Zhang, Q.-Y. Luo, W. Lu, Y.-B. Shi and D.-W. Pang, *Clin. Chem.*, 2009, **55**, 955.
181. J. B. Delehanty, I. L. Medintz, T. Pons, F. M. Brunel, P. E. Dawson and H. Mattoussi, *Bioconjugate Chem.*, 2006, **17**, 920.
182. B. C. Lagerholm, M. Wang, L. A. Ernst, D. H. Ly, H. Liu, M. P. Bruchez and A. S. Waggoner, *Nano Lett.*, 2004, **4**, 2019.
183. A. M. Derfus, W. C. W. Chan and S. N. Bhatia, *Adv. Mater.*, 2004, **16**, 961.
184. G. Ruan, A. Agrawal, A. I. Marcus and S. Nie, *J. Am. Chem. Soc.*, 2007, **129**, 14759.
185. H. Duan and S. Nie, *J. Am. Chem. Soc.*, 2007, **129**, 3333.
186. M. V. Yezhelyev, L. Qi, R. M. O'Regan, S. Nie and X. Gao, *J. Am. Chem. Soc.*, 2008, **130**, 9006.
187. B. Y. S. Kim, W. Jiang, J. Oreopoulos, C. M. Yip, J. T. Rutka and W. C. W. Chan, *Nano Lett.*, 2008, **8**, 3887.
188. A. M. Smith and S. Nie, *J. Visualized Exp.*, 2012, **68**, e4236.
189. K. Baba and K. Nishida, *Theranostics*, 2012, **2**, 655.
190. F. Pinaud, S. Clarke, A. Sittner and M. Dahan, *Nat. Methods*, 2010, **7**, 275.
191. A. M. Derfus, W. C. W. Chan and S. N. Bhatia, *Nano Lett.*, 2004, **4**, 11.
192. K. M. Tsoi, Q. Dai, B. A. Alman and W. C. W. Chan, *Acc. Chem. Res.*, 2013, **46**, 662.
193. M. C. Mancini, B. A. Kairdolf, A. M. Smith and S. Nie, *J. Am. Chem. Soc.*, 2008, **130**, 10836.
194. B. I. Ipe, M. Lehnig and C. M. Niemeyer, *Small*, 2005, **1**, 706.
195. C. E. Bradburne, J. B. Delehanty, K. Boeneman Gemmill, B. C. Mei, H. Mattoussi, K. Susumu, J. B. Blanco-Canosa, P. E. Dawson and I. L. Medintz, *Bioconjugate Chem.*, 2013, **24**, 1570.

196. F. M. Winnik and D. Maysinger, *Acc. Chem. Res.*, 2013, **46**, 672.
197. N. Ma, A. F. Marshall, S. S. Gambhir and J. Rao, *Small*, 2010, **6**, 1520.
198. Y.-P. Gu, R. Cui, Z.-L. Zhang, Z.-X. Xie and D.-W. Pang, *J. Am. Chem. Soc.*, 2012, **134**, 79.
199. Y. Zhang, G. Hong, Y. Zhang, G. Chen, F. Li, H. Dai and Q. Wang, *ACS Nano*, 2012, **6**, 3695.
200. G. Hong, J. T. Robinson, Y. Zhang, S. Diao, A. L. Antaris, Q. Wang and H. Dai, *Angew. Chem., Int. Ed.*, 2012, **51**, 9818.
201. C. Tu, X. Ma, A. House, S. M. Kauzlarich and A. Y. Louie, *ACS Med. Chem. Lett.*, 2011, **2**, 285.
202. S. Kim, Y. T. Lim, E. G. Soltesz, A. M. De Grand, J. Lee, A. Nakayama, J. A. Parker, T. Mihaljevic, R. G. Laurence, D. M. Dor, L. H. Cohn, M. G. Bawendi and J. V. Frangioni, *Nat. Biotechnol.*, 2004, **22**, 93.
203. D. R. Larson, W. R. Zipfel, R. M. Williams, S. W. Clark, M. P. Bruchez, F. W. Wise and W. W. Webb, *Science*, 2003, **300**, 1434.
204. D. Peer, J. M. Karp, S. Hong, O. C. Farokhzad, R. Margalit and R. Langer, *Nat. Nanotechnol.*, 2007, **2**, 751.
205. B. Ballou, L. A. Ernst, S. Andreko, T. Harper, J. A. J. Fitzpatrick, A. S. Waggoner and M. P. Bruchez, *Bioconjugate Chem.*, 2007, **18**, 389.
206. E. L. Bentzen, I. D. Tomlinson, J. Mason, P. Gresch, M. R. Warnement, D. Wright, E. Sanders-Bush, R. Blakely and S. J. Rosenthal, *Bioconjugate Chem.*, 2005, **16**, 1488.
207. E. Muro, T. Pons, N. Lequeux, A. Fragola, N. Sanson, Z. Lenkei and B. Dubertret, *J. Am. Chem. Soc.*, 2010, **132**, 4556.
208. K. Susumu, E. Oh, J. B. Delehanty, J. B. Blanco-Canosa, B. J. Johnson, V. Jain, W. J. t. Hervey, W. R. Algar, K. Boeneman, P. E. Dawson and I. L. Medintz, *J. Am. Chem. Soc.*, 2011, **133**, 9480.
209. H. S. Choi, W. Liu, P. Misra, E. Tanaka, J. P. Zimmer, B. Itty Ipe, M. G. Bawendi and J. V. Frangioni, *Nat. Biotechnol.*, 2007, **25**, 1165.
210. X. Wang, J. Li, Y. Wang, K. J. Cho, G. Kim, A. Gjyrezi, L. Koenig, P. Giannakakou, H. J. C. Shin, M. Tighiouart, S. Nie, Z. (G.) Chen and D. M. Shin, *ACS Nano*, 2009, **3**, 3165.
211. W. Liu, A. B. Greytak, J. Lee, C. R. Wong, J. Park, L. F. Marshall, W. Jiang, P. N. Curtin, A. Y. Ting, D. G. Nocera, D. Fukumura, R. K. Jain and M. G. Bawendi, *J. Am. Chem. Soc.*, 2010, **132**, 472.
212. W. Liu, M. Howarth, A. B. Greytak, Y. Zheng, D. G. Nocera, A. Y. Ting and M. G. Bawendi, *J. Am. Chem. Soc.*, 2008, **130**, 1274.
213. A. M. Smith and S. Nie, *J. Am. Chem. Soc.*, 2008, **130**, 11278.
214. B. Dong, C. Li, G. Chen, Y. Zhang, Y. Zhang, M. Deng and Q. Wang, *Chem. Mater.*, 2013, **25**, 2503.
215. J. H. Yu, S.-H. Kwon, Z. Petrášek, O. K. Park, S. W. Jun, K. Shin, M. Choi, Y. Il Park, K. Park, H. B. Na, N. Lee, D. W. Lee, J. H. Kim, P. Schwille and T. Hyeon, *Nat. Mater.*, 2013, **12**, 359.
216. L. Li, Y. Chen, Q. Lu, J. Ji, Y. Shen, M. Xu, R. Fei, G. Yang, K. Zhang, J. R. Zhang and J. J. Zhu, *Sci. Rep.*, 2013, **3**, 1529.

217. B. Zhang, D. Tang, I. Y. Goryacheva, R. Niessner and D. Knopp, *Chem.-Eur. J.*, 2013, **19**, 2496.
218. X. Zhu, L. Chen, P. Shen, J. Jia, D. Zhang and L. Yang, *J. Agric. Food Chem.*, 2011, **59**, 2184.
219. J. Liu, T. Song, Q. Yang, J. Tan, D. Huang and J. Chang, *J. Mater. Chem. B*, 2013, **1**, 1156.
220. C. Zhang, D. Gao, G. Zhou, L. Chen, X. A. Zhang, Z. Cui and Z. He, *Chem.-Asian J.*, 2012, **7**, 1764.
221. K. D. Wegner, S. Linden, Z. Jin, T. L. Jennings, R. E. Khoulati, P. M. van Bergen En Henegouwen and N. Hildebrandt, *Small*, 2013, **10**, 734.
222. M. C. Tu, Y. T. Chang, Y. T. Kang, H. Y. Chang, P. Chang and T. R. Yew, *Biosens. Bioelectron.*, 2012, **34**, 286.
223. K. D. Wegner, Z. Jin, S. Linden, T. L. Jennings and N. Hildebrandt, *ACS Nano*, 2013, **7**, 7411.
224. X. Wang, G. Wang, W. Li, B. Zhao, B. Xing, Y. Leng, H. Dou, K. Sun, L. Shen, X. Yuan, J. Li, K. Sun, J. Han, H. Xiao, Y. Li, P. Huang and X. Chen, *Small*, 2013, **9**, 3327.
225. C. Roh and S. K. Jo, *J. Chem. Technol. Biotechnol.*, 2011, **86**, 1475.
226. C. W. Chi, Y. H. Lao, Y. S. Li and L. C. Chen, *Biosens. Bioelectron.*, 2011, **26**, 3346.
227. T. N. Tran, J. Cui, M. R. Hartman, S. Peng, H. Funabashi, F. Duan, D. Yang, J. C. March, J. T. Lis, H. Cui and D. Luo, *J. Am. Chem. Soc.*, 2013, **135**, 14008.
228. P. Wu, L.-N. Miao, H.-F. Wang, X.-G. Shao and X.-P. Yan, *Angew. Chem., Int. Ed.*, 2011, **50**, 8118.
229. Y. Luo, X. Liu, T. Jiang, P. Liao and W. Fu, *Anal. Chem.*, 2013, **85**, 8354.
230. J. Wang, H. Han, X. Jiang, L. Huang, L. Chen and N. Li, *Anal. Chem.*, 2012, **84**, 4893.
231. C. M. Tyrakowski and P. T. Snee, *Anal. Chem.*, 2014, **86**, 2380.
232. I. L. Medintz, A. R. Clapp, H. Mattoussi, E. R. Goldman, B. Fisher and J. M. Mauro, *Nat. Mater.*, 2003, **2**, 630.
233. K. Boeneman, B. C. Mei, A. M. Dennis, G. Bao, J. R. Deschamps, H. Mattoussi and I. L. Medintz, *J. Am. Chem. Soc.*, 2009, **131**, 3828.
234. K. E. Sapsford, J. Granek, J. R. Deschamps, K. Boeneman, J. B. Blanco-Canosa, P. E. Dawson, K. Susumu, M. H. Stewart and I. L. Medintz, *ACS Nano*, 2011, **5**, 2687.
235. S. B. Lowe, J. A. G. Dick, B. E. Cohen and M. M. Stevens, *ACS Nano*, 2012, **6**, 851.
236. D. E. Prasuhn, A. Feltz, J. B. Blanco-Canosa, K. Susumu, M. H. Stewart, B. C. Mei, A. V. Yakovlev, C. Loukou, J.-M. Mallet, M. Oheim, P. E. Dawson and I. L. Medintz, *ACS Nano*, 2010, **4**, 5487.
237. W. R. Algar, D. Wegner, A. L. Huston, J. B. Blanco-Canosa, M. H. Stewart, A. Armstrong, P. E. Dawson, N. Hildebrandt and I. L. Medintz, *J. Am. Chem. Soc.*, 2012, **134**, 1876.

238. A. Moquin, F. E. Hutter, A. O. Choi, A. Khatchadourian, A. Castonguay, F. M. Winnik and D. Maysinger, *ACS Nano*, 2013, 7, 9585.

239. W. R. Algar, M. G. Ancona, A. P. Malanoski, K. Susumu and I. L. Medintz, *ACS Nano*, 2012, 6, 11044.

240. J. Wang, Y. Shan, W. W. Zhao, J. J. Xu and H. Y. Chen, *Anal. Chem.*, 2011, 83, 4004.

241. Y. Huang, S. Zhao, M. Shi, J. Chen, Z. F. Chen and H. Liang, *Anal. Chem.*, 2011, 83, 8913.

242. W. Ma, L. X. Qin, F. T. Liu, Z. Gu, J. Wang, Z. G. Pan, T. D. James and Y. T. Long, *Sci. Rep.*, 2013, 3, 1537.

243. H. Zhang, L. Zhang, R. P. Liang, J. Huang and J. D. Qiu, *Anal. Chem.*, 2013, 85, 10969.

244. X. He and N. Ma, *Small*, 2013, 9, 2527.

245. J. Li, X. Li, X. Shi, X. He, W. Wei, N. Ma and H. Chen, *ACS Appl. Mater. Interfaces*, 2013, 5, 9798.

246. M. Han, X. Gao, J. Z. Su and S. Nie, *Nat. Biotechnol.*, 2001, 19, 631.

247. C.-Y. Zhang, H.-C. Yeh, M. T. Kuroki and T.-H. Wang, *Nat. Mater.*, 2005, 4, 826.

248. J. H. Kim, D. Morikis and M. Ozkan, *Sens. Actuators, B*, 2004, 102, 315.

249. C. S. Wu, M. K. Oo, J. M. Cupps and X. Fan, *Biosens. Bioelectron.*, 2011, 26, 3870.

250. L. Xu, Y. Zhu, W. Ma, H. Kuang, L. Liu, L. Wang and C. Xu, *J. Phys. Chem. C*, 2011, 115, 16315.

251. Y. Zhang and C.-Y. Zhang, *Anal. Chem.*, 2012, 84, 224.

252. Y. Song, Y. Zhang and T. H. Wang, *Small*, 2013, 9, 1096.

253. J. Zhou, Q.-X. Wang and C.-Y. Zhang, *J. Am. Chem. Soc.*, 2013, 135, 2056.

254. D.-S. Xiang, G.-P. Zeng and Z.-K. He, *Biosens. Bioelectron.*, 2011, 26, 4405.

255. J. Hu, C.-Y. Wen, Z.-L. Zhang, M. Xie, J. Hu, M. Wu and D.-W. Pang, *Anal. Chem.*, 2013, 85, 11929.

256. X. He and N. Ma, *Anal. Chem.*, 2014, 86, 3676.

257. X. He, Z. Li, M. Chen and N. Ma, *Angew. Chem., Int. Ed.*, 2014, 53, 14447.

258. R. Cui, Y.-P. Gu, L. Bao, J.-Y. Zhao, B.-P. Qi, Z.-L. Zhang, Z.-X. Xie and D.-W. Pang, *Anal. Chem.*, 2012, 84, 8932.

259. J. Yuan, D. Wen, N. Gaponik and A. Eychmuller, *Angew. Chem., Int. Ed.*, 2013, 52, 976.

260. Q. Ma, Y. Li, Z. H. Lin, G. Tang and X. G. Su, *Nanoscale*, 2013, 5, 9726.

261. C. Kong, D. W. Li, Y. Li, R. Partovi-Nia, T. D. James, Y. T. Long and H. Tian, *Analyst*, 2012, 137, 1094.

262. Y. T. Long, C. Kong, D. W. Li, Y. Li, S. Chowdhury and H. Tian, *Small*, 2011, 7, 1624.

263. G. Jie, J. Yuan and J. Zhang, *Biosens. Bioelectron.*, 2012, 31, 69.

264. H. Zhang, B. Jiang, Y. Xiang, Y. Zhang, Y. Chai and R. Yuan, *Anal. Chim. Acta*, 2011, 688, 99.

265. P. J. Cywinski, A. J. Moro and H. G. Lohmannsroben, *Biosens. Bioelectron.*, 2014, 52, 288.

266. R. Freeman, R. Gill, I. Shweky, M. Kotler, U. Banin and I. Willner, *Angew. Chem., Int. Ed.*, 2009, **48**, 309.

267. L. Bahshi, R. Freeman, R. Gill and I. Willner, *Small*, 2009, **5**, 676.

268. X. Huang, J. Wang, H. Liu, T. Lan and J. Ren, *Talanta*, 2013, **106**, 79.

269. F. Long, C. Gu, A. Z. Gu and H. Shi, *Anal. Chem.*, 2012, **84**, 3646.

270. B. Liu, C. Tong, L. Feng, C. Wang, Y. He and C. Lu, *Chem.–Eur. J.*, 2014, **20**, 2132.

271. P. Haro-Gonzalez, L. Martinez-Maestro, I. R. Martin, J. Garcia-Sole and D. Jaque, *Small*, 2012, **8**, 2652.

272. D. Zhou, M. Lin, X. Liu, J. Li, Z. Chen, D. Yao, H. Sun, H. Zhang and B. Yang, *ACS Nano*, 2013, **7**, 2273.

273. C.-H. Hsia, A. Wuttig and H. Yang, *ACS Nano*, 2011, **5**, 9511.

274. A. M. Dennis, W. J. Rhee, D. Sotto, S. N. Dublin and G. Bao, *ACS Nano*, 2012, **6**, 2917.

275. I. L. Medintz, M. H. Stewart, S. A. Trammell, K. Susumu, J. B. Delehanty, B. C. Mei, J. S. Melinger, J. B. Blanco-Canosa, P. E. Dawson and H. Mattoussi, *Nat. Mater.*, 2010, **9**, 676.

276. A. Orte, J. M. Alvarez-Pez and M. J. Ruedas-Rama, *ACS Nano*, 2013, **7**, 6387.

277. L. X. Qin, W. Ma, D. W. Li, Y. Li, X. Chen, H. B. Kraatz, T. D. James and Y. T. Long, *Chem.–Eur. J.*, 2011, **17**, 5262.

278. E. I. Zenkevich, E. I. Sagun, V. N. Knyukshto, A. S. Stasheuski, V. A. Galievsky, A. P. Stupak, T. Blaudeck and C. von Borczyskowski, *J. Phys. Chem. C*, 2011, **115**, 21535.

279. C. M. Lemon, E. Karnas, M. G. Bawendi and D. G. Nocera, *Inorg. Chem.*, 2013, **52**, 10394.

280. L. Tan, A. Wan and H. Li, *Langmuir*, 2013, **29**, 15032.

281. L. Li, J.-F. Zhao, N. Won, H. Jin, S. Kim and J.-Y. Chen, *Nanoscale Res. Lett.*, 2012, **7**, 386.

282. J.-Y. Chen, Y.-M. Lee, D. Zhao, N.-K. Mak, R. N.-S. Wong, W.-H. Chan and N.-H. Cheung, *Photochem. Photobiol.*, 2010, **86**, 431.

283. A. C. S. Samia, X. Chen and C. Burda, *J. Am. Chem. Soc.*, 2003, **125**, 15736.

284. C. Y. Hsu, C. W. Chen, H. P. Yu, Y. F. Lin and P. S. Lai, *Biomaterials*, 2013, **34**, 1204.

285. D. Wu, G. Song, Z. Li, T. Zhang, W. Wei, M. Chen, X. He and N. Ma, *Chem. Sci.*, 2015, **6**, 3839.

286. K.-L. Chou, N. Won, J. Kwag, S. Kim and J.-Y. Chen, *J. Mater. Chem. B*, 2013, **1**, 4584.

287. E. V. Shashkov, M. Everts, E. I. Galanzha and V. P. Zharov, *Nano Lett.*, 2008, **8**, 3953.

288. S. H. Hu, Y. W. Chen, W. T. Hung, I. W. Chen and S. Y. Chen, *Adv. Mater.*, 2012, **24**, 1748.

289. H. Absalan, A. SalmanOgli, R. Rostami and A. Maghoul, *J. Therm. Biol.*, 2012, **37**, 490.

290. M. Chu, X. Pan, D. Zhang, Q. Wu, J. Peng and W. Hai, *Biomaterials*, 2012, **33**, 7071.
291. J. Xie, S. Lee and X. Chen, *Adv. Drug Delivery Rev.*, 2010, **62**, 1064.
292. J. M. Li, M. X. Zhao, H. Su, Y. Y. Wang, C. P. Tan, L. N. Ji and Z. W. Mao, *Biomaterials*, 2011, **32**, 7978.
293. J. M. Li, Y. Y. Wang, M. X. Zhao, C. P. Tan, Y. Q. Li, X. Y. Le, L. N. Ji and Z. W. Mao, *Biomaterials*, 2012, **33**, 2780.
294. J. Qian and X. Gao, *ACS Appl. Mater. Interfaces*, 2013, **5**, 2845.
295. S. Marrachea and S. Dhara, *Proc. Natl. Acad. Sci. U. S. A.*, 2012, **109**, 16288.
296. J. H. Kim, Y.-W. Noh, M. B. Heo, M. Y. Cho and Y. T. Lim, *Angew. Chem., Int. Ed.*, 2012, **51**, 9670.
297. B. Yang, Y. Li, X. Sun, X. Meng, P. Chen and N. Liu, *J. Chem. Technol. Biotechnol.*, 2013, **88**, 2169.
298. Y. Chen, H. Chen, S. Zhang, F. Chen, L. Zhang, J. Zhang, M. Zhu, H. Wu, L. Guo, J. Feng and J. Shi, *Adv. Funct. Mater.*, 2011, **21**, 270.
299. Y. S. Cho, G. Y. Lee, H. K. Sajja, W. Qian, Z. Cao, W. He, P. Karna, X. Chen, H. Mao, Y. A. Wang and L. Yang, *Small*, 2013, **9**, 1964.
300. M. X. Zhao, J. M. Li, L. Du, C. P. Tan, Q. Xia, Z. W. Mao and L. N. Ji, *Chem.–Eur. J.*, 2011, **17**, 5171.
301. V. Bagalkot and X. Gao, *ACS Nano*, 2011, **5**, 8131.
302. J. Jung, A. Solanki, K. A. Memoli, K. Kamei, H. Kim, M. A. Drahl, L. J. Williams, H. R. Tseng and K. Lee, *Angew. Chem., Int. Ed.*, 2010, **49**, 103.
303. Z.-Y. Zhang, Y.-D. Xu, Y.-Y. Ma, L.-L. Qiu, Y. Wang, J.-L. Kong and H.-M. Xiong, *Angew. Chem. Int. Ed.*, 2013, **52**, 4127.
304. V. Bagalkot, L. Zhang, E. Levy-Nissenbaum, S. Jon, P. W. Kantoff, R. Langer and O. C. Farokhzad, *Nano Lett.*, 2007, **7**, 3065.
305. Q. Jiang, C. Song, J. Nangreave, X. Liu, L. Lin, D. Qiu, Z. G. Wang, G. Zou, X. Liang, H. Yan and B. Ding, *J. Am. Chem. Soc.*, 2012, **134**, 13396.
306. L. Y. Chou, K. Zagorovsky and W. C. Chan, *Nat. Nanotechnol.*, 2014, **9**, 148.

CHAPTER 7

Upconversion Nanomaterials for Photodynamic Therapy

MARÍA J. MARÍN[a] AND DAVID A. RUSSELL*[a]

[a]School of Chemistry, University of East Anglia, Norwich Research Park, Norwich, NR4 7TJ, UK
*E-mail: D.Russell@uea.ac.uk

7.1   Introduction

Photodynamic therapy (PDT) is a treatment for cancer that uses the reactive oxygen species (ROS) generated by a photosensitiser (PS) drug following irradiation of a specific wavelength to destroy the cancerous tumour.[1] Examples of PS drugs reported as effective for PDT treatment of cancer include Rose Bengal (RB), chlorin e6 (Ce6), and porphyrin and phthalocyanine derivatives, amongst other organic and inorganic compounds.[2] PDT has been proven to show excellent clinical treatment outcomes for numerous cancer types. However, the currently used PSs have some drawbacks when used for PDT including non-specific distribution of the drug in the body and the hydrophobicity of some of the PS drugs, which require the use of delivery systems for their administration.[3]

The number of publications reporting the use of nanoparticle formulations as delivery vehicles to overcome the drawbacks of non-specificity and hydrophobicity of molecular PSs has increased dramatically during the past 10–15 years.[4] Nanoparticles can be functionalised with multiple ligands including a PS-based ligand and ligands that make the system water

RSC Nanoscience & Nanotechnology No. 40
Near Infrared Nanomaterials: Preparation, Bioimaging and Therapy Applications
Edited by Fan Zhang
© The Royal Society of Chemistry 2016
Published by the Royal Society of Chemistry, www.rsc.org

soluble. Furthermore, targeting agents such as folic acid (FA), antibodies, or lectins that recognise specific receptors on the surface of tumour cells can also be used to functionalise the nanoparticle's surface and thus render the nanosystem specific towards cancerous tissues. This specificity can also be enhanced when using nanoparticles since they accumulate in tumour tissue due the vascular enhanced permeability and retention (EPR) effect.[5] The EPR effect is characteristic of solid tumours and does not occur in normal tissue. Tumour tissues require large amounts of nutrients and oxygen for rapid growth. These demands are met by the rapid formation of new blood vessels that present anatomical defects and functional abnormalities. Tumour blood vessels are leaky, which induces the accumulation of blood plasma components, including macromolecules or nanoparticles, in the tumour tissue.[6-8] Thus, the passive accumulation of nanoparticles in the tumour increases due to the EPR effect and this phenomenon is currently used for tumour targeting in the development of new anticancer drugs.[9,10] Nanoparticles with photosensitising properties as well as liposomes, micelles, polymer-based nanoparticles, lipoprotein nanoparticles, inorganic nanoparticles, and hybrid nanoparticles containing PS drugs have been reported for both *in vitro* and *in vivo* PDT of cancer.[4,11] We have investigated the efficiency of phthalocyanine-functionalised gold nanoparticles for the *in vitro* and *in vivo* treatment of cancer using PDT.[12-18]

A possible drawback of PSs, either molecular or as part of a nanosystem, is that to generate ROS they must be excited with light, typically of visible wavelengths. Such visible light may have limited penetration depth within human tissue and so prevent the treatment of deeper tumours. Upconversion nanomaterials (UCNPs) have been reported as ideal delivery systems to overcome the 'visible light' drawback since the nanomaterials are excited by near infrared (NIR) light and then emit light of ultraviolet (UV) to visible wavelengths to excite the PS on the nanosystem. Being able to use NIR excitation for PDT of cancer is advantageous over visible light since NIR light shows minimal photodamage, exhibits low autofluorescence, and most importantly, can penetrate deeper into the tissue. Furthermore, the use of nanoparticles can overcome the problems of hydrophobicity and non-specificity towards cancerous tissues, problems exhibited by some molecular PS drugs.

Considerable efforts have been made recently to improve the synthesis and modification of UCNPs to ensure they are biocompatible and to increase their potential for biomedical applications.[19-22] The number of published articles reporting the use of UCNPs for PDT of cancer has increased dramatically over the last decade.[9,23-27] This chapter reviews the state-of-the art in the use of UCNP for PDT including both *in vitro* and *in vivo* applications.

7.2 Proof of Concept

Upconversion nanomaterials are able to absorb infrared irradiation and to emit visible light of different wavelengths. This emitted light can then be used to further excite specific PS molecules present near the upconversion nanomaterial (see Figure 7.1) resulting in the generation of ROS that destroy the cancer tissue. Using the light emitted by the UCNPs is an indirect way of exciting the PSs with a NIR light source that is able to penetrate deep into the tissue producing minimal photodamage and autofluorescence in the treated samples. Furthermore, the fact that nanomaterials are used (size <100 nm) facilitates the specific delivery of the UCNPs to the cancer cells. Several parameters have to be taken into account when designing UCNPs for PDT applications, *i.e.*:

- The luminescence spectra of UCNPs consist of sharp emission lines of different wavelengths. These spectra can be tailor-made using different rare earth dopants during the synthesis of the nanoparticles to ensure that the emission wavelengths coincide with the absorption of the PS.
- The functionalisation of the UCNPs with the PS drugs can be achieved following different protocols including: (1) the PS can be embedded within a silica layer surrounding the UCNPs; (2) the PS can be covalently attached to the surface of the nanoparticle *via* a spacer; (3) the PS can be retained near the UCNPs surface *via* hydrophobic interactions; and (4) a layer of PS material can be used to cover the UCNPs. See Figure 7.1.
- The upconversion efficiency and energy transfer from the UCNP (donor) to the PS (acceptor) has to be high to ensure the production of ROS in high yields.

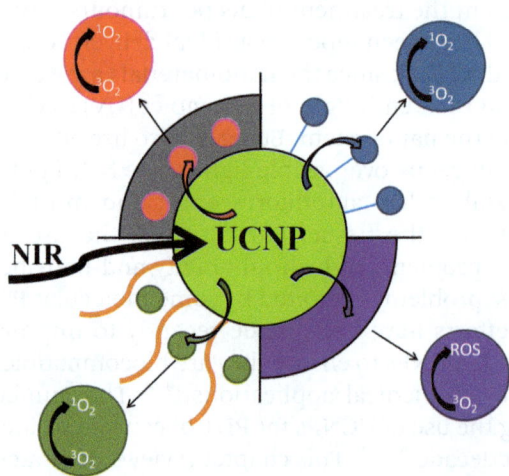

Figure 7.1   Schematic representation of the different protocols that allow the functionalisation of UCNPs with PS drugs (silica coverage, top left; covalent linkage, top right; hydrophobic interactions, bottom left; shell of PS material, bottom right). Following NIR irradiation of the UCNPs, the emitted light, of varying wavelength depending on the composition of the nanomaterial, is absorbed by the PS drug and the production of singlet oxygen or other ROS is activated.

- The nanoparticles should be biocompatible. This requires a matrix that has to be permeable to water and oxygen and has to permit the ROS to diffuse out of the matrix. Water solubility can be imparted to the nanoparticles *e.g.* by using a silica layer, polyethyleneimine, polyethylene glycol derivatives, or chitosan.
- The particles have to be delivered specifically to cancer cells without damaging healthy tissues following NIR irradiation. Specific targeting agents can be attached to the nanoparticles, such as FA or antibodies.
- The PS should not leak from the nanosystem in the biological environment.

The characterisation of UCNPs functionalised with PS drugs involves several parameters. The size of the UCNPs is measured using transmission electron microscopy (TEM) (Figure 7.2(a)) or dynamic light scattering (DLS). The upconversion luminescence of the UCNPs following NIR irradiation can be observed by eye in the dark (Figure 7.2(b)). The presence of the PS drugs on the functionalised UCNPs can be confirmed by a colour change in the solution, by the presence of the PS absorption bands in the absorption spectrum of the functionalised UCNPs (Figure 7.2(c)), or by using techniques such as nuclear magnetic resonance (NMR). The energy transfer between the UCNPs and the PS can be characterised by obtaining the steady-state upconversion luminescence (UCL) spectrum (Figure 7.2(d)) or measuring the changes in the luminescent decay lifetimes. Successful UCNPs for PDT should be able to produce ROS upon NIR irradiation. The singlet oxygen ($^1O_2$) production or the production of other ROS following NIR irradiation can be assessed using spectroscopic measurements of the singlet oxygen luminescence at 1270 nm or by using specific probes, the absorbance or fluorescence emission of which changes with the presence of ROS.

The first reports describing the potential use of UCNPs for PDT were published by Zhang *et al.* in 2007 (Figure 7.3).[28,29] Since then this field of research has received extensive attention. In the pioneering work of Zhang *et al.*, sodium yttrium fluoride ($NaYF_4$) nanoparticles codoped with ytterbium (Yb^{3+}) and erbium (Er^{3+}) of 60–120 nm were used as carriers of PS drugs.[28] Upon excitation at 974 nm, the nanoparticles emitted light at two wavelengths, ~537 and 635 nm. The $NaYF_4$:Yb,Er UCNPs were coated with a layer of silica doped with the PS merocyanine 540 (M-540) which did not affect the photon upconversion properties of the nanoparticles. The M-540–UCNPs were able to produce $^1O_2$ in a buffered solution following irradiation at 974 nm which was investigated using the disodium salt of 9,10-anthracenedipropionic acid (ADPA). The photodynamic cytotoxicity of the M-540–UCNPs was investigated using MCF-7/AZ breast cancer cells. To provide targeted uptake of the nanoparticles by the cancer cells, the M-540–UCNPs were further functionalised with a mouse monoclonal antibody, anti-MUC1/episialin, highly specific towards the MUC1 receptors overexpressed in MCF-7/AZ cells. MCF-7/AZ cells were incubated with the antibody-M-540–UCNPs and irradiated at 974 nm for 36 min. After this period, the cells displayed cell morphology

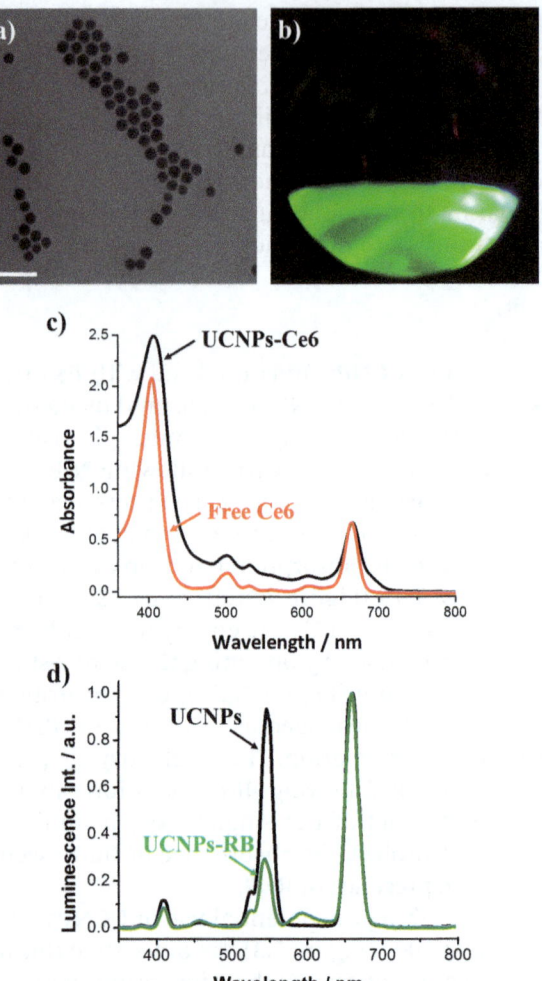

Figure 7.2 (a) TEM image of LiNaYF$_4$:Yb,Er UCNPs; scale bar = 100 nm. (b) A solution of LiNaYF$_4$:Yb,Er UCNPs irradiated with a 980 nm laser showing the green luminescence of the UCNPs. (c) Absorption spectrum of free chlorin e6 (red) and of LiNaYF$_4$:Yb,Er UCNPs covered with a silica layer where chlorin e6 is embedded (black). (d) UCL spectrum of silica-coated LiNaYF$_4$:Yb,Er UCNPs (black) and of LiNaYF$_4$:Yb,Er UCNP covered with a silica layer where Rose Bengal is embedded.

alterations, indicative of cell death, and were subsequently stained with markers of cell death such as trypan blue. Propidium iodide (PI) was also used to indicate cell death 43 min after irradiation of the UCNPs.[28] Zhang *et al.* also reported the potential use of 30–60 nm NaYF$_4$:Yb,Tm UCNPs coated with a layer of tris(bipyridine)ruthenium(II)-doped (Ru(bpy)$_3$$^{2+}$-doped) silica for PDT.[29] The UCNPs emitted blue light (~477 nm) upon irradiation at 975 nm and this photon upconversion property was not affected by the

Ru(bpy)$_3^{2+}$-doped silica layer. The functionalised UCNPs were found to generate $^1O_2$ in buffered solutions upon irradiation by a light source at 450 nm (direct illumination of the Ru(bpy)$_3^{2+}$) and at 975 nm (indirect illumination of the PS drug). The $^1O_2$ generation following irradiation with either light source was comparable.[29] The authors of these initial reports suggested that UCNP–PS nanosystems would be promising in PDT of cancer, which has been subsequently confirmed by the large number of publications now reported in this field of research. This chapter highlights the most recent work in the use of UCNPs for applications both *in vitro* and *in vivo* PDT of cancer.

Ways of improving the $^1O_2$ generation of RB–UCNPs have been investigated by Zhang and coworkers.[30,31] One of the major research interests in the field of UCNPs is to increase the UCL intensity of the UCNPs, which can be achieved by modifying the properties or the number of luminescence centres. Zhang and coworkers considered that increasing the number of emitters was the most efficient way to increase the UCL intensity.[30] This work broke away from the well-accepted optimal dopant concentration of 2 mol% for emitter Er^{3+} (in NaYF$_4$:Yb,Er UCNPs), achieving 5 mol% for Er^{3+} and resulting in an enhancement of the upconversion intensity. The strategy followed to achieve this enhancement was the spatial separation of emitter doping areas with the final nanosystem containing four parts: the core (NaYF$_4$:Yb,Er), a first separating shell (NaYF$_4$:Yb), a second illuminating shell (NaYF$_4$:Er), and finally, an active shell (NaYF$_4$:Yb). When the 5% Er^{3+}-doped separated UCNP were functionalised with the PS RB, greater singlet oxygen production following irradiation with a 980 nm laser was observed using 1,3-diphenylisobenzofuran (DPBF) as compared with the well-accepted 2% Er^{3+}-doped separated or non-separated RB–UCNPs. These results indicated that the singlet oxygen production of PS–UCNPs can be improved using spatial separation of the emitter doping area.[30] Using a similar RB-functionalised NaYF$_4$:Yb,Er@NaYF$_4$ nanosystem, Wang *et al.* determined that using core–shell structured nanoparticles is advantageous over naked cores for Förster (fluorescence)

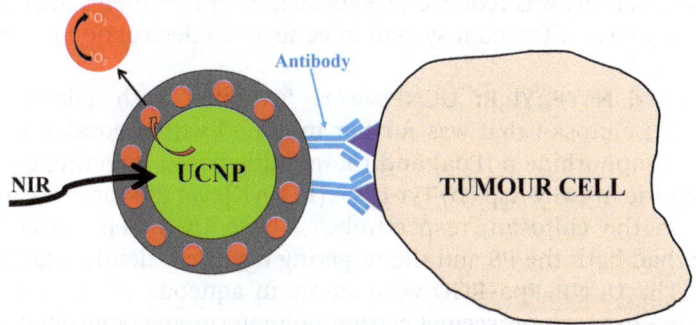

Figure 7.3   Schematic representation of the first PS-functionalised upconversion nanosystem (PS–UCNPs) reported for PDT. Figure adapted with permission from P. Zhang, *et al.*, *J. Am. Chem. Soc.*, 2007, **129**, 4526. Copyright (2007) American Chemical Society.[28]

resonance energy transfer (FRET) applications including for singlet oxygen production. This investigation concluded that nanoparticles with a core of 16 nm and a critical shell thickness of 6 nm were optimal for high singlet oxygen production (determined using DFBF following 980 nm excitation). The optimum shell thickness for singlet oxygen production was not the optimum for FRET efficiency, or for UCL. These results indicate that, the optimum core thickness is acceptor and core dependent and these parameters should be taken into consideration when designing upconversion-based nanotools for PDT.[31]

7.3 *In vitro* Applications of Upconversion Nanomaterials for PDT

7.3.1 Stability of UCNPs in Biological Media Achieved Using Polyethyleneimide

A quantitative assessment of the effectiveness of PS–UCNP systems for the targeted destruction of cancer cells was first reported by Chatterjee and Zhang using polyethyleneimine (PEI)-coated $NaYF_4$:Yb,Er UCNPs (~50 nm).[32] The PEI coat provided solubility to the system in physiological buffers and functionalised the UCNPs with amine groups for attachment of the cancer cell targeting molecule FA. To enable the use of the nanosystem for PDT, a zinc phthalocyanine (ZnPc) was physically adsorbed onto the surface of the nanoparticles with a reported encapsulation efficiency of ~97%. The singlet oxygen production following irradiation at 980 nm was determined by the photobleaching of ADPA. The potential of these ZnPc–UCNPs for PDT was evaluated *in vitro* using HT-29 human colon adenocarcinoma cells. HT-29 cells were incubated with ZnPc–UCNPs for 24 h, irradiated for 30 min with a 980 nm laser and the cell viability determined 48 h following irradiation using a 3-(4,5-dimethylthiazol-2-yl)-2,5-diphenyltetrazolium bromide (MTT) assay. Cell viability was reduced by ~80–90% after treatment, thus indicating the effectiveness of the nanosystem in cancer cell destruction following NIR irradiation.[32]

PEI-coated $NaYF_4$:Yb,Er UCNPs were 'wrapped' with a layer of *O*-carboxymethyl chitosan that was further modified with a covalently attached PS pyropheophorbide a (Ppa) and the peptide-based targeting agent cyclic pentapeptide Arg-Gly-Asp-(D)-Tyr-Lys (c(RGDyK)) *via* the carboxyl and amino groups on the chitosan, respectively.[33] These UCNP–Ppa–RGD particles, ~50 nm, had both the PS and the targeting agent covalently attached to the UCNPs. The UCNP–Ppa–RGD were stable in aqueous solution at different pH (from 5 to 8). Fluoresceinyl cypridina luciferin analogue chemiluminescence was used to confirm the ability of the nanoparticles to generate $^1O_2$. However, the excitation of the Ppa was performed directly using 635 nm excitation and not indirectly with NIR excitation. cRGDyK on the nanoparticles was used to selectively bind $\alpha_v\beta_3$ integrin, which plays an important role in tumour angiogenesis. The binding specificity of UCNP–Ppa–RGD was confirmed by incubating U87-MG human glioblastoma cancer cells (high $\alpha_v\beta_3$ integrin expression) and MCF-7 human breast cancer cells (low $\alpha_v\beta_3$ integrin expression) with the nanoparticles and then imaging the treated cells using a confocal laser scanning microscope (CLSM) exciting the Ppa directly with a 633 nm laser. Treated U87-MG cells exhibited a higher fluorescence than the treated MCF-7 cells, suggesting that the UCNP–Ppa–RGD were internalised *via* receptor-mediated endocytosis. Cell viability studies were performed with both U87-MG and MCF-7 cells treated with UCNP–Ppa–RGD using a CCK8 assay. Only the viability of U87-MG cancer cells decreased by 30–50% following NIR irradiation, confirming the strong targeting specificity and thus selective killing of the designed nanosystem.[33]

7.3.2 Stability of UCNPs in Biological Media Achieved Using Polyethylene Glycol Derivatives

Coating the UCNPs with a polyethylene glycol (PEG) derivative has also been used as a strategy to ensure biocompatibility of the nanoparticles, as first reported by Ungun *et al.*[34] NaYF$_4$:Er UCNPs with an intermediate coat of tetraphenylporphyrin (TPP) PS and an outer layer of PEG-*block*-poly(caprolactone) (PEG-*b*-PCL) of 100–300 nm were reported and the generation of $^1O_2$ by the nanosystem following NIR irradiation was confirmed.[34] The use of PEG for biocompatibility and to allow further chemical modification was suggested as research that could potentially be undertaken in the area of PDT using UCNPs.[34] The initial study by Ungun *et al.* was extended through the use of PEG-*block*-poly(DL)lactide (PEG-*b*-PLA) to stabilise the UCNPs rather than PEG-*b*-PCL since, according to the authors, PLA provided greater stability of the nanoparticles in serum.[35] 100 nm β-NaYF$_4$:Yb,Er UCNPs were coated with TPP (10 wt% loading) and stabilised with PEG-*b*-PLA. The functionalised particles were stable at 4 °C in deionised water, phosphate buffer (PB), and culture medium containing proteins for at least 3 months as indicated by DLS experiments. The efficacy of the functionalised UCNPs for PDT was investigated using HeLa cervical cancer cells. The nanoparticles exhibited no dark toxicity, but following irradiation for 45 min, significant cell death was observed by staining the cells. Quantification of cell death lead to values of 75% and 14% cell death following NIR irradiation with laser powers of 134 and 39 W cm^{-2}, respectively.[35]

A FA–PEG ester was used by Liu *et al.* to increase the targeting efficacy of NaYF$_4$:Yb,Er UCNPs in cancer cells.[36] Amino-functionalised UCNPs (~20 nm) were synthesised *via* ligand exchange where oleylamine ligands were replaced by 2-aminoethyl dihydrogen phosphate. The free amino groups were used to covalently attach the PS RB yielding UCNPs–RB (~100 molecules of RB per nanoparticle). The production of $^1O_2$ by the UCNPs–RB was investigated by measuring the photobleaching of DPBF following irradiation of the nanoparticles at 980 nm which showed a 50% consumption of DPBF after 16 min irradiation. The authors reported that the singlet oxygen production efficiency was superior to that of silica-shelled UCNPs where only 3% of the DPBF was consumed after 16 min NIR irradiation. The singlet oxygen production was also directly monitored recording the characteristic phosphorescence of $^1O_2$ at 1270 nm following NIR irradiation as well as the direct excitation of the RB. This was the first report of the direct detection of $^1O_2$ phosphorescence from a PS excited by the energy transfer from UCNPs stimulated with NIR irradiation. Steady-state UCL spectra and luminescent decay lifetimes confirmed the selective energy transfer from the UCNPs to the RB. Subsequently, the UCNPs–RB were further functionalised with a FA–PEG ester and its targeting efficacy was evaluated using JAR choriocarcinoma cells (with or without saturation with free FA) and the control NIH 3T3 non-cancerous cells. Following 980 nm irradiation in a CLSM, the luminescence from the UCNPs was only observed in JAR choriocarcinoma cells when their folate receptors (FR) were not saturated. Furthermore, cell viability experiments performed by MTT assay confirmed the photodynamic effect of the nanoparticles in JAR cells following NIR irradiation.[36]

Pérez-Prieto and coworkers employed PEG to provide water solubility to their β-NaYF$_4$:Yb,Er UCNPs, ~30 nm, that were further functionalised with a BODIPY derivative PS anchored to the surface of the nanoparticle *via* the carboxylate group and embedded in the PEG capping.[37] The reported BODIPY–PEG–UCNPs were stable both in water and in PB for 3 months and in cell culture medium for at least 1 week. The ability of the BODIPY–PEG–UCNPs to produce $^1O_2$ following irradiation at 975 nm was confirmed with a 0.61 singlet oxygen quantum yield (QY) in methanol. The efficacy of the BODIPY–PEG–UCNPs to produce $^1O_2$ following irradiation at 975 nm was also investigated chemically measuring the photoconsumption of ABDA (9,10-anthracenediyl-bis(methylene) dimalonic acid) in D_2O. A decrease in the emission of ABDA of ~50% was observed after 15 min irradiation. The cytotoxicity of the BODIPY–PEG–UCNPs was determined using human neuroblastoma SH-SY5Y cells using a XTT-assay which, in the dark, reported a cell viability of treated cells of ~90%. Following 45 min NIR irradiation of the cells treated with BODIPY–PEG–UCNP, a cell mortality of ~50% was estimated using a Cell LIVE/DEAD® Kit. The signs of cellular damage following irradiation of the treated cells were confirmed by toluidine blue-staining which showed important formation of cytosolic vacuoles, swelling, irregular nuclear membranes, picnotic cells, and large masses of heterogenous matter together with empty vacuole-like spaces.[37]

Multifunctional UCNPs coated with a PEG–phospholipid layer for aqueous solubility were reported by Chen *et al.*[38] Core–shell NaYF$_4$:Yb,Tm–NaYF$_4$:Yb,Er UCNPs (~31 nm) codoped with Er^{3+} and Tm^{3+} were chosen due to their multiple emissions at 539, 654, and 802 nm. Under 980 nm excitation, the NIR-to-visible UCL was used to excite the PS Ce6 physically adsorbed on the hydrophobic layer on the nanoparticles and therefore, to produce $^1O_2$. The NIR-to-NIR UCL was used for NIR imaging. The loading efficiency of Ce6 on the UCNP was estimated to be 23% (w/w) and no leakage of the PS was observed when the particles were kept in PB for up to 7 days. The production of $^1O_2$ by the

Ce6–UCNPs was determined by measuring the photobleaching of DPBF. The *in vitro* PDT effect of the Ce6–UCNPs was investigated using the liver cancer cells QGY-7703, and through the use of trypan as a marker of membrane damage.[38]

The potential use of PS-conjugated PEG-coated core–shell Fe$_3$O$_4$@NaYF$_4$:-Yb,Er UCNPs (~50 nm) for T2-weighted magnetic resonance imaging (MRI), UCL imaging, and PDT was reported by Zeng *et al.*[39] The PS tetrasulfonic aluminium phthalocyanine (AlPcS$_4$) was conjugated to the UCNPs *via* electrostatic interactions. Although the singlet oxygen production by the AlPcS$_4$–UCNPs was not investigated in solution, the potential use of the nanoparticles for PDT following NIR irradiation was evaluated in MCF-7 cells yielding good PDT efficacy and low cytotoxicity as measured by the MTT assay.[39]

Multiple functionality in UCNPs has been reported by Yuan *et al.*, achieving the triple combination of targeted delivery, PDT, and chemotherapy in a single nanosystem (Figure 7.4).[40] The polymeric PS poly[9,9-bis(*N*-(but-3′-ynyl)-*N*,*N*-dimethylamino)hexyl fluorenyl divinylene-*alt*-4,7-(2′,1′,3′-benzothiadiazole) dibromide (PFVBT) was attached to an α-azide-ω-carboxyl-poly(ethylene glycol) *via* the alkyne group in PFVBT reacting with the azide. The same reaction took place between the PFVBT and the azide derivative drug doxorubicin (DOX) modified with the UV-cleavable *ortho*-nitrobenzyl (NB) linker. These reactions led to a conjugated polyelectrolyte–drug conjugate (CPE–DOX) that was used to encapsulate NaYF$_4$:Yb,Tm UCNPs (~35 nm) and to provide water

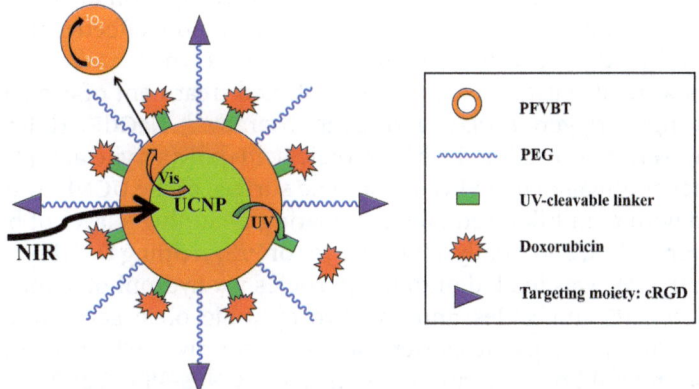

Figure 7.4 Schematic representation of the UCNPs@CPE–DOX. The UCNPs are encapsulated in a conjugated polyelectrolyte-drug system that contains the PS poly[9,9-bis(*N*-(but-3′-ynyl)-*N*,*N*-dimethylamino)hexyl fluorenyl divinylene-*alt*-4,7-(2′,1′,3′-benzothiadiazole)dibromide (PFVBT), PEG, and the chemotherapy drug doxorubicin (DOX) attached to a UV-cleavable linker. Following NIR irradiation, the PFVBT absorbs the emitted visible light and produces singlet oxygen while the UV-cleavable linker absorbs the emitted UV light to release the DOX. A targeting moiety, cyclic arginine-glycine-aspartic acid (cRGD), is also attached to the PEG to ensure specificity of the nanosystems towards certain cancer cells. Figure adapted from Yuan *et al.*[40] with permission from the Royal Society of Chemistry.

solubility to the system. The UCNPs@CPE–DOX nanoparticles were dispersible in water, PB, and cell culture medium (DMEM) for at least 7 days at 37 °C. Excitation of the particles with NIR irradiation resulted in UV light emitted at ~350 nm which overlaps with the absorption of *ortho*-nitrobenzyl and can be used to release the doxorubicin; the visible light emitted by the UCNPs following NIR irradiation was used to excite the PS PFVBT to induce the production of $^1O_2$. The ability of the UCNPs@CPE–DOX to produce $^1O_2$ following 980 nm irradiation was determined using dichlorodihydrofluorescein diacetate (DCFH-DA) and by measuring the enhancement in fluorescence of the emissive form dichlorofluorescein. 980 nm irradiation also triggered the release of doxorubicin in a laser power density related manner. To achieve a targeted delivery of the designed system, the nanoparticles were further functionalised with cyclic arginine-glycine-aspartic acid (cRGD) tripeptide for targeting cancer cells expressing integrin $\alpha_v\beta_3$. The targeting ability of the nanoparticles was investigated with two types of cancer cells that expressed different levels of integrin $\alpha_v\beta_3$: U87-MG (high levels) and MCF-7 (low levels). CLSM experiments showed higher fluorescence in the U87-MG cells than the MCF-7 cells, confirming the targeting of the cancer cells expressing high levels of integrin $\alpha_v\beta_3$. The targeting ability of the reported nanoparticles was further confirmed by performing semiquantitative analysis by flow cytometry. The UCNPs@CPE–DOX had a low dark toxicity as determined by MTT assay and the level of cell apoptosis increased following irradiation as demonstrated by FITC-tagged Annexin V assay. This investigation also proved that combined PDT and chemotherapy treatment was more effective in inducing cell death than the individual treatments on their own.[40]

UCNPs were also used to achieve a combined treatment of cancer including NIR light-triggered PDT and gene therapy.[41] NaGdF$_4$:Yb,Er UCNPs (~30 nm) were first modified with an oleic acid–poly(acrylic acid) (OA–PAA) copolymer conjugated to a PEG ligand. The surface of the UCNPs was further modified with a multilayered polymer coating including a layer of branched PEI, a layer of PAA, and finally another layer of PEI, yielding UCNPs–PEG@2×-PEI. Finally, Ce6 was loaded onto the particles *via* hydrophobic interactions between the PS molecules and the hydrophobic oleic acid layer on the UCNP's surface. The particles were stable with only 4–6% of Ce6 released in PB and in fetal bovine serum for 24 h. The UCNPs–PEG@2×PEI-Ce6 were able to generate $^1O_2$ following 980 nm irradiation as determined using the singlet oxygen sensor green (SOSG). To achieve gene therapy to silence the Polo-like kinase 1 (Plk1) oncogene overexpressed in many types of cancer cells and thus trigger cell apoptosis, a small interfering RNA (siRNA) that targets Plk1 was loaded onto the UCNPs–PEG@2×PEI-Ce6 nanoparticles (UCNPs–PEG@2×PEI-Ce6-siRNA). When HeLa cells were incubated with the UCNPs–PEG@2×PEI-Ce6-siRNA nanoparticles and imaged following 980 nm irradiation, the fluorescence emissions due to Ce6, fluorescently labelled siRNA, and UCNPs were shown to be colocalised within the cells confirming that both Ce6 and siRNA were together on the UCNPs within the cellular environment. Western blotting experiments showed that the cells treated with UCNPs–PEG@2×PEI-Ce6-siRNA experienced a Plk1 downregulation that was also confirmed by RT-qPCR. This gene silencing was not affected by the production of $^1O_2$. Finally, the efficacy of UCNPs–PEG@2×PEI-Ce6-siRNA for PDT and gene therapy was investigated performing MTT assays of HeLa cells under different treatment regimes. The UCNPs–PEG@2×PEI-Ce6-siRNA were reported to exhibit negligible dark toxicity and a greater decrease in cell viability following NIR irradiation was observed as compared to PDT or gene therapy alone.[41]

Although an important part of the work reported on UCNPs for PDT has been carried out using aromatic PSs, some research groups have designed UCNP containing fullerenes that have proved useful for their potential use for *in vitro* PDT following NIR irradiation. Liu *et al.* reported the use of multicolour UCNPs containing monomalonic fullerene ($C_{60}MA$) for NIR imaging-guided PDT of cancer cells.[42] NH_2-functionalised NaYF$_4$:Yb,Er/NaY-F$_4$:Yb,Tm were synthesised exhibiting UCL at ~450, 475, 540, 650, and 808 nm upon 980 nm irradiation. The $C_{60}MA$ was covalently attached to the UCNP *via* a cross-linking reaction between the amino group on the UCNP and the carboxylic group of the fullerene derivative. $C_{60}MA$ was able to absorb the emission at ~450, 475, 540, and 650 nm from the UCNP to generate $^1O_2$, as confirmed by measuring the chemiluminescence of fluresceinyl cypridina luciferin analogue, while the emission at 808 nm was used for NIR high-contrast imaging. The particles were further modified with PEG–succinimidyl carbonate to ensure stability of the nanoparticles in the biological environment. To enhance the cellular uptake, the targeting agent FA was covalently attached to the nanoparticles. High cellular uptake of the $C_{60}MA$–UCNP–FA was observed by HeLa cells (FR-positive) while a poor cellular uptake was exhibited by human alveolar adenocarcinoma (A549) cells (FR-negative) confirming the targeting ability of the reported nanosystem. The efficacy of PDT following NIR excitation of cells treated with the $C_{60}MA$–UCNP–FA conjugates was confirmed using HeLa cells *via* an MTT assay.[42]

7.3.3 Stability of UCNPs in Biological Media Achieved Using a Silica Layer

Another approach to the synthesis of water-dispersible UCNPs is through their encapsulation in a layer of mesoporous silica ($mSiO_2$). One of the advantages of coating the UCNPs in a silica layer is to protect the nanoparticles from degradation in the biological environment. The benefit of encapsulating the PS within a layer of $mSiO_2$ rather than within a non-porous silica coating is that, in the mesoporous silica layer, the ROS generated by the PS can be released out of the silica, thus, increasing the efficiency of PDT.[43] Qian *et al.* have coated NaYF$_4$:Yb,Er UCNPs, ~35 nm, with a thin layer of amorphous silica and a layer of $mSiO_2$ in which ZnPc was loaded. The ZnPc encapsulated within the silica layer was not released when the nanoparticles were soaked in deionised water, PB, or cell culture medium. Singlet oxygen production following irradiation of the ZnPc–UCNPs with

NIR light was confirmed using ABDA. Finally, the authors demonstrated that MB49 bladder cancer cells treated with the ZnPc–UCNPs showed reduced cell viability following NIR irradiation as compared to cells that had been treated with particles without the ZnPc PS.[43] The induced cytotoxicity on MB49 cells by the $mSiO_2$–ZnPc–UCNPs was subsequently confirmed by Guo *et al.* using an MTT assay.[44] The production of $^1O_2$ within treated MB49 cells following NIR irradiation was confirmed using 5-(and-6) carboxy-2'7'-dichlorodihydrofluorescein diacetate, and $^1O_2$-induced apoptosis was shown by changes in nuclear morphology, internucleosomal DNA fragmentation, and cytochrome *c* release.[44]

A shell of $mSiO_2$ surrounding UCNPs was also used to covalently graft a PS *via* the hydrolysis and condensation of silanol groups between the PS and the silica.[45] The covalent attachment of the PS, rather than the previously used doping methodology, could provide an improved stability to the nanosystem. Two types of multifunctional core ($NaGdF_4$:Yb,Er@CaF_2, ~10 nm)–shell (silica layer) nanomaterials were prepared by covalently grafting the PSs haematoporphyrin (HP) or silicon phthalocyanine dihydroxide (SPCD).[45] The as-synthesised nanosystems were able to produce $^1O_2$ in aqueous solutions following irradiation at 980 nm (using 9,10-anthracenediylbis(methylene) dimalonic acid, ABMD). SPCD–UCNPs produced higher levels of $^1O_2$ than the HP–UCNPs within the same irradiation time, possibly due to an enhanced overlap between the absorption spectrum of SPCD and the emission spectrum of the UCNPs. The photodynamic effect of the two types of nanoparticles was investigated in HeLa cells using calcein acetoxymethyl ester (CAM) and PI which showed a decrease in cell viability following 980 nm irradiation that was dependent on concentration, irradiation laser power density, and irradiation time. In agreement with the singlet oxygen production experiments in solution, the mortality of cells treated with SPCD–UCNPs was greater than that of HeLa cells treated with HP–UCNPs. The multifunctional utility of the designed nanosystems was extended to their use for NIR cell imaging, which resulted in low autofluorescence; and as contrast agents for MRI due to the present of Gd.[45] Zhao *et al.* also reported a multifunctional core–shell silica-coated $NaGdF_4$:Yb,Er/$NaGdF_4$ nanosystem with a phthalocyanine-derivative PS covalently incorporated inside the silica shell that could be applied for both PDT of cancer using NIR irradiation and as a contrast agent for MRI of cancer.[46]

Another multifunctional upconversion-based nanosystem for NIR imaging and PDT containing a PS covalently bound to a silica shell was reported by Wang *et al.*[47] The nanosystem contained the following components: a core of $NaYF_4$:Yb/Tm nanocrystals; a layer of silica coating the nanocrystals, to which the PS 1,8-dihydroxy-3-methylanthraquinone (DHMA) and fluorescein isothiocyanate (FITC) to provide downconversion luminescence were covalently attached; and finally, FA to target the FR overexpressed in cancer cells. When the functionalised UCNPs were irradiated with a 980 nm excitation source, they emitted light at 475 nm that could be reabsorbed by both the DHMA and the FITC on the particles and could lead to the generation of

$^1O_2$ as demonstrated using ABDA. When DHMA was only deposited onto the UCNPs covered with silica and not covalently attached, the singlet oxygen production observed was significantly lower, confirming the potential benefits of covalently attaching the PS onto the UCNPs. These results suggest that covalent binding ensures close proximity and therefore enhanced energy transfer between the UCNP and the PS. The photodynamic efficacy of the designed nanoparticles was demonstrated using an MTT assay with HeLa cells irradiated with NIR light for 30 min. The apoptotic effect induced by the designed nanoparticles was investigated using Annexin V-FITC/PI/ Hoechst 33342 staining, concluding that the nanoparticles produce early apoptosis effects rather than necrosis. Apoptosis of HeLa cells treated with the nanoparticles and irradiated with NIR light was also studied by flow cytometry which resulted in 42.75% of apoptotic cells, whereas only 4.22% of the cells became apoptotic when they were treated with nanoparticles with non-covalently-attached DHMA. The benefits of a targeted therapy were also investigated showing that the FA–DHMA–UCNPs were able to target cells expressing the FR (HeLa cells) while no binding of the particles to FR-negative cells (293T cell line, normal cells) was observed.[47] A similar study was reported by the same research group using silica-coated NaYF$_4$:Yb/Er UCNPs with the PS hypericin covalently attached to the silica shell.[48] Hypericin has the advantage of being extracted from natural herbs, which could contribute to the low cytotoxicity, determined by MTT assay, of the reported functionalised UCNP.[48]

7.3.4 Other Methods of Achieving UCNP Stability in Biological Media

Adopting a different approach, Tian *et al.* used α-cyclodextrin, a cyclic oligosaccharide, to provide water solubility to UCNPs by hydrophobic interactions between the hydrophobic cavity of α-cyclodextrin and the oleic acid on the particles.[49] Mn^{2+} ions were used as dopants of NaYF$_4$:Yb,Er UCNPs (~20–30 nm) which led to a pure dark red emission (650–700 nm) of the UCNPs following NIR irradiation that matches the absorption band of commonly used PSs.[50] The Mn^{2+}-doped UCNPs were loaded, *via* hydrophobic interactions, with Ce6 (0.158 mmol g^{-1}), ZnPc (0.165 mmol g^{-1}), or methylene blue (MB) (0.129 mmol g^{-1}) showing leakage of 10–15% in PB (pH 7.4) within 48 h.[49] The ability of the PS–UCNPs to produce $^1O_2$ following irradiation at 980 nm was investigated using DPBF. Ce6–UCNPs showed greater singlet oxygen production than the other PS–UCNPs. Neither the free PSs nor the naked UCNPs were able to produce $^1O_2$ under 980 nm irradiation. The PDT effect under 980 nm irradiation was investigated for the three different PS–UCNPs in human epithelial lung cancer A549 cells using an CCK-8 assay. The viability of cells treated with the PS–UCNPs followed by NIR irradiation decreased as compared to those untreated cells or cells treated with PS–UCNPs but not irradiated. This research was completed with the use of functionalised UCNPs for combined chemotherapy and PDT.

The Mn^{2+}-doped NaYF$_4$:Yb,Er UCNPs were loaded with the chemotherapeutic drug doxorubicin and with the PS Ce6 *via* hydrophobic interactions. The cell viability of non-irradiated A549 cells treated with the DOX–Ce6–UCNPs decreased due to the release of doxorubicin with the UCNPs used as a drug delivery system. However, the cell viability decrease was greater when the cells were treated with the DOX–Ce6–UCNPs and irradiated with NIR light, when the doxorubicin drug was delivered and the singlet oxygen produced by Ce6 was triggered by the NIR irradiation.[49]

Wang *et al.* reported the use of a functional polymeric liposome (FLP) where NaYF$_4$:Yb,Er UCNPs and the PS MC-540 were incorporated yielding (UCNP + MC-540)@FLP.[51] Three functionalised amphiphilic polymers formed the functional polymeric liposome providing unique properties to the nanosystem: (1) octadecyl-quaternised lysine-modified chitosan (OQLCS) to provide steric repulsion against particle aggregation and to allow the encapsulation of the MC-540 and the UCNP; (2) FA-grafted OQLCS (FA-OQLCS) for targeting purposes; and (3) transactivating transduction (TAT) protein-grafted OQLCS (TAT-OQLCS) to aid the nanosystem crossing the cell membrane due to the translocation capacity of TAT. The nanoparticles showed good singlet oxygen production in water following 980 nm irradiation as determined by measuring the fluorescence decay of ABDA. In human breast cancer adenocarcinoma MCF-7 cells the production of ROS following incubation with the UCNP and NIR irradiation was determined using Image-iT Live ROS. The efficacy of the PDT treatment following 980 nm for 30 min was investigated using an MTT assay confirming the superiority of nanoparticles containing the targeting agents FA and TAT (lower cell viability) as compared to those non-targeted UCNPs.[51]

Wang *et al.* reported a dual-targeting UCNP system based on the use of a fullerene derivative as the PS drug.[52] NaYF$_4$:Yb,Gd,Tm UCNPs (~16 nm) were functionalised with hyaluronated fullerene (HAC$_{60}$), to target cluster determinant 44 (CD44) receptors overexpressed on cancer cells, and with the targeting agent aminophenylboronic acid (APBA), which targets polysialic acid (PSA) receptors also overexpressed on cancer cells. Following NIR excitation, the UCL at 475, 650, and 700 nm was absorbed by the HAC$_{60}$ *via* FRET to produce $^1O_2$ as reported by the decay in intensity of DPBF. The UCL of the functionalised nanoparticles at 800 nm was used to perform high-contrast NIR luminescence imaging, while the emission at 640 nm due to the HAC$_{60}$ enables downconversion fluorescence imaging. To assess the targeting ability of the dual-targeting nanosystem, PC12 cells (overexpressing both PSA and CD44 receptors) were treated with the functionalised UCNPs and blocking assays were performed using free APBA and hyaluronic acid (HA). These experiments confirmed the benefits of incorporating the two targeting agents on the nanosystem over unique targeting agents. The PDT efficacy of the HAC$_{60}$-functionalised nanoparticles was evaluated by recording the cell viability of PC12 cells loaded with the nanoparticles and irradiated with a 980 nm laser and also by measuring the singlet oxygen production using the DCFH-DA probe and flow cytometry.[52]

The majority of the reported UCNPs used for PDT focus on UCNPs that are synthesised surrounded by hydrophobic ligands. These ligands are then replaced or modified to obtain hydrophilic nanosystems.[21] Zhou *et al.* reported a one-step nucleotide-programmed growth of UCNPs that yielded water-dispersible and biocompatible $NaYF_4$:Yb,Er UCNPs.[53] The UCNPs had a porous structure that facilitated the loading of the drug doxorubicin hydrochloride or of the PS MB. The production of $^1O_2$ by the MB–UCNP following NIR irradiation was confirmed using the DPBF probe. The PDT efficacy of the reported UCNPs was investigated *in vitro* using A549 cancer cells. A decrease in cell viability was observed in the cells treated with the UCNPs and 980 nm irradiation as compared to the viability showed by non-irradiated control cells.[53]

The majority of upconversion-based nanosystems reported for PDT are all based on the use of $NaYF_4$ cores. However, zinc oxide (ZnO)-based UCNPs have also been shown useful for MRI and NIR-trigged PDT applications and have the advantage of being dispersible in biological environments.[54] Wei *et al.* synthesised water-soluble ZnO:Er,Yb,Gd UCNPs to which the PS MB was adsorbed. Although singlet oxygen production of the MB–UCNPs was not shown in solution, a difference in cell viability between HepG2 cells incubated with the nanoparticles following NIR irradiation was observed by MTT assays.[54]

7.3.5 Upconversion Nanoparticles Containing Inorganic Photosensitisers

Inorganic PSs such as TiO_2 coupled to UCNPs have also proved useful for potential PDT treatment of cancer. Mesoporous TiO_2 ($mTiO_2$) has been used to produce ROS when deposited on the surface of UCNPs.[55] $NaGdF_4$:-Yb,Tm@$mTiO_2$ UCNPs were obtained by depositing $mTiO_2$ on the surface of silica-coated $NaGdF_4$:Yb,Tm nanoparticles. The particles were further modified with the polysaccharide HA that provided targeting capability to the system by binding to the CD44 receptor overexpressed on various tumour cells. HA is degraded by the enzyme hyaluronidase (Hyal-1) that is present in high concentrations in some malignant tumours. Following 980 nm laser irradiation, the UCNPs exhibited an emission band at 300–370 nm that overlaps with the absorption of the TiO_2 and thus induced the production of hydroxyl radicals. The $mTiO_2$ shell was loaded with the chemotherapeutic anticancer drug doxorubicin and the controlled released of the drug in the presence of enzyme Hyal-1 was investigated. The release was quicker at acidic pH, making this nanosystem suitable for the release of the drug in tumours where the pH is usually low. To confirm the effective specificity of the HA–UCNPs, breast cancer MDA-MB-231 cells (positive expression of CD44) were incubated with either HA–UCNPs or UCNPs; the breast cancer cells showed stronger fluorescence when the HA–UCNPs were loaded. No cellular uptake was seen in embryonic fibroblast NIH 3T3 cells (negative expression of CD44). The energy transfer between the UCNP and the mTiO$_2$ was confirmed by measuring the production of ROS *in vitro* using DCFH-DA. The superiority of a combined chemo–PDT treatment of cancer over individual treatments was highlighted by cell viability analyses using MTT assays.[55]

The generation of more than one ROS by a core–shell NaYF$_4$:Yb,Tm@TiO$_2$ nanosystem was investigated by Idris *et al.*[56] Investigations using a series of scavenger experiments and the SOSG dye showed that the ROS generated by the UCNPs following 980 nm were hydrogen peroxide, hydroxyl radicals, and superoxide anions. The ability of the UCNPs to produce ROS under 980 nm irradiation was maintained for over 1 month when stored at 4 °C as a dry powder or soaked in ethanol. The effectiveness of the UCNPs as PDT agents following NIR irradiation was shown by using oral squamous carcinoma cells (OSCC).[56]

7.3.6   Upconversion Nanomaterials Excited with ~808 nm Irradiation

The research reported in this chapter to this point is based on the excitation of PS–UCNPs with ~980 nm irradiation to produce ROS that can be used for PDT of cancer. However, water absorbs light at this wavelength, which could lead to cell mortality due to an overheating effect rather than through a photodynamic effect. The absorption of water at shorter wavelengths, such as ~800 nm, is minimal and thus this wavelength would be optimal for biomedical applications. Nd^{3+}-doped UCNPs have the advantage of being excitable at 808 nm and thus being more suitable for biological applications. However, Nd^{3+}-doped UCNPs also have the drawback of poor energy transfer efficiency. Wang *et al.* have recently reported the first UCNPs that, upon 808 nm irradiation, exhibit high energy transfer and can be used for simultaneous *in vitro* PDT without overheating effects and fluorescence imaging of HeLa cells.[57] Optimally designed core–shell–shell UCNPs, NaYF$_4$:Yb/Ho@NaYF$_4$:Nd@NaYF$_4$, with high efficient UCL following 808 nm irradiation were synthesised. The particles were transferred to an aqueous phase using poly(allylamine) that resulted in amino-functionalised UCNPs. In a final step, the PS RB was covalently attached to the nanoparticles *via* EDC (*N*-(3-dimethylaminopropyl)-*N*′-ethylcarbodiimide hydrochloride) coupling. Different particles with different active-shell thicknesses were synthesised and the optimum thickness for the greatest production of ^{1}O$_2$ following 808 nm irradiation (measured using DPBF) was found to be 1.5 nm. The heating effect of the nanoparticles with 808 nm irradiation was evaluated at different laser power densities and compared with the effect following 980 nm irradiation. MTT assays confirmed a higher cell survival when 808 nm irradiation was employed and concluded that a laser power density of 0.67 W cm^{-2} was appropriate for biological studies. To perform the *in vitro* bioimaging studies and PDT, the particles were further functionalised with the targeting agent FA and with PEG–SH yielding UCNPs–RB/FA. The targeting ability and bioimaging potential of the UCNPs–RB/FA was confirmed by CLSM where the internalisation of the particles was observed in FR-positive human cervix carcinoma HeLa cells but not in FR-negative human alveolar A549 cells or in HeLa cells treated with UCNPs–RB. The potential of the UCNPs–RB/FA as PDT agents *in vitro* was evaluated using the HeLa cells. Following incubation with the particles and irradiation at 808 nm (0.67 W cm^{-2}) for 10 min, a dramatic concentration-dependent change in viability was observed that reached 44.5% cell kill with a UCNPs–RB/FA concentration of 200 µg mL^{-1}.[57]

Another example of UCNPs containing Nd^{3+} has been reported, which can be excited with 808 nm irradiation and employed for *in vitro* PDT treatment.[58] This work reports an initial experiment where KB human mouth epidermal carcinoma cells and A549 human non-small-cell lung cancer cells were irradiated for up to 30 min with 808 nm (1 W cm^{-2}) and 976 nm (1 W cm^{-2}) and the cell viability was investigated. While it was reported that the cell viability was not affected by the 808 nm irradiation, 976 nm irradiation under the same conditions resulted in high cell death, confirming that 808 nm is a more appropriate wavelength for PDT. Even a laser power of up to 6 W cm^{-2} for 808 nm irradiation during 30 min induced negligible cell death. Core–shell–shell NaYbF$_4$:Nd@NaGdF$_4$:Yb/Er@NaGdF$_4$ UCNPs were synthesised that could emit UCL at 550 and 660 nm following irradiation at 808 nm and could also upconvert using 976 nm irradiation due to the NaGdF$_4$:Yb/Er. The UCNPs were amino-functionalised and the PS Ce6 was covalently conjugated to the nanoparticles *via* a carbodiimide cross-linking reaction. In a final step, the nanoparticles were capped with FA–PEG that provided water solubility and targeting ability to the FA–PEG–Ce6–UCNPs. The FRET process between the UCNPs to the Ce6 was confirmed by the significant decrease in the luminescence emission and in the decay time at ~660 nm. A 70% FRET efficiency was estimated. The PEG–Ce6–UCNPs were able to produce $^1O_2$ under 808 nm (3 W cm^{-2}) irradiation as demonstrated by the decrease in absorbance of the probe DPBF. $^1O_2$ was produced even when pork muscle tissue of 15 mm thickness was placed between the laser and the sample of PEG–Ce6–UCNPs following 5 min of 808 nm irradiation. However, no $^1O_2$ was produced under the same conditions when the sample was irradiated with a 976 nm laser. FA–PEG–Ce6–UCNPs were used for bioimaging in two different cell lines, the FR-positive human mouth epidermal carcinoma KB cell line and the FR-negative human non-small-cell lung cancer A549 cell line. The internalisation of the nanoparticles by the KB cells was greater than for A549 cells, confirming the targeted imaging ability of the reported nanosystem. The potential of the FA–PEG–Ce6–UCNPs for PDT treatment was investigated *in vitro* with KB and A549 cells. The cells were loaded with the FA–PEG–Ce6–UCNP (200 µg mL^{-1}) and irradiated with a 808 nm laser (6 W cm^{-2}) for different time intervals (1, 2, 5, and 10 min). Following 5 min irradiation, an effective cell kill was observed for KB cells *via* MTT assay (cell viability 1.6%). However, the cell viability for the A549 cells treated under the same conditions remained significantly high.[58] Table 7.1 summarises all of the UCNPs that have been developed for *in vitro* PDT.

Table 7.1 Summary of publications focusing on the use of UCNPs for *in vitro* PDT.[a]

Photosensitiser (PS)	UCNP λ_{exc} for $^1O_2$	Conjugation of the PS to the UCNP	Water solubility provided by	Targeting agent	Cell lines studied	Therapy	Ref.
M-540	NaYF$_4$:Yb,Er 974 nm	Non-covalent silica encapsulation	Silica	Antibody	MCF-7/AZ	PDT	28
ZnPc	NaYF$_4$:Yb,Er 980 nm	Non-covalent physical adsorption	PEI	Folic acid	HT-29	PDT	32
Ppa	NaYF$_4$:Yb,Er 635 nm	Covalent conjugation to chitosan	PEI	Peptide c(RGDyK)	U87-MG, MCF-7	PDT	33
TPP	NaYF$_4$:Yb,Er 978 nm	Polymeric encapsulation	PEG-*b*-PLA	n/a	HeLa	PDT	35
RB	NaYF$_4$:Yb,Er 980 nm	Covalent attachment	PEG	Folic acid	JAR, NIH 3T3	PDT	36
BODIPY derivative	NaYF$_4$:Yb,Er 975 nm	*Via* carboxylate group – PEG capping – non-covalent	PEG	n/a	SH-SY5Y	PDT	37
Ce6	NaYF$_4$:Yb,Tm/ NaYF$_4$:Yb,Er 980 nm	Non-covalent physical adsorption	PEG-phospholipid layer	n/a	QGY-7703	PDT	38
AlPcS$_4$	Fe$_3$O$_4$/ NaYF$_4$:Yb,Er 980 nm	Non-covalent electrostatic interactions	PEG	n/a	MCF-7	PDT	39
PFVBT	NaYF$_4$:Yb,Tm 980 nm	PS covalently attached to PEG and encapsulation of the UCNPs	PEG	Peptide cRGD	U87-MG, MCF-7	PDT and chemotherapy	40
Ce6	NaGdF$_4$:Yb,Er 980 nm	Non-covalent hydrophobic interactions	PEG & PEI	siRNA	HeLa	PDT	41
C$_{60}$MA	NaYF$_4$:Yb,Er/ NaYF$_4$:Yb,Tm 980 nm	Covalent attachment	PEG	Folic acid	HeLa, A549	PDT	42
ZnPc	NaYF$_4$:Yb,Er 980 nm	Non-covalent silica encapsulation	Mesoporous silica	n/a	MB49	PDT	43 and 44
HP or SPCD	NaGdF$_4$:Yb,Er 980 nm	Covalent grafting	Mesoporous silica	n/a	HeLa	PDT	45

DHMA	NaYF$_4$:Yb,Er 980 nm	Covalent attachment	Silica	Folic acid	HeLa, 293T	PDT	47
Hyperacin	NaYF$_4$:Yb,Er 980 nm	Covalent attachment	Silica	Folic acid	HeLa, HepG2	PDT	48
Ce6, ZnPc, or MB	NaYMnF$_4$:Yb,Er 980 nm	Non-covalent hydrophobic interactions	α-Cyclodextrin	n/a	A549	PDT and chemotherapy	49
MC-540	NaYF$_4$:Yb,Er 980 nm	Non-covalent encapsulation	Polymeric liposome	Folic acid and trans-activating transduction protein	MCF-7	PDT	51
HAC$_{60}$	NaYF$_4$:Yb, Gd,Tm 980 nm	Covalent attachment	HAC$_{60}$	Aminophenyl-boronic acid and hyaluronic acid	PC-12	PDT	52
MB	NaYF$_4$:Yb,Er 980 nm	Non-covalent encapsulation	Nucleotide	n/a	A549	PDT	53
TiO$_2$	NaGdF$_4$:Yb,Tm 980 nm	Deposition	Silica	Hyaluronic acid	MDA-MB-231, NIH 3T3	PDT and chemotherapy	55
TiO$_2$	NaYF$_4$:Yb,Tm 980 nm	Deposition	Silica	n/a	OSCC	PDT	56
RB	NaYF$_4$:Yb,Ho/ NaYF$_4$:Nd/ NaYF$_4$ 808 nm	Covalent attachment	Poly(allylamine) and PEG	Folic acid	A549, HeLa	PDT	57
Ce6	NaYF$_4$:Nd/ NaGdF$_4$:Yb,Er, NaGdF$_4$ 808 nm	Covalent attachment	PEG	Folic acid	A549, KB	PDT	58

[a] AlPcS$_4$, tetrasulfonic phthalocyanine aluminium; C$_{60}$MA, monomalonic fullerene; Ce6, chlorin e6; DHMA, 1,8-dihydroxy-3-methylanthraquinone; HAC$_{60}$, hyaluronated fullerene; HP, hematoporphyrin; M-540, merocyanine 540; MB, methylene blue; ^{1}O$_2$, singlet oxygen; PDT, photodynamic therapy; PEG-*b*-PLA, polyethylene glycol-*block*-poly(DL)lactide; PEG, polyethylene glycol; PEI, polyethyleneimine; PFVBT, poly[9,9-bis(*N*-(but-3,3'-ynyl)-*N*,*N*-dimethylamino)hexyl fluorenyl divinylene-*alt*-4,7-(2',1',3'-benzothiadiazole)dibromide]; Ppa, pyropheophorbide a; PS, photosensitiser; RB, Rose Bengal; siRNA, small interfering RNA; SPCD, silicon phthalocyanin dihydroxide; TPP, tetraphenylporphyrin; UCNP, upconversion nanoparticles; ZnPc, zinc phthalocyanine.

7.4 *In vivo* Applications of Upconversion Nanomaterials for PDT

7.4.1 Stability of UCNPs in Biological Media Achieved Using Polyethylene Glycol Derivatives

The first work reporting the use of UCNPs for *in vivo* PDT was published by Wang *et al.* in 2011 (Figure 7.5).[59] The PS Ce6 was loaded onto NaYF$_4$:Yb,Er UCNPs (~30 nm) functionalised with a PEG-grafted poly(maleic anhydride-*alt*-1-octadecene) amphiphilic polymer. In the supramolecular UCNPs–Ce6-complex, Ce6 was physically adsorbed (6–8% (w/w) loading) onto the surface of the PEG–UCNPs *via* hydrophobic interactions with the oleic acid layer. The as-synthesised nanoparticles were reported to be stable for more than 1 week in physiological environments such as PB, cell culture medium, and fetal bovine serum. The energy transfer between the UCNP and the Ce6 was confirmed by the quenching of the red emission at ~660 nm of the naked UCNPs following functionalisation with Ce6. The effective generation of $^1O_2$ by the UCNPs–Ce6-complex, following NIR irradiation, was determined by monitoring the photobleaching of *p*-nitroso-*N*,*N*'-dimethylaniline in the presence of histidine. The UCNPs–Ce6-complex were taken up by HeLa cells as observed by CLSM with 980 nm excitation. The use of the UCNP–Ce6-complex for NIR light-activated PDT was investigated *in vitro* using 4T1 breast cancer cells, resulting in reduced cell viability following irradiation for 10 min. *In vivo* experiments were also performed with

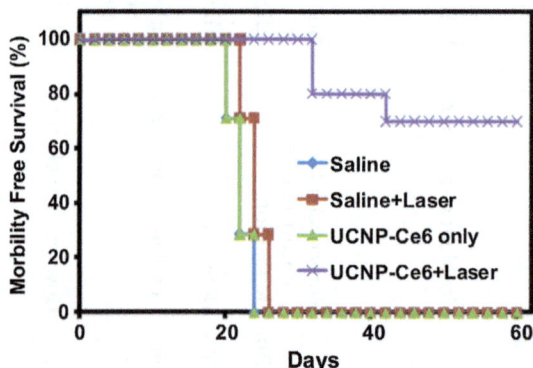

Figure 7.5   *In vivo* PDT treatment of 4T1 tumour-bearing mice. Survival curves of 4T1 breast tumour-bearing mice 60 days after various treatments: no nanoparticles and no irradiation – saline (blue); no nanoparticles and NIR irradiation – saline + laser (red); treatment with UCNP–Ce6-complex and no irradiation – UCNP–Ce6 only (green); and treatment with UCNP–Ce6-complex and NIR irradiation – UCNP–Ce6 + laser (purple). 10 mice were used in each group. Figure adapted from Wang *et al.*, Near-infrared light induced *in vivo* photodynamic therapy of cancer based on UCNPs, *Biomaterials*, 2011, **32**, 6145–6154. Copyright (2011) with permission from Elsevier.[59]

4T1 murine breast tumour-bearing BALB/c mice following intratumoral injection of the UCNPs–Ce6 complex (20 mg mL^{-1} UCNPs with 1.5 mg mL^{-1} Ce6) and NIR irradiation (0.5 W cm^{-2}) for 30 min. It is important to note that, according to the American National Standard for Safe Use of Lasers, the maximum permissible skin exposure for 980 nm continuous-wave (CW) laser irradiation is 0.73 W cm^{-2}.[60] Following injection and NIR activation of the UCNPs–Ce6 complex, the tumours of 7 out of 10 mice were eradicated in 2 weeks and the mice survived for over 60 days (Figure 7.5). These *in vivo* PDT results were complemented with quantitative biodistribution investigations that showed that, 1 day after intratumoral injection of the UCNPs–Ce6 complex, the nanoparticles had mostly accumulated in the skin on and close to the injection site, while 15 days following injection the concentration of nanoparticles in liver and spleen was higher than in the skin. 60 days following injection, the concentration of Y^{3+} in organs and tissues was barely detectable. The *in vivo* experiments concluded with an investigation on the advantage of the use of NIR for PDT over 660 nm excitation that would excite directly the Ce6 on the particles. Tumour-bearing mice were treated with the UCNPs–Ce6 complex and the tumours were blocked with 8 mm thick pork tissue during PDT treatment to simulate tumours located deep inside the body. Treatment of these mice with 980 nm light showed reduced tumour growth, which was not observed when the Ce6 was injected into the tumours and treated with 660 nm irradiation under the same experimental set up.[59] Following this initial work, the use of different PS–UCNPs for *in vivo* PDT has been reported by the same research group and by other authors.

Wang *et al.* reported an improved UCNP nanosystem for *in vivo* PDT treatment in 2013.[61] The red emission of NaYF$_4$:Yb,Er UCNPs was enhanced by doping the nanomaterial with Mn^{2+}. The Mn^{2+}-doped UCNPs (~20 nm) exhibited an intense red luminescence at ~650–670 nm that was used for the excitation of Ce6 loaded on the nanoparticles. The Ce6 was loaded on the nanoparticles following a layer-by-layer self-assembly protocol that yielded UCNPs@2×Ce6 and led to an enhancement of the PDT efficiency of the nanosystem. Furthermore, the presence of Mn^{2+} ions on the nanoparticles enabled their use as T1-weighted MRI contrast agents. Following 980 nm excitation, the UCNPs@2×Ce6 were able to produce $^1O_2$, as determined by the photobleaching of *N,N'*-dimethyl-4-nitrosoaniline. The production of singlet oxygen by the UCNPs@2×Ce6 was greater than that of the UCNPs@3×Ce6, possibly due to a self-quenching effect. The UCNPs@2×Ce6 were modified with a pH-sensitive polymer synthesised by cografting poly(allylamine hydrochloride) (PAH) with PEG and dimethylmaleic acid (DMMA) to achieve a more efficient internalisation by cancer cells. The surface of the UCNPs@2×Ce6–DMMA–PEG was negatively charged at pH 7.4 but it became positively charged without the PEG layer at pH 6.8. The internalisation of the pH-sensitive nanoparticles by HeLa human epithelial carcinoma cells and the efficacy of the PDT treatment under NIR light (as determined by calcein-AM/PI double staining) was greater at pH 6.8 than at pH 7.4 due to the positive charge on the surface of the nanoparticles at the slightly acidic pH. The PDT efficacy of the UCNPs@2×Ce6–DMMA–PEG was investigated *in vivo* using BALB/c mice bearing a 4T1 murine breast cancer tumour. The UCNPs@2×Ce6–DMMA–PEG (10 mg mL^{-1}) accumulated in the tumour and retarded the tumour growth following NIR irradiation (0.5 W cm^{-2} for 30 min) when the UCN-Ps@2×Ce6–DMMA–PEG were intratumorally injected. However, when the nanoparticles were administered *via* intravenous injection, the majority of the UCNPs@2×Ce6–DMMA–PEG were retained in the liver and spleen and the efficacy of the PDT treatment was reported as unsatisfactory.[61]

PEG–phospholipids were used to provide water solubility to NaYF$_4$:Yb,Er/NaGdF$_4$ core–shell particles (~42 nm) that could be used for simultaneous *in vivo* dual-mode imaging (MR and fluorescence imaging) and PDT.[62] The PS Ce6 was physically adsorbed within the hydrophobic phospholipid layers surrounding the UCNPs and chemically conjugated through the carboxy group to the amine functional group on the UCNPs, leading to an enhanced loading of Ce6 on the nanoparticles (>10^3 Ce6 per particle). The Ce6–UCNPs were stable in aqueous media and did not degrade at pH 4.5 (the pH value of acidic intracellular environments). The production of ^{1}O$_2$ by the Ce6–UCNPs following NIR irradiation was observed using 9,10-dimethylanthracene (DMA). In live U87-MG glioblastoma cells, where the dark cytotoxicity of the Ce6–UCNPs was negligible as observed using an MTT assay, the singlet oxygen production following NIR irradiation was confirmed using 5-(and 6-)carboxy-2′,7′-dichlorodihydrofluorescein diacetate and the cell viability was evaluated using the MTT assay. The Ce6–UCNPs were investigated *in vivo* using BALB/c mice bearing a U87-MG glioblastoma tumour. Following tail vein injection of the Ce6–UCNPs (150 µL, 0.1 mg rare earth metal), the nanoparticles accumulated in the tumour through the EPR effect. Following 980 nm irradiation (0.6 W cm^{-2} for 5 min), the tumour growth was inhibited for 14 days and tumour necrosis was induced as confirmed by the histological examination of the treated mice. The authors reported that the Ce6–UCNPs could be also used to image the tumour using UCL and MRI.[62]

Low laser power doses (0.39 W cm^{-2} for 15 min) and low nanoparticle concentrations (10 mg mL^{-1}) were used by Xia *et al.* to achieve the *in vivo* PDT treatment of a liver tumour and to perform image-guided PDT using a PEG-modified nanosystem.[63] The efficiency of the singlet oxygen production by the reported nanosystem was achieved by enhancing the red upconversion emission (at 660 nm) using NaYF$_4$:Yb,Er with a 25% Yb^{3+} doping. Furthermore, the PS ZnPc(COOH)$_4$ was covalently bound to the UCNPs surface which shortened the distance between the donor and the acceptor and thus enhanced the energy transfer between the nanoparticle and the PS. NaYF$_4$:25%Yb, 2%Er oleic acid-coated nanoparticles (~30 nm) were synthesised and modified *via* a ligand exchange process after which the nanoparticles were coated with poly(allylamine), resulting in hydrophilic NH$_2$–UCNPs. The amino groups were used to covalently link the PS ZnPc(COOH)$_4$ and a PEG derivative that provides stability to the nanosystem in the biological media. The nanosystem was able to produce ^{1}O$_2$ following 980 nm irradiation (as measured by DPBF, 74% consumed after 4 min irradiation) in a much higher yield than the UCNPs containing the PS physically absorbed (10% DPBF consumed after 4 min irradiation) confirming the superiority of the covalent attachment of the PS to the UCNP's surface. The energy transfer between the donor UCNPs and the acceptor $ZnPc(COOH)_4$ (estimated to be 80.9%) was observed by the quenching of the UCL emission band and the changes in the average decay time at 660 nm of the UCNPs in the presence of the acceptor. These changes were less noticeable for ZnPc–UCNPs where the PS was physically absorbed. To achieve *in vitro* and *in vivo* targeting imaging and PDT, the UCNPs were further functionalised with a covalently linked FA. The targeting ability of the ZnPc–UCNPs–FA was confirmed using HeLa cells (FR-positive) and human alveolar adenocarcinoma A549 cells as a control (FR-negative). A greater uptake of the ZnPc–UCNP–FA was observed in the HeLa cells. Following NIR irradiation, the cell viability (as determined by MTT assay) of treated HeLa cells decreased, confirming the potential use of the nanosystem for image-guided PDT. The ZnPc–UCNPs–FA were also used for *in vivo* PDT in Hepa1-6 tumour-bearing C57/6J mice. The particles were intratumorally injected and the tumours irradiated with a 980 nm laser (0.39 W cm^{-2}) for 15 min. The tumour growth was slowed for 14 days for the treated mice with a tumour inhibition ratio of ~80.1%. Histological analysis of tumour, liver, heart, spleen, lung, and kidney were performed 14 days after treatment revealing no pathological changes in the heart, lung, kidney, liver, or spleen and markedly increased apoptotic and necrotic tumour cells.[63]

7.4.2 Stability of UCNPs in Biological Media Achieved Using Polyethyleneimide

PEI-coated $NaYF_4$:Yb,Er UCNPs (~50 nm) containing ZnPc bound to the surface *via* physical adsorption have been reported as useful for the photodynamic inactivation of viruses (DENV2 and adenovirus type 5) in suspension, in DENV2-infected HepG2 human hepatocellular carcinoma cells, and in BALB/c mice innoculated with DENV2 virus.[64] Mice treated with the ZnPc–UCNPs and NIR irradiation remained healthy during the 15 day observation period. Furthermore, the amine functional group of the PEI was used to conjugate an antibody specific for the DENV2 envelope protein released to the surface of the host cell during virus replication. The antibody-conjugated UCNP showed improved specificity, demonstrating the potential antiviral strategy of PS–UCNPs.[64]

7.4.3 Stability of UCNPs in Biological Media Achieved Using a Silica Layer

Enhanced *in vivo*-targeted NIR PDT was achieved by Idris *et al.* using a dual-PS approach.[65] The low upconversion efficiency of the majority of PS–UCNP-based nanosystems is a potential drawback in the use of these tools for PDT. Idris *et al.* reported the synthesis of $NaYF_4$:Yb,Er coated with $mSiO_2$

in which the PSs MC-540 and ZnPc were encapsulated yielding MC-540–UCNPs–ZnPc. Upon excitation at 980 nm, the UCNPs emitted light at two wavelengths, 540 nm and 660 nm, which could be reabsorbed by MC-540 and ZnPc, respectively, inducing the generation of cytotoxic $^1O_2$. The ability of the MC-540–UCNPs–ZnPc to produce $^1O_2$ in aqueous solutions was evaluated by measuring the consumption of the ABDA dye following 980 nm irradiation for 80 min at 20 min intervals. The consumption of ABDA and thus the singlet oxygen production was statistically significantly greater for MC-540–UCNPs–ZnPc than for those UCNPs functionalised only with one of the PSs, either MC-540–UCNPs or ZnPc–UCNPs. The production of ROS was also examined in B16-F0 melanoma cells using the marker 6-carboxy-2′,7′-dichlorodihydro-fluorescein diacetate, which emits bright fluorescence in the presence of ROS. For the *in vitro* situation, those cells treated with MC-540–UCNPs–ZnPc showed greater fluorescence than cells treated with either MC-540–UCNPs or ZnPc–UCNPs. The efficacy of the PDT treatment after irradiation with NIR light for 40 min was demonstrated using a MTS assay. To demonstrate the efficacy of the MC-540–UCNPs–ZnPc for *in vivo* PDT treatment, B16-F0 mela-noma cells loaded with the MC-540–UCNPs–ZnPc (500 µg mL^{-1}) were injected under the skin of C57BL/6 mice and then irradiated with a 980 nm laser (415 mW cm^{-2} for 2 h) which induced an inhibition of the tumour growth and an enhancement in the population of apoptotic cells as compared to control experiments. Following a second approach to *in vivo* experiments, MC-540–UCNPs–ZnPc were intratumourally injected into C57BL/6 mice bearing mel-anoma tumours followed by 980 nm irradiation for 1 h. The PDT treatment slowed the tumour growth during the 11 day period when the mice were monitored. Finally, the MC-540–UCNPs–ZnPc were further functionalised with PEG, to enhance the circulatory lifetime of the nanoparticles, and with FA, to target the FR overexpressed on B16-F0 cells. When the nanoparticles were intravenously injected in mice bearing B16-F0 melanoma tumours and following NIR irradiation, a significantly greater reduction in tumour growth was observed as compared to the controls. However, the complete regression of the tumours was not achieved with the as-synthesised nanosystem.[65]

A mSiO$_2$ layer also provided aqueous solubility to Mn^{2+}-doped UCNPs that were used for a dual cancer treatment.[66] NaY(Mn^{2+})F$_4$:Yb,Er UCNPs (20 nm) were synthesised and encapsulated by a mSiO$_2$ layer in which a hydrocar-booctadecyltrimethoxysilane chain (C18) and Ce6 were grafted. The Ce6–C18–UCNPs@SiO$_2$ were able to produce $^1O_2$ following 980 nm irradiation as confirmed by the photobleaching of N,N'-dimethyl-4-nitrosoaniline. Once the production of $^1O_2$ was confirmed, an amphiphilic copolymer containing 9,10-dialkoxyanthracene was used to encapsulate the UCNPs forming the outermost layer and the anticancer drug doxorubicin was then loaded. When exposed to $^1O_2$, the dialkoxyanthracene group degrades, becoming detached from the nanoparticle surface and hence releasing the anticancer drug in a controlled manner. The *in vitro* drug release was confirmed in KB cells by CLSM where, following NIR irradiation, an increase in the fluorescence due to doxorubicin was observed over time. The dark cellular cytotoxicity of the particles was negligible as confirmed by MTT assays while cytotoxicity was enhanced following NIR irradiation due to the singlet oxygen production and the release of the doxorubicin. The combined therapy was shown to be superior to the single therapies. The efficacy of the combined therapy was also shown *in vivo* using athymic nude mice bearing KB cells treated by intratumoral injection of the nanoparticles (10 mg mL^{-1}) and exposed to 980 nm irradiation (0.5 W cm^{-2}) for 30 min. The tumour growth was slowed for up to 23 days and the histology studies indicated that little impact had been caused to the liver or to other major organs. The UCNPs were also used for tumour imaging even 3 days after treatment.[66]

mSiO$_2$-coated UCNPs were used by Fan *et al.* to design a nanotheranostic system that could be used for synergetic chemo-/radio-/photodynamic therapy and simultaneous MR/UCL imaging.[67] Core–shell NaYF$_4$:Yb/Er/Tm@NaGdF$_4$ nanoparticles were synthesised emitting at ~350–370 nm, 460–500 nm, 510–530 nm, 530–570 nm, 630–700 nm, and 770–810 nm. These characteristics facilitate the use of the NaYF$_4$:Yb/Er/Tm@NaGdF$_4$ UCNPs for NIR-to-visible and NIR-to-NIR *in vitro* and *in vivo* imaging and the presence of the Gd has potential as an MRI contrast agent. The nanoparticles were coated with a non-porous layer of silica followed by the deposition of a layer of mSiO$_2$ shell, yielding Gd–UCNPs@SiO$_2$@mSiO$_2$. As a final step, the intermediate non-porous silica layer was etched and the surface of the particles was coated with polyvinylpyrrolidone. The nanoparticles showed significant endocytosis by cancerous cells and were successfully used for MR/UCL bimodal imaging *in vivo* following intravenous injection in 4T1 tumour-bearing mice. Passive targeting induced the accumulation of the nanoparticles in the tumours. To achieve the synergetic chem-/radio-/photodynamic therapy, the radiosensitising/chemodrug docetaxel (Dtxl) was loaded into the cavity left between the core and the mSiO$_2$ layer. Dtxl can be released from the nanoparticles in hydrophobic environments such as the cell membrane. Furthermore, the radiosensitising PS haematoporphyrin (HP) was loaded in the mSiO$_2$ shell by physical absorption and by covalent grafting. The Dtxl–HP–UCNPs were able to produce $^1O_2$ as indicated by the decrease in absorption of DPBF. *In vitro* experiments confirmed the superiority of the combined therapy when the particles were irradiated with both NIR and X-ray irradiation as compared to the individual therapy modalities. The utility of this complex system was also highlighted *in vivo* treating 4T1 tumour-bearing mice by intratumoral injection of the particles (30 mg mL^{-1}, 150 mL) followed by NIR (2.5 W cm^{-2} for 30 min) and X-ray irradiation. After several days following the treatment, the tumours were eradicated (without reappearing over the following 120 days) due to the combined effect of singlet oxygen production, chemotherapy, and radiotherapy. The *in vivo* treatment was also attempted following intravenous injection of the UCNPs, but this treatment was shown to be less effective than intratumoral injection.[67]

Another multifunctional platform for multimodal imaging and synergistic therapy of cancer using NIR irradiation was recently reported by Lv *et al.*[68] GdOF:Ln@SiO$_2$ UCNPs were synthesised, conjugated with the PS ZnPc and the surface decorated with the photothermal therapy (PTT) agent carbon dots. The nanoparticles were further functionalised with the targeting agent FA. The generation of $^1O_2$ by the nanosystem was shown using DPBF, and the enhanced temperature (~15 °C) following NIR irradiation was demonstrated *via in vivo* IR thermal imaging of tumour-bearing mice. The reported nanosystem was biocompatible, as demonstrated by measuring cell viability of treated L929 fibroblast cells and *in vivo* haemolysis. To increase the multifunctionality of the nanosystem, the anticancer drug doxorubicin (DOX) was loaded onto the nanoparticle using the $mSiO_2$ shell. DOX was released from the particles at acidic pH, characteristic of cancer cells. A synergistic effect of photothermal, photodynamic therapy, and chemotherapy was observed when HeLa cells were incubated with the nanoparticles and irradiated with 980 nm irradiation as demonstrated by MTT assay and by Calcein AM/PI staining. CLSM was used to demonstrate the uptake of the nanoparticles by the cells and the release of DOX. The synergistic therapy was also demonstrated *in vivo* when H22 tumour-bearing mice were injected intratumorally with the nanoparticles and treated with NIR irradiation with a markedly increased number of apoptotic and necrotic cells as compared to the controls. Haematoxylin and eosin staining of the heart, lung, kidney, liver, and spleen showed no pathological changes in the analysed organs. The behaviour of the nanoparticles *in vivo* following intravenous injection was also investigated showing the targeting of the tumour due to the presence of FA on the nanoparticles, and effective PDT treatment following NIR irradiation. Biodistribution and blood circulation studies indicated the accumulation of the nanoparticles in liver, spleen and lung with the Gd concentrations low in the heart and kidney. The nanoparticles were eliminated through the bile–gut pathway. The designed UCNPs also showed potential as multimodal imaging systems including UCL imaging, CT imaging and MRI.[68]

Local hypoxia, responsible for lower effectiveness of PDT, can be mitigated by the combination of PDT with bioreductive prodrugs that can be activated in hypoxic environments. This was the strategy followed by Liu *et al.* in the design of $NaYF_4$:Yb,Er,Gd UCNPs that were loaded with the PS drug SPCD and with the bioreductive prodrug tirapazamine (TPZ) and used for a synergetic tumour therapy both *in vitro* and *in vivo*.[69] The UCNPs were coated with a shell of non-porous silica where the SPCD was loaded and a shell of $mSiO_2$ where the TPZ was loaded, resulting in UCNPs–SPCD–TPZ. The nanoparticles did not show cytotoxicity in HeLa cancer cells or in healthy NRK-52E and BRL cells; and no pathological changes were observed, after 7 and 30 days, with histological analyses of mice injected with the nanoparticles (100 mg kg^{-1}). In solution, singlet oxygen production by the UCNPs–SPCD under 980 nm irradiation was confirmed by measuring the fluorescence decay rate of DPBF. *In vitro*, the singlet oxygen production following NIR irradiation of the cells loaded with UCNPs–SPCD was investigated using flow cytometry to detect the emission of dichlorofluorescein and CLSM to observe the emission of fluorescein isothiocyanate. The anticancer efficacy of the UCNPs–SPCD–TPZ under 980 nm irradiation in HeLa cells was highlighted using a CCK-8 assay.

In vivo experiments were also performed with the UCNPs–SPCD–TPZ. In an initial experiment, the hypoxic subvolumes in tumours were quantified using ^{18}F-labelled MISO positron emission tomography which confirmed that a tumour preinjected with the UCNPs–SPCD–TPZ (150 μL, 5.33 mg mL^{-1}) and treated with 980 nm irradiation (1.4 W cm^{-2} for 15 min) exhibited strong local hypoxia. Finally, the potential of the UCNPs–SPCD–TPZ for NIR-induced synergetic therapy of tumours was confirmed in tumour-bearing BALB/c nude mice, tumour growth being suppressed for 16 days following treatment.[69]

7.4.4 Other Methods of Achieving UCNP Stability in Biological Media

Chitosan has been used to provide water solubility to ZnPc–UCNPs that were used for *in vivo* PDT by Cui *et al.*[70] Hydrophilic NaYF$_4$:Yb,Er UCNPs (~35 nm) were obtained by coating the surface of the oleic acid-capped NaYF$_4$ nanoparticles with amphiphilic chitosan. These nanoparticles were further functionalised by loading the PS ZnPc onto the UCNP *via* hydrophobic interactions, yielding ZnPc–chitosan–UCNPs. The production of $^1O_2$ by the ZnPc–chitosan–UCNP, following irradiation at 980 nm, was evaluated using DPBF and a ~45% decrease in the absorption of the probe was observed within the first 30 min following NIR irradiation. The cytotoxicity of the reported ZnPc–chitosan–UCNPs was low in human embryonic lung fibroblast (HELF) cells and human breast adenocarcinoma (MCF-7) cells as determined using an MTT assay. The intracellular uptake of ZnPc–chitosan–UCNPs by MCF-7 cells was evaluated using CLSM, obtaining the emission due to the UCNPs following irradiation at 980 nm and the emission due to the ZnPc using direct excitation at 650 nm. The PDT effect of the ZnPc–chitosan–UCNPs following NIR irradiation (10 min) was confirmed in MCF-7 cells using an MTT cell viability assay; and cell apoptosis was investigated using an Annexin V-FITC/PI apoptosis detection kit. Finally, the ZnPc–chitosan–UCNPs were used for *in vivo* PDT treatment of female Kunming mice bearing a S180 sarcoma tumour. 14 days after intratumoral injection of the ZnPc–chitosan–UCNPs, the nanoparticles were only found in the tumour tissue with no traces of the UCNPs present in other organs such as liver, lung, heart, spleen, intestine, and kidney as observed by fluorescence imaging of these organs. NIR PDT treatment of the mice injected with the ZnPc–chitosan–UCNPs induced significant reduction of the tumour volume as compared to mice treated only with NIR light or non-treated mice.[70] To enhance the tumour selectivity of ZnPc–chitosan–UCNPs, the surface of the nanoparticles was further functionalised with a folate-modified amphiphilic chitosan (FASOC) layer in which the ZnPc was encapsulated yielding FASOC–ZnPc–UCNP.[71] The red emission of the UCNP following NIR irradiation was used to activate the PS and to produce $^1O_2$ (measured using DPBF) while the green UCL was employed for cell imaging. The production of $^1O_2$ was higher when the particles were directly irradiated with 660 nm light (~68%) than when they were indirectly irradiated with NIR light (~53%). However, the advantage of NIR irradiation was highlighted when a thick slice of pork tissue was placed between the irradiation source and the nanoparticle solution. *In vitro* cytotoxicity of the FASOC–ZnPc–UCNPs was investigated in HELF human embryo lung cells and in human breast carcinoma MDA-MB-231 cells showing a low toxicity in both cell lines. The target ability of the FASOC–ZnPc–UCNPs was assessed in cancer cells that expressed different levels of FR. Human hepatocellular carcinoma Bel-7402 and MDA-MB-231 cells, both FR-positive, showed higher uptake of nanoparticles than human lung adenocarcinoma A549 cells (FR-negative), confirming the target efficacy of the reported nanosystem. The FR-mediated cellular uptake was confirmed by receptor blocking experiments loading the cells with an excess of FA. The intracellular production of $^1O_2$ in cells loaded with the FASOC–ZnPc–UCNPs irradiated with NIR light was confirmed using the probe DCFH-DA. Active-tumour targeting of FASOC–ZnPc–UCNP was demonstrated *in vivo* in mice bearing FR-positive hepatocellular Bel-7402 and sarcoma S180 tumours. The uptake of the subcutaneously injected particles (33 mg kg^{-1} containing 3.86 mg kg^{-1} ZnPc) was observed 4 h after injection with a peak in the tumour to skin ratio after 24 h that was still high at 96 h after injection. The *in vivo* PDT efficacy of the FASOC–ZnPc–UCNPs was investigated following irradiation at both 980 nm (0.4 W cm^{-2} for 15 min) and 660 nm. For subcutaneous tumours, 660 nm irradiation led to a higher tumour inhibition than 980 nm irradiation. However, when a 10 mm thick pork tissue was placed over the tumour during irradiation, the mice treated with 980 nm irradiation showed a greater decrease in tumour growth than those irradiated with the 660 nm light; thus demonstrating the superiority of NIR for PDT treatment of deeper lying tumours.[71]

An interesting approach was described by Chen *et al.* that reported the use of bovine serum albumin (BSA)-coated UCNP for a combined NIR-induced PDT and PTT to perform synergistic *in vivo* tumour treatment.[72] Polyacrylic acid-coated NaGdF$_4$:Yb,Er UCNPs (~60 nm) were further coated by covalently conjugating BSA *via* amide bond formation (UCNPs@BSA) which provided water stability to the nanoparticles. The PS RB and the NIR-absorbing dye IR825 were loaded onto the UCNPs@BSA using the hydrophobic domains of the BSA. The dual-dye nanoparticles could be used as a dual modal MR (due to the Gd) and upconversion optical imaging probe and as a dual therapeutic system for combined PDT and PTT using NIR irradiation. The singlet oxygen production by the functionalised nanoparticles under 980 nm irradiation was investigated using SOSG and the photothermal effect upon 808 nm irradiation was confirmed by a rise of temperature that was shown to be negligible following 980 nm irradiation. The designed nanoparticles were not cytotoxic in the dark when incubated with 4T1 murine breast cancer cells, as confirmed by MTT assay; however, the cells' viability was considerably reduced following 980 nm and 808 nm irradiation of the treated cells. The research was concluded with the *in vivo* application of the reported nanoparticles for dual imaging and dual therapy using mice bearing 4T1 tumours. Following intratumoral injection of the nanoparticles and 808 nm irradiation, changes in temperature of up to 15 °C were observed. Furthermore, the efficacy of the combined PDT (980 nm, 0.4 W cm^{-2} for 30 min) and PTT (808 nm, 0.5 W cm^{-2} for 5 min) therapies was also investigated and the mice showed a remarkable delay in tumour growth 14 days after treatment, which was not observed when only one of the therapies was applied.[72]

The vast majority of UCNP-based nanosystems reported for PDT use NaYF$_4$ as the host material. However, LiYF$_4$ can also be used as a host material, to be doped with Ln^{3+}, and the first report that describes the use of a LiYF$_4$-based UCNP for both *in vivo* and *in vitro* PDT has been recently reported by Wang *et al.*[73] LiYF$_4$:Yb,Er UCNPs (47 nm) were synthesised and the oleate ligand removed, resulting in positive Ln^{3+} ions on the surface of the particles that could interact with electronegative groups of ZnPc–COOH *via* electrostatic interactions. The UCNPs were further modified with polyvinylpyrrolidone, which provides excellent water solubility to the nanoparticles. The nanoparticles exhibited negligible dark toxicity in HELF cells as determined using the CCK-8 assay and were able to produce high levels of $^1O_2$ following 980 nm irradiation (91% decrease in the absorbance of DPBF after 26 min). The successful energy transfer between the donor and the acceptor was confirmed by the quenching of the red UCL and the changes in the lifetime at 669 nm. The energy transfer efficiency was determined to be 96.3%. The *in vitro* PDT efficacy of the UCNP following NIR irradiation was examined in MDA-MB-231 breast cancer cells and showed a significant decrease in cell viability. *In vivo* PDT was also performed by intratumorally injecting (0.5 mg mL^{-1}) H22 tumour-bearing mice. Following irradiation at 0.5 W cm^{-2} for 20 min the tumour growth was attenuated during a 14 day period during which the tumour volumes were measured. The reported nanoparticles were also used for cell imaging.[73]

7.4.5 Upconversion Nanoparticles Containing Prodrugs

All of the research reported to date using UCNP for PDT *in vivo* has been based on the use of PS molecules such as RB, MB, ZnPc, or Ce6. The first work describing the use of the clinically used PDT prodrug 5-aminolevulinic acid (ALA) for UCNP-based PDT was recently reported by Punjabi *et al.*[74] With the aim of increasing the deep-tissue (>12 mm) treatment of tumours following PDT, the authors reported a UCNP-based nanosystem with amplified red emission, since red emission penetrates deeper into tissues than green emission. To achieve this enhanced red emission, α-NaYF$_4$:Yb,Er(2%)@CaF$_2$ UCNPs (~26 nm) were synthesised containing an optimal core Yb^{3+} ratio of 80% which led to nanomaterials with a QY of ~3.2% under 10 W cm^{-2} irradiation. The prodrug ALA was conjugated to the UCNP *via* a covalent pH-sensitive hydrazone linkage, yielding a nanosystem that was able to produce PpIX. The ability of the nanoparticles for *in vitro* PDT treatment was investigated using HeLa cells. Following irradiation of treated cells with 980 nm light for 20 min, almost 70% of the cells were killed as measured by MTT assay. The results obtained using the α-NaYF$_4$:Yb(80%),Er(2%)@CaF$_2$–ALA–UCNPs were compared with those obtained using β-NaYF$_4$:Yb(20%),Er(2%)@β-NaYF$_4$–ALA nanoparticles, and the superiority of the α-NaYF$_4$:Yb(80%),Er(2%)@CaF$_2$–ALA in killing cancer cells following NIR irradiation was confirmed. The *in vitro* singlet oxygen production when the cells were incubated with the ALA–UCNPs and irradiated with a 980 nm laser was confirmed by two methods: (1) using DCFH-DA to stain the cells and observing the changes in emission using a fluorescence microscope, and (2) quantifying the formation of 2′,7′-dichlorofluorescein using a microtitre plate reader. The ability of the nanosystem to perform *in vitro* deep-tumour PDT treatment was investigated by placing a piece of pork between the NIR laser and the HeLa cells. α-NaYF$_4$:Yb(80%),Er(2%)@CaF$_2$–ALA were able to induce 30% cell death even with a 12 mm piece of pork tissue between the laser and the cells; however, the β-NaYF$_4$:Yb(20%),Er(2%)@β-NaYF$_4$–ALA nanoparticles were not able to induce cell death under the same conditions. The *in vivo* efficiency of the ALA–UCNPs was proven under different conditions. Female BALB/c mice subcutaneously injected into the back with 4T1 breast cancer cells were intratumorally treated with the α-NaYF$_4$:Yb(80%),Er(2%)@CaF$_2$–ALA nanoparticles (20 mg mL^{-1}) and then irradiated with either red light (635–685 nm) to directly activate ALA or with 980 nm laser to activate the UCNPs. Both treatments slowed the tumour growth with no statistically significant differences found between them. When a piece of pork tissue (1.2 cm) was placed between the tumour and the laser, 980 nm irradiation (0.5 W cm^{-2} for 40 min) resulted in a ~150 mm^3 tumour reduction while irradiation with red light did not produce a therapeutic effect, thus confirming the superiority of NIR irradiation for deep-tissue PDT treatment.[74]

7.4.6   UCNPs Containing Inorganic Photosensitisers

Inorganic PSs such as TiO$_2$ coupled to UCNPs have also been used for *in vivo* PDT treatment of cancers.[75] NaGdF$_4$:Yb,Tm UCNPs (~50–80 nm) were synthesised, coated with a silica shell (~20 nm) and then a shell of TiO$_2$ nanoparticles was deposited yielding NaGdF$_4$:Yb,Tm@SiO$_2$@TiO$_2$ UCNPs (Gd-Si-Ti–UCNPs). These nanoparticles were further modified to include amino groups on the surface that were used to conjugate the targeting agent FA, resulting in FA–Gd-Si-Ti–UCNPs. The production of ROS by the FA–Gd-Si-Ti–UCNPs following 980 nm irradiation was not investigated in solution. The dark cytotoxicity of FA–Gd-Si-Ti–UCNPs was low in both human breast cancer MCF-7 cells and human cervix carcinoma HeLa cells even at high concentrations (500 μg mL^{-1}) as evaluated by MTT assay. The low cytotoxicity was also observed *in vivo* following tail vein injection of the nanoparticles in healthy female BALB/c nude mice and haematoxylin and eosin staining of the major organs. The efficacy of the FA–Gd-Si-Ti–UCNPs for *in vitro* PDT following 980 nm irradiation (0.6 W cm^{-2}, 20 min) was investigated for HeLa and MCF-7 cells using an MTT assay and CAM and PI staining. Although cell viabilities decreased to ~30% for both cell lines, it is important to note that a decrease in cell viability of ~70% was also observed when the cells were treated in the presence of NaGdF$_4$:Yb/Tm UCNPs under the same conditions. The potential application of the FA–Gd–Si–Ti–UCNPs for *in vivo* PDT was investigated with MCF-7 tumour-bearing nude mice. The particles (100 μL, 500 μg mL^{-1}) were intratumourally injected and the mice treated with 980 nm irradiation (0.6 W cm^{-2}, 20 min) resulting in a growth inhibition ratio of ~88.6% 2 weeks after treatment. Histological analysis was performed of the tumour, heart, liver, spleen, lung, and kidney by haematoxylin and eosin staining showing extensive coagulative necrosis in tumour cells and no pathological changes or necrosis in the major organs investigated. With the presence of the Gd ions, the nanoparticles were used to perform *in vivo* MRI studies.[75]

TiO$_2$ has been also used by Lucky *et al.* to report a UCNP-based nanosystem with good therapeutic efficacy both *in vitro* and *in vivo*.[76] A core of NaYF$_4$:-Yb,Tm UCNPs (~25 nm) was first coated with a thin layer of silica followed by grafting of (3-aminopropyl)-trimethoxysilane and binding of a Ti precursor that yielded a thin layer of TiO$_2$ on the nanoparticles. The nanoparticles were further modified with maleimide–PEG–silane resulting in Mal–PEG–TiO$_2$–UCNPs. The nanoparticles were able to generate a significant amount of ROS in PB following irradiation at 980 nm even in the presence of a 10 mm tissue (phantom). The ROS were not observed when direct excitation of the TiO$_2$ shell was attempted using UV light. The cellular uptake of the Mal–PEG–TiO$_2$–UCNPs was investigated showing a more reduced cellular uptake than TiO$_2$–UCNPs in mouse macrophage cells. However, the behaviour of Mal–PEG–TiO$_2$–UCNPs and TiO$_2$–UCNPs in OSCC was the opposite. Furthermore, an enhancement in cellular uptake of Mal–PEG–TiO$_2$–UCNPs was also observed when compared to PEG–TiO$_2$–UCNPs indicating a potential interaction between maleimide on the nanoparticles and cell-surface thiols. The dark toxicity of the Mal–PEG–TiO$_2$–UCNPs and TiO$_2$–UCNPs, investigated using an MTS proliferation assay and Trypan Blue dye exclusion method, was found negligible up to concentrations of 1 mM for both types of nanoparticles. Haemolysis assays were also performed with both nanoparticles and concluded a good haemocompatibility and suitability for *in vivo* applications. After optimisation of the PDT parameters, the nanoparticles (1 mM) were evaluated for *in vitro* PDT following 980 nm irradiation (~2.1 W cm^{-2}, 5 min). Mal–PEG–TiO$_2$–UCNPs showed a greater induction of cell death as compared to TiO$_2$–UCNPs and the mechanism of cell death was found to be necrosis. The generation of ROS intracellularly was quantified using DCFH-DA. Interestingly, following incubation of Mal–PEG–TiO$_2$–UCN for 1 h, the nanoparticles were found in the cell membrane while after 6 h incubation they were mostly located in the cytoplasm. Despite the different location of the nanoparticles, cell death was observed upon irradiation following either of the two different incubation times. *In vivo*, a homogeneous distribution of the Mal–PEG–TiO$_2$–UCNPs was found within the tumour 4 h after intratumoral injection. Following NIR irradiation of the treated mice, necrotic areas were found within the treated area and severe destruction of tumour cells was indicated by haematoxylin and eosin staining. The efficacy of the nanoparticles as PDT agents was investigated by intratumoural treatment of mice followed by NIR irradiation. A delay in tumour growth was observed for both Mal–PEG–TiO$_2$–UCNPs and TiO$_2$–UCNPs. However, although the survival rate was reduced for mice treated with TiO$_2$–UCNPs, no mice died following treatment with Mal–PEG–TiO$_2$–UCNPs confirming the superiority of the Mal–PEG containing TiO$_2$–UCNPs.[76]

TiO$_2$ inorganic PSs have been also used in the design of a bifunctional anticancer UCNP-based nanosystem that combined chemotherapy with PDT for the treatment of drug-resistant breast cancers.[77] NaYF$_4$:Yb/Tm–TiO$_2$ were synthesised and their surface modified with polymer–PEG that provided dispensability and biocompatibility to the nanoparticles. FA was coupled to the nanoparticles which were further modified by loading the anticancer drug DOX. The generation of ROS in solution following 980 nm irradiation of the nanoparticles was confirmed using ABDA. The release of DOX from the nanoparticles was pH dependent (with a higher release at lower pH) but also irradiation dependent. Flow cytometry and bio-TEM were used to study the uptake of the nanoparticles by two different types of cells, human breast cancer MCF-7 cells and Adriamycin-resistant human breast cancer (MCF-7/AR) cells. These investigations showed a greater uptake of the particles by MCF-7, hence confirming the targeting activity of FA on the nanoparticles. CLSM confirmed the enhanced DOX release in MCF-7 treated with FA–UCNPs. The nanoparticles did not show dark cytotoxicity either *in vitro* (*via* MTT assay) or *in vivo* following tail vein injection (*via* haematoxylin and eosin staining). The superiority of the combined chemo- and NIR (500 mW cm^{-2}) PDT was confirmed *in vitro* using both MCF-7 and MCF-7/AR cells and by undertaking MTT assays. The potential use of the designed nanoparticles for *in vivo* chemo-NIR PDT treatment to overcome multidrug resistance was also investigated for 15 days using MCF-7 or MCF-7/AR-bearing female BALB/c nude mice. The nanoparticles were intravenously injected (100 μg mL^{-1}) and, following NIR irradiation (0.5 W cm^{-2} for 10 min), the combined therapy showed disappearance of the tumour by 15 days after treatment, a result that is superior to the single-mode treatments. Histological analysis *via* haematoxylin and eosin staining of the tumour and major organs of the treated mice showed apoptotic and necrotic tumour cells and no pathological changes in the major organs such as heart, liver, spleen, lung, and kidney.[77]

Another multifunctional anticancer nanosystem based on TiO$_2$–UCNP was reported by Lv *et al.* and used NIR irradiation to achieve both PDT and PTT.[78] The authors reported an uncommon core–shell–shell structure, TiO$_2$@Y$_2$Ti$_2$O$_7$@YOF:Yb,Tm, which emitted blue light following irradiation at 980 nm. The blue light induced the Y$_2$Ti$_2$O$_7$ to produce ROS (confirmed using the DPBF probe) and, during the energy transfer process, a thermal effect also occurred. The nanoparticles were further modified by loading the anticancer drug DOX. The nanoparticles (<500 μg mL^{-1}) were nontoxic in the dark, as indicated by the good biocompatibility shown by L929 fibroblast cells using an MTT assay and the negligible haemolysis of red blood cells. The efficacy of the nanoparticles for the treatment of cancer cells using NIR irradiation was demonstrated by MTT assay, Trypan Blue, and AM/PI marked HeLa cells *in vitro*. The application of the nanoparticles for *in vivo* treatment of cancer was investigated using H22 hepatocarcinoma tumour-bearing mice. Following intravenous injection of the nanoparticles and 980 nm irradiation (0.72 W cm^{-2}, 10 min), a reduction in the tumour size was observed and tumour histologic sections indicated an elevated number of apoptotic and necrotic cells. Furthermore, no pathological changes were observed in the major organs such as liver, lung, and kidney.[78] Table 7.2 summarises all of the reported UCNPs used for *in vivo* PDT.

7.5 Conclusions

Based on the increasing number of publications that report the use of PS–UCNPs for the effective *in vitro* and *in vivo* treatment of cancer following PDT, these nanoparticle-based tools can be thought of as the next generation of nanodrugs for PDT. The PS–UCNPs nanotools exhibit several advantages for PDT cancer therapy including the following.

- Photosensitiser–UCNPs can be excited with NIR light. This facilitates the treatment of cancers in deep-lying tissue while minimising any photodamage.
- Following NIR irradiation, the UCNPs emit light at different wavelengths that can be reabsorbed by the vast majority of currently used PSs to induce production of ROS.
- UCNPs can be used as delivery vehicles for *in vivo* PDT due to the ease of functionalisation of the nanomaterial *e.g.* with PEG, PEI, silica layer, or BSA.
- The specific delivery of UCNPs to tumours can easily be achieved by the EPR effect due to the size of the nanoparticle. However, the surface of the UCNP can also be readily functionalised with targeting agents including FA, antibodies, or peptides.
- Multifunctional UCNPs can be designed that enable multimodal imaging, such as MRI and UCL imaging, in combination with multitherapy including photodynamic therapy, photothermal therapy, chemotherapy, and radiotherapy. Multifunctional UCNPs have been shown to enhance tumour destruction *via* combined therapies and allow image-guided therapies; being therefore suitable tools for theranostics (therapy + diagnosis).

A potential negative aspect of the UCNP systems reported to date is the requirement for intratumoral injection of nanoparticles for successful *in vivo* PDT. However, this drawback may have been overcome with the introduction of Nd^{3+}-doped UCNPs that can be excited with 808 nm irradiation, thus avoiding the water absorption of the typically used 980 nm light. Consequently, it is apparent that these innovative UCNPs will find utility in the future developments of PDT cancer therapy, particularly for tumours that are difficult to treat.

Table 7.2 Summary of publications focusing on the use of UCNPs for *in vivo* PDT.[a]

Photosensitiser (PS)	UCNP λ_{exc} for $^1O_2$	Conjugation of the PS to the UCNP	Water solubility provided by	Targeting agent	Tumour studied	Therapy	Reference
Ce6	NaYF$_4$:Yb,Er 980 nm	Non-covalent physical adsorption	PEG-amphiphilic polymer	n/a	4T1	PDT	59
Ce6	NaYMnF$_4$:Yb,Er 980 nm	Non-covalent physical adsorption	PEG-polymer	n/a	4T1	PDT	61
Ce6	Core-shell NaYF$_4$:Yb,Er/ NaGdF$_4$ 980 nm	Non-covalent physical adsorption and covalent attachment	PEG-phospholipid	n/a	U87-MG	PDT	62
ZnPc(COOH)$_4$	NaYF$_4$:Yb,Er 980 nm	Covalent attachment	PEG	Folic acid	Hepa1-6	PDT	63
MC-540 and ZnPc	NaYF$_4$:Yb,Er 980 nm	Non-covalent silica encapsulation	Mesoporous silica and PEG	Folic acid	B16-F0	PDT	65
Ce6	NaYMnF$_4$:Yb,Er 980 nm	Non-covalent silica encapsulation	Mesoporous silica	n/a	KB	PDT and chemotherapy	66
HP	NaGdF$_4$:Yb,Er,Tm/ NaGdF$_4$ 980 nm	Non-covalent physical adsorption and covalent grafting	Mesoporous silica	n/a	4T1	PDT, radiotherapy, and chemotherapy	67
ZnPc	GdOF:Ln 980 nm	Covalent attachment	Silica	Folic acid	H22	PDT, PTT, and chemotherapy	68
SPCD	NaYF$_4$:Yb,Er,Gd 980 nm	Non-covalent silica encapsulation	Silica	n/a	HeLa	PDT and chemotherapy	69

ZnPc	NaYF$_4$:Yb,Er 980 nm	Non-covalent hydro-phobic interactions	Amphiphilic chitosan	n/a	S180	PDT	70
ZnPc	NaYF$_4$:Yb,Er 980 nm	Non-covalent hydro-phobic interactions	Amphiphilic chitosan	Folic acid	Bel-7402	PDT	71
RB	NaGdF$_4$:Yb,Er 980 nm	Non-covalent hydro-phobic interactions	BSA	n/a	4T1	PDT and PTT	72
ZnPc–COOH	LiYF$_4$:Yb,Er 980 nm	Non-covalent electro-static interactions	Polyvinylpyrro-lidone	n/a	H22	PDT	73
ALA	NaYF$_4$:Yb,Er 980 nm	Covalent attachment	PAA	n/a	4T1	PDT	74
TiO$_2$	NaGdF$_4$:Yb,Tm 980 nm	Deposition	Silica	Folic acid	MCF-7	PDT	75
TiO$_2$	NaYF$_4$:Yb,Tm 980 nm	Deposition	Silica and PEG	PEG–maleimide	OSCC	PDT	76
TiO$_2$	NaYF$_4$:Yb,Tm 980 nm	Deposition	Polymer–PEG	Folic acid	MCF-7, MCF-7/AR	PDT and chemotherapy	77
TiO$_2$	Y$_2$Ti$_2$O$_7$, YOF:Yb,Tm 980 nm	Deposition	TiO$_2$	Folic acid	H22	PDT and PTT	78

[a]ALA, 5-aminolevulinic acid; BSA, bovine serum albumin; Ce6, chlorin e6; HP, hematoporphyrin; M-540, merocyanine 540; ^{1}O$_2$, singlet oxygen; PAA, poly(acrylic acid); PDT, photodynamic therapy; PEG, polyethylene glycol; PS, photosensitiser; PTT, photothermal therapy; RB, Rose Bengal; SPCD, silicon phthalocyanine dihydroxide; UCNP, upconversion nanoparticles; ZnPc, zinc phthalocyanine.

Acknowledgement

The authors are grateful to the School of Chemistry, University of East Anglia, for financial support.

References

1. T. J. Dougherty, C. J. Gomer, B. W. Henderson, G. Jori, D. Kessel, M. Korbelik, J. Moan and Q. Peng, *J. Natl. Cancer Inst.*, 1998, **90**, 889.
2. M. C. DeRosa and R. J. Crutchley, *Coord. Chem. Rev.*, 2002, **233–234**, 351.
3. A. B. Ormond and H. S. Freeman, *Materials*, 2013, **6**, 817.
4. G. Obaid and D. A. Russell, in *Handbook of Photomedicine*, ed. M. R. Hamblin and Y.-Y. Huang, Taylor & Francis, CRC Press, Boca Raton, FL, 2013, pp. 367–378.
5. A. K. Iyer, G. Khaled, J. Fang and H. Maeda, *Drug Discovery Today*, 2006, **11**, 812.
6. H. Maeda, J. Wu, T. Sawa, Y. Matsumura and K. Hori, *J. Controlled Release*, 2000, **65**, 271.
7. H. Maeda, J. Fang, T. Inutsuka and Y. Kitamoto, *Int. Immunopharmacol.*, 2003, **3**, 319.
8. J. Fang, H. Nakamura and H. Maeda, *Adv. Drug Delivery Rev.*, 2011, **63**, 136.
9. D. K. Chatterjee, L. S. Fong and Y. Zhang, *Adv. Drug Delivery Rev.*, 2008, **60**, 1627.
10. G. F. Paciotti, D. G. I. Kingston and L. Tamarkin, *Drug Dev. Res.*, 2006, **67**, 47.
11. S. H. Voon, L. V. Kiew, H. B. Lee, S. H. Lim, M. I. Noordin, A. Kamkaew, K. Burgess and L. Y. Chung, *Small*, 2014, **10**, 4993.
12. G. Obaid, I. Chambrier, M. J. Cook and D. A. Russell, *Photochem. Photobiol. Sci.*, 2015, **14**, 737.
13. M. Camerin, M. Moreno, M. J. Marín, C. L. Schofield, I. Chambrier, M. J. Cook, O. Coppellotti, G. Jori and D. A. Russell, *Photochem. Photobiol. Sci.*, 2016, DOI: 10.1039/c5pp00463b.
14. G. Obaid, I. Chambrier, M. J. Cook and D. A. Russell, *Angew. Chem., Int. Ed.*, 2012, **51**, 6158.
15. T. Stuchinskaya, M. Moreno, M. J. Cook, D. R. Edwards and D. A. Russell, *Photochem. Photobiol. Sci.*, 2011, **10**, 822.
16. M. Camerin, M. Magaraggia, M. Soncin, G. Jori, M. Moreno, I. Chambrier, M. J. Cook and D. A. Russell, *Eur. J. Cancer*, 2010, **46**, 1910.
17. M. E. Wieder, D. C. Hone, M. J. Cook, M. M. Handsley, J. Gavrilovic and D. A. Russell, *Photochem. Photobiol. Sci.*, 2006, **5**, 727.
18. D. C. Hone, P. I. Walker, R. Evans-Gowing, S. FitzGerald, A. Beeby, I. Chambrier, M. J. Cook and D. A. Russell, *Langmuir*, 2002, **18**, 2985.
19. A. Sedlmeier and H. H. Gorris, *Chem. Soc. Rev.*, 2015, **44**, 1526.
20. V. Biju, *Chem. Soc. Rev.*, 2014, **43**, 744.

21. V. Muhr, S. Wilhelm, T. Hirsch and O. S. Wolfbeis, *Acc. Chem. Res.*, 2014, **47**, 3481.
22. M. Haase and H. Schäfer, *Angew. Chem., Int. Ed.*, 2011, **50**, 5808.
23. V. Shanmugam, S. Selvakumar and C.-S. Yeh, *Chem. Soc. Rev.*, 2014, **43**, 6254.
24. A. G. Arguinzoniz, E. Ruggiero, A. Habtemariam, J. Hernández-Gil, L. Salassa and J. C. Mareque-Rivas, *Part. Part. Syst. Charact.*, 2014, **31**, 46.
25. L. Cheng, C. Wang and Z. Liu, *Nanoscale*, 2013, **5**, 23.
26. S. Jiang, M. K. Gnanasammandhan and Y. Zhang, *J. R. Soc., Interface*, 2010, **7**, 3.
27. C. Wang, L. Cheng and Z. Liu, *Theranostics*, 2013, **3**, 317.
28. P. Zhang, W. Steelant, M. Kumar and M. Scholfield, *J. Am. Chem. Soc.*, 2007, **129**, 4526.
29. Y. Guo, M. Kumar and P. Zhang, *Chem. Mater.*, 2007, **19**, 6071.
30. X. Liu, X. Kong, Y. Zhang, L. Tu, Y. Wang, Q. Zeng, C. Li, Z. Shi and H. Zhang, *Chem. Commun.*, 2011, **47**, 11957.
31. Y. Wang, K. Liu, X. Liu, K. I. Dohnalová, T. Gregorkiewicz, X. Kong, M. C. G. Aalders, W. J. Buma and H. Zhang, *J. Phys. Chem. Lett.*, 2011, **2**, 2083.
32. D. K. Chatterjee and Y. Zhang, *Nanomedicine*, 2008, **3**, 73.
33. A. Zhou, Y. Wei, B. Wu, Q. Chen and D. Xing, *Mol. Pharmaceutics*, 2012, **9**, 1580.
34. B. Ungun, R. K. Prud'homme, S. J. Budijon, J. Shan, S. F. Lim, Y. Ju and R. Austin, *Opt. Express*, 2009, **17**, 80.
35. J. Shan, S. J. Budijono, G. Hu, N. Yao, Y. Kang, Y. Ju and R. K. Prud'homme, *Adv. Funct. Mater.*, 2011, **21**, 2488.
36. K. Liu, X. Liu, Q. Zeng, Y. Zhang, L. Tu, T. Liu, X. Kong, Y. Wang, F. Cao, S. A. G. Lambrechts, M. C. G. Aalders and H. Zhang, *ACS Nano*, 2012, **6**, 4054.
37. M. Gonzalez-Bejar, M. Liras, L. Frances-Soriano, V. Voliani, V. Herranz-Perez, M. Duran-Moreno, J. M. Garcia-Verdugo, E. I. Alarcon, J. C. Scaiano and J. Pérez-Prieto, *J. Mater. Chem. B*, 2014, **2**, 4554.
38. X. Chen, Z. Zhao, M. Jiang, D. Que, S. Shi and N. Zheng, *New J. Chem.*, 2013, **37**, 1782.
39. L. Y. Zeng, L. C. Xiang, W. Z. Ren, J. J. Zheng, T. H. Li, B. Chen, J. C. Zhang, C. W. Mao, A. G. Li and A. G. Wu, *RSC Adv.*, 2013, **3**, 13915.
40. Y. Yuan, Y. Min, Q. Hu, B. Xing and B. Liu, *Nanoscale*, 2014, **6**, 11259.
41. X. Wang, K. Liu, G. Yang, L. Cheng, L. He, Y. Liu, Y. Li, L. Guo and Z. Liu, *Nanoscale*, 2014, **6**, 9198.
42. X. Liu, M. Zheng, X. Kong, Y. Zhang, Q. Zeng, Z. Sun, W. J. Buma and H. Zhang, *Chem. Commun.*, 2013, **49**, 3224.
43. H. S. Qian, H. C. Guo, P. C.-L. Ho, R. Mahendran and Y. Zhang, *Small*, 2009, **5**, 2285.
44. H. Guo, H. Qian, N. M. Idris and Y. Zhang, *Nanomed. Nanotechnol. Biol. Med.*, 2010, **6**, 486.
45. X.-F. Qiao, J.-C. Zhou, J.-W. Xiao, Y.-F. Wang, L.-D. Sun and C.-H. Yan, *Nanoscale*, 2012, **4**, 4611.

46. Z. Zhao, Y. Han, C. Lin, D. Hu, F. Wang, X. Chen, Z. Chen and N. Zheng, *Chem. Asian J.*, 2012, **7**, 830.
47. F. Wang, X. Yang, L. Ma, B. Huang, N. Na, Y. E, D. He and J. Ouyang, *J. Mater. Chem.*, 2012, **22**, 24597.
48. X. Yang, Q. Xiao, C. Niu, N. Jin, J. Ouyang, X. Xiao and D. He, *J. Mater. Chem. B*, 2013, **1**, 2757.
49. G. Tian, W. Ren, L. Yan, S. Jian, Z. Gu, L. Zhou, S. Jin, W. Yin, S. Li and Y. Zhao, *Small*, 2013, **9**, 1929.
50. G. Tian, Z. J. Gu, L. J. Zhou, W. Y. Yin, X. X. Liu, L. Yan, S. Jin, W. L. Ren, G. M. Xing, S. J. Li and Y. L. Zhao, *Adv. Mater.*, 2012, **24**, 1226.
51. H. Wang, Z. Liu, S. Wang, C. Dong, X. Gong, P. Zhao and J. Chang, *ACS Appl. Mater. Interfaces*, 2014, **6**, 3219.
52. X. Wang, C.-X. Yang, J.-T. Chen and X.-P. Yan, *Anal. Chem.*, 2014, **86**, 3263.
53. L. Zhou, Z. Li, Z. Liu, M. Yin, J. Ren and X. Qu, *Nanoscale*, 2014, **6**, 1445.
54. X. Wei, W. Wang and K. Chen, *J. Phys. Chem. C*, 2013, **117**, 23716.
55. M. Yin, E. Ju, Z. Chen, Z. Li, J. Ren and X. Qu, *Chem.–Eur. J.*, 2014, **20**, 14012.
56. N. M. Idris, S. S. Lucky, Z. Li, K. Huang and Y. Zhang, *J. Mater. Chem. B*, 2014, **2**, 7017.
57. D. Wang, B. Xue, X. Kong, L. Tu, X. Liu, Y. Zhang, Y. Chang, Y. Luo, H. Zhao and H. Zhang, *Nanoscale*, 2015, **7**, 190.
58. F. Ai, Q. Ju, X. Zhang, X. Chen, F. Wang and G. Zhu, *Sci. Rep.*, 2015, **5**, 10785.
59. C. Wang, H. Tao, L. Cheng and Z. Liu, *Biomaterials*, 2011, **32**, 6145.
60. G. Chen, H. Qiu, P. N. Prasad and X. Chen, *Chem. Rev.*, 2014, **114**, 5161.
61. C. Wang, L. Cheng, Y. Liu, X. Wang, X. Ma, Z. Deng, Y. Li and Z. Liu, *Adv. Funct. Mater.*, 2013, **23**, 3077.
62. Y. I. Park, H. M. Kim, J. H. Kim, K. C. Moon, B. Yoo, K. T. Lee, N. Lee, Y. Choi, W. Park, D. Ling, K. Na, W. K. Moon, S. H. Choi, H. S. Park, S.-Y. Yoon, Y. D. Suh, S. H. Lee and T. Hyeon, *Adv. Mater.*, 2012, **24**, 5755.
63. L. Xia, X. Kong, X. Liu, L. Tu, Y. Zhang, Y. Chang, K. Liu, D. Shen, H. Zhao and H. Zhang, *Biomaterials*, 2014, **35**, 4146.
64. M. E. Lim, Y.-l. Lee, Y. Zhang and J. J. H. Chu, *Biomaterials*, 2012, **33**, 1912.
65. N. M. Idris, M. K. Gnanasammandhan, J. Zhang, P. C. Ho, R. Mahendran and Y. Zhang, *Nat. Med.*, 2012, **18**, 1580.
66. S. Yang, N. Li, Z. Liu, W. Sha, D. Chen, Q. Xu and J. Lu, *Nanoscale*, 2014, **6**, 14903.
67. W. Fan, B. Shen, W. Bu, F. Chen, Q. He, K. Zhao, S. Zhang, L. Zhou, W. Peng, Q. Xiao, D. Ni, J. Liu and J. Shi, *Biomaterials*, 2014, **35**, 8992.
68. R. Lv, P. Yang, F. He, S. Gai, C. Li, Y. Dai, G. Yang and J. Lin, *ACS Nano*, 2015, **9**, 1630.
69. Y. Liu, Y. Liu, W. Bu, C. Cheng, C. Zuo, Q. Xiao, Y. Sun, D. Ni, C. Zhang, J. Liu and J. Shi, *Angew. Chem., Int. Ed.*, 2015, **54**, 8105.
70. S. Cui, H. Chen, H. Zhu, J. Tian, X. Chi, Z. Qian, S. Achilefu and Y. Gu, *J. Mater. Chem.*, 2012, **22**, 4861.

71. S. Cui, D. Yin, Y. Chen, Y. Di, H. Chen, Y. Ma, S. Achilefu and Y. Gu, *ACS Nano*, 2013, 7, 676.
72. Q. Chen, C. Wang, L. Cheng, W. He, Z. Cheng and Z. Liu, *Biomaterials*, 2014, **35**, 2915.
73. M. Wang, Z. Chen, W. Zheng, H. Zhu, S. Lu, E. Ma, D. Tu, S. Zhou, M. Huang and X. Chen, *Nanoscale*, 2014, **6**, 8274.
74. A. Punjabi, X. Wu, A. Tokatli-Apollon, M. El-Rifai, H. Lee, Y. Zhang, C. Wang, Z. Liu, E. M. Chan, C. Duan and G. Han, *ACS Nano*, 2014, **8**, 10621.
75. L. Zhang, L. Zeng, Y. Pan, S. Luo, W. Ren, A. Gong, X. Ma, H. Liang, G. Lu and A. Wu, *Biomaterials*, 2015, **44**, 82.
76. S. S. Lucky, N. M. Idris, Z. Li, K. Huang, K. C. Soo and Y. Zhang, *ACS Nano*, 2015, **9**, 191.
77. L. Y. Zeng, Y. W. Pan, Y. Tian, X. Wang, W. Z. Ren, S. J. Wang, G. M. Lu and A. G. Wu, *Biomaterials*, 2015, **57**, 93.
78. R. Lv, C. Zhong, R. Li, P. Yang, F. He, S. Gai, Z. Hou, G. Yang and J. Lin, *Chem. Mater.*, 2015, **27**, 1751.

CHAPTER 8

Near Infrared Nanomaterials for Triggered Drug and Gene Delivery

BEI LIU[a,b], CHUNXIA LI*[a], AND JUN LIN*[a]

[a]State Key Laboratory of Rare Earth Resource Utilization, Changchun Institute of Applied Chemistry, Chinese Academy of Sciences, Changchun, 130022, P. R. China; [b]University of Chinese Academy of Sciences, Beijing, 100049, P. R. China
*E-mail: jlin@ciac.ac.cn, cxli@ciac.ac.cn

8.1 General Introduction

Recently, the treatment of cancer with nanoparticle (NP)-based therapeutics has become a hotspot because nanocarrier-mediated drug delivery can effectively overcome the intrinsic defects of conventional therapeutic drugs, such as poor water solubility, lack of targeting capability, non-specific distribution, systemic toxicity, and low therapeutic index. However, most NP-based therapeutics release their loaded cargos in a sustained and continuous way, resulting in decreased drug accumulation in the targeted cancer cells/tissues. Therefore, smart control over the release patterns and profiles in response to physiological conditions or external stimuli is urgently needed: on one hand, this can effectively increase the drug accumulation at targeted cancer cells/tissues, drastically decrease the systemic toxicity, and optimize therapeutic

RSC Nanoscience & Nanotechnology No. 40
Near Infrared Nanomaterials: Preparation, Bioimaging and Therapy Applications
Edited by Fan Zhang
© The Royal Society of Chemistry 2016
Published by the Royal Society of Chemistry, www.rsc.org

outcomes for the cancer therapy; on the other hand, a delivery system that responds rapidly to a stimulus can be employed for real-time manipulation of drug dosage to potentially avoid under- or over-dosing. With respect to the biological system, the stimuli used to trigger the release of biomolecules can be broadly classified as either internal (*e.g.*, pH value, ion concentration, small molecules, enzymes)[1] or external (*e.g.*, light, ultrasound, electric field, magnetic field, heating).[2] Among the stimulus-sensitive drug or gene delivery nanocarriers, light-responsive vehicles are particularly attractive because of their non-invasive nature and the high spatial resolution of light. In recent years, researchers have reported many light-responsive vehicles for the delivery of drugs or genes for cancer treatment. For example, Mal successfully illustrated the release of drugs under ultraviolet (UV) irradiation using coumarin dimers as photocleavable molecular gates.[3] However, one of the major hurdles of light-responsive vehicles is their low penetration and potential damage to normal tissues, since most light-responsive nanocarriers need UV/Vis light for controllable drug release. The strong light absorption and scattering of UV/Vis light has seriously limited *in vivo* applications. In order to solve this problem, near infrared (NIR) light is preferred as the excitation source to trigger the release of the loaded drugs or genes owing to its high spatiotemporal resolution, high tissue penetration ability, and minimal photodamage. The exploration of NIR-responsive drug delivery systems opens up a new and exciting possibility for nanomedicine. Here, different kinds of nanocarriers or methods are illustrated for NIR-triggered drug or gene delivery, including the overall introduction of nanocarriers (Section 8.2); three types of strategies to achieve NIR light-based control over the delivery process: photoresponsive nanocarriers (Section 8.3), photocaging of bioactive cargos (Section 8.4), and photothermal transduction for NIR-triggered nanocarriers (Section 8.5). Finally we discuss current challenges for NIR nanomaterials and potential solutions.

8.2 Nanocarriers for NIR-Triggered Drug or Gene Release

8.2.1 Introduction

Over the past several decades, remarkable progress has been made in the development and application of engineered drug or gene nanocarriers with optimal sizes, shapes, and surface properties. These nanocarriers could not only help to increase the solubility and prolong the circulation half-life, but also effectively improve the biodistribution and reduce the immunogenicity of their cargos. To date, many materials, *e.g.*, organic nanomaterials, such as liposomes, block copolymers, dendrimers; and inorganic nanomaterials, such as microporous silica (MSN), Au, Ag, CuS NPs; and some innovative organic–inorganic hybrid composites, have been successfully utilized as drug or gene carriers. Each kind of nanomaterial has its own advantages. For example, organic nanomaterials have been used for decades because of their low toxicity and some of them have been approved for human clinical use by the United States Food and Drug Administration (FDA). Inorganic-based vectors have also been widely investigated because of their possible enhanced stability in biological systems. Organic–inorganic hybrid composites can combine the respective merits of organic and inorganic materials. Some nanocarriers, with sizes controlled typically in the range of 1–100 nm and compositions/structures engineered to load biomolecules in a variety of configurations, have been approved for cancer therapy *in vitro/in vivo*, and are poised to significantly improve the treatment outcomes for cancer diseases.

Drug or gene delivery nanocarriers with good performances for NIR-triggered release should have two key features: (1) good stability for secure protection and delivery of the loaded cargos and (2) the ability to release the encapsulated cargos upon NIR irradiation. To realize these two key features, it is necessary to produce smart vesicles. The smart vesicle membrane should have good integrity, but with specific channels to facilitate the release of the encapsulated active species under NIR light. The following is a brief introduction to nanocarriers for NIR-triggered release.

8.2.2 Organic Nanomaterials

Light-triggered drug or gene delivery systems using polymeric materials have been intensively investigated for applications in biomedicine and tissue engineering, owing to their versatility. Biocompatible systems that are capable of storing and gradually releasing active molecules are mainly organic. There are many kinds of organic nanomaterials suitable for drug or gene delivery, such as micelles, liposomes, cyclodextrins, hydrogels, or polymeric NPs. In order to construct organic nanocarriers with good performance for light-triggered release of biomolecules, some photoresponsive functional groups, such as azobenzene, pyrene, nitrobenzene, cinnamoyl, or spirobenzopyran,[4] which can change their conformations or other properties upon irradiation by light of a suitable wavelength,[5] must be introduced to improve the photosensitivity. After proper modification of photoresponsive functional groups in the organic nanomaterials, these vehicles will generally be photoresponsive in themselves. On irradiation by light, the nanomaterial vehicle will disrupt and release the loaded biomolecules. A well-known example is the reversible molecular switching of the azobenzene group (and its derivatives), which can undergo phase transformation from *trans* to *cis* when irradiated by UV light (300–380 nm) and from *cis* to *trans* when irradiated by visible light, allowing for photoregulated control of biomolecule release.[6]

Four different kinds of NIR-responsive organic nanomaterials—micelles, biodegradable polypeptides, liposomes, and hydrogels—are discussed in Section 8.4 and their drug or gene release profiles upon NIR irradiation are considered in detail.

8.2.3   Inorganic Nanomaterials

Inorganic nanomaterials have been employed for storage of active molecules because of their rich variety, precision in size/shape control, excellent physicochemical properties, and multifunctionality, although their inability to degrade *in vivo* has somewhat limited their scope of application. There are many kinds of inorganic drug or gene delivery systems, such as MSNs, hollow Y_2O_3, Fe_3O_4 NPs, Au nanostructures, and so on.

Although organic nanocarriers have progressed to later stages of clinical tests or applications, and a few of them, such as liposomes, polymer micelles, and protein NPs, have been approved by the FDA, most of the nanomedicines based on inorganic nanocarriers are still at the preclinical or basic research stage. However, it is worth noting that Au nanostructures and MSNs have already received FDA approval for clinical trials. In fact, these two are currently the most promising inorganic nanobiomaterials and much research has focused on them.[7] In what follows, we consider Au nanostructures and MSNs as examples to illustrate the typical properties of inorganic nanocarriers. The further construction of NIR-triggered drug or gene delivery systems based on Au or MSNs is also presented here.

Au nanostructures have gained a lot of attention in the field of nanomedicine because of their advantageous properties. For example, unlike the normal drug or gene nanocarriers, Au nanostructures can not only load the biomolecules, but also possess pronounced photothermal properties for direct cancer therapy, in addition to a variety of optical properties useful for diagnostics (*e.g.*, fluorescence for Au nanoclusters,[8] multiphoton luminescence for Au nanorods and nanocages,[9] and strong optical absorption for all of them). In order to load the biomolecules, solid Au nanostructures (*e.g.*, nanospheres, nanostars, and nanorods)[10] must be further modified while Au structures with hollow interiors (*e.g.*, nanoshells and nanocages)[11] can efficiently encapsulate drugs within their cavities. The construction of NIR-triggered nanocarriers generally takes advantage of the capability of Au nanostructures to transduce the absorbed NIR light into heat, which can be used to enhance the rapid release of drug molecules.[12] A brief discussion of this system can be found in Section 8.5.

Since their introduction by Kuroda's and Kresge's groups in the 1990s,[13] MSNs have experienced many developments owing to their unique features, such as large surface area and pore volume, high chemical and thermal stability, excellent biocompatibility, and versatile chemistry for further functionalization. Moreover, these NPs, composed of highly ordered mesoporous silica structures with uniform but adjustable pore size, can provide a physical encapsulation that protects the entrapped cargos from degradation and denaturization. Many researchers have developed different kinds of MSNs as candidates for the accommodation of guest molecules, such as MSN nanospheres, hollow nanospheres, asymmetric single-hole mesoporous nanocages, and hollow ellipsoidal MSNs, as shown in Figure 8.1a–d. These types of MSN nanomedicines have exhibited great potential for enhancing

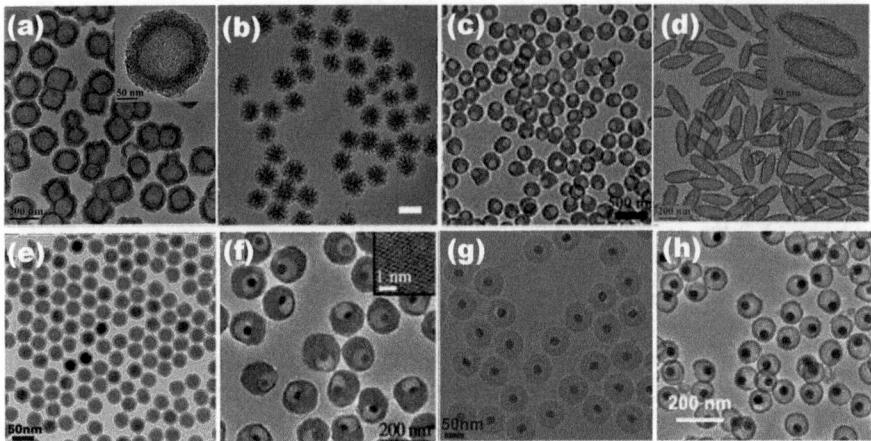

Figure 8.1   Transmission electron microscopy (TEM) images of various types of MSN and UCNPs@mSiO$_2$ nanostructures: (a) hollow MSN nanospheres, (b) MSN nanospheres, (c) asymmetric single-hole mesoporous nanocages, (d) hollow ellipsoidal MSN NPs, (e) UCNP@SiO$_2$ with a thin and dense silica layer, (f) eccentric single-hole UCNP@mSiO$_2$ nanorattles, (g) centric UCNPs@mSiO$_2$ nanospheres, (h) yolk–shell structured UCNPs@mSiO$_2$ nanospheres. Figures (a) and (d) reprinted from ref. 13a with permission from the Royal Society of Chemistry; figure (b) reprinted from ref. 13b with permission from the Royal Society of Chemistry; figures (c) and (f) reprinted with permission from X. Li, L. Zhou, Y. Wei, A. M. El-Toni, F. Zhang and D. Zhao, *J. Am. Chem. Soc.*, 2015, **137**, 5903. Copyright (2015) American Chemical Society;[13c] figure (e) reproduced from ref. 14a with permission from John Wiley and Sons. Copyright © 2014 Wiley-VCH GmbH & Co. KGaA, Weinheim; figure (g) reproduced from ref. 14b with permission from John Wiley and Sons. Copyright © 2013 Wiley-VCH Verlag GmbH & Co. KGaA, Weinheim; figure (h) reproduced from L. Zhao, J. Peng, M. Chen, Y. Liu, L. Yao, W. Feng and F. Li, *ACS Appl. Mater. Interfaces*, 2014, **6**, 11190.[14c] Copyright (2014) American Chemical Society.

anticancer efficacy and decreasing the toxic side effects of chemotherapeutic drugs. In particular, multifunctional composites of upconversion NPs (UCNPs) and MSNs have also been designed, synthesized, and applied in biomolecule delivery by taking advantage of the attractive properties of UCNPs and MSNs. Much research[14] has been done to synthesize different kinds of uniform core–shell UCNPs@mSiO$_2$ NPs, as shown in Figure 8.1e–h. Our group has also successfully construct an anticancer drug nanocarrier NaYF$_4$:Yb,Er@NaGdF$_4$:Yb@mSiO$_2$, in which the mesoporous silica is directly coated onto the UCNPs. Although MSN or UCNPs@mSiO$_2$ have been developed as good delivery nanocarriers, further functionalization is needed in order to obtain a NIR-triggered drug delivery system. To date, NIR delivery nancarriers based on MSNs or UCNPs@mSiO$_2$ can be realized by functionalizing them with light-responsive pore blocking caps, such as NPs, supermolecular assemblies, and large molecules, as discussed in Section 8.3.

8.2.4   Organic–Inorganic Hybrid Composites

As a major requirement for the design of NP carriers, the composition of vehicles needs to be precisely engineered and optimized to achieve drug release at therapeutically optimal rates and dose regimes. As discussed above, NP carriers have been prepared and tested using a wide variety of organic or inorganic materials. However, these nanocarriers still have some limitations. For example, the inorganic carriers have good thermal or chemical stability but poor biocompatibility, while the organic carriers have good biocompatibility but poor mechanical properties.[15] Combining their respective merits in the rational design and construction of multifunctional organic–inorganic hybrid composites is of great importance.

There are three main kinds of strategies for building the organic–inorganic hybrid NPs to act as NIR-triggered nanocarriers:

- Photoresponsive polymer molecules can be modified inside the mesoporous external walls of inorganic NPs.
- The external mesoporous shells of inorganic NPs can be functionalized with photoresponsive organic groups to reduce the free diameter of the pores, thus controlling the delivery of the bioactive drugs.
- The formation of a novel organosilica shell (organic fragments bonded covalently to silica units) could allow the controlled release of loaded biomolecules.

These strategies allow the preparation of new multifunctional organic–inorganic hybrid cargo delivery vehicles with greatly improved as-prepared properties, in which the polymer component improves hydrophilicity and controls the biomolecule release profiles, while the inorganic component supports the polymer and supplies good mechanical properties.

8.3   Photoresponsive Nanocarriers

8.3.1   Introduction

To construct a NIR-controlled NP-based delivery system for the treatment of cancer, the most direct solution is to use a nanocarrier that is itself photoresponsive. Upon NIR irradiation, the photoresponsive nanocarrier is disrupted and releases its anticancer drug cargo. NIR light not only penetrates deeper into tissue and has less detrimental effect on healthy cells than UV/Vis light, but can also provide greater selectivity in terms of control over the time and the location of nanocarrier disruption. Generally speaking, photoresponsive or photocontrol nanocarriers can be realized by modifying them with photosensitive moieties (*e.g.*, azobenzene, pyrene, nitrobenzene, cinnamoyl, spirobenzopyran) that can change their conformation or other properties upon light irradiation, making the resulting particles photoresponsive. After suitable modification, photochemical reactions, such as *trans–cis* isomerization, molecular dimerization, and bond cleavage, can lead to the disruption of polymeric NPs as a result of these photoinduced structural and/or property changes. Much exciting work has been done to encapsulate drug or gene molecules in this kind of light-responsive polymeric biomaterials. Here, we have divided those NIR photoresponsive or photocontrol nanocarriers into four main species: NIR-responsive micelle materials (Section 8.3.2); NIR-responsive biodegradable polypeptide materials (Section 8.3.3); NIR-responsive liposomes (Section 8.3.4); and NIR-responsive hydrogels (Section 8.3.5).

8.3.2   NIR-Responsive Micelle Materials

Micelles, which are generally formed by the self-aggregation of amphiphilic block copolymer in an aqueous medium, have emerged as a novel drug delivery system with good prospects. In fact, they have found extensive applications in cargo delivery because of their ability to encapsulate and retain hydrophobic cargos in their self-assembled interior. Encapsulating biomolecules inside polymeric micelles has many advantages: it can enhance the efficacy of the loaded drugs or genes and improve their pharmacokinetics *in vivo*; it can also protect protein drugs or genes from enzymatic digestion, thus leading to a long blood circulation time. Despite significant advances in drug or gene delivery, micelle therapeutics still have some hurdles to overcome before they can be used for *in vivo* bioapplications. One of the main problems is their extensive distribution throughout the body. To solve this problem, NIR-responsive micelles have been developed that can control the stability of the drug complex and release drugs at the tumor site by NIR triggering. In this section we discuss NIR-triggered functional polymeric micelles, which have been increasingly investigated as drug/gene delivery systems.

In order to make polymeric micelles NIR-sensitive, the polymer must contain photochromic groups with a photoreaction that increases the polarity of the polymer and shifts the hydrophilic–hydrophobic balance toward micellar disruption. In other words, when exposed to NIR irradiation, the hydrophobic tail of molecules constituting the micelle becomes hydrophilic and the micelle is destroyed, releasing the loaded cargos. Several approaches to regulating the structure of micelles using light have been proved effective, but there remains a major common concern which hampers their potential use in biomedical applications: most of the explored photoreactions require high-energy UV/Vis light, rather than longer-wavelength NIR light that penetrates deeper into tissue and is less detrimental to healthy cells. Much research has been devoted to finding alternative methods.

One possibility is to use micelles with high two-photon absorption of NIR light. In order to obtain two-photon absorption micelles, some moieties with two-photon absorption of NIR light need to be modified into the micelles. For example, Zhao and his coworkers have reported the syntheses of two kinds of amphiphilic block copolymer (BCP), of which the hydrophobic block is a polymethacrylate that contains either *o*-nitrobenzyl or coumarin moieties exhibiting two-photon absorption of NIR light.[16] The strong NIR absorption in these polymers induced photolysis of the chromophores, resulting in their removal from the polymethacrylate and its conversion to a hydrophilic poly(methacrylic acid), which finally led to the irreversible disruption of BCP micelles in aqueous solution, as shown in Figure 8.2a–d. The *o*-nitrobenzyl or coumarin chromophore, with its large two-photon absorption cross-section, plays an important role in this successful disruption of micelles under two-photon NIR irradiation.

In addition to the irreversible disruption triggered by NIR light, reversible photoisomerization reactions of some chromophores, such as the photodimerization of coumarin,[17] have also been explored in order to design photocontrollable BCP micelles. The use of reversible photoisomerization or irreversible photocleavage reactions of various chromophores helps to design reversible or irreversible light-dissociable BCP micelles.

However, the photoreactions activated by two-photon absorption of NIR light are generally slow and inefficient because the two-photon-absorbing cross-sections of the chromophores are typically low. Moreover, the simultaneous absorption of two photons necessitates high laser power density and thus requires the use of a femtosecond pulse laser. An attractive alternative way of using NIR light is based on lanthanide-doped UCNPs, which can absorb NIR light and convert it to higher-energy photons in the UV, visible, or NIR regions. In contrast to two-photon absorption, the excitation of UCNPs by NIR light occurs *via* sequential multiple absorptions with real energy levels, which requires much lower power density, so that a continuous-wave (CW) diode NIR laser is sufficient as the excitation source. UCNPs can be used as NIR-triggered delivery vehicles based on NIR-responsive micelle materials by coating the UV/Vis-regulated BCP polymers on the NP surface. In a detailed example,[18] Bin Yan *et al.* demonstrate a novel strategy enabling the use of a NIR laser to disrupt BCP micelles and trigger the release of their payloads by combining the UCNPs with BCP, as shown in Figure 8.2e–h. In their paper, $NaYF_4$:Yb,Tm UCNPs were encapsulated inside the micelles of poly(ethylene oxide)-*block*-poly(4,5-dimethoxy-2-nitro-benzylmethacrylate) to form a NIR-responsive payload delivery system. When the composite NPs are exposed to 980 nm light, the UCNPs emit photons in the UV region, which can be in turn absorbed by *o*-nitrobenzyl groups on the micelle core-forming block, activating the photocleavage reaction and leading to the dissociation of the micelles. In this way, the release of the coloaded hydrophobic species can be triggered. It is worth noting that the UCNPs can offer additional benefits as well as providing a UV/Vis light source for the breakdown of the polymers: the surface of UCNPs can also be loaded with multiple therapeutic agents to offer synergistic effects, or with other ligands to provide targeting and selectivity, and the luminescent nature of the UCNPs makes them ideal for providing information about where and when the release event has taken place ("release and report"). In fact, UCNPs based on lanthanide-doped $NaYF_4$ NPs have already been widely investigated not only in the field of drug delivery systems, but also in other biological applications.

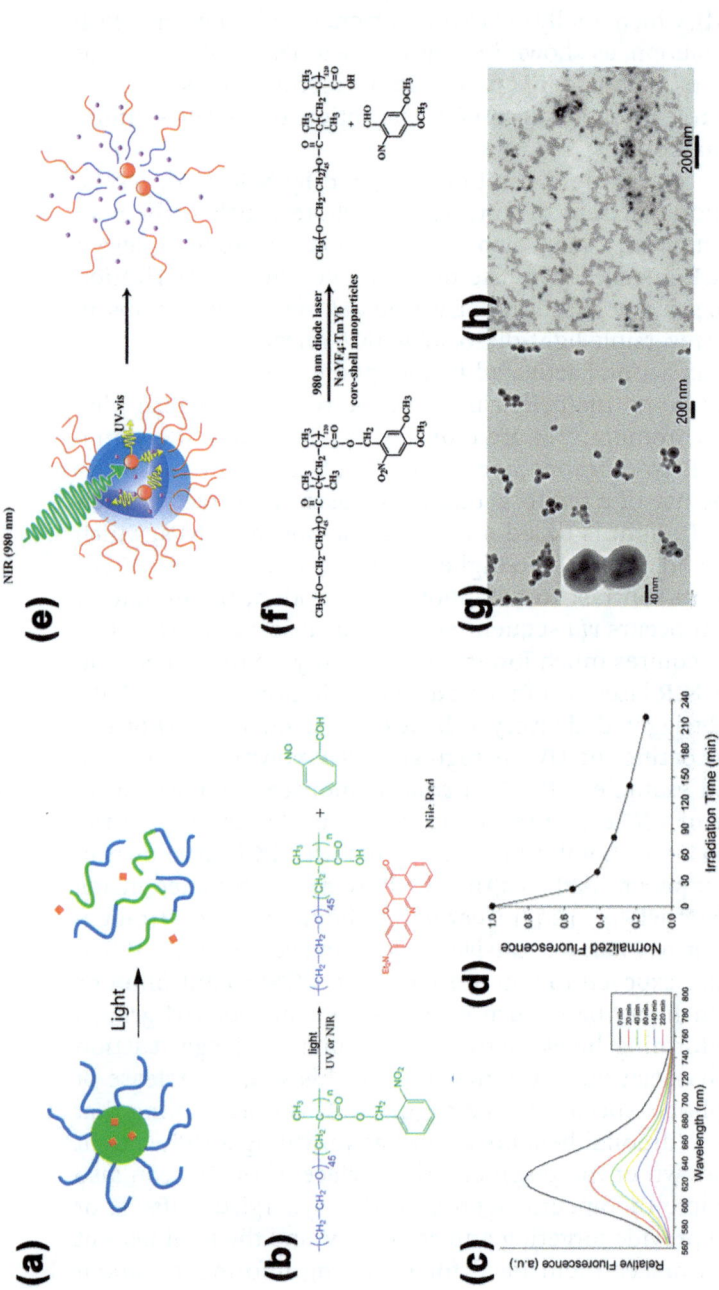

Figure 8.2 (a) Schematic illustration of the photo-controlled release of an encapsulated agent as a result of the photoinduced dissociation of the polymer micelles. (b) Chemical structure and the photolysis of the 2-nitrobenzyl-containing amphiphilic block copolymer; and chemical structure of Nile Red (labeled as NR). (c) Fluorescence spectra of NR-loaded micelles of p3 in aqueous solution upon NIR irradiation (700 nm), showing the decrease and red shift of fluorescence emission due to exposure (release) of NR molecules to water. (d) The plot of normalized fluorescence *vs.* irradiation time. (e) Schematic illustration of using NIR excitation of UCNPs to trigger dissociation of micelles. (f) Chemical structure and NIR-triggered photoreaction of the PEO-*b*-PNBMA. (g) TEM image of UCNPs-loaded micelles before NIR irradiation, the inset showing two magnified micelles containing several NPs. (h) TEM image of the same micellar solution after NIR irradiation (5 W, 4 h), showing the disintegration of micelles. Reprinted with permission from ref. 16a and 18. Figures (a)–(d) reproduced from J. Jiang, X. Tong, D. Morrisand and Y. Zhao, *Macromolecules*, 2006, **39**, 4633.[16a] Copyright (2006) American Chemical Society. Figures (e)–(h) reproduced from B. Yan, J. C. Boyer, N. R. Branda and Y. Zhao, *J. Am. Chem. Soc.*, 2011, **133**, 19714.[18] Copyright (2011) American Chemical Society.

Although increasing efforts have been put into NIR-sensitive NPs for drug/gene/small interfering RNA (siRNA) delivery, these polymeric micelles still have some limitations, such as the lack of highly efficient tumor-targeting properties and NIR sensitivity. Many researchers have focused on developing more effective NIR-triggered polymeric micelles for cancer treatment. For example, Dong and his coworkers introduced a new strategy that simultaneously combines sugar-triggered targeting (active targeting) and two-photon sensitivity to yield a degradable and dendritic micellar nanocarrier, in which the desired sugar residues and light-responsive groups can be modularly conjugated and/or altered.[19] They loaded the clinical anticancer drug doxorubicin (DOX) into the micelles and controlled its release by changing the light irradiation time, which induces the gradual disruption of the micelles in aqueous solution. These sugar-targeted NIR-triggered drug delivery vesicles may have potential for current cancer therapy and nanomedicine.

8.3.3 NIR-Responsive Liposomes

Liposomes, generally consisting of a lipid bilayer with a typical diameter of 50–500 nm, can be formed when lipids are emulsified in an aqueous medium.[20] Interactions between water molecules and amphiphilic lipid molecules allow liposomes to form spontaneously, trapping an aqueous volume within the core of each liposome. This gives liposomes the capability to selectively sequester solutes for encapsulation, forming the basis for biomolecule delivery.[21] Of the all nanomaterial-based drug or gene delivery systems used for anticancer therapy, the liposome-based delivery system is considered as one of the most established and successful platform technologies, owing to its obvious advantages, such as the ability to encapsulate highly toxic or poorly soluble pharmaceuticals and the easy modification of the liposome surface with better efficacy and fewer systemic side effects. Several liposome-based drug and gene delivery systems have been approved for clinical use by the FDA.[22] However, the passive targeting mechanism of the liposomes, *i.e.*, the so-called enhanced permeation and retention (EPR) effect in cancerous tissues, results in slow release or poor availability of the encapsulated drug. Therefore, the appropriate functionalization of liposomes is essential in order to render them simultaneously resistant to drug leakage in the circulation and able to rapidly release their contents at the site of interest. Many of the current strategies to enhance temporal or spatial control of drug release focus on incorporating components into the liposome membranes to achieve thermal, pH, photochemical, or enzymatically triggered release. Here we focus mainly on NIR-responsive liposomes for drug or gene delivery. In order to construct such NIR-responsive liposomes, appropriate components need to be incorporated into the liposome membranes. This still presents several challenges:

- Easily synthesized and biocompatible triggering agents with a strong NIR absorption but small NP size are required.

- The liposomes coupled with the NP triggers should retain their lipid membrane integrity to avoid premature release or chemical degradation of the drug contents.
- Triggering of the composite NPs should require only short bursts of NIR irradiation so that the NPs remain localized.

Because of these limitations, relatively little work on controlled release using NIR light has been reported, although liposomes were the first type of NPs in clinical use.

A commonly used strategy to prepare a liposome-based NIR-responsive vehicle, while optimizing liposome composition and structure to enhance circulation time and drug retention, is to incorporate Au nanostructures into the thermally sensitized liposome membranes. On NIR irradiation, Au nanostructures, such as nanoshells, nanorods or NPs, convert NIR light into heat, increasing the permeability of the liposome membrane *via* a photothermal conversion process. Finally, the anticancer drugs or genes are released from the liposomes.

Generally speaking, there are two ways of preparing Au nanostructure-based liposomes. The first and also the most commonly used method is to combine small Au nanostructures with liposomes loaded with cargo molecules (*e.g.*, by adsorbing on liposomes, forming large aggregates with them, or embedding inside them).[23] There are many examples of this concept. For example, Wu and his coworkers successfully synthesized small hollow Au nanoshells (HGNs) which are either encapsulated within liposomes (by an interdigitation–fusion method),[24] or tethered to the liposome membrane with an Au-SH–PEG–lipid linker.[25] HGNs were selected due to their ease of synthesis and small dimensions compared with other NIR-absorbing nanostructures. After the construction, femtosecond pulses of NIR light could induce release of liposome contents within seconds. Another approach is to assemble the amphiphilic AuNPs (with mixed polymer brush coatings) into a plasmonic vesicular nanostructure.[26] In this synthesis method, successful nanocrystal synthesis and controlled surface-initiated polymerization can open a wealth of possibilities for integrating different types of nanocrystals into multifunctional vesicles.

8.3.4 NIR-Responsive Hydrogels

In recent years, hydrogels with a three-dimensional cross-linked network of hydrophilic (water-soluble) polymers have been developed owing to their unique advantages, such as biocompatibility, adjustability of pores in hydrogel, and degradability under appropriate conditions.[27] Naturally occurring polymers used for hydrogels include chitosan, hyaluronic acid, dextran, alginate, collagen, and gelatin. Synthetic hydrogel polymers include poly(2-hydroxyethyl methacrylate) (PHEMA), poly(2-hydroxypropyl methacrylate) (PHPMA), PAAm, poly(vinyl alcohol) (PVA), and poly(ethylene oxide) (PEO).[28]

Stimulus-responsive moieties must be incorporated into the hydrogel networks to render them degradable under appropriate conditions. As-obtained smart hydrogels can be sensitive to physicochemical changes such as pH,[29] molecular recognition,[30] temperature,[31] or light. Once the relevant environmental change occurs, the elastic networks shrink or degrade, resulting in controlled release of the preloaded drugs or genes.

Here we focus mainly on NIR-responsive hybrid nanogels. Normally, these hydrogels proceed through a gel–sol phase transition or gel volume contraction–expansion process of a thermoresponsive poly(NIPAM-AAm)-based gel by introducing some photothermal constituents, such as dyes, AuNPs, carbon nanotubes (CNTs), or graphene oxide (GO). This kind of NIR-responsive nanogels was discussed in detail in the section of 8.5. There is also another way to construct NIR-triggered nanocarriers based on hydrogels, in which the hydrogel nanocarrier itself is made light-sensitive by introducing light-responsive moieties into the hydrogel network. For example, a novel NIR-triggered release system has been constructed by Zhao and his coworkers from hydrogels loaded with UCNPs. In those NIR nanomaterials, the cargos can be preloaded inside a polymer hydrogel, effectively shutting down their bioactivity. The chemical structure of the hydrogel has a cross-linked hybrid polyacrylamide–poly(ethylene glycol) (PEG) structure held together by photoresponsive *o*-nitrobenzyl groups. When exposed to NIR light, as shown in Figure 8.3a and b, the UCNPs within this cross-linked system generate UV light, which can cleave the *o*-nitrobenzyl groups by a typical photo-to-oxidation process, resulting in the breakdown of the entire gel (gel–sol transition) and triggering the release of the entrapped biomacromolecules (Figure 8.3c and d).[32]

8.3.5 NIR-Responsive Biodegradable Polypeptide Materials

As mentioned earlier, NIR-responsive polymers are of particular interest since the light allows remote control of the drug release process in a manner that is both spatially and temporally controlled. However, most photoresponsive copolymers have undergone limited *in vivo* and clinical trials owing to their poor biodegradability. Therefore, discovering NIR-responsive biodegradable polymers is very important.

In recent years, polypeptides have been intensively studied as stimulus-responsive polymers owing to their obvious attractive features such as their inherent ionizable groups (*i.e.*, NH_2 and COOH), hierarchical self-assembly, and multiple secondary or tertiary structures. The biocompatibility and biodegradability of hybrid polypeptides or proteins have attracted great attention for possible clinical use; some have entered clinical trials.

The >20 natural amino acids and their synthetic derivatives constitute an abundant monomer source for the construction of multifunctional polypeptides. Like the micelles, liposomes, and hydrogels described above, photoresponsive polypeptides can be successfully constructed by introducing some photoresponsive moieties such as coumarin, 2-nitrobenzyl, cinnamyl, and

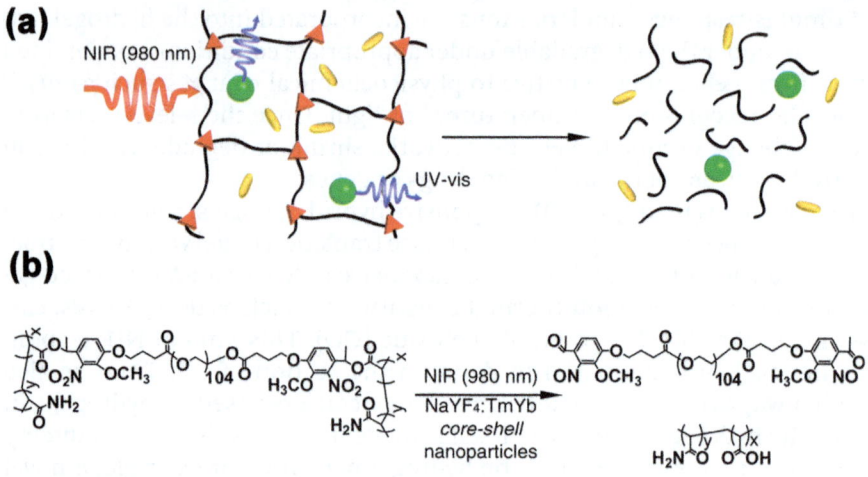

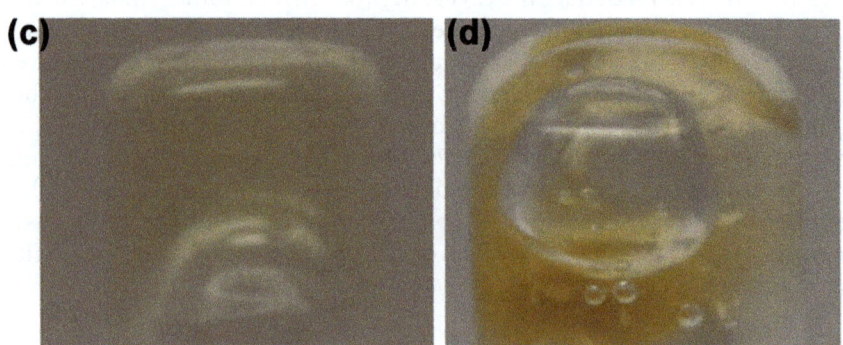

Figure 8.3   (a) Schematic illustration of the NIR-triggered degradation of a photo-sensitive hydrogel using the UV light generated by encapsulated UCNPs. The polymeric components are depicted as black lines, the photocleavable cross-links as red triangles, the UCNPs as green spheres, and the trapped biomacromolecules as yellow rods. (b) Chemical structure of the hydrogel containing photocleavable *o*-nitrobenzyl moieties in the cross-linker and the NIR-induced photoreaction of the hydrogel *via* UV light emitted by loaded NaYF$_4$:TmYb core–shell UCNPs. (c and d) Photographs of a UCNP loaded hydrogel (~0.08 mL) (c) before and (d) after irradiation with 980 nm light (5 W, 195 min), showing the NIR-induced gel–sol transition of the hydrogel. Reprinted with permission from B. Yan, J. C. Boyer, D. Habault, N. R. Branda and Y. Zhao, *J. Am. Chem. Soc.*, 2012, **134**, 16558.[32] Copyright (2012) American Chemical Society.

spiropyran into polypeptides, which are then used to construct a phototrig-gered drug or gene delivery system. For example, Dong and coworkers[33] designed a photoresponsive *S*-(*o*-nitrobenzyl)-ʟ-cysteine *N*-carboxyanhy-dride (NBC–NCA) monomer, then synthesized the related PNBC-*b*-PEO block copolymers from the ring-opening polymerization of NBC–NCA. The final as-obtained micelles formed a phototriggered drug release system: upon UV irradiation, progressive decrease of the hydrophobic interaction between DOX and the PNBC block promoted the release of DOX due to the photo-cleavage of pendant *o*-nitrobenzyl groups. However, as discussed above, the irradiation source is high-energy UV light. In fact, most of the light-responsive photochromic moieties incorporated into polypeptide structures absorb high-energy UV/Vis light. Some kinds of NIR-responsive polypeptides *via* two-photon or upconversion absorption have been reported in recent years. For example, Zhao and his coworkers[34] synthesized a NIR-responsive poly-peptide by incorporating 6-bromo-7-hydroxycoumarin-4-ylmethyl groups into PEG–poly(L-glutamic acid). The resulting polypeptide block can be dis-rupted by the shifted hydrophilic/hydrophobic balance under 794 nm NIR irradiation, causing the cleavage of coumarin groups. An antibacterial drug (rifampicin) and an anticancer drug (paclitaxel) were loaded into the photo-sensitive BCP micelles, and the NIR-triggered drug release was studied in a series of experiments.

8.4 Photocaging of Bioactive Cargos

8.4.1 Introduction

NIR photocontrolled NP-based delivery systems can achieved not only by using nanocarriers that are themselves photoresponsive, as described in Section 8.3, but also by modification with a photolabile "caging" moiety on the surface of hollow or mesoporous nanocarriers. This is called the pho-tocaging of bioactive cargos. In this kind of NIR-triggered delivery system, the nanocarrier itself does not show any photoresponsive or photocon-trol behaviors. The bioactive molecules such as drugs, nucleic acids, and other small molecules, can be temporarily "inactivated" by blocking key functional groups (carboxyl groups, amino groups, phosphate moieties, hydroxyl groups, *etc.*) using photolabile molecules, or by being locked in the pores of the nanocarrier. Upon irradiation with a suitable NIR wave-length, this NIR-photolabile caging is released, rendering the biomolecule active again.

The NIR photocaging of bioactive cargos can be achieved either by direct modification of the bioactive cargo with a photolabile caging moiety, or by coating NIR-responsive polymers, NPs, or DNA onto the surface of hollow or mesoporous nanocarriers. However, to construct this kind of NIR-triggered delivery system, the chemical modifications of nanocarriers must meet a number of requirements:

- The modification should introduce the ability to absorb NIR light effectively (*i.e.*, a high molar extinction coefficient or two-photon cross-section).
- The photochemistry of the modification must be highly efficient, so that the NIR dosage required to trigger drug release is not too high.

- The modification must alter the carrier in a manner that leads to a substantial change in drug delivery activity upon NIR light excitation.
- The modification should be stable to biological conditions and inert (non-toxic) both before and after irradiation.[35]

Photocaging drug or gene delivery systems can deploy three main strategies to achieve NIR-based control over the delivery process: direct blocking of bioactive cargos with a photolabile caging moiety (Section 8.4.2); using polymeric molecular nanostructures as light-triggered gatekeepers (Section 8.4.3); or using nucleic acids as light-triggered gatekeepers (Section 8.4.4).

8.4.2   Direct Blocking of Bioactive Cargos with a Photolabile Caging Moiety

In this strategy, the surface of the nanocarrier is modified using photolinkers to which the bioactive cargo is attached. These photolinkers can be cleaved upon NIR irradiation, resulting in the release of the linked cargos. In this way, the carriers loaded with the inactivated cargos can be activated in the region of interest by NIR irradiation alone. In this type of delivery system, the procedure of covalently linking the photolabile group to biomolecules is termed "caging", and the linked biomolecules are said to be "caged".

Numerous photolabile molecules have been developed for the modification of bioactive cargos, which can be divided into four groups depending on the mechanism of photolysis:

- *o*-nitrobenzyl and related groups, such as nitrophenyl ethyl (NPE), *o*-nitrobenzyl (NB), and 1-(4,5-dimethoxy-2-nitrophenyl) diazoethane (DMNPE)
- coumarin-4-ylmethyl and related groups, such as 7-methoxycoumarin-4-ylmethyl (MCM)
- *p*-hydroxyphenacyl (pHP) group, which is a promising alternative to the nitrobenzyl-based groups
- other miscellaneous groups, such as nitroindolinyl (NI) and 4-methoxyl-7-nitroindolinyl (MNI).

These "phototriggers" are all small organic molecules that perform two important functions: precise control over drug release and acting as a linker between the NPs and the drug.

There are many examples of this kind of nanocarriers, such as Fe/Si NPs as a nanocarrier with fluorescent 7-hydroxy coumarin as a phototrigger,[36] mesoporous silica grafted with a coumarin-based phototrigger,[37] and an MSN-based drug carrier with nitroveratryl moieties.[38] Although these direct-blocking nanocarriers are effective, they are mainly limited by the predominant use of UV/Vis light, to which most photolabile groups are sensitive. As already mentioned, NIR-triggered nanomaterials are greatly preferable for drug or gene delivery and two strategies have been developed to achieve this goal: upconversion excitation and two-photon excitation.

Numerous efforts have been made to take advantages of Yb/Tm/Er/Ho codoped UCNPs as an effective UV source by multiphoton absorption of NIR light. UCNPs have also been used to construct NIR-triggered delivery carriers by direct blocking of the bioactive cargo with a photolabile caging moiety. In this way, the UCNPs, which can convert NIR to UV/Vis light, have been developed to serve as nanotransducers. For example, Krull and his coworkers[39] have described an investigation of photocleavage at the surface of UCNPs to release the well-known chemotherapy drug 5-fluorouracil (5-FU) by utilizing the popular photolabile protecting group *o*-nitrobenzyl, as shown in Figure 8.4a. In this investigation, core–shell UCNPs composed of a β-NaYF$_4$:4.95% Yb, 0.08% Tm core, and a β-NaYF$_4$ shell were coated with *o*-phosphorylethanolamine ligands and coupled to an *o*-nitrobenzyl derivative of 5-FU. Under NIR irradiation, the photoluminescence emission bands centered at 365, 455, and 485 nm from the UCNPs were in resonance with the absorption band of the ONB–FU derivative, resulting in photocleavage and subsequent release of the 5-FU from this delivery vehicle. The results revealed that the release of 5-FU was complete in <14 min, using a NIR laser source centered at 980 nm that operated at a power of <100 mW. The efficiency of the triggered release was as high as 77% of the total ONB–FU conjugate, and the rate of release could be tuned with the laser power output. Our research group has also reported a new kind of strategy focusing on the use of UCNPs as the carriers to deliver a light-activated prodrug *via* an NIR-to-UV strategy. By using upconverted UV emission from core–shell structured NaYF$_4$:Yb/Tm@ NaGdF$_4$:Yb UCNPs, a *trans*-Pt(IV) prodrug, *trans*-, *trans*-, *trans*-[Pt(N$_3$)$_2$(NH$_3$) (py)(O$_2$CCH$_2$CH$_2$COOH)$_2$] (DPP) was effectively released and further activated to the highly toxic Pt(II) derivative for the treatment of cancer.[40]

Two-photon absorption-induced excitation is also a promising approach for building NIR-triggered carriers. Here, the nanocarrier is directly conjugated to the active biomolecule by a photosensitive linker. Under NIR irradiation, an uphill energy conversion happens through the use of two-photon absorbing chromophores. A subsequent energy transfer to the photosensitive linker then helps to release the caged cargo. For example, researchers[41] have established a NIR phototriggered-on-demand drug release system by utilizing a two-photon absorbing rhodamine B fluorescent dye to graft onto the surface of gum arabic-modified Fe$_3$O$_4$ magnetic NPs. The results show that this drug release system has a steady drug release profile on exposure to NIR, whereas no detectable release was observed in the dark.

Apart from the above-mentioned situations, in which the bioactive drugs or prodrugs are caged with a NIR-photolabile caging moiety and released under NIR irradiation, other bioactive molecules, such as targeting ligands, can also act as bioactive cargos to be caged. Once NIR-irradiated, the caged targeting ligands become active again to allow site-specific labeling. For example, Chien *et al.*[42] used a photolabile nitrobenzyl (NBz) group for uncaging folate (FA) targeting ligands in antitumor applications. In this paper, FA

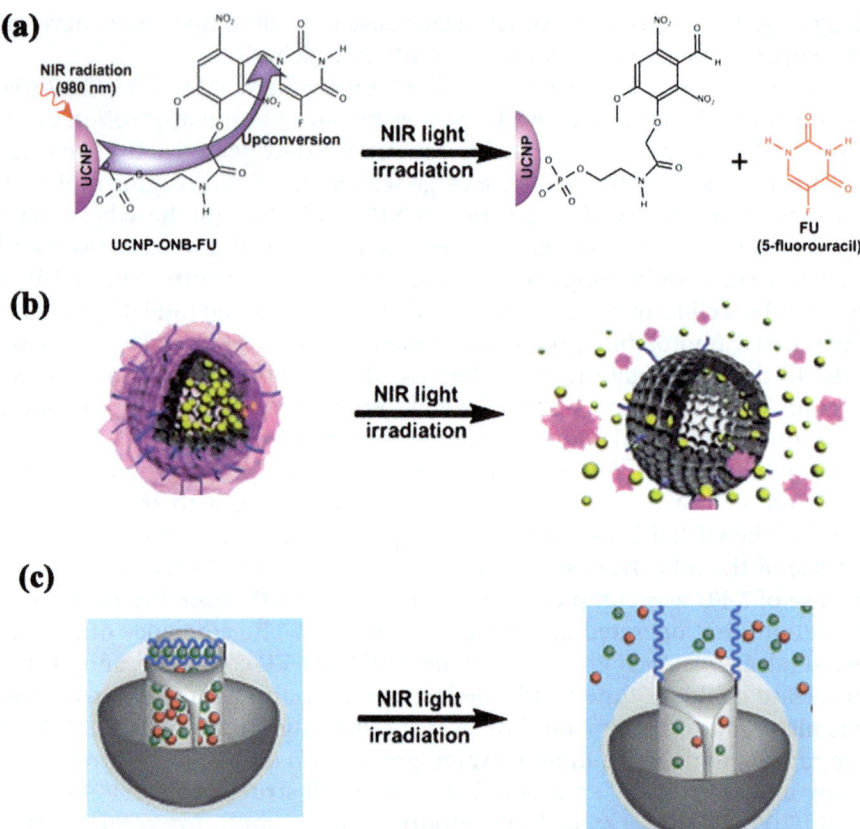

Figure 8.4   Schematic depiction of (a) NIR controlled release of 5-fluorouracil from the UCNP surface through direct breaking the photolabile "caging" moiety; (b) NIR-controlled release of cargos from HMS@C18@HAMA-FA-*b*-DDACMM by degrading the polymer molecules upon NIR exposure; (c) NIR-controlled release of loaded cargos from coralyne/ICG@poly(A)–MSN nanocarriers through opening up the nucleic acid "gatekeepers" under NIR laser stimuli. Figure (a) reprinted with permission from L. L. Fedoryshin, A. J. Tavares, E. Petryayeva, S. Doughan, and U. J. Krull, *ACS Appl. Mater. Interfaces*, 2014, **6**, 13600.[39] Copyright (2014) American Chemical Society; figure (b) reprinted from ref. 45 with permission from the Royal Society of Chemistry; figure (c) reprinted from ref. 54 with permission from the Royal Society of Chemistry.

ligands were caged using the NBz group through covalent attachment to the carbohydrate moiety of the FA molecule. Laser excitation at 980 nm was used to produce UCNP photoluminescence at 360 nm, and this short wavelength resulted in uncaging of FA by excitation of the NBz group. The system was then used for targeted tumor binding to cell-surface FA receptors. DOX was coupled to the UCNP surface *via* a labile disulfide bond and released intracellularly through cleavage by lysosomal enzymes after internalization of NPs, and inhibition of tumor growth was noted.

8.4.3 Polymer Molecular Nanostructures As Light-Triggered Gatekeepers

In the above situations, the biomolecules were covalently linked to the surface of nanocarriers modified with photolabile groups, which may easily be affected by the surrounding environment. To solve this problem, a new kind of light-triggered drug nanovehicle has been designed, in which the active cargos are loaded into mesoporous inorganic scaffolds (*e.g.* SiO_2, Fe_3O_4, Au) while the photolabile caging moieties are modified on the surface of the inorganic nanocarriers. Upon light irradiation, the caging moieties can be "opened" and the cargos released. Compared with other light-triggered biomolecular nanocarriers, such as polymeric micelles, the mesoporous inorganic scaffolds used here have well-controlled morphology, excellent biocompatibility, good monodispersity, and high drug loading. Therefore, this is an attractive and suitable alternative approach to combine polymers with nanoscopic inorganic solids in order to enhance the functionality of nanocarriers.

Generally speaking, gated nanocarriers are prepared by grafting photolabile group caps onto the external surface of mesoporous inorganic scaffolds which have been loaded with some particular cargos. There are usually two important parts in these nanocarriers: the molecular caps, which are the switchable entities controlling the NIR-triggered release of confined guests, and a suitable inorganic support acting as a nanocontainer (for loading the guests), to which gate-like molecules can be easily grafted. Both components are important, and their selection determines the NIR-controlled release.

Gated nanocarriers have developed quickly in recent years and many kinds of caps or gates have been developed, including polymer molecular nanostructures which can achieve conformational changes or cleavage of chemical bonds under UV, visible, or NIR light illumination. Such light-triggered gatekeepers, such as semi-rotaxane pore-capping nanostructures,[43] "photomechanical" azobenzene groups,[44] or copolymers with high two-photon absorption cross-section,[45] can be covalently grafted onto the preorganized or pre-existing stable nanoscopic supports, resulting in cooperative functional polymer molecular behaviors that are not found in unanchored molecules or in the unfunctionalized solids alone (Figure 8.4b). Such molecules are sometimes described as "phototriggers," which can undergo irreversible photochemistry: once uncaged, the drugs are released and the drug carrier remains open until it is removed in some other manner. Alternately, they may be "photoswitches", which can undergo reversible photochemistry so that the drug carrier system can produce many rounds of drug release/unrelease states. Since the motion-based functional processes of these functional polymer molecular materials, such as translocation, reversible mass movement, or controlled molecular transport, can be exactly controlled by light after their successful assembly into the inorganic scaffold, these molecules can achieve the goal of light-controlled drug delivery by undergoing phototriggering or photoswitching behavior, satisfying most of the required criteria for nanocarriers.

Some researchers have also investigated NIR-controlled gene delivery systems. For example, Jayakumar and his coworkers tried loading caged green fluorescent protein (GFP) plasmids into the pores of mesoporous silica-coated $NaYF_4$ UCNPs doped with Yb^{3+} and Tm^{3+}. Their results showed that gene expression was restricted to NIR-irradiated cells.[46] After that, they further investigated the simultaneous delivery and photoactivation of photomorpholinos and TPPS2a, a photosensitizer used for photochemical internalization, using these particles. This resulted in endosomal escape and enhanced gene knockdown in targeted cells *in vitro* and *in vivo* in a murine melanoma model.[47]

Most photolabile caging moieties are induced by UV/Vis light, but two-photon absorption or UCNP-assisted photochemistry can be used to convert the NIR irradiation to higher-energy photons in the UV/Vis regions in order to achieve NIR-triggered biomolecule release. However, the reported excitation intensity for either two-photon absorption or UCNP-assisted photochemistry is usually several to several hundred W cm^{-2}. Since an intensity of several W cm^{-2} is dangerous according to the American National Standard for Safe Use of Lasers, it is highly desirable to reduce the NIR intensity for drug delivery systems. Much research has been devoted to developing the efficiency of NIR light utilization in order to reduce the required NIR intensity to a medically harmless dose. For example, Wu and his coworkers[48] demonstrated a strategy to trigger drug release under 974 nm irradiation at only 0.35 W cm^{-2}, based on UCNP-assisted photochemistry. They fabricated a drug delivery system by loading mesoporous silica-coated UCNPs with DOX; blue-light-cleavable Ru complexes were then successfully gated on the surface of the mesoporous silica shells. The required NIR intensity was so low because a three-photon process is sufficient to create the required blue photons and induce the photocleavage of Ru complexes.

In addition to light-controlled or -responsive functional polymer molecular materials that can be coated on the surface of inorganic scaffolds for NIR-triggered drug or gene delivery, some other stimulus-sensitive polymer molecules, such as the thermosensitive or singlet oxygen ($^1O_2$)-sensitive molecules can also be used as coatings. However, this requires the additional introduction of some nanomaterial that can absorb NIR in order to produce the special stimulus that the caging molecules require. For example, if we coat thermosensitive molecules on the inorganic scaffold, a nanomaterial that can convert NIR light to heat needs to be introduced. For example, AuNPs or AuNRs can be coated with thermoresponsive polymers for phototriggered drug release *via* a photothermal conversion process. This is discussed in Section 8.5. Here, another example which involves $^1O_2$-sensitive molecules is illustrated. Lu and his coworkers reported a new core–shell nanocomposite which was fabricated for NIR-controlled release of anticancer drugs accompanied by photodynamic therapy (PDT). In this case, the higher-energy visible red light (660 nm) produced by the UCNP core upon NIR irradiation (980 nm) could excite Ce6 to produce $^1O_2$. Upon excitation by $^1O_2$, the 9,10-dialkoxyanthracene (DN) groups in the amphiphilic copolymer outside the core were converted into 9,10-anthra-quinone (AQ), causing degradation of the polymer and resulting in the release of the loaded biomolecules. As a result, the prepared nanocomposite could potentially be employed for NIR-triggered release of anticancer drugs and PDT, as well as *in vivo* bioimaging.[49]

8.4.4 Nucleic Acids As Light-Triggered Gatekeepers

As mentioned in Section 8.4.3, many reports have been limited to the use of polymer molecular nanostructures as light-triggered gatekeepers. In addition to polymer molecules, nucleic acids can also act as controllable nanovalves on the surface of some preorganized or pre-existing stable nanoscopic supports. Compared to polymer molecules, nucleic acids have their own unique advantages, such as their remarkable molecular self-recognition capabilities and unique structural motif properties. The resulting DNA nanomachines have therefore also attracted much attention for practical applications.

DNA nanomachines that can respond to external stimuli such as pH, temperature, or enzymes to control pore opening/closing have been successfully fabricated and characterized by many researchers. For example, Bein and his coworkers exploited a thermal-sensitive drug delivery system based on biotin-labeled double-stranded (DS) DNA: the DNA double strands were attached to the pores of mesoporous NPs. Once the temperature changed, the hybridization of the dsDNA could be controlled to open or close in order to manipulate the release of the loaded biomolecules.[50] Ren and his coworkers also reported a pH-sensitive nanocarrier based on quadruplex DNA, in which the release of the loaded biomolecules can be controlled by changing the pH of the solution.[51] Furthermore, a further DNA duplex coating on the pores of as-obtained NPs made this system endonuclease-sensitive by using endonuclease to degrade the DNA for pore opening.[52]

Since light as a trigger has significant advantages for easy control of movement and conformation compared with temperature, pH, or other external stimuli, light-triggered gatekeepers based on nucleic acids have attracted a lot of attention. To construct light-triggered nanocarriers, generally the dsDNA should incorporate some photon-sensitive molecules to form so-called photon-manipulated nucleic acids. The as-obtained photon-manipulated nucleic acids, to be utilized as an important building block, were then coated on the pore area of hollow or porous inorganic structures. For example, Tan and his coworkers[53] reported a photoresponsive DNA/mesoporous silica hybrid formed by introducing azobenzene moieties into the DNA molecule. The azobenzene-incorporated dsDNA was immobilized on the surface of MSN NPs. UV/Vis irradiation induced the photoisomerization of azobenzene, resulting in the dehybridization/hybridization switching of the complementary DNA strands, thereby causing uncapping/capping of the pore gates of the mesoporous silica.

There is also an indirect way to construct NIR-triggered nanocarriers based on DNA gatekeepers, rather than directly constructing photon-manipulated nucleic acids: NIR-sensitive biomolecular nanocarriers can be constructed by coating other stimulus-sensitive nucleic acids on the surface of the inorganic scaffolds. However, these systems must contain some nanomaterials that can absorb NIR light to produce the special stimulus that the caging DNA gatekeeper requires. For example, He and his coworkers[54] coated temperature-sensitive nucleic acids on the surface of MSNs by the cooperative binding of poly(A) strands with silica, as shown in Figure 8.4c. Indocyanine green (ICG), which has high efficiency for the conversion of NIR light into heat, was also coloaded into the MSNs. Upon NIR irradiation, the heat generated from the ICG could make non-Watson–Crick secondary structures unstable, leading to the dehybridization of the cooperative binding structure and an open-gate state. Tang and his coworkers[55] also synthesized a kind of NIR-responsive nanocarrier which has a AuNR core and mesoporous silica shell, capped with reversible single-stranded DNA valves. Upon NIR irradiation, the AuNR core converted light into heat through the photothermal effect. The heat dissipated into the surroundings, destroying the electrostatic interaction between DNA and the silicon shell and leading to an "on" state of the valves and release of the cargo molecules from the nanocarrier.

8.5 Photothermal Transduction for NIR-Triggered Nanocarriers

8.5.1 Introduction

Recently, NPs exhibiting photothermal properties have been widely exploited in photothermal treatment of cancer (PTT), which has achieved encouraging levels of therapeutic efficacy in many *in vivo* animal studies. Ideally photothermal agents should exhibit strong absorbance in the NIR region, and efficiently transfer the absorbed NIR optical energy into heat, mainly because of their intense surface plasmon resonance (SPR), a collective oscillation of free electrons upon interaction with light, and low quantum yield. Nanomaterials with these properties may be inorganic (*e.g.* various noble metals,[56] carbon nanomaterials,[57] Pd nanosheets,[58] CuS NPs,[59] and a few other newly reported ones[60]), or organic (*e.g.* NIR-absorbing conjugated polymers,[61] porphysomes,[62] and nanomicelles encapsulated NIR dyes[63]). Photothermal agents absorb the light radiation and convert it to thermal energy, heating up their surroundings. The heat can not only be used to directly burn cancer cells, but also helps to trigger and/or enhance drug or gene release if well-designed smart nanoplatforms are employed.

In the following four sections we discuss NIR-triggered photothermal therapy (Section 8.5.2); photothermally enhanced drug delivery (Section 8.5.3); NIR-controllable drug release based on thermosensitive polymers (Section 8.5.4); and NIR-controllable drug release through decreasing the binding affinity (Section 8.5.5).

8.5.2 NIR-Triggered Photothermal Therapy

Thermal treatment is considered to increase the temperature of part or all of the body above its normal temperature for a defined period of time, which can have positive effects in patients with an ongoing disease such as cancer. PTT, mainly using NPs capable of efficient heat generation under laser illumination, has attracted attention for some time, but there are two main limitations in its development: animal tissues have strong extinction coefficients in the visible range of the optical spectrum, and the laser energy will be absorbed by both healthy and cancerous tissues. These two limitations not only reduce the efficacy of heat delivery within the tumor but also increase non-specific damage to adjacent tissues, so PTT is not considered a reliable technique. To overcome those problems, NIR-triggered PTT is attracting considerable attention since the NIR could deliver high thermal energy to cancer tissue with little collateral damage to normal tissue.

Several nanomaterials with strong NIR-absorbing capability have shown encouraging PTT effects both *in vitro* and *in vivo*. Among them, Au nanostructures (AuNSs) have attracted much attention owing to their unique features, such as straightforward synthesis, tunable size and morphology, good biocompatibility, chemical inertness, well-defined SPR absorption, and easy surface modification. The well-defined SPR absorption of AuNSs plays a particularly important role in NIR-triggered PTT. Many researchers have focused on developing complex geometries, such as Au nanoshells, nanorods, nanocages and even nanostars, instead of simple nanospheres (Figure 8.5a–c), since the dynamics of plasmonic oscillations is strongly modified when the shape of the AuNP deviates from a highly symmetric spherical shape. The influence of shape in the optical extinction of metallic NPs can be explained by the so-called Gans model, which is a formal extension of Mie's theory and explains the extinction cross-sections of dilute solutions of randomly oriented AuNPs with ellipsoidal geometries. This model yields good results in describing and predicting the optical properties of elliptic metallic NPs, but it cannot be applied to more complex geometries. To do that, advanced simulation methods (such as discrete dipole approximation, DDA) are needed. Indeed, there are several publications demonstrating how it is now possible to simulate the optical properties of metallic NPs with almost any arbitrary geometry.[64]

Among the various non-spherical metallic NPs, Au nanorods (AuNRs) have attracted a great deal of attention due to the large number of synthesis methods available, the high achievable monodispersity, and the rational control over the aspect ratio, which is primarily responsible for the change in their optical properties. As shown in Figure 8.5d–g, the aspect ratio of AuNPs can be changed at will. As a result, the extinction spectra of AuNRs with different aspect ratios can be tuned. As shown in Figure 8.5h, for all nanorods, the NIR extinction bands are strongly red-shifted when the aspect ratio is increased, thus allowing for spectral tunability into the biological windows by simply changing the aspect ratio of the AuNRs. The AuNPs widely used

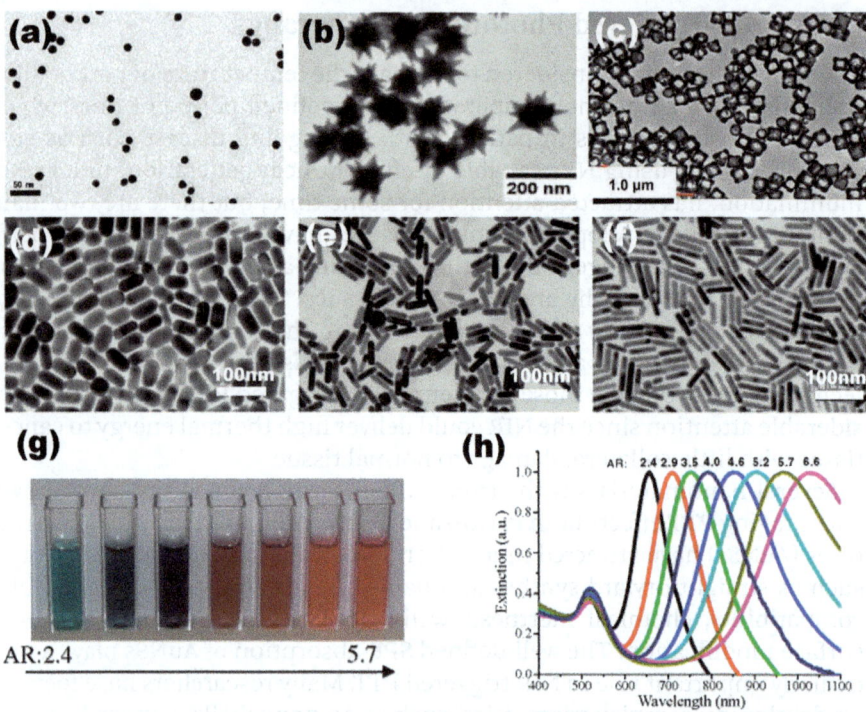

Figure 8.5 TEM images of Au nanostructures with different geometries: (a) Au NPs, (b) Au nanostars, (c) Au nanocages, and (d–f) Au nanorods with various aspect ratios. The different colors (g) and different SPR wavelengths (h) of Au nanorods with various aspect ratios. Figure (a) reprinted from J. Kim *et al.*, Transfection and intracellular trafficking properties of carbon dot-Au NP molecular assembly conjugated with PEI-pDNA, *Biomaterials*, 2013, **34**, Copyright (2013) with permission from Elsevier.[10a] Figure (b) reproduced from ref. 10b © IOP Publishing. Reproduced with permission. All rights reserved. Figure (c) reproduced from R. Vankayala *et al.*, Gold nanoshells-mediated bimodal photodynamic and photothermal cancer treatment using ultra-low doses of near infra-red light, *Biomaterials*, 2014, **35**, Copyright (2014) with permission from Elsevier.[11e] Figures (d)–(h) reproduced from X. Huang *et al.*, Gold NPs:Optical properties and implementations in cancer diagnosis and photothermal therapy, *J. Adv. Res.*, 2010, **1**, Copyright (2010) with permission from Elsevier.[10c]

as photothermal agents have an aspect ratio of 4, which shows a longitudinal absorbance at approximately 800 nm (Figure 8.5h). This phenomenon makes AuNRs particularly suitable for PTT. For example, Van Maltzahn *et al.* reported PEG-coated AuNRs (AuNR–PEG) used as an efficient photothermal nanoheater.[65] Tong *et al.* also investigated the photothermal effects of FA-conjugated AuNRs to human malignant nasopharyngeal carcinoma (KB) cells by both CW NIR laser and femtosecond-pulsed laser irradiation,[66] and the investigation results show an effective antitumor effect. Au nanoshells have also been widely investigated since they can be easily synthesized by seed-mediated shell growth. Hirsch *et al.*[67] first demonstrated nanoshell-based therapy by irradiation of nanoshells injected into tumors grown in mice, achieving temperature increases >30 °C and significant cell death. Subsequently many other works have investigated nanoshell-based hyperthermia, as described in recent reviews.[68]

Although AuNSs are widely used in the PTT field, their practical applications are limited by their high cost. There are also many other nanomaterials that could be used for PTT, such as carbon-based materials, organics, and semiconductors. One of the most recent promising candidates is Cu_9S_5 NPs, owing to their advanced properties such as low cost, high stability, low cytotoxicity, and the intrinsic NIR absorption derived from d–d energy band transitions. Recently, our group[69] also reported DOX-loaded core–shell structured Cu_9S_5@mSiO_2 nanofibrous fabrics (labeled as DOX–Cu_9S_5@mSiO_2–PG) made by an electrospinning process. The as-obtained multifunctional spun fabric can be surgically implanted directly to the tumor site of mice to deliver synergistic orthotopic therapy, combining chemotherapy by the controlled release of DOX from the mesoporous SiO_2 with PTT through the photothermal transformation of Cu_9S_5 under NIR irradiation. The results showed that the as-obtained DOX–Cu_9S_5@mSiO_2–PG has a great inhibitory effect on tumor growth, especially for unresectable tumors or metastases.

The NIR photothermal agents mentioned above are mainly inorganic nanomaterials. Very recently, NIR-absorbing conjugated polymers, such as polyaniline, poly(3,4-ethylenedioxythiophene)poly(styrenesulfonate) (PEDOT:PSS) and polypyrrole (PPy) have also been widely reported to act as photothermal agents.[70] For example, Dai *et al.*[71] reported a novel fabrication of PEGylated polypyrrole NPs conjugating gadolinium chelates (Gd–PEG–PPy NPs) for dual-mode magnetic resonance imaging (MRI)/photoacoustic imaging-guided PTT of cancer. Tumor growth was effectively inhibited after treatment with Gd–PEG–PPy NPs in combination with NIR irradiation, indicating the great potential of PPy NPs as photothermal agents for localized tumor PTT because of their good *in vitro* and *in vivo* biocompatibility, significant photothermal conversion efficiency, and remarkable photostability.[72]

8.5.3 NIR-Controllable Drug Release Through Increasing the Diffusion Speed

As discussed earlier, many kinds of highly efficient NIR-absorbing nanomaterials have been intensively explored for cancer PTT.[73] Apart from being directly utilized for photothermal ablation of cancer cells, the photothermal effect of NIR-absorbing nanomaterials can also be exploited for remotely controlled drug or gene release.[74] One of the simplest and commonly used ways of realizing this idea is to increase the diffusion speed of bioactive cargos by increasing the temperature generated by photothermal NPs. The decreased viscosity of the surrounding fluid when the temperature rises can also help to enhance the release of biomolecules, favoring diffusion of the drug away from the nanocarriers.

The direct acceleration of biomolecule release by increasing the temperature of the surrounding environment has many advantages, such as easy fabrication, convenient control, and the increased drug payload. As far as we know, the construction of this kind of delivery vehicle is mainly achieved either by embedding photothermal NPs within a structure containing the drug (*e.g.* drug-loaded hydrogels or liposomes), or by directly fabricating a photothermal mesoporous structure that can load the drug inside (*e.g.* drug-loaded hollow Au nanospheres or Au/SiO$_2$). Hollow Au or CuS spheres or core–shell structures such as Au@mSiO$_2$, CuS@mSiO$_2$, and Ag$_2$S@mSiO$_2$ NPs are all widely investigated as nanocarriers for this kind of photothermally enhanced drug delivery.[75] For example, Wu *et al.*[75a] successfully synthesized PEG-modified DOX-loaded MSN@CuS nanohybrids as efficient NIR-triggered drug delivery carriers. In this system, as shown in Figure 8.6a–c, the NIR irradiation significantly elevated the local temperature inside the illumination zone, and the higher temperature enhanced the Brownian motion of particles, which promoted drug release from the MSNs. In addition, both Zhang *et al.*[76] and Zhao *et al.*[77] have also developed mesoporous silica shell-coated AuNPs and loaded the drug within the porous interior. Tang *et al.*[78] also developed a coating of mesoporous Au nanoshells on a mesoporous silica nanorattle core. In these chemothermal therapy systems, the AuNPs can convert NIR light into heat. The heat is transferred from the surface of the AuNSs to the surrounding environment and can be used not only for hyperthermia cancer therapy, but also to trigger the release of chemotherapeutic agents attached to the AuNSs, achieving the concept of NIR-triggered drug delivery.

Apart from the direct enhancement of desorption or release of biomolecules by increasing the temperature of the surrounding environment to accelerate the diffusion rate, a number of researchers have found that the cell membrane permeability is significantly enhanced as the environmental temperature gradually increases from 37 to 43 °C, which could accelerate the cellular uptake of NPs to improve the effect of antitumor therapy.[79] This phenomenon may be ascribed to the fact that many NPs, although with different sizes, shapes, and surface functionalization, are engulfed by cells mainly *via* energy-dependent pathways.[80] Nanocarriers that combine photothermal NPs and drugs could therefore exhibit increased cellular uptake upon laser irradiation, effectively helping to treat the tumor. For example, Sherlock *et al.* showed that DOX-loaded FeCo/graphitic carbon shell (FeCo/GC) nanocrystals have enhanced cellular uptake upon NIR irradiation.[81] Liu *et al.* also used functionalized nano-GO for photothermally enhanced intracellular delivery of Ce6, DNA plasmid, and siRNA, successfully combining PDT with PTT and gene therapy.[82]

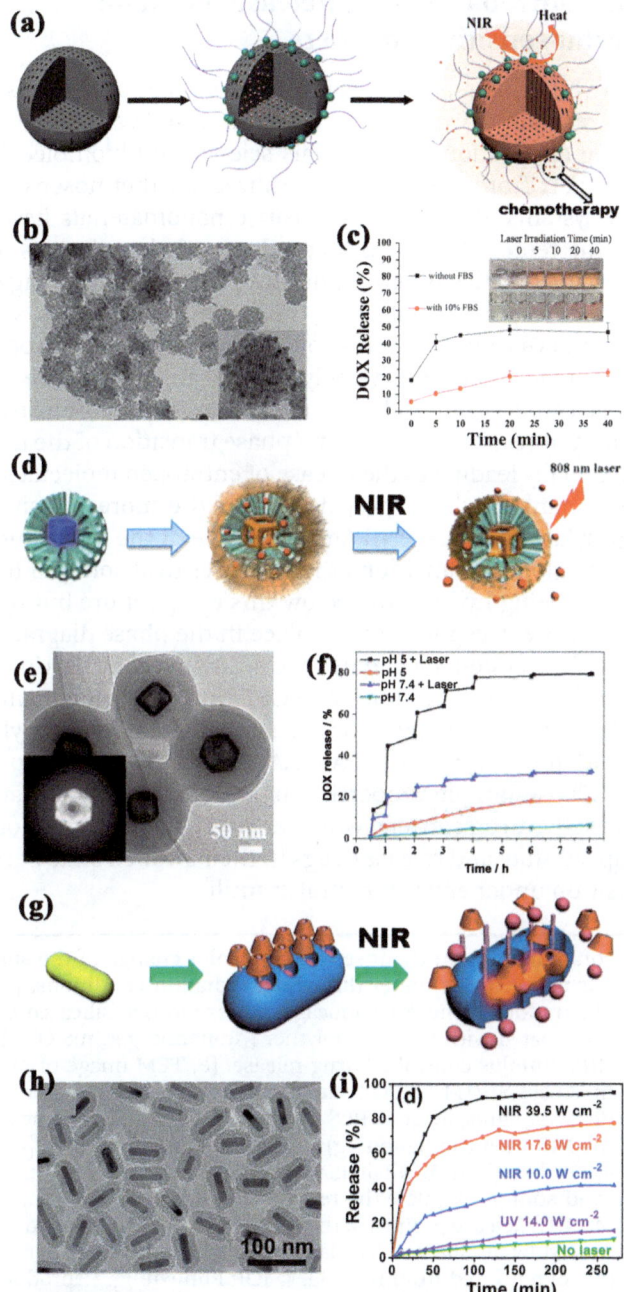

Figure 8.6 (a) Schematic view of the as-obtained PEG-modified DOX-loading MSN@CuS nanohybrids and NIR-triggered drug release. (b) Low- and high-magnification (inset) TEM images of as-synthesized PEG-modified MSN@CuS. (c) NIR laser-induced release behavior of the PEG-modified DOX-loading MSN@CuS nanohybrids in pure water or in 10% FBS (v/v)

8.5.4 NIR-Controllable Drug Release Based on Thermosensitive Polymers

Thermosensitive polymers, which can exhibit structural changes or volume transitions in respond to external temperature stimuli, are highly valued and have promising applications in materials science and biomolecular delivery applications. Apart from being directly utilized for thermosensitive drug or gene delivery systems, those thermosensitive nanomaterials have also been exploited for the remote NIR-controlled release of biomolecules. In order to harness the characteristics of NIR light and produce heat to trigger the disruption of thermoresponsive polymers, photosensitive moieties, such as dyes, AuNPs, CNTs, and GO NPs that can strongly absorb in the NIR spectral range, are embedded in thermoresponsive polymer matrices. The conversion of light energy to heat through non-radiative relaxation of photosensitive moieties causes heating and results in the volume phase transition of the thermosensitive polymers, finally leading to the release of entrapped molecules.

Among the many kinds of polymers with thermoresponsive behavior, poly(N-isopropylacrylamide) (PNIPAm) gel is one of the most studied. It has an obvious coil–globule transition at its lower critical solution temperature (LCST) of 32 °C, being hydrophilic below this temperature but hydrophobic above it. The LCST corresponds to the place in the phase diagram where the entropic gain of the system overcomes the enthalpic contribution associated with hydrogen bonds. In addition, the LCST can be raised at will by polymerizing PNIPAm with a more hydrophilic copolymer, such as acrylamide. For example, a 95:5 molar ratio of NIPAm:AAm has been shown to result in a LCST >37 °C. Those unique properties make PNIPAm-based materials suitable to use as active drug carriers, which are designed to retain loaded drugs in their collapsed state and release drugs in their swollen state because of the volume transition under environmental stimuli.

---

containing water; the inset was the color change of the supernatant of our nanohybrids under different irradiation conditions. (d) Schematic illustration of the Au nanocages@mesoporous silica core–shell structure coated with a smart polymer (Au-nanocage@mSiO$_2$@PNIPAM) for NIR stimulus controlled drug release. (e) TEM image of Au-nanocage@mSiO$_2$@PNIPAM. (f) DOX release curves from the Au-nanocage@mSiO$_2$@PNIPAM nanocarrier in PBS buffer at pH 7.4 and pH 5, with or without the NIR laser irradiation. (g) Schematic illustration of SC[4]A-QAS nanovalves based on AuNR@MSN. The nanovalves can be operated by NIR irradiation to regulate the release of drugs. (h) TEM images of AuNR@MSN. (i) Release profiles of the RhB-loaded, SC[4]A-capped AuNR@MSNs caused by an 808 nm NIR laser with different power densities. Figures (a)-(c) reprinted from ref. 75a. © IOP Publishing. Reproduced with permission. All rights reserved. Figures (d)-(f) Reprinted with permission from J. Yang, D. Shen, L. Zhou, W. Li, X. Li, C. Yao, R. Wang, A. M. El-Toni, F. Zhang and D. Zhao, *Chem. Mater.*, 2013, **25**, 3030.[84] Copyright (2013) American Chemical Society. Figures (g)-(i) reproduced from ref. 89 with permission from the Royal Society of Chemistry.

Several recent studies have utilized the thermosensitive PNIPAm and its copolymers to decorate AuNCs or AuNRs for drug loading. A burst release of the loaded drugs was observed when the nanocarriers were exposed to NIR light, due to the photothermally induced phase decomposition of the coating PNIPAm polymer in which the drug molecules were encapsulated. For instance, Halas and West employed optically active nanoparticles, SiO_2 particles coated with thin Au shells in a bulk NIPAAm-*co*-AAm hydrogel. The relation between the diameter of the SiO_2 core and the Au shell thickness was varied to shift the SPR of the Au to the spectral range 700–1050 nm. Following 1064 nm irradiation, release of the loaded drug was triggered by the shrinkage of the hydrogel.[83] Similarly, Zhao *et al.* have also successfully developed a novel multifunctional NIR-controlled drug release system based on Au nanocages as photothermal cores, mesoporous silica shells as supporters to increase the anticancer drug loading, and thermally responsive PNIPAm as NIR-stimulated gatekeepers (Au-nanocage@mSiO_2@PNIPAm), as shown in Figure 8.6d–f.[84] Except for the AuNPs as the photosensitive moieties, the combination of GO or CNTs with PNIPAm is also promising for NIR-responsive drug delivery applications, such as the water-soluble graphene sheets–PNIPAm,[85] GO sheets–poly(NIPAM-*co*-AA) microgel,[86] and CNT–pNIPAM composite hydrogels.[87] In addition to PNIPAm and its derivatives, genetically engineered polypeptides, particularly those based on elastin, also show similar thermoresponsive behaviors and have been used as NIR-triggered drug delivery vehicles. For example, Charati *et al.*[88] formulated a composite polypeptide hydrogel with AuNRs, which exhibits unique features suitable for on-demand changes in gels, potentially applicable as a NIR-sensitive drug release device. This kind of hydrogel is formed from a genetically engineered multiblock polypeptide that exhibits a temperature-dependent transition from a solid to liquid state.

The construction principle of this kind of NIR-triggered nanocarrier can be described as follows: various NIR absorbing nanostructures (*e.g.*, AuNRs, Pd@Ag NP, CuS nPs, graphene nanosheets) as well as the bioactive molecules (*e.g.*, drugs and genes) are appropriately contained in some kind of thermal-responsive polymer matrixes. Once exposed to NIR irradiation, the NIR-absorbing nanostructures induce local heating and, consequently, melting of the polymers, resulting in the controlled release of entrapped molecules. The release profiles are controlled by the extent and timing of the light exposure, including stepwise release with intermittent light. The NIR-triggered dissociation of these thermal responsive polymers provides a drug release technique that could be performed by transdermal exposure. However, this approach still has some constraints: (1) generally it is applicable only to thermosensitive polymers that have a hydration–dehydration phase transition temperature; and (2) photoinduced disruption of the thermal responsive polymers cannot be retained after the NIR irradiation is turned off, because the heating effect disappears and the initial solution temperature recovers. Further investigations are needed to improve the feasibility of this kind of nanocarriers.

8.5.5 NIR-Triggered Release Through Destroying the Binding Affinity

As we have seen, different kinds of efficient NIR-absorbing nanomaterials with suitable surface modifications have been extensively explored for NIR-triggered drug or gene delivery. Apart from the situations mentioned above, the hyperthermia from photothermal agents can also efficiently release cargos by decreasing or even destroying the binding affinity between a thermosensitive caging moiety and the nanocarrier's loaded active biomolecules. The active biomolecules are loaded into photothermal mesoporous structures (*e.g.* Au@mSiO$_2$, CuS@mSiO$_2$, and Ag$_2$S@mSiO$_2$ NPs or hollow Au or CuS spheres) with well-controlled morphologies, and a suitable thermosensitive caging moiety is modified onto the surface of the nanocarriers, resulting in a smart drug nanovehicle. Upon NIR irradiation, the heat generated by photothermal agents opens the thermosensitive caps or gates by decreasing or destroying the binding affinity, resulting in the release of the loaded biomolecules. A typical example is shown in Figure 8.6g–i. A biocompatible NIR-responsive nanovalve system was successfully designed and synthesized based on the modification of sulfonatocalix[4]arene (SC[4]A) supramolecular switches on the surface of a mesoporous silica-coated AuNR (AuNR@MSN).[89] In this system, the negatively charged SC[4]A rings encircle the quaternary ammonium salt (QAS) stalks on the surfaces of AuNR@MSN *via* host–guest complexation. When plasmonic heating was generated from the NIR-stimulated AuNP cores, it decreased the ring–stalk binding affinity, leading to the dissociation of SC[4]A rings from the stalks, thus opening the nanovalves and releasing the cargos. In this way, a NIR-controlled release based on photothermal agents was successfully achieved through decreasing the binding affinity between the gating caps and the pores.

In addition to these, several research groups[90] have also exploited DNA-coated NIR-absorbing nanostructures for drug delivery, in which duplex or quadruplex DNA served as a gatekeeper with a smart response to temperature. Similarly, upon NIR irradiation, the photothermal phenomenon could trigger the "melting" of dsDNA (or dsDNA–RNA hybrid) attached to nanocarriers, resulting in the release of drug molecules from the nanocarriers. For example, Qu *et al.*[91] demonstrated a novel NIR-responsive drug delivery platform which combined Au@mSiO$_2$ NPs with a unique temperature-dependent assembly property of duplex DNA. Upon NIR irradiation, the photothermal effect of the AuNRs led to a rapid rise in the local temperature, resulting in the dehybrization of the linkage DNA duplex that anchored the G-quadruplex DNA cap to the surface of the mesoporous silica-based materials, allowing the release of the entrapped guest.

Since noble metal NPs exhibit localized SPR, these plasmon resonant particles can not only have excellent light to heat conversion efficiencies, as discussed above, but also exhibit some non-thermal phenomena that could modulate the interaction between the carrier molecule and the bioactive molecule to be delivered. For example, Halas and his coworkers demonstrate an engineered AuNS-based therapeutic oligonucleotide delivery vehicle that released its cargo on demand upon NIR laser illumination through this non-thermal mechanism. In this gene delivery system, "hot" or excited electrons transferred from the metal to the carrier nucleic acid strand, increasing the electrostatic repulsion between the carrier and therapeutic nucleic acid, thereby resulting in dehybridization and release.[92] Chen and his coworkers[93] also demonstrated a mild synthetic method of synthesizing Fe_3O_4@C@Ag core–shell NPs, in which the drug could be incorporated within the carbon shell, while drug release can be externally triggered by the localized SPR of AgNPs upon excitation with an 800 nm laser.

8.6 Current Challenges and Potential Solutions

8.6.1 Introduction

As discussed above, the development of drug delivery systems with the capacity to respond to NIR light has attracted much attention owing to their dramatically enhanced therapeutic efficacy. However, though NIR-controlled release has been reported for many kinds of functional NPs, the reported examples still have many disadvantages, such as irreversibility of the cargo release mechanism, not operating in aqueous environments, monotonicity of the excitation wavelength, limited NIR excitation light harvesting ability, low sensitivity and targeting accuracy, as well as limited functionalities for imaging and therapy. Improvement of NIR-sensitive NP platforms is greatly needed. Here, we discuss four considerations for the improvement of NP platforms: multichannel controlled drug or gene delivery systems (Section 8.6.2); NIR nanocarriers based on 808 nm excited UCNPs (Section 8.6.3); multimodality imaging-assisted NIR nanocarriers (Section 8.6.4); and NIR-triggered combined therapy (Section 8.6.5). We hope that these suggestions will be helpful in the construction of better nanoplatforms for the next generation of NIR-triggered nanomaterials.

8.6.2 Multichannel Controlled Drug or Gene Delivery Systems

In most cases, the as-obtained nanocarriers can release the loaded biomolecules in response to only one kind of external stimulus, such as light, pH, or temperature, and realize a monoresponsive nanochannel. However, the sensitivity of these nanocarriers is not very efficient. Even for NIR-triggered phototherapy, the effective penetration depth of NIR light is still limited to no more than 1 cm and its sensitivity and accuracy to treat tumors located deep inside the body is thus limited. Therefore, delivery systems triggered by at least two different inputs have recently been considered by researchers. Generally, in order to get dual-responsive nanocarriers, two kinds of smart materials or one material with two functional groups must be included in the same NP. For example, two kinds of responsive materials, such as PAA and PNIPAm,[94] can be simultaneously employed in the a pH- and temperature- dual-responsive nanocarrier, or polymers with two functional groups, such as poly[2-(dimethylamino) ethyl methacrylate] (PDMAEMA)[95] and poly(NIPAM-*co*-AA),[96] can also be used. In this way, cargo release from the nanocarrier system can be controlled more conveniently and at will.

Some of the present authors have also reported dual-channel drug or gene delivery systems controlled by NIR light and some other chemical input. For example, Guillem and his coworkers[97] reported a NIR- and pH-switched gate-like functional hybrid material consisting of a nanoscopic MCM-41 matrix and nanoscopic caps (Figure 8.7a). In this system, preorganized nanoscopic solid structures can be used to load the biomolecules while the AuNPs attached to the surface of the inorganic supports act as NIR- or pH-switched gate cappers. When the pH is high and there is no NIR light excitation, no biomolecules were detected owing to the effective pore blockage by the AuNPs. However, as shown in Figure 8.7b, when the pH changed from 5 to 3, this hybrid material was able to release the entrapped cargos effectively, which can be explained by the Higuchi model. Furthermore, upon NIR irradiation, as shown in Figure 8.7c, the strong absorption of AuNPs led to the thermal cleavage of the boronic ester linkage that anchors the AuNPs to the surface

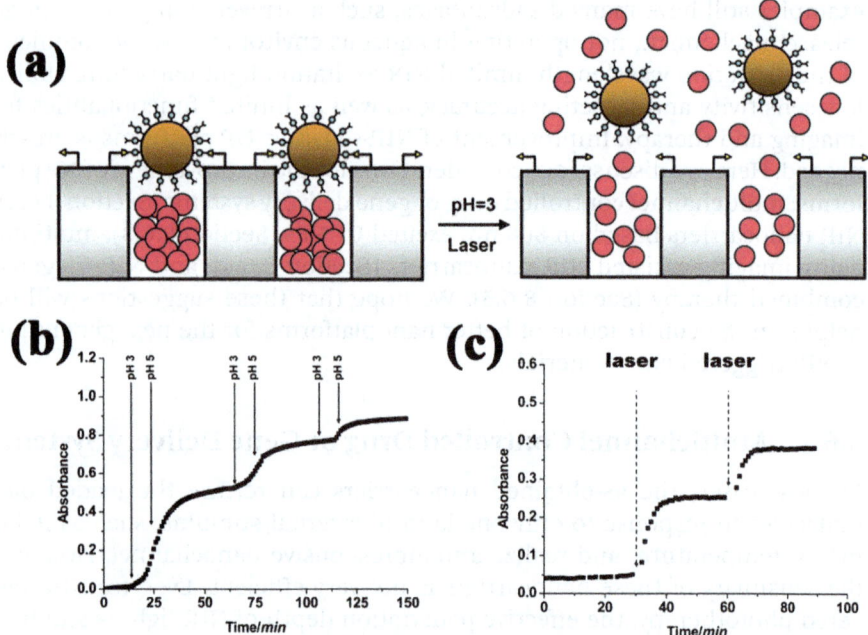

Figure 8.7 (a) Release of the entrapped guest from mesoporous silica supports (labeled as solid S2), simultaneously triggered by pH and laser light. (b) Partial guest release from solid S2 as function of pH variations. (c) Partial guest release from solid S2 controlled by irradiation with laser light. Reprinted with permission from E. Aznar, M. D. Marcos, R. Martínez-Máñez, F. Sancenón, J. Soto, P. Amorós and C. Guillem, *J. Am. Chem. Soc.*, 2009, **131**, 6833.[97] Copyright (2009) American Chemical Society.

of the mesoporous silica-based material, thus resulting in an effective way to release the entrapped cargos. In this way, a bis-switching gate-like system controllable by both NIR light and pH using AuNPs as nanoscopic caps has been successfully synthesized.

8.6.3 NIR Nanocarriers Based on 808 nm Excited UCNPs

As mentioned earlier, most photosensitive groups, such as coumarin,[98] 2-nitrobenzyl,[99] and 7-nitroindoline,[100] are sensitive to UV/Vis light with low tissue penetration, which restricts the practical applications of light-triggered drug release systems in biomedicine. To overcome this problem, transducers that can convert longer-wavelength light, such as NIR, into the UV/Vis region are required. As mentioned in Sections 8.3 and 8.4, this can be achieved by upconversion excitation. In fact, UCNPs have emerged as appealing candidate donors in NIR-triggered drug or gene delivery systems.

Many groups have successfully taken advantage of UV/Vis upconversion emission for NIR phototriggered release of loaded drugs and genes or photo-activation of macromolecules such as some prodrugs, D-luciferin, or DNA.[101] However, there is still a problem: lanthanide UCNPs codoped with Yb^{3+} or Tm^{3+}/Er^{3+} are commonly used to convert 980 nm CW NIR light into UV/Vis, but this can produce undesirable heat that will damage biological tissue when long irradiation times or strong irradiation intensity are involved, mainly because of the noticeable water absorption at this wavelength. To address this problem, many researchers have tried to design UCNPs that can absorb a more biocompatible wavelength of 808 nm, since water is much more transparent for NIR light at 808 nm.

In recent years, core–shell cascade sensitized $Nd^{3+}/Yb^{3+}/Er^{3+}$ (Tm^{3+}) tri-doped UCNPs with 808 nm excitation wavelength have been widely developed because of the large absorption cross-section of Nd^{3+} at 808 nm and the high energy transfer efficiency from Nd^{3+} to Yb^{3+} (up to 70%). Shen and his coworkers[102] first introduced Nd^{3+} as a sensitizer to construct the 808 nm excited UCNPs. In this case, successful doping of Nd^{3+} with Yb^{3+} and activator ions Er^{3+}/Tm^{3+} in a UCNP core allows a successful energy transfer: $Nd^{3+} \rightarrow Yb^{3+} \rightarrow Er^{3+}/Tm^{3+}$. However, the biggest challenge is the deleterious energy transfer from activator ions to Nd^{3+} *via* cross-relaxations, which could lead to poor NIR absorption and consequently weak upconverted emission. To solve this problem, a new core–shell structure that separates the activator Er^{3+}/Tm^{3+} and the sensitizer Nd^{3+} into different core or shell layers has been achieved.[103] In this way, enhanced upconversion luminescence (UC) can be obtained compared with triply doped $NaYF_4$:Yb,Er,Nd UCNPs.

Since Nd^{3+}-sensitized UCNPs have the potential to be the next generation of luminophores, NIR nanocarriers based on 808 nm excited UCNPs are also being developed. The major strategies immobilizing the biomolecules onto the 808 nm excited UCNPs are similar to those for the 980 nm excited UCNPs mentioned above, such as using adsorbed or tethered polymers or the interior porosity of MSNs. Although the 808 nm excited UCNPs tackle the existing problem of 980 nm excited UCNPs by reducing undesirable heat, they have yet to be applied in NIR nanomaterials. It remains to be see how far this new luminophore advances NIR-triggered therapeutic capabilities.

8.6.4 Multimodality Imaging-Assisted NIR Nanocarriers

In recent years, multifunctional nanocarriers that can integrate both diagnosis and therapy into a single multimodal nanoplatform to construct a so-called "theranostic" platform have increasingly gained attention in biomedical applications. An appropriate medical imaging modality can not only help to diagnose disease and visualize NP accumulation, but also facilitate the evaluation of treatment effects. For the NIR-triggered cancer therapy systems discussed here, appropriate medical imaging is extremely meaningful and important because of the following three points:

- Before NIR irradiation, multimodality imaging can help to identify tumor size, shape, and location to make sure that the planned NIR exposure will cover the whole tumor.
- Imaging can detect when the accumulation of phototherapeutic nanocarrier agent in the tumor is at its highest in order to achieve the greatest phototherapeutic efficacy.
- After the NIR irradiation treatment, advanced imaging techniques are also necessary to assess its effectiveness as early as possible.

Many researchers have focused on imaging-assisted NIR nanocarriers for precise cancer diagnosis and location of the tumor site to guide the external laser irradiation without damaging the surrounding healthy tissues. For example, Dai and his coworkers[104] fabricated a theranostic agent with Au nanoshells coated around poly(lactic acid) NPs entrapping DOX, followed by linking a Mn-porphyrin derivative onto the surface in order to endow a greatly improved relaxivity for MRI diagnosis. As could be expected, the results show that the grafted Mn-porphyrin derivative favors accurate cancer diagnosis and tumor location to guide NIR irradiation for tumor therapy. Recently, they have also successfully synthesized PEGylated polypyrrole NPs conjugating gadolinium chelates (Gd–PEG–PPy NPs),[105] which similarly have high MRI contrast capability to guide the laser irradiation for photothermal ablation of tumors.

Recently, various nanostructures have been designed to achieve MRI, near infrared fluorescence, photoacoustics (PA), computed tomography (CT), or positron emission tomography for tumor detection. Each imaging modality has its own advantages and intrinsic limitations, and monomodal imaging of NPs cannot satisfy ideal imaging demands such as integrity of spatial resolution and ultrasensitivity, owing to the respective drawbacks of each technique. For instance, MRI provides spatial resolution, but often limited sensitivity; UCL imaging has excellent sensitivity, but lacks spatial and anatomical resolution. In order to solve this problem, multifunctional NPs have been explored for multimodal imaging that can provide multiple complementary imaging features. For example, several theranostic NPs integrating MRI with UCL or CT exhibit considerable advantageous imaging features for tumor detection, treatment guidance, and monitoring. Several imaging agents such as MRI contrast agents, fluorescent dyes, rare earth complexes, and radioactive agents can be encapsulated or conjugated within the NIR-triggered nanocarriers to achieve their respective functions. In this way, complementary information can be acquired and lead to more accurate diagnoses for the NIR nanocarriers. Tan and his coworkers[106] reported accurate diagnosis and targeted therapy with dual molecular imaging (MRI/optical imaging)-assisted NIR-triggered therapy. In this case, a drug-loaded Fe_3O_4@Au nanorose platform was constructed, whose inner Fe_3O_4 core could function as an MRI agent while loaded DOX can be monitored by its fluorescence. Our group[40] also reported a multifunctional NIR-triggered drug delivery system combining UCL/MRI/CT trimodality imaging and NIR-activated platinum prodrug delivery, as shown in Figure 8.8.

Although there are many examples of multimodality imaging-assisted NIR nanocarriers, the reported imaging and therapy procedures are basically implemented separately, failing to obtain the therapeutic feedback for precise treatment in a timely way. If the characteristic biochemical changes during the whole therapy process could be monitored in real time, over- or under-treatment would be effectively prevented to improve the therapeutic and economic efficiency. Although urgently needed, this kind of real-time therapeutic monitoring remains a challenge. Yang and his coworkers[107] have successfully integrated a newly synthesized pH-activatable fluorescent probe, BDP-688, into multifunctional nanomicelles for the self-feedback of therapeutic efficacy. In this case, the perfect self-feedback function comes from the fast and reversible fluorescence response of BDP-688 to pH, which helps to achieve the real-time visualization of lysosomal pH gradient change and precise therapy. In this way, the strategy provides a powerful tool for real-time evaluation of cancer treatment.

8.6.5 NIR-Triggered Combined Therapy

Cancer remains one of the most devastating diseases in modern society, causing millions of death every year, which may be attributed to many causes such as rapid metastasis, development of drug resistance, and individual differences in cancer patients. Although some current therapeutic methods, such as chemotherapy and radiotherapy, can efficiently kill cancer cells by directly breaking down their DNA structures, their therapeutic effects are still far from satisfactory. Combination therapy, which uses two or more therapeutic approaches together, has shown great potential in numerous preclinical and clinical studies. Similarly, in order to improve the effectiveness of anticancer therapy, NIR-triggered drug or gene therapy may be integrated with other therapeutic methods for combination therapy.

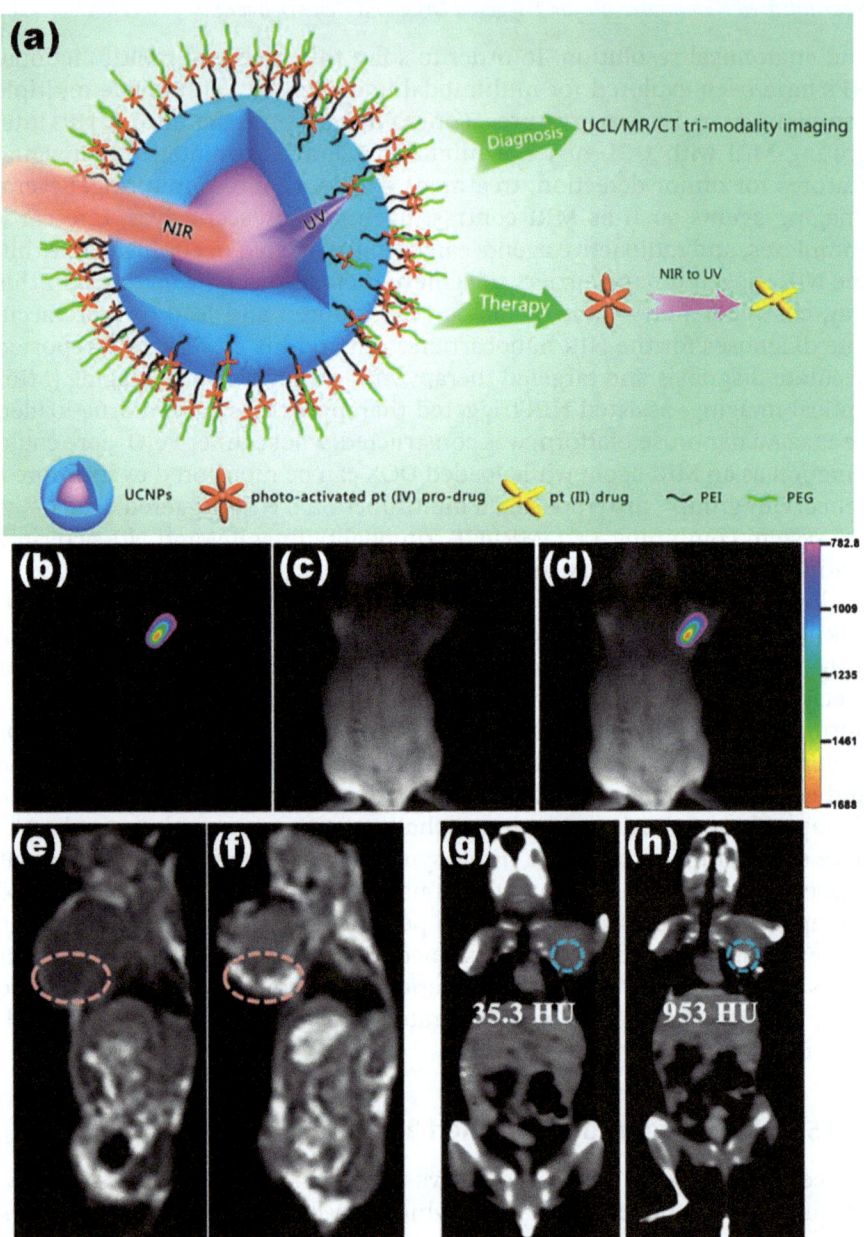

Figure 8.8   Schematic illustration of the characterization of UCNPs–DPP–PEG NPs (a). *In vivo* UCL imaging of a tumor-bearing Balb/c mouse after injection of NPs at the tumor site: upconversion luminescence (b), bright field (c), and overlay images (d). T_1-weighted MRI of a tumor-bearing Balb/c mouse: preinjection (e) and after injection (f) *in situ*. CT imagings of a tumor-bearing Balb/c mouse: preinjection (g) and after injection (h) *in situ*. Reprinted with permission from Y. L. Dai, H. H. Xiao, J. H. Liu, Q. H. Yuan, P. A. Ma, D. M. Yang, C. X. Li, Z. Y. Cheng, Z. Y. Hou, P. P. Yang and J. Lin. *J. Am. Chem. Soc.*, 2013, **135**, 18920.[40] Copyright (2013) American Chemical Society.

In fact, the combination of chemotherapy and PTT has received tremendous interest in recent years.[108] In Section 8.5 we described some commonly employed mechanisms using photothermal effect to promote drug or gene release. In these systems, the combination of NIR-triggered PTT and chemotherapy has been realized and could significantly improve therapeutic efficacy together with minimal side effects for *in vitro/in vivo* applications. Various photothermal agents such as AuNSs, Cu_9S_5 NPs, and grapheme nanosheets with suitable surface modifications have been extensively explored.[109] For example, coating AuNSs individually with a mesoporous silica shell can simply achieve combined PTT and chemotherapy, in which the mesoporous silica cavities are utilized as reservoirs for chemotherapeutics, while AuNRs serve as a local heat generator to induce phototherapy and trigger drug release.[110] Instead of mesoporous silica shell, some other materials, such as DNA or polymers, can also be coated on the surface of AuNR for the construction of NIR-triggered chemo/photothermal therapy materials.[111] The resulting combination therapy shows outstanding potential for further exploration.

Recently, PDT, which can generate cytotoxic $^1O_2$ under photoirradiation, has attracted a lot of attention because of its highly localized therapy effects, lower toxic side effects, no cumulative toxicity, minimal invasiveness, cost-effectiveness, and short treatment time. Admittedly, PDT has yet to gain widespread acceptance as a frontline cancer therapy. One of its major limitations is poor tissue penetration. Therefore, much effort has been made to shift the maximum absorption of photosensitizers to NIR wavelengths, including the synthesis of nanomaterials with strong NIR absorbance or exploitation of UCNPs as a transducer to convert NIR to UV/Vis light. If NIR-triggered PDT is combined with chemotherapy or PTT, a single NIR laser irradiation could trigger PDT and chemotherapy/PTT simultaneously, with improved operational convenience and patient comfort. In addition, the achievement of combination therapy would allow these therapeutic modes to cooperate and compensate each other efficiently, thus leading to the complete inhibition of DNA repair and causing permanent cell death. Many drug delivery systems that are capable of delivering photodynamic photosensitizers, chemotherapy drugs, or photothermal molecules have been explored by different groups. For example, various NIR-absorbing nanostructures (*e.g.*, AuNRs, AuNFs) can be used as photosensitizer delivery vehicles for carrying the photosensitizers (*e.g.*, ICG, Ce6) with strong absorption in the NIR region. In this way, excellent PDT and PTT synergistic therapeutic effects can be achieved under a single NIR laser irradiation.[112] Our group[113] also developed a mild and rational route to synthesize novel multifunctional GdOF:Ln@SiO$_2$ mesoporous capsules as drug carrier for NIR-triggered multiple anticancer therapies (PDT, PTT, and chemotherapy). As shown in Figure 8.9a and b, these yolk-like microcapsules have large hollow cavities between the core and shell, which can incorporate the PDT agent (ZnPc), PTT agent (carbon dots), and DOX into the framework to integrate PDT with PTT and chemotherapy for enhanced antitumor efficiency. The anticancer synergistic therapeutic effects are shown in Figure 8.9c–f, revealing that combination therapy shows great potential for cancer treatment. Apart from utilizing different NIR-absorbing nanostructures as

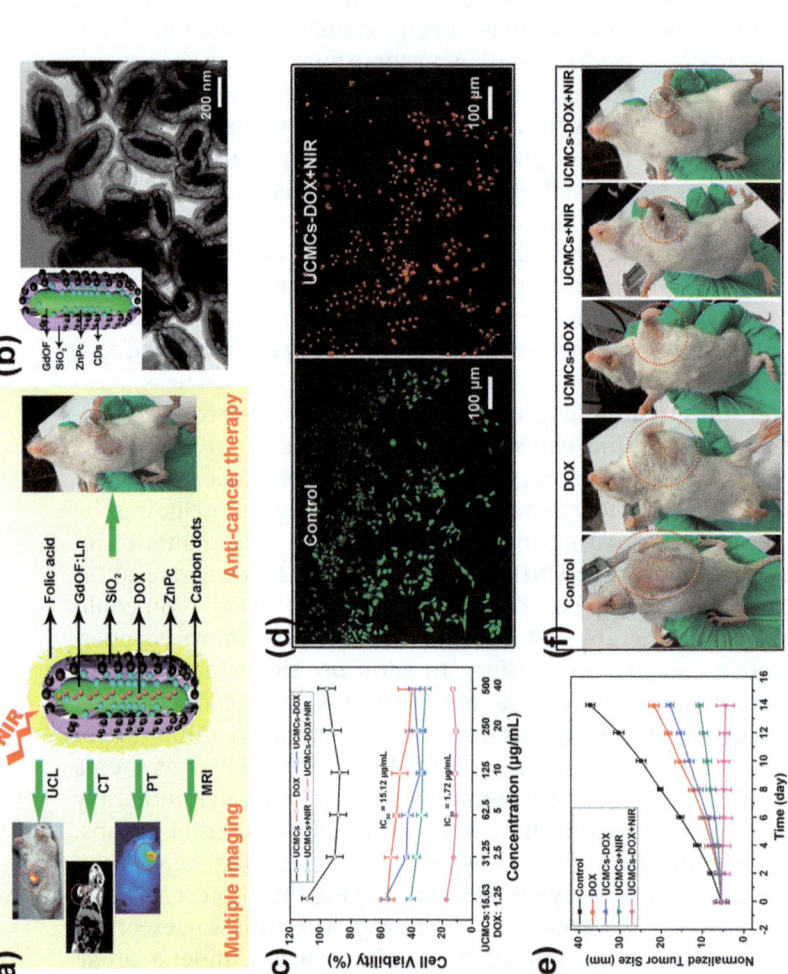

Figure 8.9 (a) Schematic illustration of the bioapplications of the as-obtained GdOF:Ln@SiO₂–ZnPc–CDs microcapsule for multiple imaging and anti-tumor combined therapy under NIR irradiation. (b) TEM images and morphology sketch (the inset) of GdOF:Ln@SiO₂–ZnPc–CDs microcapsule. (c) *In vitro* viability of HeLa cells incubated for 24 h with DOX, UCMCs, UCMCs–DOX at different concentrations with or without NIR laser irradiation. (d) Confocal laser scanning microscope (CLSM) images of HeLa cells incubated with culture without irradiation and incubated with UCMCs–DOX with NIR irradiation dyed with calcium AM and PI. (e) The tumor size of H22 tumor in different groups after treatment. (f) Representative photographs of mice after various intratumoral treatments. Reprinted with permission from R. Lv, P. Yang, F. He, S. Gai, C. Li, Y. Dai, G. Yang and I. Lin. *ACS Nano*, 2015, **9**, 1630.[113] Copyright (2015) American Chemical Society.

the heat generator and photosensitizer for combined PTT and PDT separately, there are currently some nanomaterials that could not only sensitize the formation of $^1O_2$ to exert PDT effects, but also simultaneously generate the heat to "cook" tumors. For example, Hwang *et al.* reported the preparation of a unique Au nanoechinus structure with exceptionally high extinction coefficients in the NIR region, which could be used as a dual-mode PDT and PTT reagent for the complete destruction of solid tumors using NIR light in the first and second biological windows.[114] Peng and colleagues also exploited a traditional hydrophobic photosensitizer (phthalocyanine) as both reactive oxygen species (ROS) and heat generator for combined PDT and PTT under 730 nm laser irradiation after being loaded in hollow silica NPs for an effective combined therapeutic effect.[115]

8.7 Summary

NIR control of drug or gene delivery is considered crucial to boost local effective accumulation while minimizing side effects, resulting in improved therapeutic efficacy. The exploration of NIR light-responsive drug or gene delivery systems opens up new and exciting possibilities for nanomedicine. In this chapter, we have primarily summarized different kinds of NIR-triggered drug or gene delivery platforms, including their various formation mechanisms, synthetic procedures, structural characterizations, and potential bioapplications. Despite successes so far, the properties of NIR nanocarriers are in need of further optimization in order to fulfill the requirements of practical applications. Therefore, current challenges and potential solutions for NIR nanocarriers were also discussed. This highly dynamic research field will certainly continue to produce breakthrough discoveries for cancer treatment of in the near future.

References

1. (a) S. Mura, J. Nicolas and P. Couvreur, *Nat. Mater.*, 2013, **12**, 991; (b) R. Mo, T. Jiang, R. DiSanto, W. Tai and Z. Gu, *Nat. Commun.*, 2014, **5**, 3364; (c) B. Khorsand, G. Lapointe, C. Brett and J. K. Oh, *Biomacromolecules*, 2013, **14**, 2103; (d) R. Duan, F. Xia and L. Jiang, *ACS Nano*, 2013, **7**, 8344; (e) Y. Zhang, H. F. Chan and K. W. Leong, *Adv. Drug Delivery Rev.*, 2013, **65**, 104.
2. (a) X. Zhao, J. Kim, C. A. Cezar, N. Huebsch, K. Lee, K. Bouhadir and D. J. Mooney, *Proc. Natl. Acad. Sci. U. S. A.*, 2011, **108**, 67; (b) R. Cheng, F. Meng, C. Deng, H. A. Klok and Z. Zhong, *Biomaterials*, 2013, **34**, 3647; (c) S. R. Sirsi and M. A. Borden, *Adv. Drug Delivery Rev.*, 2013, **72**, 3; (d) V. Pillay, T. S. Tsai, Y. E. Choonara, L. C. du Toit, P. Kumar, G. Modi, D. Naidoo, L. K. Tomar, C. Tyagi and V. M. K. Ndesendo, *J. Biomed. Mater. Res., Part A*, 2014, **102**, 2039.
3. (a) N. K. Mal, M. Fujiwara and Y. Tanaka, *Nature*, 2003, **421**, 350; (b) N. K. Mal, M. Fujiwara, Y. Tanaka, T. Taguchi and M. Matsukata, *Chem. Mater.*, 2003, **15**, 3385.

4. (a) X. H. Xia, M. X. Yang, L. K. Oetjen, Y. Zhang, Q. G. Li, J. Y. Chen and Y. Xia, *Nanoscale*, 2011, **3**, 950; (b) C. Alvarez-Lorenzo, L. Bromberg and A. Concheiro, *Photochem. Photobiol.*, 2009, **85**, 848; (c) S. Sortino, *J. Mater. Chem.*, 2012, **22**, 301.

5. (a) N. Fomina, C. McFearin, M. Sermsakdi, O. Edigin and A. Almutairi, *J. Am. Chem. Soc.*, 2010, **132**, 9540; (b) J. S. Katz and J. A. Burdick, *Macromol. Biosci.*, 2010, **10**, 339; (c) Y. Zhang, Q. Yin, L. Yin, L. Ma, L. Tang and J. Cheng, *Angew. Chem., Int. Ed.*, 2013, **125**, 6563; (d) Y. Zhang, Q. Yin, L. Yin, L. Ma, L. Tang and J. Cheng, *Angew. Chem., Int. Ed.*, 2013, **52**, 6435.

6. S. Mura, J. Nicolas and P. Couvreur, *Nat. Mater.*, 2013, **12**, 991.

7. M. Benezra, O. Penate-Medina, P. B. Zanzonico, D. Schaer, H. Ow, A. Burns, E. DeStanchina, V. Longo, E. Herz, S. Iyer, J. Wolchok, S. M. Larson, U. Wiesner and M. S. Bradbury, *J. Clin. Invest.*, 2011, **121**, 2768.

8. (a) J. Zheng, C. Zhang and R. M. Dickson, *Phys. Rev. Lett.*, 2004, **93**, 077402; (b) C. A. J. Lin, T. Y. Yang, C. H. Lee, S. H. Huang, R. A. Sperling, M. Zanella, J. K. Li, J. L. Shen, H. H. Wang, H. I. Yeh, W. J. Parak and W. H. Chang, *ACS Nano*, 2009, **3**, 395.

9. (a) H. Wang, T. B. Huff, D. A. Zweifel, W. He, P. S. Low, A. Wei and J. X. Cheng, *Proc. Natl. Acad. Sci. U. S. A.*, 2005, **102**, 15752; (b) L. Au, Q. Zhang, C. M. Cobley, M. Gidding, A. G. Schwartz, J. Y. Chen and Y. Xia, *ACS Nano*, 2010, **4**, 35; (c) L. Tong, C. M. Cobley, J. Y. Chen, Y. Xia and J. X. Cheng, *Angew. Chem., Int. Ed.*, 2010, **122**, 3563; (d) A. Srivatsan, S. V. Jenkins, M. Jeon, Z. J. Wu, C. Kim, J. Y. Chen and R. K. Pandey, *Theranostics*, 2014, **4**, 163.

10. (a) J. Kim, J. Park, H. Kim, K. Singha and W. J. Kim, *Biomaterials*, 2013, **34**, 7168; (b) J. R. G. Navarro, D. Manchon, F. Lerouge, N. P. Blanchard, S. Marotte, Y. Leverrier, J. Marvel, F. Chaput, G. Micouin, A.-M. Gabudean, A. Mosset, E. Cottancin, P. L. Baldeck, K. Kamada and S. Parola, *Nanotechnology*, 2012, **23**, 465602; (c) X. Huang and M. A. El-Sayed, *J. Adv. Res.*, 2010, **1**, 13.

11. (a) J. Yang, J. Lee, J. Kang, S. J. Oh, H. J. Ko, J. H. Son, K. Lee, J. S. Suh, Y. M. Huh and S. Haam, *Adv. Mater.*, 2009, **21**, 4339; (b) K. Nagpal, S. K. Singh and D. N. Mishra, *Chem. Pharm. Bull.*, 2010, **58**, 1423; (c) R. C. Mundargi, V. R. Babu, V. Rangaswamy, P. Patel and T. M. Aminabhavi, *J. Controlled Release*, 2008, **125**, 193; (d) E. Leo, B. Brina, F. Forni and M. A. Vandelli, *Int. J. Pharm.*, 2004, **278**, 133; (e) R. Vankayala, C.-C. Lin, P. Kalluru, C.-S. Chiang and K. C. Hwang, *Biomaterials*, 2014, **35**, 5527.

12. (a) J. You, R. Zhang, C. Xiong, M. Zhong, M. Melancon, S. Gupta, A. M. Nick, A. K. Sood and C. Li, *Cancer Res.*, 2012, **72**, 4777; (b) Z. Xiao, C. Ji, J. Shi, E. M. Pridgen, J. Frieder, J. Wu and O. C. Farokhzad, *Angew. Chem., Int. Ed.*, 2012, **124**, 12023; (c) Y. T. Chang, P. Y. Liao, H. S. Sheu, Y. J. Tseng, F. Y. Cheng and C. S. Yeh, *Adv. Mater.*, 2012, **24**, 3309; (d) S. J. Leung and M. Romanowski, *Theranostics*, 2012, **2**, 1020.

13. (a) W. R. Zhao, M. D. Lang, Y. S. Li, L. Li and J. L. Shi, *J. Mater. Chem.*, 2009, **19**, 2778; (b) L. Xiong, X. Du, B. Y. Shi, J. X. Bi, F. Kleitzb and S. Z. Qiao, *J. Mater. Chem. B*, 2015, **3**, 1712; (c) X. Li, L. Zhou, Y. Wei, A. M. El-Toni, F. Zhang and D. Zhao, *J. Am. Chem. Soc.*, 2015, **137**, 5903.

14. (a) B. Liu, C. Li, D. Yang, Z. Hou, P. a. Ma, Z. Cheng, H. Lian, S. Huang and J. Lin, *Eur. J. Inorg. Chem.*, 2014, **2014**, 1906; (b) C. X. Li, D. M. Yang, P. A. Ma, Y. Y. Chen, Y. Wu, Z. Y. Hou, Y. L. Dai, J. H. Zhao, C. P. Sui and J. Lin, *Small*, 2013, **9**, 4150; (c) L. Zhao, J. Peng, M. Chen, Y. Liu, L. Yao, W. Feng and F. Li, *ACS Appl. Mater. Interfaces*, 2014, **6**, 11190.

15. (a) A. Corma, U. Díaz, M. Arrica, E. Fernández and Í. Ortega, *Angew. Chem., Int. Ed.*, 2009, **48**, 6247; (b) E. R. Gillies and J. M. J. Frechet, *Bioconjugate Chem.*, 2005, **16**, 361.

16. (a) J. Jiang, X. Tong, D. Morrisand and Y. Zhao, *Macromolecules*, 2006, **39**, 4633; (b) J. Babin, M. Pelletier, M. Lepage, J. F.Allard, D. Morris and Y. Zhao, *Angew. Chem., Int. Ed.*, 2009, **121**, 3379.

17. (a) J. Jiang, B. Qi, M. Lepage and Y. Zhao, *Macromolecules*, 2006, **39**, 4633; (b) J. Babin, M. Lepage and Y. Zhao, *Macromolecules*, 2008, **41**, 1246.

18. B. Yan, J. C. Boyer, N. R. Branda and Y. Zhao, *J. Am. Chem. Soc.*, 2011, **133**, 19714.

19. L. Sun, Y. Yang, C. M. Dong and Y. Wei, *small*, 2011, **7**, 401.

20. (a) G. Gregoriadis, *Trends Biotechnol.*, 1995, **13**, 527; (b) A. Sharma and U. S. Sharma, *Int. J. Pharm.*, 1997, **154**, 123; (c) Y. Malam, M. Loizidou and A. M. Seifalian, *Trends Pharmacol. Sci.*, 2009, **30**, 592.

21. (a) A. D. Bangham, M. M. Standish and J. C. Watkins, *J. Mol. Biol.*, 1965, **13**, 238; (b) G. Gregoriadis, *N. Engl. J. Med.*, 1976, **295**, 765; (c) T. Nii and F. Ishii, *Int. J. Pharm.*, 2005, **298**, 198; (d) D. Zucker, D. Marcus, Y. Barenholz and A. Goldblum, *J. Controlled Release*, 2009, **139**, 73.

22. A. Wagner, M. Platzgummer, G. Kreismayr, H. Quendler, G. Stiegler, B. Ferko, G. Vecera, K. Vorauer-Uhl and H. Katinger, *J. Liposome Res.*, 2006, **16**, 311.

23. Y. D. Jin and X. H. Gao, *J. Am. Chem. Soc.*, 2009, **131**, 17774.

24. C. Boyer and J. A. Zasadzinski, *ACS Nano*, 2007, **1**, 176.

25. G. Wu, A. Mikhailovsky, H. A. Khant, C. Fu, W. Chiu and J. A. Zasadzinski, *J. Am. Chem. Soc.*, 2008, **130**, 8175.

26. J. B. Song, L. Cheng, A. P. Liu, J. Yin, M. Kuang and H. W. Duan, *J. Am. Chem. Soc.*, 2011, **133**, 10760.

27. (a) Y. Qiu and K. Park, *Adv. Drug Delivery Rev.*, 2001, **53**, 321; (b) N. A. Peppas, *Curr. Opin. Colloid Interface Sci.*, 1997, **2**, 531; (c) M. Hamidi, A. Azadi and P. Rafiei, *Adv. Drug Delivery Rev.*, 2008, **60**, 1638; (d) T. R. Hoare and D. S. Kohane, *Polymer*, 2008, **49**, 1993.

28. J. K. Oh, D. I. Lee and J. M. Park, *Prog. Polym. Sci.*, 2009, **34**, 1261.

29. R. A. Siegel and B. A. Firestone, *Macromolecules*, 1988, **21**, 3254.

30. T. Miyata, N. Asami and T. Uragami, *Nature*, 1999, **399**, 766.

31. W. A. Petka, J. L. Harden, K. P. McGrath, D. Wirtz and D. A. Tirrell, *Science*, 1998, **281**, 389.

32. B. Yan, J. C. Boyer, D. Habault, N. R. Branda and Y. Zhao, *J. Am. Chem. Soc.*, 2012, **134**, 16558.

33. G. Liu and C. M. Dong, *Biomacromolecules*, 2012, **13**, 1573.

34. S. Kumar, J. F. Allard, D. Morris, Y. L. Dory, M. Lepage and Y. Zhao, *J. Mater. Chem.*, 2012, **22**, 7252.

35. A. A. Beharry and G. A. Woolley, *Chem. Soc. Rev.*, 2011, **40**, 4422.
36. S. Karthik, N. Puvvada, B. N. Prashanth Kumar, S. Rajput, A. Pathak, M. Mandal and N. D. Pradeep Singh, *ACS Appl. Mater. Interfaces*, 2013, **5**, 5232.
37. Q. N. Lin, Q. Huang, C. Y. Li, C. Y. Bao, Z. Z. Liu, F. Y. Li and L. Y. Zhu, *J. Am. Chem. Soc.*, 2010, **132**, 10645.
38. N. Z. Knez evic, B. G. Trewyn and V. S. Y. Lin, *Chem.–Eur. J.*, 2011, **17**, 3338.
39. L. L. Fedoryshin, A. J. Tavares, E. Petryayeva, S. Doughan and U. J. Krull, *ACS Appl. Mater. Interfaces*, 2014, **6**, 13600.
40. Y. L. Dai, H. H. Xiao, J. H. Liu, Q. H. Yuan, P. A. Ma, D. M. Yang, C. X. Li, Z. Y. Cheng, Z. Y. Hou, P. P. Yang and J. Lin, *J. Am. Chem. Soc.*, 2013, **135**, 18920.
41. S. S. Banerjee and D. H. Chen, *Nanotechnology*, 2009, **20**, 185103.
42. Y. H. Chien, Y. L. Chou, S. W. Wang, S. T. Hung, M. C. Liau, Y. J. Chao, C. H. Su and C. S. Yeh, *ACS Nano*, 2013, **7**, 8516.
43. (a) S. Saha, K. C. F. Leung, T. D. Nguyen, J. F. Stoddart and J. I. Zink, *Adv. Funct. Mater.*, 2007, **17**, 685; (b) K. C. F. Leung, T. D. Nguyen, J. F. Stoddart and J. I. Zink, *Chem. Mater.*, 2006, **18**, 5919.
44. J. N. Liu, W. B. Bu, L. M. Pan and J. L. Shi, *Angew. Chem., Int. Ed.*, 2013, **52**, 4375.
45. W. D. Ji, N. J. Li, D. Y. Chen, X. X. Qi, W. W. Sha, Y. Jiao, Q. F. Xu and J. M. Lu, *J. Mater. Chem. B*, 2013, **1**, 5942.
46. M. K. Jayakumar, N. M. Idris and Y. Zhang, *Proc. Natl. Acad. Sci. U. S. A.*, 2012, **109**, 8483.
47. M. K. G. Jayakumar, A. Bansal, K. Huang, R. Yao, B. N. Li and Y. Zhang, *ACS Nano*, 2014, **8**, 4848.
48. S. Q. He, K. Krippes, S. Ritz, Z. J. Chen, A. Best, H. J. Butt, V. Mailander and S. Wu, *Chem. Commun.*, 2015, **51**, 431.
49. S. Yang, N. J. Li, Z. Liu, W. W. Sha, D. Y. Chen, Q. F. Xu and J. M. Lu, *Nanoscale*, 2014, **6**, 14903.
50. A. Schlossbauer, S. Warncke, P. M. E. Gramlich, J. Kecht, A. Manetto, T. Carell and T. Bein, *Angew. Chem., Int. Ed.*, 2010, **49**, 4734.
51. C. Chem, F. Pu, Z. Z. Huang, Z. Liu, J. S. Ren and X. G. Qu, *Nucleic Acids Res.*, 2011, **39**, 1638.
52. C. Chen, J. Geng, F. Pu, X. Yang, J. Ren and X. Qu, *Angew. Chem., Int. Ed.*, 2011, **50**, 882.
53. Q. Yuan, Y. F. Zhang, T. Chen, D. Q. Lu, Z. L. Zhao, X. B. Zhang, Z. X. Li, C. H. Yan and W. H. Tan, *ACS Nano*, 2012, **6**, 6337.
54. M. Chen, S. N. Yang, X. X. He, K. M. Wang, P. C. Qiu and D. G. He, *J. Mater. Chem. B*, 2014, **2**, 6064.
55. N. Li, Z. Z. Yu, W. Pan, Y. Y. Han, T. T. Zhang and B. Tang, *Adv. Funct. Mater.*, 2013, **23**, 2255.
56. (a) B. Nikoobakht and M. A. El-Sayed, *Chem. Mater.*, 2003, **15**, 1957; (b) Y. Xia, W. Li, C. M. Cobley, J. Chen, X. Xia, Q. Zhang, M. Yang, E. C. Cho and P. K. Brown, *Acc. Chem. Res.*, 2011, **44**, 914.
57. K. Yang, L. Feng, X. Shi and Z. Liu, *Chem. Soc. Rev.*, 2013, **42**, 530.
58. X. Huang, S. Tang, X. Mu, Y. Dai, G. Chen, Z. Zhou, F. Ruan, Z. Yang and N. Zheng, *Nat. Nanotechnol.*, 2011, **6**, 28.

59. (a) M. Zhou, R. Zhang, M. Huang, W. Lu, S. Song, M. P. Melancon, M. Tian, D. Liang and C. Li, *J. Am. Chem. Soc.*, 2010, **132**, 15351; (b) Q. Tian, M. Tang, Y. Sun, R. Zou, Z. Chen, M. Zhu, S. Yang, J. Wang, J. Wang and J. Hu, *Adv. Mater.*, 2011, **23**, 3542.

60. (a) J. Li, F. Jiang, B. Yang, X. R. Song, Y. Liu, H. H. Yang, D. R. Cao, W. R. Shi and G. N. Chen, *Sci. Rep.*, 2013, **3**, 1998; (b) S. S. Chou, B. Kaehr, J. Kim, B. M. Foley, M. De, P. E. Hopkins, J. Huang, C. J. Brinker and V. P. Dravid, *Angew. Chem., Int. Ed.*, 2013, **52**, 4160.

61. (a) K. Yang, H. Xu, L. Cheng, C. Sun, J. Wang and Z. Liu, *Adv. Mater.*, 2012, **24**, 5586; (b) L. Cheng, K. Yang, Q. Chen and Z. Liu, *ACS Nano*, 2012, **6**, 5605.

62. J. F. Lovell, C. S. Jin, E. Huynh, H. Jin, C. Kim, J. L. Rubinstein, W. C. W. Chan, W. Cao, L. V. Wang and G. Zheng, *Nat. Mater.*, 2011, **10**, 324.

63. L. Cheng, W. He, H. Gong, C. Wang, Q. Chen, Z. Cheng and Z. Liu, *Adv. Funct. Mater.*, 2013, **23**, 5893.

64. (a) E. Carbo-Argibay, B. Rodriguez-Gonzalez, J. Pacifico, I. Pastoriza-Santos, J. Perez-Juste and L. M. Liz-Marzan, *Angew. Chem., Int. Ed.*, 2007, **46**, 8983; (b) J. Pérez-Juste, I. Pastoriza-Santos, L. M. Liz-Marzán and P. Mulvaney, *Coord. Chem. Rev.*, 2005, **249**, 1870; (c) P. K. Jain, K. S. Lee, I. H. El-Sayed and M. A. El-Sayed, *J. Phys. Chem. B*, 2006, **110**, 7238.

65. (a) G. v. Maltzahn, J. H. Park, A. Agrawal, N. K. Bandaru, S. K. Das, M. J. Sailor and S. N. Bhatia, *Cancer Res.*, 2009, **69**, 3892; (b) E. B. Dickerson, E. C. Dreaden, X. Huang, I. H. El-Sayed, H. Chu, S. Pushpanketh, J. F. McDonald and M. A. El-Sayed, *Cancer Lett.*, 2008, **269**, 57.

66. L. Tong, Y. Zhao, T. B. Huff, M. N. Hansen, A. Wei and J. X. Cheng, *Adv. Mater.*, 2007, **19**, 3136.

67. L. R. Hirsch, R. J. Stafford, J. A. Bankson, S. R. Sershen, B. Rivera, R. E. Price, J. D. Hazle, N. J. Halas and J. L. West, *Proc. Natl. Acad. Sci. U. S. A.*, 2003, **100**, 13549.

68. (a) L. R. Hirsch, A. M. Gobin, A. R. Lowery, F. Tam, R. A. Drezek, N. J. Halas and J. L. West, *Ann. Biomed. Eng.*, 2006, **34**, 15; (b) S. Lal, S. E. Clare and N. J. Halas, *Acc. Chem. Res.*, 2008, **41**, 1842.

69. Y. Y. Chen, Z. Y. Hou, B. Liu, S. S. Huang, C. X. Li and J. Lin, *Dalton Trans*, 2015, **44**, 3118.

70. (a) J. Yang, J. Choi, D. Bang, E. Kim, E. K. Lim, H. Park, J. S. Suh, K. Lee, K. H. Yoo, E. K. Kim, Y. M. Huh and S. Haam, *Angew. Chem., Int. Ed.*, 2011, **50**, 441; (b) H. Chong, C. Nie, C. Zhu, Q. Yang, L. Liu, F. Lv and S. Wang, *Langmuir*, 2012, **28**, 2091; (c) K. Yang, H. Xu, L. Cheng, C. Sun, J. Wang and Z. Liu, *Adv. Mater.*, 2012, **24**, 5586; (d) Z. Zha, X. Yue, Q. Ren and Z. Dai, *Adv. Mater.*, 2012, **25**, 777; (e) M. Chen, X. Fang, S. Tang and N. Zheng, *Chem. Commun.*, 2012, **48**, 8934; (f) L. Cheng, K. Yang, Q. Chen and Z. Liu, *ACS Nano*, 2012, **6**, 5605; (g) L. Feng, C. Zhu, H. Yuan, L. Liu, F. Lv and S. Wang, *Chem. Soc. Rev.*, 2013, **42**, 6620.

71. X. Liang, Y. Li, X. Li, L. Jing, Z. Deng, X. Yue, C. Li and Z. Dai, *Adv. Funct. Mater.*, 2015, **25**, 1451.

72. (a) Z. Zha, X. Yue, Q. Ren and Z. Dai, *Adv. Mater.*, 2013, **25**, 777; (b) Z. Zha, J. Wang, E. Qu, S. Zhang, Y. Jin, S. Wang and Z. Dai, *Sci. Rep.*, 2013, **3**, 2360.
73. (a) M. P. Melancon, M. Zhou and C. Li, *Acc. Chem. Res.*, 2011, **44**, 947; (b) S. Lal, S. E. Clare and N. J. Halas, *Acc. Chem. Res.*, 2008, **41**, 1842.
74. (a) G. Wu, A. Mikhailovsky, H. A. Khant, C. Fu, W. Chiu and J. A. Zasadzinski, *J. Am. Chem. Soc.*, 2008, **130**, 8175; (b) S. J. Leung, X. M. Kachur, M. C. Bobnick and M. Romanowski, *Adv. Funct. Mater.*, 2011, **21**, 1113; (c) J. Croissant and J. I. Zink, *J. Am. Chem. Soc.*, 2012, **134**, 7628.
75. (a) L. Wu, M. Wu, Y. Zeng, D. Zhang, A. Zheng, X. Liu and J. Liu, *Nanotechnology*, 2015, **26**, 025102; (b) X. Liu, F. Fu, K. Xu, R. Zou, J. Yang, Q. Wang, Q. Liu, Z. Xiao and J. Hu, *J. Mater. Chem. B*, 2014, **2**, 5358; (c) D. Jaque, L. Martinez Maestro, B. del Rosal, P. Haro-Gonzalez, A. Benayas, J. L. Plaza, E. Martin Rodriguez and J. Garcia Sole, *Nanoscale*, 2014, **6**, 9494.
76. Z. Zhang, L. Wang, J. Wang, X. Jiang, X. Li, Z. Hu, Y. Ji, X. Wu and C. Chen, *Adv. Mater.*, 2012, **24**, 1418.
77. J. Yang, D. Shen, L. Zhou, W. Li, X. Li, C. Yao, R. Wang, A. M. EI-Toni, F. Zhang and D. Zhao, *Chem. Mater.*, 2013, **25**, 3030.
78. H. Liu, D. Chen, L. Li, T. Liu, L. Tan, X. Wu and F. Tang, *Angew. Chem., Int. Ed.*, 2011, **50**, 891.
79. (a) S. Sherlock and H. Dai, *Nano Res.*, 2011, **4**, 1248; (b) G. Hong, J. Z. Wu, J. T. Robinson, H. Wang, B. Zhang and H. Dai, *Nat. Commun.*, 2012, **3**, 700; (c) L. Feng, X. Yang, X. Shi, X. Tan, R. Peng, J. Wang and Z. Liu, *Small*, 2013, **9**, 1989; (d) S. Shi, X. Zhu, Z. Zhao, W. Fang, M. Chen, Y. Huang and X. Chen, *J. Mater. Chem. B*, 2013, **1**, 1133.
80. (a) B. D. Chithrani and W. C. W. Chan, *Nano Lett.*, 2007, 7, 1542; (b) S. E. A. Gratton, P. A. Ropp, P. D. Pohlhaus, J. C. Luft, V. J. Madden, M. E. Napier and J. M. DeSimone, *Proc. Natl. Acad. Sci. U. S. A.*, 2008, **105**, 11613; (c) N. W. S. Kam, Z. Liu and H. Dai, *Angew. Chem., Int. Ed.*, 2006, **118**, 591.
81. S. P. Sherlock, S. M. Tabakman, L. Xie and H. Dai, *ACS Nano*, 2011, **5**, 1505.
82. (a) B. Tian, C. Wang, S. Zhang, L. Feng and Z. Liu, *ACS Nano*, 2011, **5**, 7000; (b) H. Kim and W. J. Kim, *Small*, 2014, **10**, 117.
83. S. R. Sershen, S. L. Westcott, N. J. Halas and J. L. West, *J. Biomed. Mater. Res.*, 2000, **51**, 293.
84. J. Yang, D. Shen, L. Zhou, W. Li, X. Li, C. Yao, R. Wang, A. M. El-Toni, F. Zhang and D. Zhao, *Chem. Mater.*, 2013, **25**, 3030.
85. Y. Pan, H. Bao, N. G. Sahoo, T. Wu and L. Li, *Adv. Funct. Mater.*, 2011, **21**, 2754.
86. S. Sun and P. Wu, *J. Mater. Chem.*, 2011, **21**, 4095.
87. X. Zhang, C. L. Pint, M. H. Lee, B. E. Schubert, A. Jamshidi, H. Takei, H. Ko, A. Gillies, R. Bardhan, J. J. Urban, M. Wu, R. Fearing and A. Javey, *Nano Lett.*, 2011, **11**, 3239.
88. M. B. Charati, I. Lee, K. C. Hribar and J. A. Burdick, *Small*, 2010, **6**, 1608.

89. H. Li, L. L. Tan, P. Jia, Q. L. Li, Y. L. Sun, J. Zhang, Y. Q. Ning, J. H. Yu and Y. W. Yang, *Chem. Sci.*, 2014, **5**, 2804.

90. (a) Y. L. Luo, Y. S. Shiao and Y. F. Huang, *ACS Nano*, 2011, **5**, 7796; (b) G. B. Braun, A. Pallaoro, G. Wu, D. Missirlis, J. A. Zasadzinski, M. Tirrell and N. O. Reich, *ACS Nano*, 2009, **3**, 2007; (c) C. C. Chen, Y. P. Lin, C. W. Wang, H. C. Tzeng, C. H. Wu, Y. C. Chen, C. P. Chen, L. C. Chen and Y. C. Wu, *J. Am. Chem. Soc.*, 2006, **128**, 3709.

91. X. Yang, X. Liu, Z. Liu, F. Pu, J. Ren and X. Qu, *Adv. Mater.*, 2012, **24**, 2890.

92. R. Huschka, A. Barhoumi, Q. Liu, J. A. Roth, L. Ji and N. J. Halas, *ACS Nano*, 2012, **6**, 7681.

93. J. Chen, Z. Guo, H. B. Wang, M. Gong, X. K. Kong, P. Xia and Q. W. Chen, *Biomaterials*, 2013, **34**, 571.

94. X. Hou, F. Yang, L. Li, Y. Song, L. Jiang and D. Zhu, *J. Am. Chem. Soc.*, 2010, **132**, 11736.

95. L. X. Zhang, S. L. Cai, Y. B. Zheng, X. H. Cao and Y. Q. Li, *Adv. Funct. Mater.*, 2011, **21**, 2103.

96. W. Guo, H. Xia, L. Cao, F. Xia, S. Wang, G. Zhang, Y. Song, Y. Wang, L. Jiang and D. Zhu, *Adv. Funct. Mater.*, 2010, **20**, 3561.

97. E. Aznar, M. D. Marcos, R. Martínez-Máñez, F. Sancenón, J. Soto, P. Amorós and C. Guillem, *J. Am. Chem. Soc.*, 2009, **131**, 6833.

98. (a) N. K. Mal, M. Fujiwara and Y. Tanaka, *Nature*, 2003, **421**, 350; (b) Q. N. Lin, C. Y. Bao, S. Y. Cheng, Y. L. Yang, W. Ji and L. Y. Zhu, *J. Am. Chem. Soc.*, 2012, **134**, 5052; (c) Q. N. Lin, C. Y. Bao, Y. L. Yang, Q. N. Liang, D. S. Zhang, S. Y. Cheng and L. Y. Zhu, *Adv. Mater.*, 2013, **25**, 1981.

99. P. Neveu, I. Aujard, C. Benbrahim, T. Le Saux, J. F. Allemand, S. Vriz, D. Bensimon and L. Jullien, *Angew. Chem., Int. Ed.*, 2008, **47**, 3744.

100. G. Papageorgiou, D. C. Ogden, A. Barth and J. E. T. Corrie, *J. Am. Chem. Soc.*, 1999, **121**, 6503.

101. (a) Y. M. Yang, Q. Shao, R. R. Deng, C. Wang, X. Teng, K. Cheng, Z. Cheng, L. Huang, Z. Liu, X. G. Liu and B. G. Xing, *Angew. Chem., Int. Ed.*, 2012, **51**, 3125; (b) Y. M. Yang, B. Velmurugan, X. G. Liu and B. G. Xing, *Small*, 2013, **9**, 2937; (c) M. K. G. Jayakumar, N. M. Idris and Y. Zhang, *Proc. Natl. Acad. Sci. U. S. A.*, 2012, **109**, 8483; (d) Y. M. Yang, F. Liu, X. G. Liu and B. G. Xing, *Nanoscale*, 2013, **5**, 231.

102. J. Shen, *et al.*, *Adv. Opt. Mater.*, 2013, **1**, 644.

103. (a) Y. F. Wang, G. Y. Liu, L. D. Sun, J. W. Xiao, J. C. Zhou and C. H. Yan, *ACS Nano*, 2013, **7**, 7200; (b) X. Xie, N. Gao, R. Deng, Q. Sun, Q. H. Xu and X. Liu, *J. Am. Chem. Soc.*, 2013, **135**, 12608.

104. L. Jing, X. Liang, X. Li, L. Lin, Y. Yang, X. Yue and Z. Dai, *Theranostics*, 2014, **4**, 858.

105. X. Liang, Y. Li, X. Li, L. Jing, Z. Deng, X. Yue, C. Li and Z. Dai, *Adv. Funct. Mater.*, 2015, **25**, 1451.

106. C. Li, T. Chen, I. Ocsoy, G. Zhu, E. Yasun, M. You, C. Wu, J. Zheng, E. Song, C. Z. Huang and W. Tan, *Adv. Funct. Mater.*, 2014, **24**, 1772.

107. J. Tian, L. Ding, H. Ju, Y. Yang, X. Li, Z. Shen, Z. Zhu, J.-S. Yu and C. J. Yang, *Angew. Chem., Int. Ed.*, 2014, **53**, 9544.
108. (a) J. C. Y. Kah, R. C. Y. Wan, K. Y. Wong, S. Mhaisalkar, C. J. R. Sheppard and M. Olivo, *Lasers Surg. Med.*, 2008, **40**, 584; (b) W. Chen, R. Carubelli, H. Liu and R. Nordquist, *Mol. Biotechnol.*, 2003, **25**, 37; (c) S. Sherlock and H. Dai, *Nano Res.*, 2011, **4**, 1248; (d) Y. Matsushita-Ishiodori and T. Ohtsuki, *Acc. Chem. Res.*, 2012, **45**, 1039; (e) M. P. Melancon, M. Zhou and C. Li, *Acc. Chem. Res.*, 2011, **44**, 947; (f) S. M. Waldow, P. R. Morrison and L. I. Grossweiner, *Lasers Surg. Med.*, 1988, **8**, 510.
109. (a) Z. Zhang, J. Wang and C. Chen, *Adv. Mater.*, 2013, **25**, 3869; (b) A. M. Alkilany, L. B. Thompson, S. P. Boulos, P. N. Sisco and C. J. Murphy, *Adv. Drug Delivery Rev.*, 2012, **64**, 190; (c) Y. T. Chang, P. Y. Liao, H. S. Sheu, Y. J. Tseng, F. Y. Cheng and C. S. Yeh, *Adv. Mater.*, 2012, **24**, 3309; (d) J. Yang, D. Shen, L. Zhou, W. Li, X. Li, C. Yao, R. Wang, A. M. El-Toni, F. Zhang and D. Zhao, *Chem. Mater.*, 2013, **25**, 3030; (e) P. Huang, L. Bao, C. Zhang, J. Lin, T. Luo, D. Yang, M. He, Z. Li, G. Gao, B. Gao, S. Fu and D. Cui, *Biomaterials*, 2011, **32**, 9796; (f) Y. Wang, K. Wang, J. Zhao, X. Liu, J. Bu, X. Yan and R. Huang, *J. Am. Chem. Soc.*, 2013, **135**, 4799.
110. (a) Z. Zhang, L. Wang, J. Wang, X. Jiang, X. Li, Z. Hu, Y. Ji, X. Wu and C. Chen, *Adv. Mater.*, 2012, **24**, 1418; (b) X. Yang, X. Liu, Z. Liu, F. Pu, J. Ren and X. Qu, *Adv. Mater.*, 2012, **24**, 2890; (c) X. Yang, Z. Liu, Z. Li, F. Pu, J. Ren and X. Qu, *Chem.–Eur. J.*, 2013, **19**, 10388.
111. (a) A. R. Kim, S. W. Shin, S. W. Cho, J. Y. Lee, D. I. Kim and S. H. Um, *Adv. Healthcare Mater.*, 2013, **2**, 1252; (b) Y. Zhong, C. Wang, L. Cheng, F. Meng, Z. Zhong and Z. Liu, *Biomacromolecules*, 2013, **14**, 2411; (c) F. Y. Cheng, C. H. Su, P. C. Wu and C. S. Yeh, *Chem. Commun.*, 2010, **46**, 3167.
112. (a) W. S. Kuo, C. N. Chang, Y. T. Chang, M. H. Yang, Y. H. Chien, S. J. Chen and C. S. Yeh, *Angew. Chem., Int. Ed.*, 2010, **122**, 2771; (b) R. Chen, X. Zheng, H. Qian, X. Wang, J. Wang and X. Jiang, *Biomater. Sci.*, 2013, **1**, 285; (c) S. Wang, P. Huang, L. Nie, R. Xing, D. Liu, Z. Wang, J. Lin, S. Chen, G. Niu, G. Lu and X. Chen, *Adv. Mater.*, 2013, **25**, 3055.
113. R. Lv, P. Yang, F. He, S. Gai, C. Li, Y. Dai, G. Yang and J. Lin, *ACS Nano*, 2015, **9**, 1630.
114. P. Vijayaraghavan, C. H. Liu, R. Vankayala, C. S. Chiang and K. C. Hwang, *Adv. Mater.*, 2014, **26**, 6689.
115. J. Peng, L. Zhao, X. Zhu, Y. Sun, W. Feng, Y. Gao, L. Wang and F. Li, *Biomaterials*, 2013, **34**, 7905.

CHAPTER 9

Near Infrared Nanomaterials for Photothermal Therapy

ZIXIAO LIU[a] AND ZHIGANG CHEN*[a]

[a]State Key Laboratory for Modification of Chemical Fibers and Polymer Materials, College of Materials Science and Engineering, Donghua University, Shanghai, 201620, P. R. China
*E-mail: zgchen@dhu.edu.cn

9.1   Introduction

Cancer is a leading cause of death worldwide, accounting for 8.2 million deaths (around 13% of all deaths) in 2012.[1] For cancer treatment, several conventional therapeutic techniques have been developed, including surgery, radiotherapy, chemotherapy, immunotherapy, and traditional Chinese medicine. In the past several decades, those therapeutic techniques have indeed achieved certain progress. However, they also have many disadvantages, such as harming healthy cells and destroying the immune system, resulting in an increased incidence of secondary cancers. These disadvantages cause pain and distress to patients and reduce the cure rate of cancer. Therefore, it is still essential to develop new and efficient therapy methods with minimal damage to patients for cancer therapy.

The absorbance of near infrared (NIR) light in the 700–1000 nm wavelength range is much lower in biological tissues than visible light, which gives it a deep penetrating property. For example, the typical penetration depth of 980 nm light in biological tissue can be several centimeters.[2–4] As a result, NIR

RSC Nanoscience & Nanotechnology No. 40
Near Infrared Nanomaterials: Preparation, Bioimaging and Therapy Applications
Edited by Fan Zhang
© The Royal Society of Chemistry 2016
Published by the Royal Society of Chemistry, www.rsc.org

laser-induced photothermal ablation therapy (PAT) has attracted increasing interest as a minimally invasive and potentially more effective alternative to conventional approaches for cancer treatment.[5] In a typical PAT process, the NIR photothermal agent is injected either directly into a tumor or intravenously. After intravenous injection, the agent can be aggregated into tumors by the enhanced permeability and retention (EPR) effect or the conjugated targeting factor. If a NIR laser is then used to irradiate the tumor, the photothermal agent *in vivo* can absorb the laser light and convert it into thermal energy, thus heating the tumor. As the temperature increases (>42 °C), cancer cells *in vivo* can be destroyed by thermal ablation.

Obviously, a prerequisite for the development of NIR laser-induced PAT is to find low-cost and biocompatible photothermal agents with high photothermal efficiency and photostability. Currently, there are four main types of photothermal agents. The first type is organic compounds, including organic dyes and polymers. Some of the organic photothermal agents are highly biocompatible and can be readily degraded in the human body. However, they may suffer from limitations such as photobleaching and/or unsatisfactory photothermal conversion efficiency. The second type is metal-based photothermal agents, including Au and Pd nanomaterials. These noble metal-based nanostructures exhibit intense NIR photoabsorption and are the most studied photothermal agents, but they are very expensive and can easily reshape at high temperature during the photothermal conversion process.[6,7] The third type is carbon-based photothermal agents, including carbon nanotubes (CNTs) and graphene. Carbon-based photothermal agents do not melt at high temperature but have a relatively low absorption coefficient in the NIR region. The last type is semiconductor photothermal agents. Usually, these low-cost agents have excellent NIR-absorbing properties and relatively high photostabilities, giving them several advantages over other photothermal agents. However, the semiconductor photothermal agents are still under investigation for lower cytotoxicity and higher biocompatibility.

For evaluating photothermal agents, one of the key parameters is photothermal conversion efficiency. In this chapter, we first introduce the measurement method for photothermal conversion efficiency, which is discussed by an example from our group. Subsequently, we summarize the progress of research on the four kinds of photothermal agents as well as the combination of PAT with other nanobiotechnology. The future prospects and challenges of this novel approach to cancer therapy are addressed in the final section of this chapter.

9.2 Measurement Method for Photothermal Conversion Efficiency

Photothermal conversion efficiency (PCE) is an important parameter that describes the photothermal performances of materials. The classical calculation method for PCE developed by Roper *et al.* in 2007 has since been widely used in PCE calculation.[8]

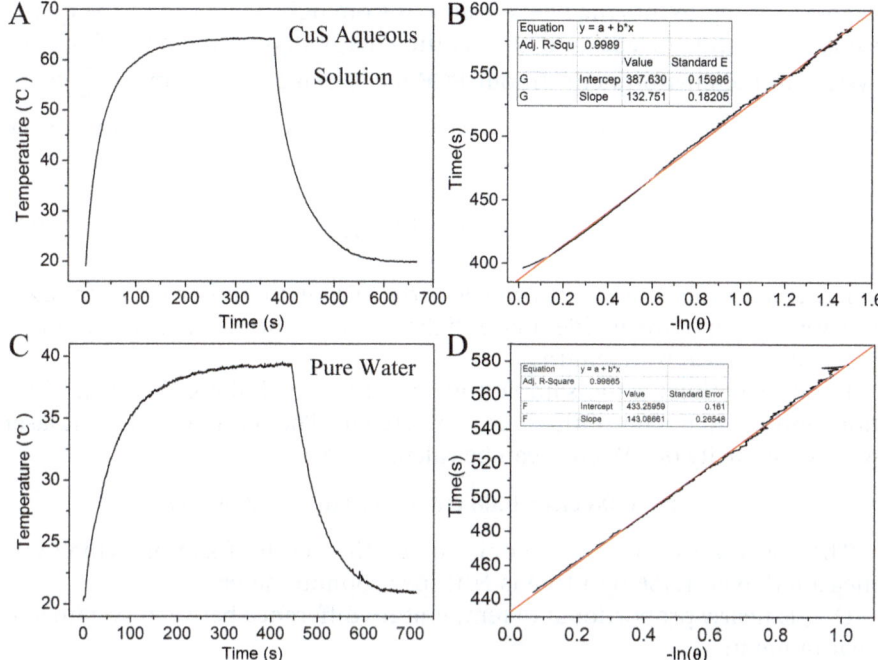

Figure 9.1 Temperature elevation and natural cooling of 0.765 mg mL^{-1} CuS aqueous solution (A) and pure water (C) under the irradiation of 980 nm laser with intensity 3 W cm^{-2}. Linear fitting of the function between time and negative natural logarithm of driving force temperature of CuS aqueous solution (B) and water (D), corresponding to the cooling period of (A) and (C) respectively.

In a typical PCE measurement, we first need to heat the aqueous solution containing photothermal agents by laser irradiation. When the temperature reaches a stable maximum, the laser is shut off and the sample cools down naturally to environmental temperature. Thus, we can obtain a heating–cooling curve by recording the time-dependent temperature change by means of a thermocouple or IR thermal imager. Taking the aqueous solution of CuS nanoplates with a concentration of 0.765 mg mL^{-1} as an example (as used by our group), the heating–cooling curve is obtained by 980 nm laser irradiation with power density 3 W cm^{-2} (Figure 9.1A).

Following the Roper method, the whole system exists in energy balance and this can be expressed as follows:

$$\sum_i m_i C_{p,i} \frac{dT}{dt} = Q_{PAs} + Q_{Dis} - Q_{Surr} \qquad (9.1)$$

In eqn (9.1), m_i and $C_{p,i}$ are the mass and heat capacity of system components (photothermal agent, solvent, quartz sample cell, and so on), T is system temperature, t is time, Q_{PAs} is the heating power caused by photothermal agents, Q_{Dis} represents the heat dissipated from light absorbed by the quartz sample cell and solvent, and Q_{Surr} is the energy dissipating away from the system to the environment. In our PCE measurement experiment, $\sum_i m_i C_{p,i}$ was determined to be 0.201 J °C^{-1}. Q_{PAs} can be calculated according to the following equation:

$$Q_{PAs} = P(1 - 10^{-A_\lambda})\eta_T \tag{9.2}$$

where P is the incident laser power, η_T is the PCE of photothermal agents, λ is the wavelength of the incident laser light, and A_λ is the absorbance of photothermal agents at wavelength λ.

In our experiment, the sample tube was irradiated along the axial direction, and the diameter of the tube was 0.16 cm. The power of incident laser with an intensity of 3 W cm^{-2} can be calculated as:

$$P = \pi \times 0.08 \text{ cm} \times 0.08 \text{ cm} \times 3 \text{ W cm}^{-2} = 0.0603 \text{ W}$$

The absorbance value, A_λ, should match the depth of sample which was measured to be 1.486 by a UV-Vis-NIR spectrophotometer.

Q_{Surr} is nearly proportional to temperature difference between system and environment:

$$Q_{Surr} = hS(T - T_{Surr}) \tag{9.3}$$

where T_{Surr} is the surrounding temperature, S is the surface area of the container, and h is a heat transfer coefficient.

Irradiated by a single-wavelength laser, the heat input, $Q_{NC} + Q_{Dis}$, is definite. The heat output, Q_{Surr}, increases with the enhancement of T, according to eqn (9.3). When the system temperature keeps a stable maximum, $dT/dt = 0$, there is a balance between heat input and heat output:

$$\sum_i m_i C_{p,i} \frac{dT}{dt} = Q_{PAs} + Q_{Dis} - Q_{Surr} = 0 \tag{9.4}$$

Substituting eqn (9.2) and eqn (9.3) for Q_{PAs} and Q_{Surr} into eqn (9.4):

$$P(1 - 10^{-A_\lambda})\eta_T + Q_{Dis} - hS(T_{Max} - T_{Surr}) = 0 \tag{9.5}$$

After rearranging, the PCE can be determined:

$$\eta_T = \frac{hS(T_{Max} - T_{Surr}) - Q_{Dis}}{P(1 - 10^{-A_\lambda})} \tag{9.6}$$

In our experiment, the T_{Max} and T_{Surr} were 64.1 °C and 16.8 °C, respectively. Now just two parameters, hS and Q_{Dis}, remain to be determined.

In order to get hS, a dimensionless driving force temperature, θ, needs to be introduced by using the maximum system temperature (T_{Max}):

$$\theta = \frac{T - T_{\mathrm{Surr}}}{T_{\mathrm{Max}} - T_{\mathrm{Surr}}} \tag{9.7}$$

Furthermore, a system time constant τ_s can be denoted as:

$$\tau_s = \frac{\sum_i m_i C_{p,i}}{hS} \tag{9.8}$$

Substituting eqn (9.2), eqn (9.3), eqn (9.7), and eqn (9.8) into eqn (9.1) and rearranging yields:

$$\frac{d\theta}{dt} = \frac{1}{\tau_s}\left[\frac{Q_{\mathrm{PAs}} + Q_{\mathrm{Dis}}}{hS(T_{\mathrm{Max}} - T_{\mathrm{Surr}})} - \theta\right] \tag{9.9}$$

When the laser irradiation is shut off, $Q_{\mathrm{PAs}} + Q_{\mathrm{Dis}} = 0$, the system is cooling, reducing eqn (9.9) to:

$$dt = -\tau_s \frac{d\theta}{\theta} \tag{9.10}$$

Integrating eqn (9.10) yields:

$$t = -\tau_s \ln \theta \tag{9.11}$$

Therefore, the time constant τ_s can be determined by the slope of the linear fitting between cooling time (t) and the negative natural logarithm of driving force temperature $(-\ln \theta)$. As shown in Figure 9.1B, the slope is 132.75 s. Then hS can be calculated according to eqn (9.8):

$$hS = \frac{\sum_i m_i C_{p,i}}{\tau_s} = \frac{0.201 \, \mathrm{J\,°C^{-1}}}{132.75 \, \mathrm{s}} = 0.00151 \, \mathrm{W\,°C^{-1}}$$

On the other hand, Q_{Dis} can be measured independently using the same sample cell containing pure solvent. The same heating process needs to be repeated for the solvent sample, as shown in Figure 9.1C. $\sum_i m_i C_{p,i}$ was measured to be 0.199 J °C^{-1}. The same linear fitting between cooling time and $-\ln(\theta)$ of the solvent sample should also be carried out (Figure 9.1D). τ_s' of the solvent sample was determined to be 143.09 s, and $h'S'$ can be calculated as follow:

$$h'S' = \frac{\sum_i m_i C_{p,i}}{\tau_s'} = \frac{0.199 \, \mathrm{J\,°C^{-1}}}{143.09 \, \mathrm{s}} = 0.00139 \, \mathrm{W\,°C^{-1}}$$

Without the photothermal agent, the energy balance of the solvent system can be described by:

$$\sum_i m_i C_{p,i} \frac{\mathrm{d}T}{\mathrm{d}t} = Q_{\text{Dis}} - Q_{\text{Surr}} \tag{9.12}$$

When the system temperature goes to a stable maximum, $\mathrm{d}T/\mathrm{d}t = 0$, substituting eqn (9.3) for Q_{Surr} into eqn (9.12) and rearranging:

$$Q_{\text{Dis}} = Q_{\text{surr}} = h'S'(T'_{\text{Max}} - T_{\text{Surr}}) \tag{9.13}$$

T_{Surr} remained at 16.8 °C and T_{Max} was 39.2 °C, according to the temperature recording. Q_{Dis} can be calculated as follow:

$$Q_{\text{Dis}} = Q_{\text{Surr}} = h'S'(T'_{\text{Max}} - T_{\text{Surr}})$$
$$= 0.00139 \text{ W } °C^{-1} \times (39.2 °C - 16.8 °C) = 0.0311 \text{ W}$$

The Q_{Dis} of the solvent sample is similar to the real Q_{Dis} of the aqueous CuS sample. Thus, after substituting Q_{Dis} and hS into eqn (9.6), the PCE can be calculated as follows:

$$\eta_{\text{T}} = \frac{hS(T_{\text{Max}} - T_{\text{Surr}}) - Q_{\text{Dis}}}{P(1 - 10^{-A_\lambda})}$$
$$= \frac{0.00151 \text{ W } °C^{-1} \times (64.1 °C - 16.8 °C) - 0.0311 \text{ W}}{0.0603 \text{ W} \times (1 - 10^{-1.486})} = 0.691$$

Thus, the PCE of our CuS sample was measured to be 69.1%.

9.3 Organic Photothermal Agents

NIR-absorbing organic compounds are promising for the applications of PAT of cancer, and have been the subject of study for some time. Organic photothermal agents can be synthesized in the laboratory (such as NIR-absorbing dyes and polymers) and can also be made from organic compounds which can be found in nature. Classical organic photothermal agents include dyes, polymer nanoparticles, and natural organic compounds. The following section discusses important findings in this area.

9.3.1 Organic Dyes

Organic dyes with high quantum yield (QY) can efficiently emit absorbed optical energy in the form of fluorescence but only a small part of the absorbed energy is transferred into heat, indicating low photothermal efficiency. On the other hand, low-QY organic dyes with high NIR absorption may offer high photothermal efficiency. In general, most of research has been based on indocyanine green (ICG) and ICG-like heptamethine dyes.

Among all the photothermal organic dyes, ICG is the only one that has been approved by the United States Food and Drug Administration (FDA) for clinical applications. As a result, ICG has been widely researched for PAT of cancer in recent years.[9-13] For example, Yu *et al.* synthesized ICG-containing capsules (ICG-capsules) with an average size of 120 nm by means of a three-step room-temperature synthesis.[10] The ICG-capsules exhibited two relatively strong broad peaks in the NIR absorbance spectrum (>750 nm), which were red-shifted from those in the free ICG spectrum. The ICG-capsules were then coated with anti-EGFR to target cancer cells and the targeting effect was proved in an *in vitro* experiment: the anti-EGFR coated capsules selectively attached to cancerous cells and caused ~45% and ~95% cell death under 808 nm laser irradiation at an intensity of 3 W cm^{-2} and 6 W cm^{-2}, respectively.

Constrained by the absorbing wavelength and photobleaching, ICG is not a perfect therapeutic agent for cancer. In order to modify the defects of ICG, heptamethine NIR dyes have been synthesized, which have a similar structure to ICG. Those heptamethine dyes can be used as NIR imaging probes and photothermal agents.[14-18] For example, Cheng *et al.* synthesized a novel NIR-absorbing heptamethine indocyanine dye, IR825, showing a super-high NIR absorption peak at about 825 nm and a rather low QY (<0.1%).[19] The IR825 was then modified by a polyethylene glycol (PEG)-grafted amphiphilic polymer, forming IR825–PEG micelle nanoparticles (Figure 9.2A). The as-synthesized IR825–PEG nanoparticles also have high absorption peaks in the NIR region (Figure 9.2B). More importantly, the NIR absorption of IR825–PEG remained unchanged after 10 min of laser irradiation (808 nm, 0.5 W cm^{-2}), indicating the dramatically improved photothermal stability compared to ICG. With strong NIR absorbance, the IR825–PEG showed better photothermal performance than ICG and obvious an concentration-dependent temperature increase under 808 nm laser irradiation (Figure 9.2C and D). After being labeled by Cy5.5 and intravenously injected into 4T1 tumor-bearing mice, the IR825–PEG nanoparticles were found to be enriched in the tumor by *in vivo* fluorescence imaging, which could be due to the EPR effect. The tumor uptake was measured to be as high as 22.5% ID g^{-1}. Subsequently, the mice were anesthetized and exposed to 808 nm laser irradiation (0.5 W cm^{-2}) for 5 min. The of tumor surface temperature increased from ~30 to ~60 °C, while the mice without IR825–PEG injection showed little change under the same irradiation. The IR825–PEG nanoparticles show a good therapeutic effect in mice, but this report did not give direct data about the degradation behavior of IR825–PEG. Compared with PEG, human serum albumin (HSA) is a safer natural carrier, abundant in the human body. The same group prepared an IR825–HSA nanocomplex by simply complexing IR825 to HSA.[20] The IR825–HSA exhibited no appreciable dark toxicity and could undergo rapid renal excretion. Compared to IR825–PEG, IR825–HSA is one step closer to clinical use.

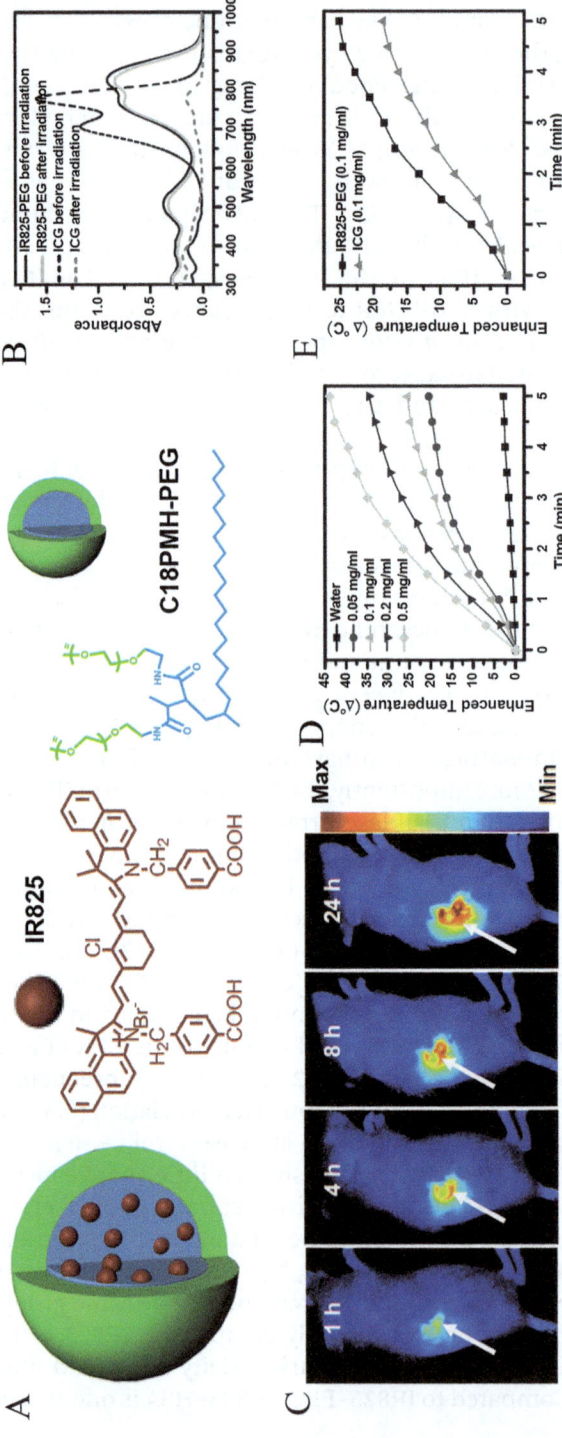

Figure 9.2 (A) A scheme illustrating the composition of IR825–PEG nanoparticles. (B) Absorbance spectra of IR825–PEG (0.01 mg mL^{-1}) and ICG molecules (0.01 mg mL^{-1}) before and after 10 min of laser irradiation. (C) *In vivo* fluorescence images of 4T1 tumor-bearing mice at different time points after the injection of IR825–PEG–Cy5.5. (D) The heating curves of water and IR825–PEG solutions under 808 nm laser irradiation (0.5 W cm^{-2}). (E) The comparison of heating curves between IR825–PEG and ICG solutions under the irradiation of 808 nm at power density 0.5 W cm^{-2}. Reproduced from ref. 19 with permission from John Wiley and Sons. Copyright © 2013 Wiley-VCH Verlag GmbH & Co. KGaA, Weinheim.

9.3.2 Polymer Nanoparticles

NIR-absorbing polymers are another kind of organic photothermal agent. In recent years, many groups have explored the applications of conjugated polymers for PAT, as a result of their high NIR-absorbing property derived from the conjugated molecular structure.[21-29] At present, research on polymer photothermal agents focus on polyaniline (PANI), polypyrrole (PPy), and poly(3,4-ethylenedioxythiophene):poly(4-styrenesulfonate) (PEDOT:PSS).

PANI has an intriguing optical absorbing ability, which means that its absorbance peak can be red-shifted to the NIR region by doping from emeralidine base (EB) to emeralidine salt (ES) state. In 2011, Yang *et al.* synthesized ES PANI nanoparticles (ES PANPs) by an oxidative polymerization process, using anilinium salts protonated by hydrochloride (HCl) and ammonium persulfate as an oxidant.[21] The synthesized ES PANPs can be transited to be EB PANPs by doping them with NaOH in order to homogenize the particle size of PANI. Those EB PANPs are spherical with a size of 115.6 ± 16.3 nm and exhibit a low absorption in the NIR region. Interestingly, the EB PANPs can be doped by biological dopants (intracellular protons and oxidative species from mitochondria) and form ES PANPs again, which have higher NIR absorbance and cause a temperature elevation to 54.8 °C under 808 nm laser irradiation (2.45 W cm^{-2}) in 5 min. This property showed an excellent photothermal therapeutic effect in both *in vitro* and *in vivo* experiments on 808 nm laser irradiation of EB PANPs in A431 cells. These nanoparticles show promise for cancer therapy in future.

PPy nanoparticles have been confirmed in previous studies to have good biocompatibility and low long-term cytotoxicity.[24,25] In 2011, Armes *et al.* reported the synthesis of PPy nanoparticles as a potential optical coherence tomography contrast agent for cancer imaging. They found that PPy nanoparticles exhibited strong NIR photoabsorption but they did not investigate the photothermal effects.[26] Subsequently, Liu's group[23] and Dai's group[27] almost simultaneously reported the use of PPy nanoparticles as photothermal agents.

PPy was polymerized from pyrrole through a microemulsion method in the aqueous phase by using Fe^{3+} as the catalyst and poly(vinyl alcohol) (PVA) as the stabilizer (Figure 9.3A). The samples consisted of spherical nanoparticles with diameter of ~46 nm[27] or 60 nm,[23] and they exhibited a broad NIR absorption band from 700 to 1200 nm (Figure 9.3B and C). Importantly, the RPMI-1640 culture medium containing PPy nanoparticles (30 µg mL^{-1}) exhibited an obvious temperature elevation from 21.3 to 55.8 °C, indicating the efficient photothermal effect. As a result, PPy nanoparticles resulted in significant death of HeLa cells, as observed in both a fluorescence staining assay and an MTT assay.[27] By intratumoral injection of PPy, Liu *et al.* further realized excellent tumor treatment efficacy using an ultra-low-power NIR laser irradiation at 0.25 W cm^{-2} (75 J cm^{-2}) and achieved 100% tumor elimination, without observing significant toxic side effects after treatment.[23]

In 2012, Liu *et al.* prepared polymer-modified PEDOT:PSS *via* a layer-by-layer assembly method.[29] The negatively charged PEDOT:PSS nanoparticles (Figure 9.4A) were first coated with positively charged poly(allylamine

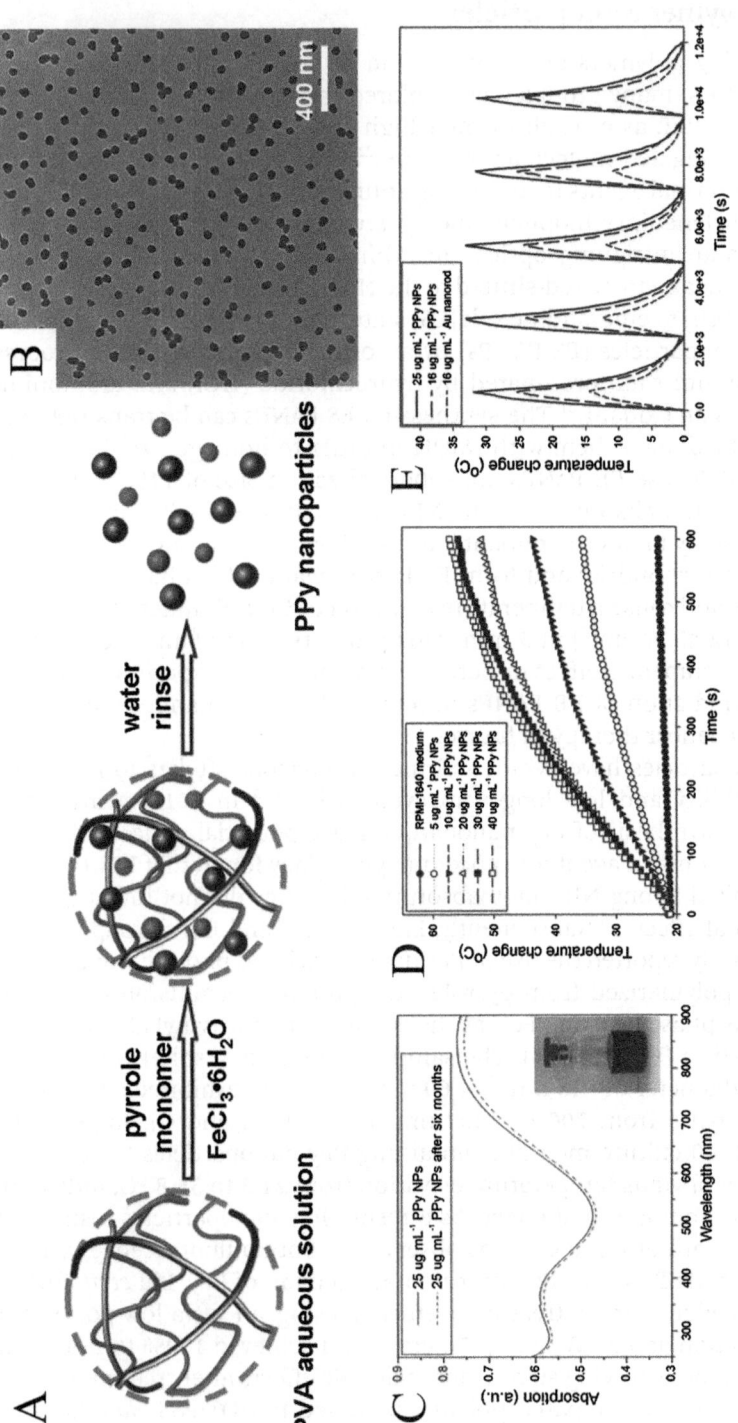

Figure 9.3 (A) Schematic illustration of PPy nanoparticles prepared in an aqueous dispersion of water-soluble polymer/metal cation complexes. (B) Transmission electron microscopy (TEM) image of the as-prepared PPy nanoparticles. (C) UV-Vis-NIR absorption spectrum of PVA stabilized PPy nanoparticles dispersed in water and stored at 4 °C for 6 months (inset photograph is the as-prepared PPy sample). Heating curves of PPy at various concentrations (D) and the comparison between PPy NPs and Au nanorods over five cycles of NIR laser irradiation (E). Reproduced from ref. 27 with permission from John Wiley and Sons. Copyright © 2013 Wiley-VCH Verlag GmbH& Co. KGaA, Weinheim.

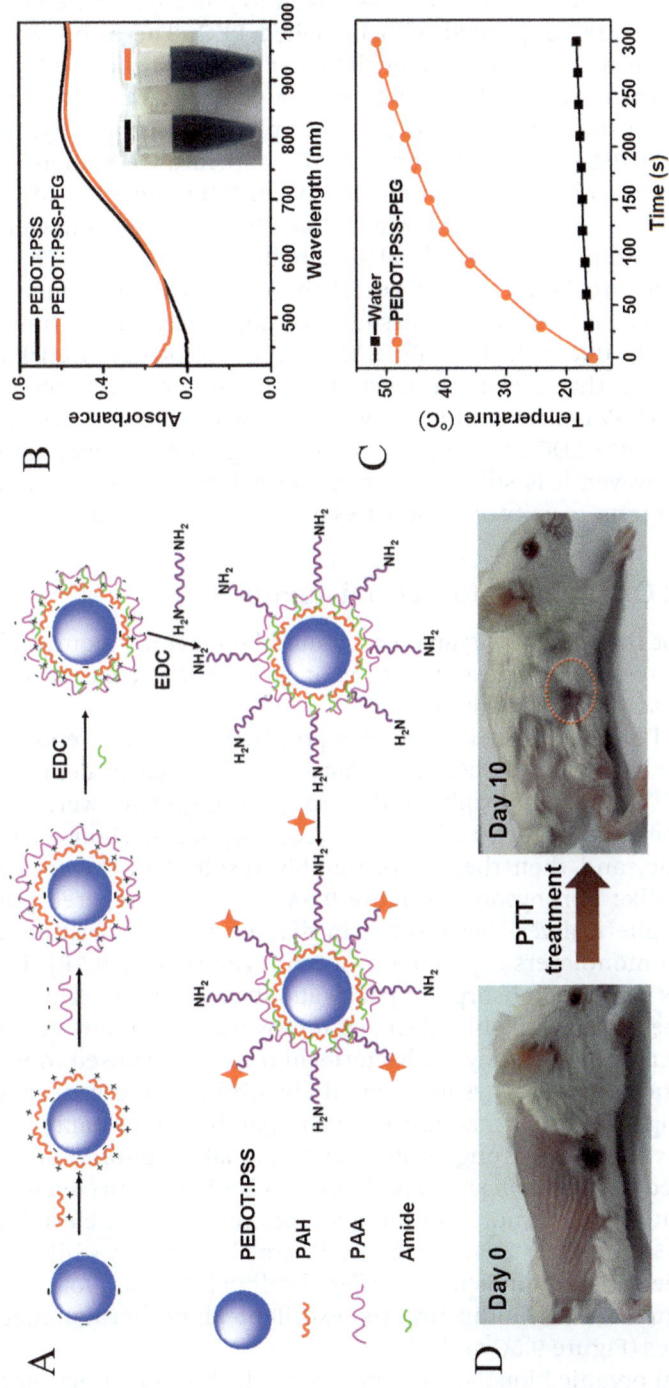

Figure 9.4 (A) Schematic representation showing the fabrication process of PEDOT:PSS–PEG. (B) The absorbance spectra of PEDOT:PSS and PEDOT:PSS–PEG solutions. Inset: Photos of PEDOT:PSS (left) and PEDOT:PSS–PEG (right) aqueous solutions. (C) Heating curves of pure water and PEDOT:PSS–PEG (0.1 mg mL^{-1}) under 808 nm laser irradiation at power density of 1 W cm^{-2}. (D) Representative photos of a mouse injected by PEDOT:PSS–PEG at day 0 before PAT treatment and at day 10 after treatment. Reproduced with permission from L. Cheng, K. Yang, Q. Chen and Z. Liu, *ACS Nano*, 2012, **6**, 5605. Copyright (2012) American Chemical Society.[29]

hydrochloride) (PAH) and then negatively charged poly(acrylic acid) (PAA). After that, the two coating layers were cross-linked by amide formation and then the particles were conjugated with branched PEG. The as-prepared PEDOT:PSS nanoparticles had an average diameter of 80 nm and exhibited high optical absorbance in the NIR region with a peak at 830 nm (Figure 9.4B). This strong absorbance led to an excellent photothermal ability: the 0.1 mg mL^{-1} PEDOT:PSS solution showed a >35 °C temperature elevation in 5 min under 808 nm laser irradiation (intensity 1 W cm^{-2}) (Figure 9.4C). Moreover, the PEDOT:PSS nanoparticles showed a "stealth-like" behavior, which means a long second-phase blood circulation half-life of 21.4 ± 3.1 h, as proved by the blood circulation test. This excellent property allows it enough time to be accumulated in cancerous tissue, showing 28.02% ID g^{-1} tumor uptake, which may be due to the EPR effect. The accumulation of PEDOT:PSS in tumors resulted in the efficient destruction of the tumor under 808 nm laser irradiation (0.5 W cm^{-2}) for 5 min, as vividly shown in Figure 9.4D. The excellent PAT effect of PEDOT:PSS make it a promising photothermal agent for the future. However, it is still necessary to investigate the degradation behavior and long-term toxicity of PEDOT:PSS nanoparticles *in vivo*.

9.3.3 Natural Organic Photothermal Agents

There is no doubt that natural organic PAs should be ideal for biomedical applications. The research interest on natural organic PAs has concentrated mainly on porphysomes and melanin.

Porphyrin–lipid based nanodevices, called porphysomes, were reported for the first time in 2011 and since then have been studied in depth by Zheng's group.[30–34] In 2011, phospholipid–porphyrin conjugates were synthesized by an acylation reaction between lysophosphatidylcholine and pyropheophorbide, and then their self-assembly resulted in the formation of liposome-like porphysomes (Figure 9.5A).[30] The as-prepared porphysomes were spherical vesicles 100 nm in diameter and consisting of by two porphyrin monolayers separated by a 2 nm gap (Figure 9.5B). The spherical vesicles exhibited absorption peaks at 480 and 680 nm. The 680 nm peak could be further red-shifted to 760 nm by using subunits generated from another type of porphyrin, bacteriochlorophyll. Exposed to 637 nm laser irradiation, the porphysomes could absorb light and convert it to heat, with a high photothermal efficiency comparable to Au nanorods. The porphysomes showed a strong photoacoustic signal for photoacoustic imaging, as proved in further *in vivo* experiments. After being injected with porphysome solution, KB tumors exhibited a rapid temperature elevation to 60 °C under 658 nm laser irradiation (1.9 W cm^{-2}) for 1 min while the control group injected with phosphate-buffered saline (PBS) had low temperature elevation to 40 °C, indicating the excellent photothermal effect from porphysomes (Figure 9.5C and D).

Another natural organic biopolymer is melanin, which has distinct functions and is found in many organisms. Melanin can protect humans and

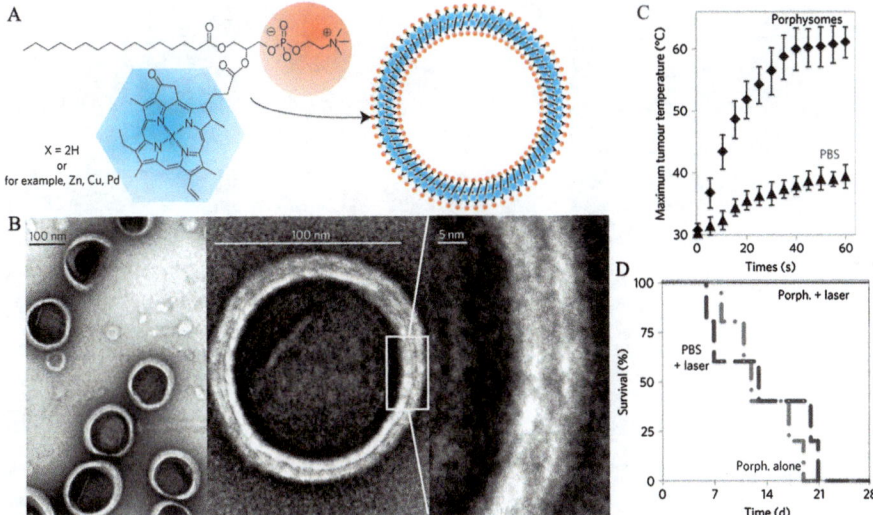

Figure 9.5   (A) Schematic showing of a pyropheophorbide–lipid porphysome. The phospholipid headgroup (red) and porphyrin (blue) are highlighted in the subunit (left) and assembled nanovesicle (right). (B) TEM images of negatively stained porphysomes. (C) Maximum tumor temperature during 60 s laser irradiation. (D) Survival plot of tumor-bearing mice treated with the indicated conditions. Reproduced by permission from Macmillan Publishers Ltd.: *Nat. Mater.*[30] Copyright (2011).

animals from ultraviolet light.[35,36] Moreover, its absorption spectrum contains NIR regions, which has inspired the development of melanin-based PAs. Lu *et al.* synthesized dopamine–melanin colloidal nanospheres (Dpa-melanin CNSs) by the oxidation and self-polymerization of dopamine in a water/ethanol/ammonia solution at room temperature.[36] Dpa-melanin CNSs consisted of spherical particles with diameter ~160 nm, and exhibiting a broad absorption ranging from ultraviolet to NIR wavelengths (Figure 9.6A and B). The CNSs showed a temperature elevation of 33.6 °C in 500 s under 808 nm laser irradiation (2 W cm^{-2}), and the PCE can be calculated as 40%. Furthermore, they could be dispersed in 10% blood serum solution and remain stable after 24 h, revealing their high potential for *in vivo* applications. In the *in vivo* experiments, the prepared CNSs exhibited excellent biodegradability and high median lethal dose, without obvious long-term toxicity during their retention in rats.[36] In the *in vitro* experiment, cancer cells (4T1 and HeLa) were incubated with Dpa-melanin and exposed to 808 nm laser irradiation at 2 W cm^{-2} for 5 min. Then the cells were stained with calcein AM and propidium iodide. The live cells could be distinguished by the green fluorescence of calcein AM and dead cells by the red fluorescence of propidium iodide. The majority of cancer cells within the laser spot were killed, while the remaining regions showed negligible cell death (Figure 9.6C–F), proving the death of cancer cells caused by photothermal ablation of Dpa-melanin CNSs. More

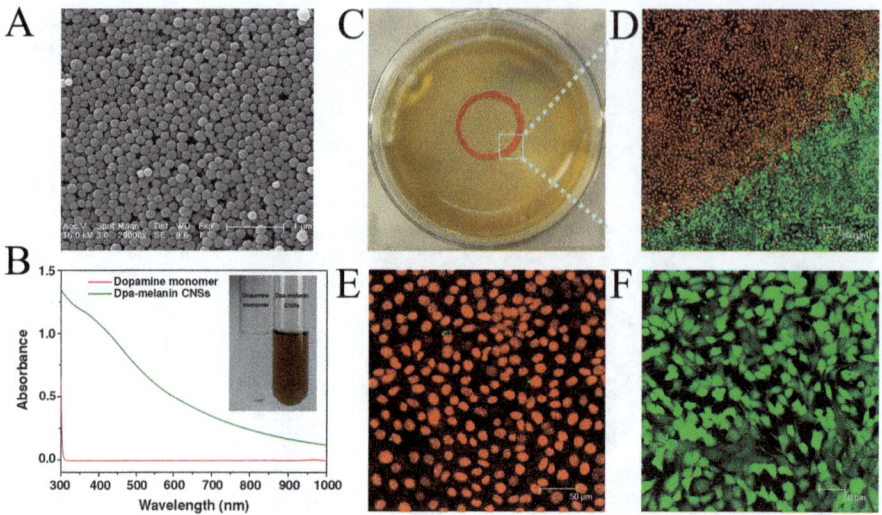

Figure 9.6   (A) Scanning electron microscopy (SEM) image of Dpa-melanin CNSs. (B) Comparison of absorption spectra between dopamine monomer and Dpa-melanin CNSs. (C) Digital photo of the culture dish after incubating 4T1 cell with Dpa-melanin CNSs. The red circle shows the laser spot. (D–F) Confocal images of calcein AM (green, live cells) and propidium iodide (red, dead cells) costained 4T1 cells after laser irradiation. Reproduced from Y. Liu, K. Ai, J. Liu, M. Deng, Y. He and L. Lu, *Adv. Mater.* 2013, **25**, 1353 with permission from John Wiley and Sons.[36] Copyright © 2013 Wiley-VCH Verlag GmbH & Co. KGaA, Weinheim.

interestingly, Dpa-melanin CNSs can also be conjugated with other biofunctional components, providing a future platform for simultaneous cancer diagnosis and treatment.

To summarize briefly, organic PAs with strong NIR absorption show promise for PAT of cancer. Compared to inorganic nanomaterials, they exhibit excellent biocompatibility, biodegradability and low long-term toxicity, which are favorable for biomedical applications. However, they also have some disadvantages, such as relatively low photostability. Therefore, it is still necessary to improve organic PAs for future biomedical applications.

9.4   Metal-Based Photothermal Agents

Metal-based nanomaterials have been regarded as the most widely and deeply studied photothermal agents. The metal nanoparticles exhibit excellent localized surface plasmon resonance (LSPR) properties, caused by nanoscale oscillations of free electrons on the surface of metal particles. The intriguing LSPR property makes metal nanoparticles absorb light in visible and NIR regions and convert the optical energy to heat. This characteristic of metal nanoparticles has greatly motivated the research on metal-based photothermal agents. At present, the studies of metal-based photothermal agents focus mainly on Au and Pd nanomaterials.

9.4.1 Au Nanomaterials

Au nanomaterials are the most typical metal photothermal agents, and their LSPR is closely related to their shapes and structures. Currently, several Au nanomaterials with different shapes have been well developed as excellent photothermal agents, including Au nanoparticles (AuNPs), Au nanorods (AuNRs), Au nanoshells (AuNSs), and Au nanocages (AuNCs).

9.4.1.1 AuNPs

AuNPs are simplest of all the Au nanostructures. In most conditions, AuNPs show strong visible absorption and the absorbing wavelengths are closely related to their size and aggregation state. The absorbed light energy can be converted to heat by the LSPR effect of AuNPs, and the heat can be used for cancer PAT. It is worthy of notice that the absorption cross-section of AuNPs is much larger than that of small organic dyes, which is an advantage for practical applications.

EI-Sayed and coworkers[37] synthesized 40 nm AuNPs by citrate reduction of $HAuCl_4$ and then conjugated them with anti-EGFR antibody for selective attachment to cancer cells. The as-prepared AuNPs showed a measured absorption maximum of 530 nm. The AuNPs were then incubated with two oral squamous carcinoma cell lines (HSC 313 and HOC 3 Clone 8) and one benign epithelial cell line (HaCaT) for 40 min. The cells were rinsed with PBS buffer and then exposed to continuous-wave (CW) argon ion laser at various power densities (64, 57, 51, 45, 38, 32, 25, 19 and 13 W cm^{-2}, 514 nm, 4 min). The results showed that it is sufficient to use lower laser energy (19 W cm^{-2}) to destroy malignant cells, leaving the benign ones intact. Another group also reported targeted PAT of breast cancer cells (Hs578T) by transferrin-conjugated AuNPs under the of visible CW laser irradiation.[38]

Research on dispersed AuNPs in the domain of nanomedicine has remained *in vitro*, due to the relatively narrow absorption wavelength in the visible band (nearly 520 nm), which can hardly penetrate biological tissues. These tissues demonstrate strong absorption and scattering in visible wavelengths and thus restrict the penetration depth of laser irradiation around 520 nm. In order to overcome this shortcoming, there have been many reports on modifying the absorption of AuNPs from visible to longer wavelengths induced by aggregation.[39-46] Those aggregated AuNPs were also used for cancer PAT. For example, He *et al.*[46] demonstrated the self-assembly of poly(ethyl oxide)-*b*-polystyrene (amphiphilic block copolymer) coated AuNPs (14, 20, 30, and 40 nm) in different solvents and got NIR-absorbed, micelle-like aggregations of AuNPs (Figure 9.7A and B).

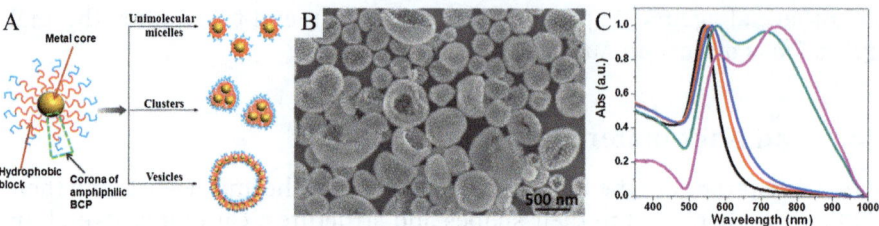

Figure 9.7   (A) Schematic illustration of micelle-like AuNPs and the assembly structures. (B) SEM images of vesicular assemblies of 40 nm Au. (C) UV-Vis
spectra of 40 nm Au using amphiphilic block copolymer of different
molecular weights (increasing from left to right, along with the plasmonic peaks). Reproduced with permission from J. He, X. Huang, Y.-C.
Li, Y. Liu, T. Babu, M. A. Aronova, S. Wang, Z. Lu, X. Chen and Z. Nie, *J.
Am. Chem. Soc.*, 2013, **135**, 7974.[46] Copyright (2013) American Chemical
Society.

The various assembled superstructures, several hundred nanometers in
size, showed strong photoabsorption with large red shift to the NIR range
(Figure 9.7C), as a result of the remarkable plasmonic coupling of the Au
cores. The assemblies of 40 nm AuNPs reached a temperature >60 °C in
5 min under 808 nm laser irradiation (1 W cm^{-2}). The 808 nm irradiation
can penetrate biological tissues, so the assembled AuNPs can move us one
step closer to *in vivo* experiments. The AuNPs assemblies were injected
intratumorally into nude mice bearing 4T1 tumors and the tumors were
exposed to 808 nm laser irradiation (1 W cm^{-2}). Tumors were heated to
55–60 °C in 1 min which is high enough to ablate tumor *in vivo*. Tumors
treated with these aggregated AuNPs and laser irradiation were effectively
ablated without showing recurrence.

As well as self-assembling in different solvents, AuNPs can also aggregate
according to the surrounding pH level. The microenvironment in tumors is
known to be mild acidic, and this can lead to the aggregation of AuNPs.[43–45]
Liu *et al.*[45] facilely prepared ~16 nm AuNPs by surface modification of mixed
self-assembled monolayers of weak electrolytic 11-mercaptoundecanoic acid
and strong electrolytic (10-mercaptodecyl)trimethylammonium bromide
(Figure 9.8A). The zwitterionic AuNPs can be stable at the pH of blood and
normal tissues (pH 7.4), but aggregate instantly in response to the acidic
extracellular environment (pH 6.5) of solid tumors (Figure 9.8B). It takes only
a few seconds for the mixed-charge system to respond to the change of pH.
This intriguing phenomenon can also be observed by the absorbance spectra
and transmission electron microscopy (TEM) images (Figure 9.8C and D). It
is worthy of notice that the aggregated AuNPs exhibit NIR absorbance (Figure
9.8C), which can be used for PAT of cancer *in vivo*. After intravenous administration into normal ICR mice, the AuNPs showed excellent stealth ability to
resist uptake by macrophages and no apparent histopathological abnormalities, lesions, or noticeable toxic effect in liver, spleen, and kidney at 24 h after
NP administration. After 24 h post-injection, the tumors were irradiated by

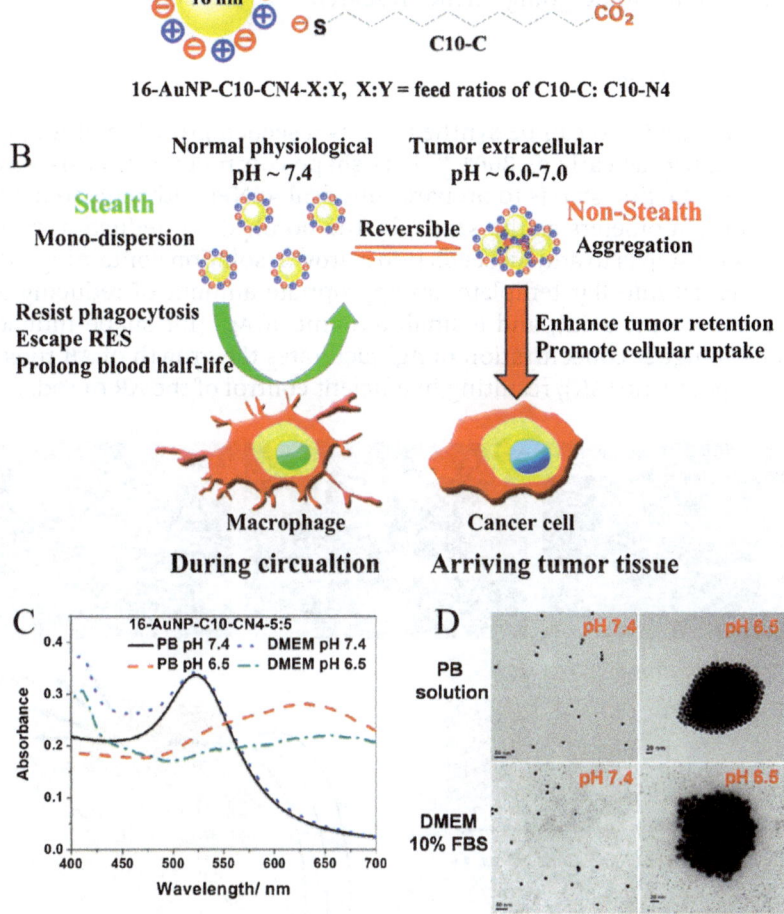

Figure 9.8 (A) Schematics of pH-responsive zwitterionic AuNPs modified with mixed-charge thiols. (B) Schematic illustration of the acidic tumor selectivity of pH-responsive zwitterionic AuNPs. (C) UV-Vis spectra of AuNPs incubated in PB solution (50 mM) and DMEM culture medium with 10% FBS at pH 7.4 and 6.5. (D) Representative TEM images of AuNPs (scale bar is 50 nm upper images, 20 nm in lower images). Reproduced with permission from X. Liu, Y. Chen, H. Li, N. Huang, Q. Jin, K. Ren and J. Ji, *ACS Nano* 2013, 7, 6244.[45] Copyright (2013) American Chemical Society.

an 808 nm laser once every 2 days up to the sixth day after the first treatment. Tumor weight after the treatment was obviously lower than other control groups. Compared to non-sensitive PEGylated AuNPs, the total accumulation, retention, and cellular uptake of AuNPs were significantly enhanced by the pH-induced aggregation effect. This can be a universal strategy to obtain stealth properties and pH sensitivity at the same time. The self-assembled AuNPs with red-shifted plasmonic absorption show potential for *in vivo* PAT and is currently undergoing further research.

9.4.1.2   AuNRs

AuNRs (Figure 9.9A) can be synthesized by a seeded growth method, which was developed as early as 2001.[47] This shape-controlled synthesis involves two steps. The first step is to prepare spherical AuNPs with a mean diameter of several nanometers as the seeds, by the borohydride reduction method. The second step is to add Au seeds into a growth solution containing Au salt, a rod-shaped micellar template, an appropriate amount of reducing agent (such as ascorbic acid), and a small amount of Ag^+ (for shape induction). Usually, a higher concentration of Ag^+ facilitates the growth of Au rods with larger aspect ratio (AR), resulting in efficient control of the AR of rods.

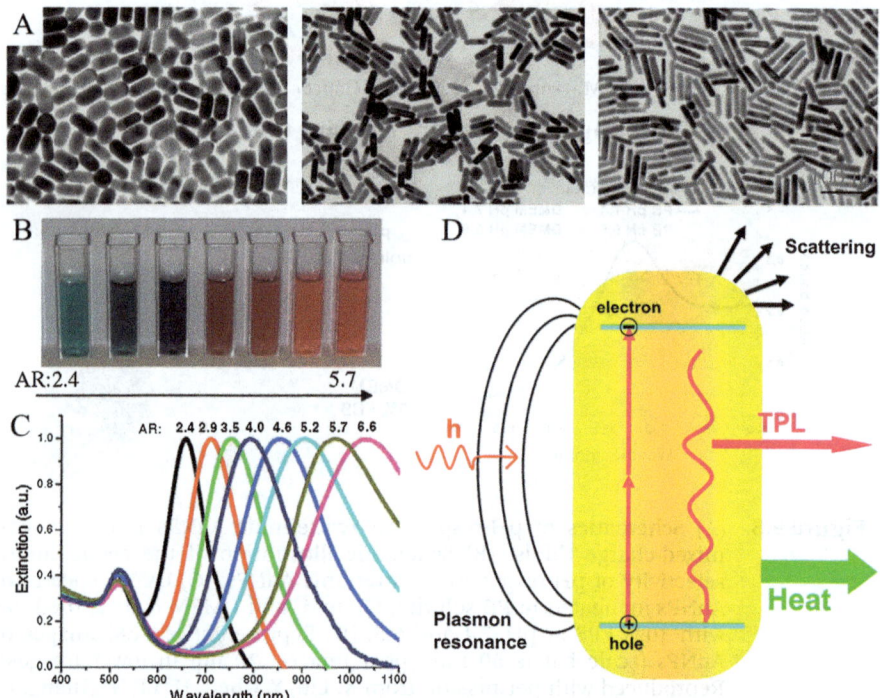

Figure 9.9   TEM images (A), differing colors (B), and absorbance spectra (C) of AuNRs with various aspect ratios. (D) Schematic illustration of the photophysical processes in gold nanorods. Reproduced from X. Huang *et al.*, *J. Adv. Res.*, **1**, Gold nanoparticles: optical properties and implementation in cancer diagnosis and photothermal therapy, 2010, 16–13 Copyright (2010) with permission from Elsevier[51] and ref. 56 with permission from John Wiley and Sons. Copyright © 2009 The Authors. Journal Compilation. The American Society of Photobiology.

Compared with AuNPs, AuNRs offer significant advantages for biomedical applications. AuNRs show two plasmon modes, longitudinal and transverse LSPR modes, associated with electron oscillations along the length axis and the transverse direction respectively.[48-50] This intriguing property immediately gives AuNRs a weak short-wavelength absorbance (~520 nm) and a strong long-wavelength absorbance in NIR, unlike AuNPs which need to be aggregated in order to have NIR absorbance. NIR excites the longitudinal plasmon resonance mode of AuNRs, resulting in both absorption and resonant light scattering. The two-photon absorption of NIR light leads to an electronic transition from the d band to the sp band. At the same time, electron–hole pairs generate. The jumping of excited electrons results in two-photon luminescence emission, while the electron–phonon collisions generate heat (Figure 9.9D). Moreover, as their AR changes, the absorption wavelength of AuNRs can be synthetically tuned and expanded to a broad range, covering visible and NIR regions (Figure 9.9A–C).

So far, AuNRs have been extensively explored for PAT applications.[51-61] For example, Von Maltzahn *et al.*[53] coated AuNRs (axial sizes of 12.7 ± 3.4 and 47 ± 9.3 nm) with amino-PEG–thiol polymer to get PEGylated AuNRs (AuNRs–PEG). After PEGylation, AuNRs were rendered highly stable *in vitro*, showing minimal spectral shifting (which would indicate particle destabilization and aggregation) even after 1000 h in 0.15 mol L^{-1} NaCl or 10% human serum aqueous solution.[53] Then, the AuNRs–PEG in PBS were injected into mice with MDA-MB-435 human tumors through the tail vein (20 mg Au kg^{-1}) and were found to exhibit blood half-lives of about 17 h, allowing passive accumulation into the xenograft tumors (~7% ID g^{-1} tumor uptake). During this time, the AuNRs–PEG maintained their longitudinal plasmon resonance around 810 nm, allowing spectrophotometric detection in serum over time.[53] The tumor-bearing mice were exposed to NIR irradiation (810 nm, 2 W cm^{-2}, 5 min) at 72 h post-injection and tumors were rapidly heated to temperatures of over 70 °C, whereas saline-injected mice displayed less focal temperature increases with maximum surface temperatures of ~40 °C.[53] The experimental mice group subjected to 810 nm laser irradiation showed a survival time >50 days, compared to 33 days for control groups. Moreover, the AuNRs–PEG could also act as dense X-ray absorbing agents for X-ray computed tomography (CT). It exhibited approximately twofold amplified X-ray contrast compared with Isovue-370, a clinical iodine standard for X-ray CT.

The main driving force for accumulating AuNRs–PEG in tumors is the EPR effect, which somehow limits the tumor uptake of PAs. In order to break this restriction, researchers tried to explore the targeted or intracellular therapy by conjugating AuNRs with biomolecules. For example, El-sayed *et al.*[55] conjugated AuNRs with anti-epidermal growth factor receptor (anti-EGFR) monoclonal antibodies. Those AuNRs scatter different colored light (depending on their size and shape) in dark-field microscopy. The anti-EGFR-conjugated AuNRs could bind specifically to malignant cells and give them a distinguishable imaging contrast, while individual non-cancerous cells are hardly identifiable, because anti-EGFR conjugated AuNRs have no specific interactions with normal cells.[55] After incubation with anti-EGFR conjugated AuNRs for 30 min, cells were exposed to 800 nm laser irradiation at power values of 40, 80, 120, 160, and 200 mW for 4 min each with a focus spot of 1 mm in diameter. The cancer cells died (detected by the cell viability test with trypan blue) when exposed to the laser at ≥80 mW, corresponding to 10 W cm^{-2}. However, almost no non-cancerous cells died under the same conditions, so the highly targeted affinity of anti-EGFR led to a selective photothermal ablation of tumor cells without the destruction of non-malignant cells. In addition to anti-EGFR, there are still other targeted biomolecules that have been conjugated to AuNRs for targeted cancer PAT, such as folate acid, arginine-glycine-aspartic acid (RGD), chitosan, transferrin, and bacterial pathogens.[57–61]

9.4.1.3   *AuNSs*

AuNSs can be synthesized by attaching AuNPs onto the surfaces of spherical nanocores, which is called the seed-mediated method. Generally, those spherical nanocores can be dielectric nanoparticles, such as silica and polymer nanoparticles.[62,63] AuNSs are promising for PAT *in vivo*. Firstly, the absorbance ability of AuNSs in NIR region is six times more than that of ICG, which makes it a strong NIR absorber.[64] Secondly, AuNSs have the contrast ability for dark-field imaging and optical coherence tomography (OCT).[63,65] In 2007, Gobin *et al.* synthesized AuNSs (12 nm in thickness) on silica cores (119 ± 11 nm in diameter) and then coated the AuNSs with PEG.[63] The PEG-modified nanoshells were injected intravenously into tumor-bearing mice and passively accumulated in the tumor tissue due to the leakiness of the tumor vasculature, which dramatically enhanced the NIR scattering, increasing the OCT contrast. After treating with 808 nm laser irradiation at a power density of 4 W cm^{-2} and a spot size of 5 mm in diameter for 3 min, the tumor were completely regressed and the treated mice survived >6 weeks.

AuNSs can also be grown on polymer cores. Liu *et al.* synthesized AuNSs by assembling AuNSs (10–20 nm) on the surface of carboxylated polystyrene spheres (200 nm in size).[66] The as-prepared AuNSs showed a narrower plasmon resonance absorption peak as a result of the higher refractive index of polystyrene compared to silica. This narrowed absorption peak enhanced the absorbing efficiency at NIR wavelengths, and the 1.2 mg mL^{-1} sample caused >30 °C temperature elevation in <10 min under 808 nm laser irradiation (4 W cm^{-2}). In another study, Ke *et al.* developed a multifunctional theranostic agent based on gold-nanoshelled microcapsules (AuNS–MCs) by electrostatic adsorption of AuNPs as seeds onto the polymeric microcapsule surfaces, followed by the formation of gold nanoshells by using a surface seeding method.[67] The seeding process made the attached AuNPs large enough to cluster and form rough edges with a more compact shell. The prepared spherical AuNSs with a mean size of 2.32 ± 1.07 μm showed a strong absorbing peak ranging from 650 to 900 nm (Figure 9.10B) and could reach 55 °C in 10 min under laser irradiation (808 nm, 2 W, 10 min, Figure 9.10C).

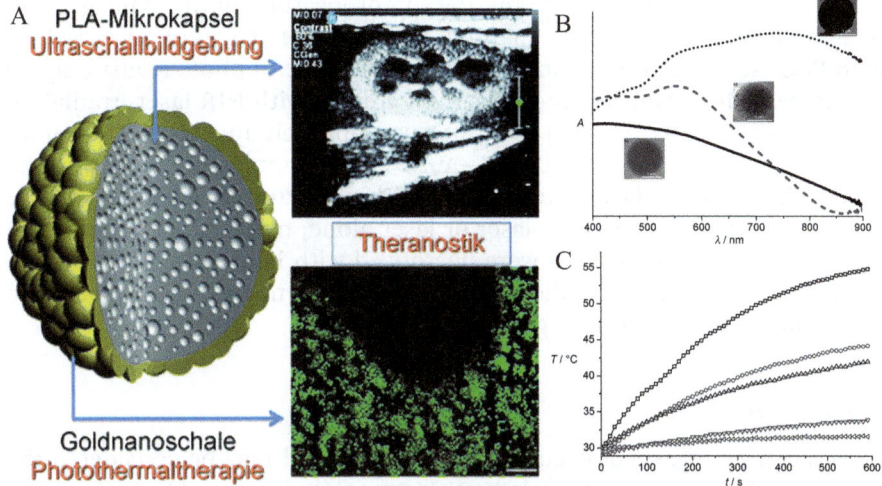

Figure 9.10 (A) Schematic illustration of the structure and functions of AuNS–MCs. (B) UV-Vis spectra of blank microcapsules, AuNPs–MCs, and AuNSs–MCs (from left to right). Insets: TEM images of each respective sample. (C) Temperature elevating curves of AuNS–MCs agent at different concentrations: 0 mg mL^{-1}, 0.05 mg mL^{-1}, 0.15 mg mL^{-1}, 0.3 mg mL^{-1}, and 0.5 mg mL^{-1} (from the bottom up). Reproduced from H. Ke, J. Wang, Z. Dai, Y. Jin, E. Qu, Z. Xing, C. Guo, X. Yue and J. Liu, *Angew. Chem.* 2011, **123**, 3073 with permission from John Wiley and Sons.[67] Copyright © 2011 Wiley-VCH Verlag GmbH & Co. KGaA, Weinheim.

After incubating the AuNSs with HeLa cells (cultured in six-well plates) for 1 h, the cells were illuminated by an NIR laser (808 nm, 8 W cm^{-2} for 10 min). The intracellular proteins in the HeLa cells were denatured, inhibiting their normal cellular growth and proliferation, by the NIR light-induced thermal effect of AuNSs. This can be observed by fluorescent cell staining under an inverted fluorescence microscope. Importantly, the polymeric microcapsule cores were made by the water-in-oil-in-water (W/O/W) double-emulsion method, leaving a small hollow space inside the microcapsule (Figure 9.10A). These hollow spaces are the basis for ultrasound-responsive properties. In this way, this multifunctional theranostic agent can be used for both photothermal therapy and photoacoustic imaging of cancer.

Furthermore, some ligands were also conjugated to AuNSs for targeted cancer PAT.[68-71] For example, Lu *et al.* synthesized PEGylated AuNSs (PEG–AuNSs) and then functionalized them with cyclic RGD peptide (RGD–PEG–AuNSs).[71] The spherical hollow AuNSs had an average diameter of approximately 40 nm and showed excellent absorption around 800 nm. The RGD–PEG–AuNSs and PEG–AuNSs were then injected intravenously into mice with U87 human glioma in brain. Cyclic RGD peptides have high binding affinity to integrin receptors on glioma cells, so the RGD–PEG–AuNSs could attach to the tumor selectively. The photoacoustic tomography images clearly revealed brain tumor for 24 h after injection, and the tumor location on those images correlated well with mouse brain anatomy. However, the images of mice injected with PEG–AuNSs exhibited no significant difference in photoacoustic signal intensities. Moreover, the mice were also treated with NIR laser irradiation (16 W cm^{-2}, 3 min, 808 nm) 24 h after nanoparticle injection. The tumor cells in mice treated with RGD–PEG–AuNSs were completely ablated with no discernible residue of viable tumor cells within the tumor periphery. In mice treated with PEG–AuNSs plus laser or laser alone, however, ~45% and 30% respectively of tumor tissues were necrotized with large numbers of viable tumor cells within the periphery. Mice in photothermally treated group also lived longer than the control group.

9.4.1.4 AuNCs

AuNCs can be regarded as cube-shaped AuNSs and may be considered the most complex Au nanomaterials. AuNCs were first developed by Xia's group, using an Ag template-engaged galvanic replacement reaction with aqueous HAuCl$_4$ solution.[72-74] That is, Ag nanoparticles react with HAuCl$_4$ solution to generate epitaxial gold atoms and form an Au–Ag alloy. A dealloying process then selectively removes silver atoms and forms AuNCs.[75]

In studies by Xia's group,[76] poly(vinyl pyrrolidone) (PVP) was found to preferentially interact with the {1 0 0} (six side faces) rather than {1 1 1} facets (eight corners) of Ag nanocubes. This mechanism can be used in the synthesis of nanocages. When Ag nanocubes (Figure 9.11A) with truncated corners (Figure 9.11B) were added to an aqueous solution of HAuCl$_4$, Ag atoms were dissolved at the corners without PVP and galvanic replacement reactions occurred on the six side faces, thus forming nanocages with controllable pores (Figure 9.11C and D) and tunable absorbing peaks (Figure 9.11E).[77] In another experiment, Xia *et al.* further functionalized the nanocages with thiolated PEG and then conjugated the PEGylated nanocages with monoclonal anti-HER2 antibodies to selectively target the antigens (EGFR2) on SK-BR-3 cells.[78] These AuNCs have a surface plasmon resonance (SPR) peak around 812 nm. When incubated with cancer cells, the AuNCs selectively attached to the surfaces of SK-BR-3 cells. The cells were then irradiated by a femtosecond laser (diameter ~2 mm, intensity 1.5 W cm^{-2}, duration 5 min). The cells were then stained with calcein AM and ethidium homodimer 1 (EthD-1). The colorless calcein AM was enzymatically converted to green-fluorescencing calcein only in viable cells, while EthD-1 can penetrate through cells, stain DNA, and emit red fluorescence only when the cell membrane is irreversibly ablated by photothermal agents (*i.e.*, the cells are dead).[78] The circular laser-irradiated area shows almost no green calcein fluorescence (Figure 9.11E), while the void region in calcein AM-stained image is nicely filled by the red fluorescence of EthD-1 (Figure 9.11F), proving the cell death was caused by the NIR laser irradiation after incubation with AuNCs. Interestingly, the experimentally measured photothermal damage area went up with the increase of the irradiation power density.

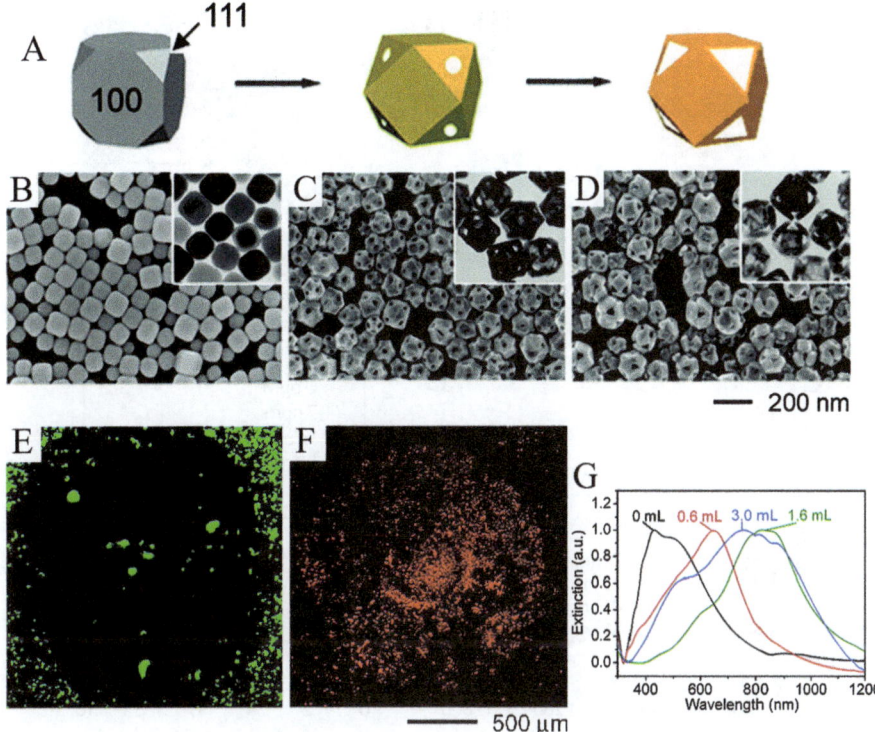

Figure 9.11 (A) Schematic illustration detailing all major steps involved in the formation of Au–Ag nanocages with well-controlled pores at the corners. SEM images (B–D) and absorption curves (G) of the Ag nanocubes with truncated corners titrated with 0.1 mM HAuCl₄, 0, 1.6 and 3.0 mL, respectively. Insets in (B–D): TEM images of each respective sample. *In vitro* experiments of SK-BR-3 breast cancer cells treated by immuno Au nanocages and 810 nm laser: (E) calcein AM assay (where green fluorescence indicates live cells), and (F) ethidium homodimer-1 (EthD-1) assay (where red fluorescence indicates dead cells). Reproduced with permission from J. Chen, J. M. McLellan, A. Siekkinen, Y. Xiong, Z.-Y. Li and Y. Xia, *J. Am. Chem. Soc.* 2006, **128**, 14776.[77] and J. Chen, D. Wang, J. Xi, L. Au, A. Siekkinen, A. Warsen, Z.-Y. Li, H. Zhang, Y. Xia and X. Li, *Nano. Lett.* 2007, 7, 1318.[78] Copyright (2007) American Chemical Society.

9.4.2 Pd Nanosheets

Pd nanosheets are another kind of noble metal nanostructure with tunable NIR plasmon absorbance to have appeared in recent years. In 2011, Zheng *et al.* demonstrated the facile synthesis of freestanding hexagonal Pd nanosheets that are <10 atomic layers thick (Figure 9.12A and B), using carbon monoxide as a confining agent.[79] The blue-colored Pd nanosheets showed a well-defined and tunable (826–1068 nm) strong SPR peak in the NIR regions (Figure 9.12C). The temperature of a solution containing 27 ppm

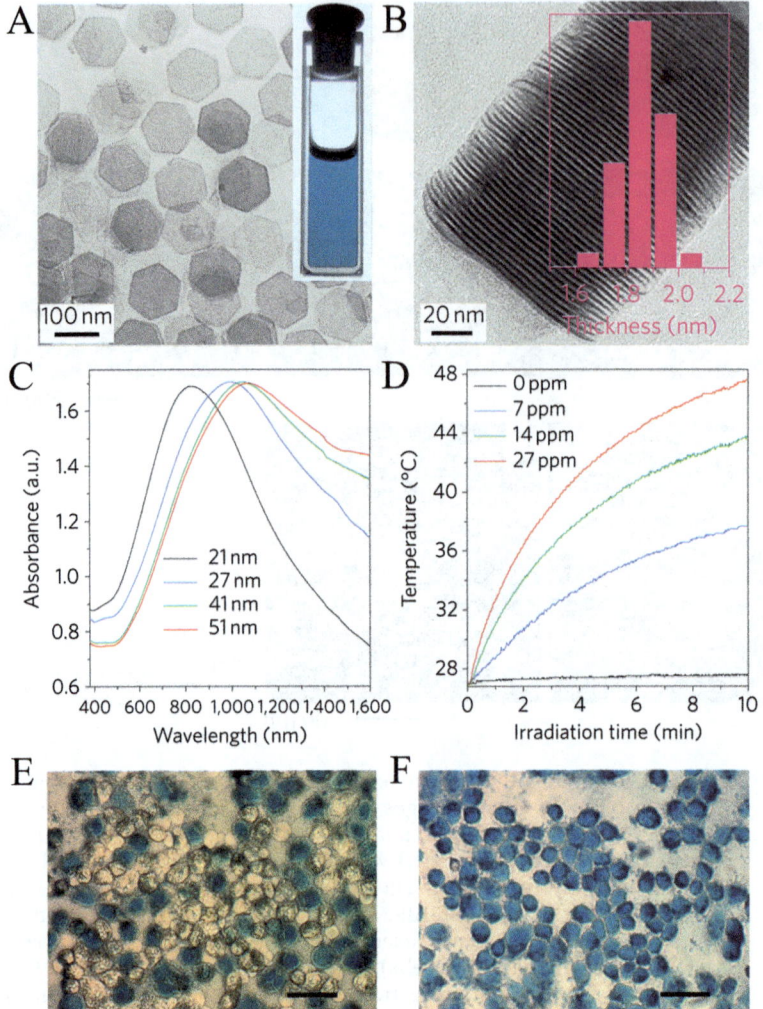

Figure 9.12 (A) TEM image of the Pd nanosheets. Inset: photograph of an ethanol dispersion of the as-prepared Pd nanosheets in a cuvette. (B) TEM image of the assembly of Pd nanosheets perpendicular to the TEM grid. Inset: thickness distribution of the Pd nanosheets. The absorption spectra (C) and heating curves (D) of as-synthesized Pd nanosheets. Image of cancer cells corresponding to 2 min (E) and 5 min (F) 808 nm laser irradiation. Dead cells are stained with Trypan Blue. Reproduced by permission from Macmillan Publishers Ltd.: *Nat. Nanotechnol.*[79] Copyright (2010).

Pd nanosheets rose by 20.7 °C in 10 min under NIR laser irradiation (808 nm, 1 W), while the control group showed only 0.5 °C elevation under the same NIR laser irradiation (Figure 9.12D). Meanwhile, upon 808 nm irradiation (2 W) for 30 min, the sheet-like structure of the Pd nanosheets was well retained, exhibiting higher photothermal stability than Au nanostructures.

After incubating polyethyleneimine-exchanged Pd nanosheets with liver cancer cells and irradiating the cells with an 808 nm laser (1.4 W cm^{-2}), ~50% of the cells were killed in 2 min and ~100% died in 5 min (Figure 9.12E and F).

To some extent, the ultrathin nature of 1.8 nm Pd nanosheets prevents their effective uptake by cancer cells. In order to promote cellular uptake of Pd nanosheets, Zheng *et al.* coated the nanosheets with silica to increase the sheet thickness and give the sheet surface a positive charge.[80] The silica-coated Pd nanosheets had an average thickness of ~32 nm and a 13-fold enhancement in cell uptake after functionalization with amino groups, while the uptake of Pd nanosheets without any surface modification was only 0.1% (0.11 ppm). Furthermore, the silica-coated Pd nanosheets exhibited an enhanced photothermal destructive efficacy for cancer cells *in vitro*. In another study, they synthesized Pd@Ag core–shell bimetallic nanoplates to enhance the photothermal stability.[81] The Pd@Ag nanoplates were prepared by using uniform hexagonal Pd nanoplates, 1.8 nm thick, as seeds to direct the epitaxial growth of Ag. Importantly, Pd@Ag core–shell nanoplates retained their 2D structure well after 808 nm laser irradiation (2 W, 30 min), whereas the 2D structure of Pd nanosheets changed after the same irradiation.

9.5   Carbon-Based Photothermal Agents

Carbon nanomaterials are a large class of low-dimensional materials that has attracted a great deal of interest in the 30 years since the first member of the class, fullerene (Figure 9.13 left), arrived on the global academic stage in 1985.[82] The discoveries of CNTs and graphene (Figure 9.13 middle & right), in 1991 and 2004 respectively, further enriched the carbon family.[83,84] Carbon nanomaterials show broad absorption in the visible and NIR regions and can convert energy from light to heat, showing great prospects for PAT of cancer. Carbon-based PAs exhibit excellent photostability, which means that they unlikely to melt or decompose due to the high temperature caused

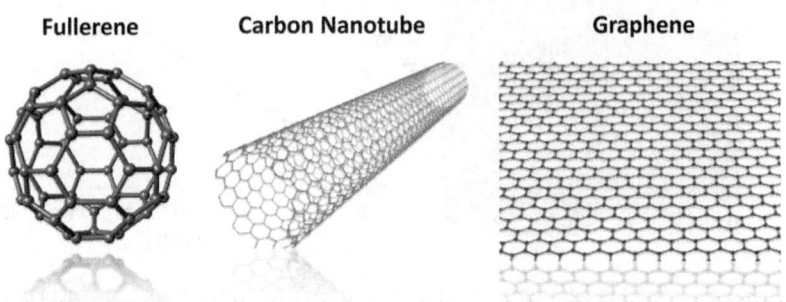

Figure 9.13   Structure schematic illustration of various carbon nanomaterials. Reproduced with permission from Hong, G.; Diao, S.; Antaris, A. L.; Dai, H. *Chem. Rev.* 2015, **115**, 10816. Copyright (2015) American Chemical Society.[96]

by LSPR, unlike metal PAs. This is a huge advantage for constructing highly stable photothermal agents. At present, the main research interest focuses on CNTs and graphenes.

9.5.1   Carbon Nanotubes

The classic carbon nanomaterial, CNTs were soon demonstrated to have photothermal properties. Early research into CNT-based PAT was limited to *in vitro* experiments. For example, Dai *et al.* modified single-walled CNTs (SWNTs) with DNA (Figure 9.14A and C).[85] DNA-coated SWNTs (DNA–SWNTs) showed wide absorbance in the visible and NIR regions (Figure 9.14B). The temperature of DNA–SWNTs in aqueous solution (25 mg L^{-1}) can be heated to 70 °C under 808 nm laser irradiation (1.4 W cm^{-2}, 2 min, Figure 9.14D). The DNA–SWNTs were then incubated with HeLa cells and irradiated by 808 nm laser (1.4 W cm^{-2}) for 2 min. Many of the HeLa cells were killed. After conjugation with folic acid (FA), the SWNTs exhibited targeted PAT to FA-receptor-positive cancer cells *in vitro* without harming normal cells. Many other

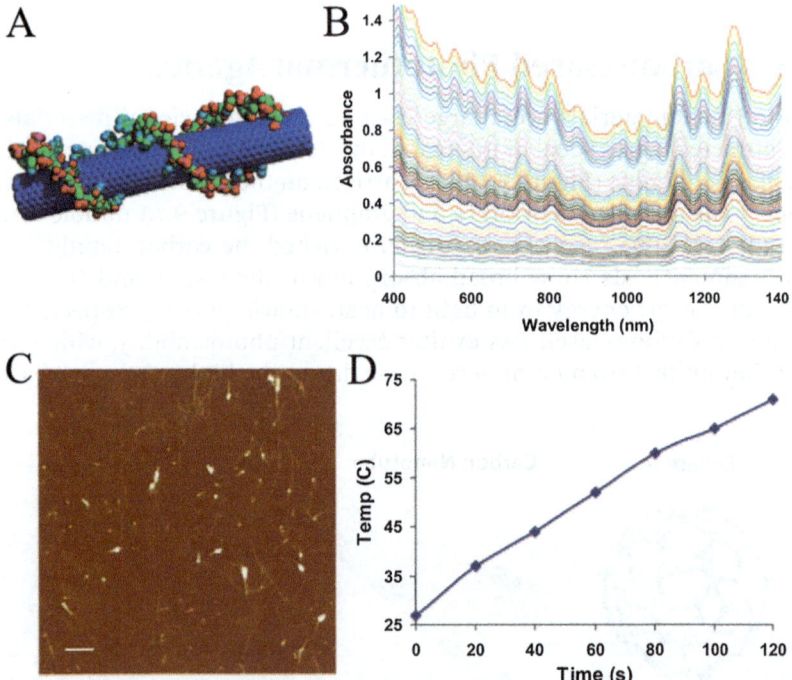

Figure 9.14   Schematic (A) and AFM image (C) of individual DNA–SWCNTs. (B) UV-Vis spectra of DNA–SWCNT solutions. (D) Temperature evolution of a DNA–SWCNT solution during continuous radiation by a 808 nm laser. Reproduced with permission from ref. 85. Copyright (2005) National Academy of Sciences, USA.

groups also studied CNT-based PAT of cancer cells *in vitro*, *e.g.* targeting peptide or antibody-coupled CNTs.[86–88]

Three different groups studied *in vivo* PAT by CNTs in 2009.[89–91] Moon *et al.* injected the PEGylated SWNTs intratumorally into carcinoma KB tumors.[89] The tumors were then irradiated by 808 nm laser (76 W cm^{-2}, 3 min), leading to complete destruction of tumor cells without side effects over 6 months.

Burke *et al.* injected pluronic F127 functionalized multiwalled CNTs (MWNTs) into kidney tumors and then ablated the tumors by 1064 nm laser irradiation (3 W cm^{-2}, 3 min).[90] The temperature of the tumor tissue was increased to 76 °C and the tumors were totally destroyed. In another report, Ghosh and coworkers demonstrated that DNA-encased MWNTs could also be used to safely eradicate PC3 xenograft tumors *in vivo* under 1064 nm laser irradiation (2.5 W cm^{-2}, 70 s).[91]

Several reports proved that intratumorally injected CNTs can kill tumor cells by NIR irradiation. However, intratumoral injection is not an ideal route for clinical applications. Intravenous injection is more convenient, but with the prerequisite that the CNTs should disperse in the bloodstream and then accumulate in tumors. Thus, the surface of CNTs needs to be modified for highly efficient tumor uptake. In a study by Liu and coworkers, SWNTs were coated with PEGylated PMHC$_{18}$ polymers in different degrees and were then injected intravenously into mice with 4T1 tumors.[92] They found that heavily PEGylated SWNTs showed 12–13 h blood circulation half-time and high uptake in the tumor cells (the highest was ~23% ID g^{-1}). However, insufficiently coated SWNTs exhibited a blood half-life of 2.5 h and were easily cleared out from the blood with little tumor uptake. After 808 nm laser irradiation (1 W cm^{-2}, 5 min), the surface temperature of tumors containing SWNTs reached ~50 °C and the tumors were partially damaged, growing more slowly after treatment.

Besides the photothermal property, SWNTs also exhibit autofluorescence in the 1.1–1.7 μm region. Thus, SWNTs can also be used as a photoluminescence-guided photothermal agent. The first report by Dai and coworkers appeared in 2010.[93] In their report, SWNTs were first coated with C$_{18}$–PMH–mPEG and thus can be dispersed in water (Figure 9.15A and B). At the same mass concentration (0.35 mg mL^{-1}), PEG-coated SWNTs exhibited threefold higher optical absorbance than AuNRs at 808 nm (Figure 9.15C). After that, they intravenously injected PEGylated SWNTs into 4T1 tumor-bearing mice. The PEGylated SWNTs showed a blood circulation half-life of 18.9 h and were passively taken up in the tumors through the EPR effect. As the SWNTs accumulated, the tumors fluoresced brightly in the 1.1–1.4 μm range (Figure 9.15D and E). Guided by the fluorescent imaging, the tumors were subjected to 808 nm laser irradiation (0.6 W cm^{-2}, 5 min). The temperature of the tumor tissue increased to 52.9 °C, causing complete tumor destruction. Therefore, this group successfully achieved fluorescence-guided PAT with CNTs.

Based on the photoluminescence property, SWNTs can also be used to detect cancer metastasis, which directly or indirectly causes more than 90% of deaths from cancer.[5,94] In their latest work, Liang *et al.*[94] intratumorally injected C$_{18}$–PMH–PEG-coated SWNTs in 4T1 tumor-bearing mice. The NIR-II

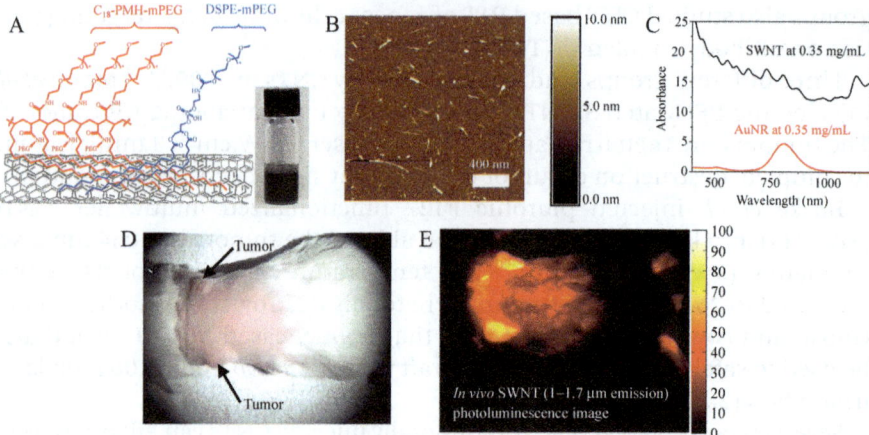

Figure 9.15 (A) Schematic illustration of the C_{18}–PMH–mPEG and phospholipid DSPE–mPEG functionalized SWCNT. Inset: photo of an aqueous suspension of functionalized SWCNTs. (B) AFM image of SWCNTs. (C) Absorption curves of SWCNTs and AuNRs with the same mass concentration of 0.35 mg mL^{-1}. (D) Optical image of a mouse with two 4T1 tumors (indicated by arrows). (E) An NIR photoluminescent image 48 h after injection. Reproduced from ref. 93 with kind permission from Springer Science+Business Media.

fluorescence imaging (1000–1700 nm) of SWNTs showed up in the popliteal sentinel lymph node (SLN) at 20 min post-injection, and the signals reached their highest at about 90 min. This indicated the presence of metastases in the SLN, derived from the primary tumor. The primary tumor and SLN were then laser irradiated (808 nm, 0.5 W cm^{-2} and 0.8 W cm^{-2}, respectively). The temperature of the primary tumor and the SLN rapidly rose to 55 and 45 °C, respectively. PAT at both primary tumors and lymph nodes greatly prolonged the survival of tumor-bearing mice, compared with mice treated by surgery or irradiated at the primary tumor only.

9.5.2 Graphene

The photothermal application of one-dimensional CNTs scored great success in cancer therapy. Inspired by that, two-dimensional graphene has also attracted attention and brings more opportunities in nanobiomedicine because of its unique physical and chemical properties.[95,96]

Based on its wide absorbance in the visible and NIR regions, graphene has been explored for photothermal applications in recent years. In 2010, a pioneering study by Liu *et al.*[97] demonstrated the application of PEGylated nanographene sheets (NGS–PEG) for PAT of cancer *in vivo*. Graphene oxide (GO) sheets were conjugated with amine-terminated six-arm branched PEG by amide formation and the sizes were in the range of 10–50 nm (Figure 9.16A and B). Meanwhile, the NGS–PEG exhibited high optical

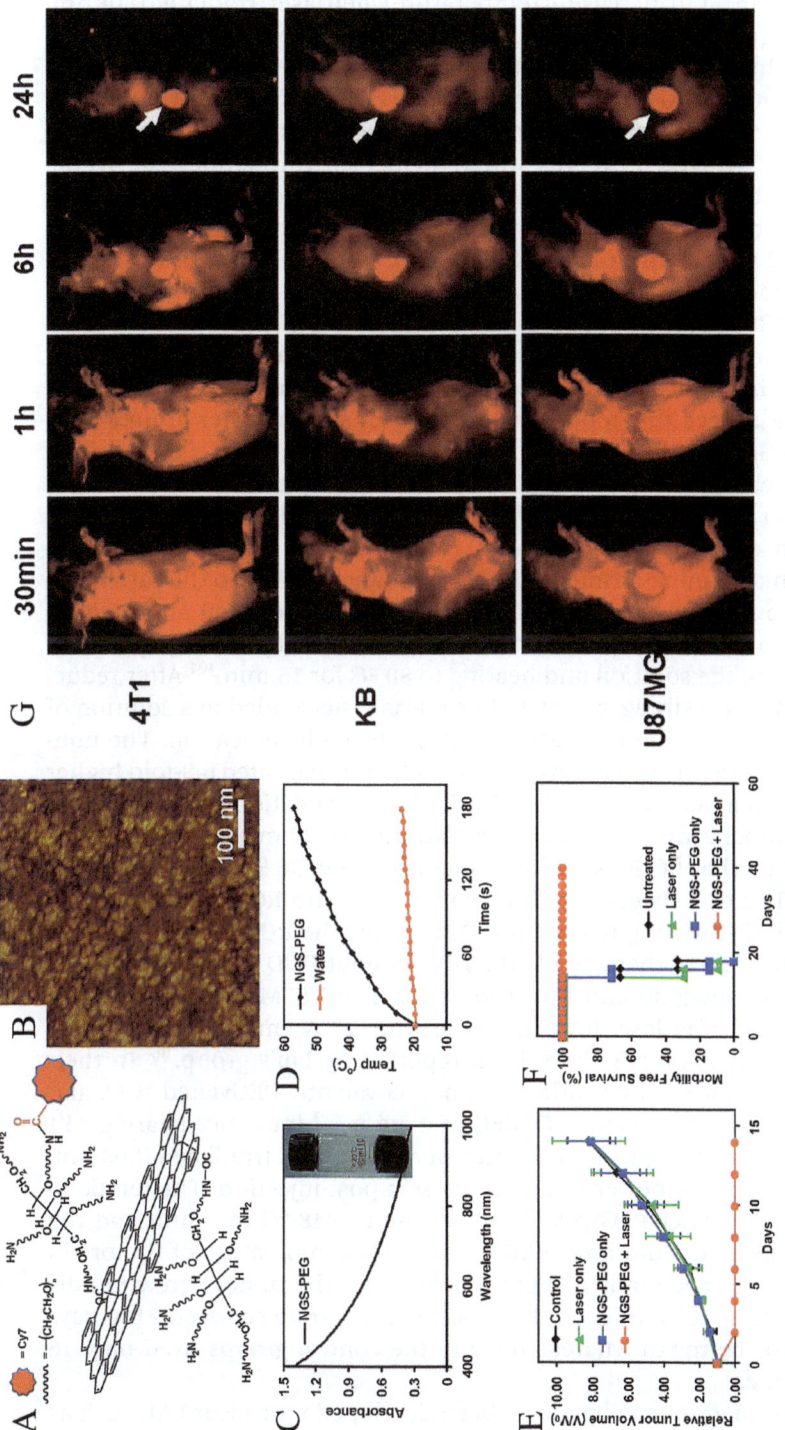

Figure 9.16 (A) Structure scheme of Cy7-labeled nGO–PEG. The AFM image (B), UV-Vis-NIR spectrum (C), and heating curves under 808 nm laser (D) of as-prepared nGO–PEG. (E) Tumor growth curves of mice with 4T1 tumor after various treatments indicated. (F) Survival curves of mice with 4T1, KB, and U87MG tumors after injection of Cy7-labeled nGO–PEG at different time points. High tumor uptake was observed for all three tumor models. Reproduced with permission from K. Yang, S. Zhang, G. Zhang, X. Sun, S.-T. Lee, Z. Liu, *Nano. Lett.* 2010, **10**, 3318. Copyright (2010) American Chemical Society.[97]

absorption and a rapid increase of temperature when laser irradiated (808 nm, 2 W cm^{-2}; Figure 9.16C and D). In order to study the *in vivo* behavior of NGS, researchers labeled NGS–PEG with Cy7 and injected the NGS–PEG–Cy7 intravenously into mice bearing 4T1 murine breast cancer tumors, KB human epidermoid carcinoma tumors, and U87MG human glioblastoma tumors. The Cy7–NGS–PEG was widely dispersed throughout the whole animal at the beginning and then enriched in the tumors over time, showing prominent tumor uptake in 24 h (Figure 9.16G). After that, the tumors were exposed to laser irradiation (808 nm, 2 W cm^{-2}). After laser irradiation the surface temperature of tumors reached ~50 °C, and tumors on mice with NGS–PEG disappeared. More importantly, mice in the treated group were tumor-free after PAT and survived >40 days without a single death, while the three control groups showed average life spans of ~16 days (Figure 9.16F). Since this report was published, several other groups have also studied GO-based PAT.[98-100]

GO can be chemically reduced to rGO, showing a much higher NIR absorbance. Higher NIR absorbance means higher PCE and a better photothermal ablation effect in PAT of cancer. Based on this, Dai and coworkers covalently linked amine-terminated, six-arm branched PEG to the carboxylic acid groups on GO in order to stabilize GO in buffer solution and break it into ~20 nm pieces. The GO was then reduced to rGO by adding hydrazine monohydrate to the solution and heating to 80 °C for 15 min.[101] After reduction, the rGO was easily aggregated; thus it was resuspended in a solution of the amphiphilic polymer C$_{18}$–PMH–mPEG$_{5000}$ by bath sonication. The non-covalently functionalized rGO (average size ~20 nm) exhibited 6.8-fold higher NIR absorption at 808 nm than GO. Under laser irradiation (808 nm, 0.6 W cm^{-2}), the temperature of rGO at a concentration of 20 mg L^{-1} rose to 55 °C in 5 min, while that of GO solution remained below 36 °C under the same conditions. The rGO sheets were then conjugated with RGD-based peptides and Cy5. After incubating rGO with U87MG cells, the RGD-conjugated rGO selectively attached to them, while the rGO without RGD did not, as proved by fluorescence imaging and flow cytometry. The U87MG cells were completely destroyed after laser irradiation (808 nm, 15 W cm^{-2}, 8 min). Further *in vivo* experiments on rGO have been reported by Liu's group.[102] In their report, they optimized the synthesis of non-covalently PEGylated rGO, and intravenously injected 200 μl rGO–PEG at 2 mg mL^{-1} into mice bearing 4T1 tumors (a dose of 20 mg kg^{-1}). The mice were then laser irradiated (808 nm, 0.15 W cm^{-2}; ultralow power intensity) at 48 h post-injection. The temperature of tumors in rGO–PEG treated mice rose to ~48 °C in 5 min and the tumors disappeared 1 day after PAT, whereas the temperature of tumors in control groups showed only a slight increase and the tumors grew rapidly after treatment. Finally, mice in the experimental group survived >100 days without tumor regrowth whereas mice in the control groups lived only 16 days on average.

Other kinds of graphene have also been developed for cancer PAT, such as GO nanoribbons[100] and nanomeshes.[103]

9.6 Semiconductor Photothermal Agents

Semiconductor photothermal agents have many advantages, such as low cost, high NIR absorption, high photothermal ability, and relatively high stability. In general, the studies of semiconductor photothermal agents mainly focus on Cu, W, and some other metal sulfides.

9.6.1 Cu-Based Photothermal Agents

One of the most studied semiconductor photothermal agents is CuS nanocrystals, a chalcogenide-based semiconductor with an excellent LSPR absorption band in the NIR range.[104] The strong LSPR property arises from p-type carriers in vacancy-doped CuS nanocrystals.[105] In 2010, Chen and coworkers reported the first photothermal application of CuS nanoparticles (CuS NPs), which was also the first report on the use of semiconductor nanoparticles for cancer PAT.[106] They synthesized, CuS NPs with an average size of ~3 nm (Figure 9.17A) that showed an increased absorption band in the NIR regions (Figure 9.17B), with maximum absorption at 900 nm. Under laser irradiation

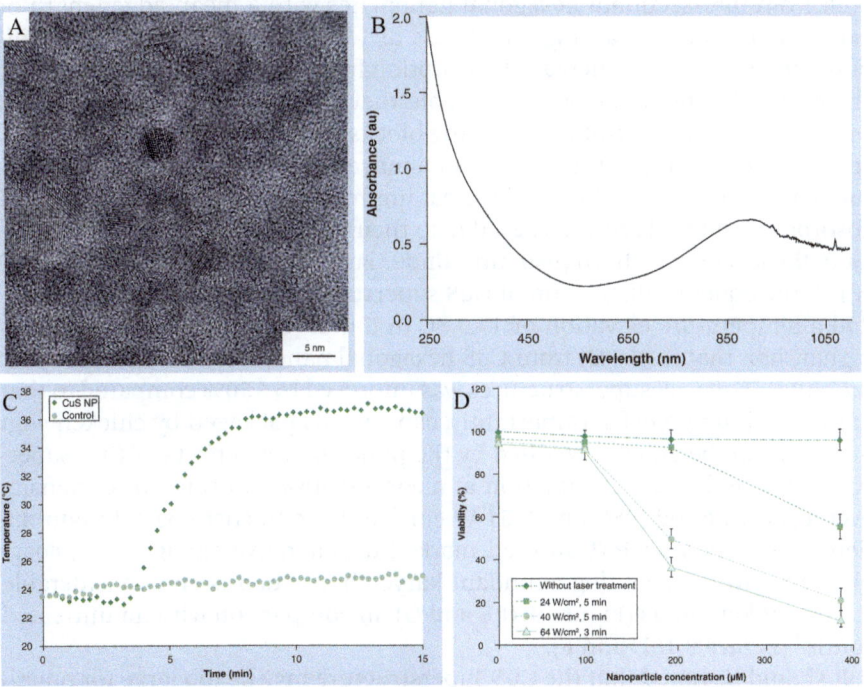

Figure 9.17 (A) HRTEM image of CuS NPs. (B) Optical absorption spectrum of CuS NPs. (C) Temperature elevating curves of CuS NPs and pure water under 808 nm NIR (24 W cm^{-2}) for 15 min. (D) Cell viability of the cells treated as indicated. Reproduced with permission from ref. 106. Copyright (2010) Future Medicine Ltd.

(808 nm, 24 W cm^{-2}), the aqueous dispersion of CuS NPs (770 µM) exhibited a rapid temperature increase of 12.7 °C in 5 min (Figure 9.17C). In their *in vitro* experiments, HeLa cells were incubated with CuS NPs for 2 h and laser irradiated (808 nm, 24 W cm^{-2}, 5 min). The percentage of viable cells was 55.6 ± 5.8% when the concentration of CuS NPs was 384 µM, and decreased to 21.2 ± 5.6% and 12.2 ± 3.7% when the laser power was increased to 40 W cm^{-2} for 5 min and 64 W cm^{-2} for 3 min, respectively (Figure 9.17D).[106] In an MTT assay, CuS NPs displayed minimal cytotoxic effects with a profile similar to 20 nm AuNPs, which the researchers also synthesized. The NIR absorption, small size, low cost, and low cytotoxicity make CuS NPs a promising photo-thermal agent for cancer PAT.

In this first study, the power intensity needed (>24 W cm^{-2}) is approximately 72 times higher than the conservative limit (~0.33 W cm^{-2})[93] for human skin exposure to 808 nm laser light, which is a great barrier to *in vivo* applications. To address this problem, our group has developed several kinds of CuS nanomaterials with better photothermal performances.[107–109] In 2011, we reported a kind of hydrophilic flower-like CuS superstructure, which was synthesized by a controllable hydrothermal route assisted with PVP at 180 °C for 48 h.[107] These CuS superstructures were ~1 µm in size, and were in fact built from intersectional hexagonal nanoplates with a mean edge length of ~500–800 nm and an average thickness of ~50 nm (Figure 9.18A). In addition, they also showed increased absorption in the near-IR regions, and their absorption intensity was almost double that of CuS hexagonal nanoplates with the same concentration (building blocks of superstructures) across the spectrum, especially in the NIR. This enhanced NIR photoabsorption of CuS superstructures resulted from the great improvement of their reflection and absorption ability (Figure 9.18B), due to their size (~1000 nm) which is similar to the laser wavelength (980 nm). Under laser irradiation (980 nm, 0.51 W cm^{-2}), the aqueous dispersion of CuS superstructures (0.25 mg mL^{-1}) exhibited a temperature elevation of 17.3 °C in 5 min (Figure 9.18C), which was higher than that (11.6 °C) from CuS hexagonal nanoplates. It is evident that the NIR PCE of CuS superstructures was improved by ~50% compared to that of their building blocks. Importantly, cancer cells packaged by chicken skin or *in vivo* can be efficiently killed by the photothermal effects of CuS super-structures under laser irradiation at a conservative and safe power density over a short period (980 nm, 0.51 W cm^{-2}, 5–10 min. Histological examination of tumors after PAT showed marked degenerative changes, *i.e.*, coag-ulative necrosis, including abundant karyorrhectic debris and considerable regions of karyolysis (Figure 9.18E and G), in comparison with an untreated control (Figure 9.18D and F).

It should be noted that the CuS superstructure may be too large for practical biological applications. To solve this problem, we synthesized hydrophilic Cu$_9$S$_5$ plate-like nanocrystals with a mean size of ~70 × 13 nm by a two-step route of thermal decomposition and ligand exchange.[108] The aqueous dispersion of Cu$_9$S$_5$ nanocrystals exhibited an enhanced absorption with the increase of wavelength in NIR regions, where the molar extinction coefficient

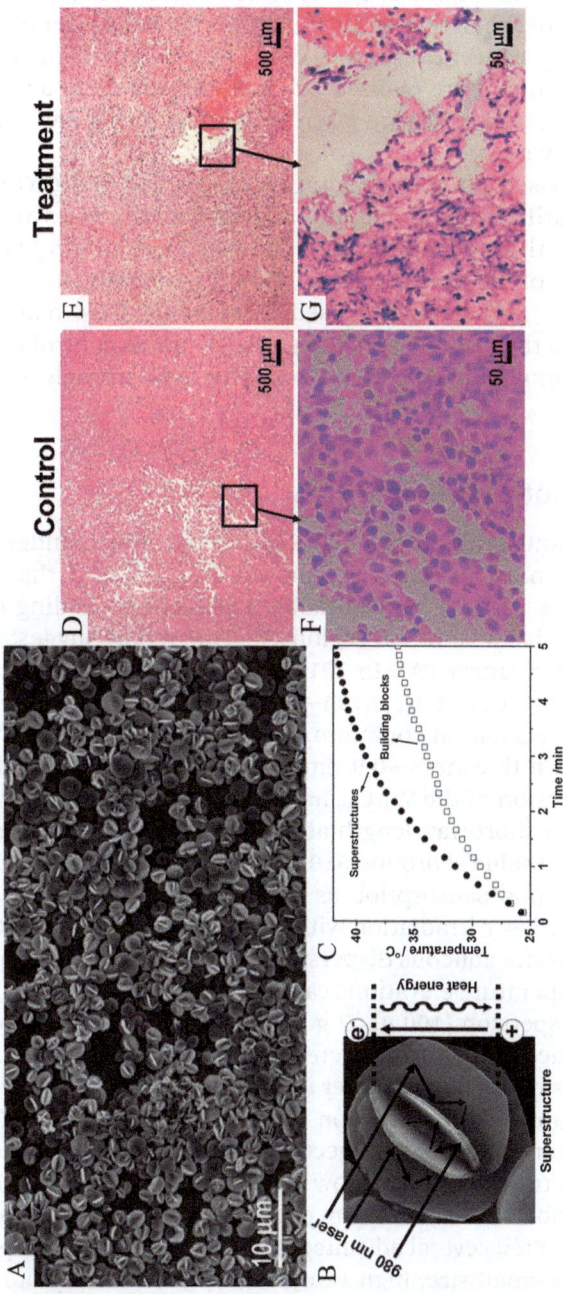

Figure 9.18 (A) Low-magnification SEM image of flower-like CuS superstructure. (B) Mechanism scheme of a CuS superstructure serving as 980 nm laser-cavity mirror and its photothermal conversion. (C) Temperature elevation between the aqueous dispersion of CuS superstructures and CuS building blocks. (D–G) Representative H&E-stained histological images of *ex vivo* tumor sections treated by 980 nm laser irradiation with power density 0.51 W cm^{-2} over a period of 10 min, injected with: (D and F) water; (E and G) CuS superstructure. Reproduced from Q. W. Tian, M. H. Tang, Y. G. Sun, R. J. Zou, Z. G. Chen, M. F. Zhu, S. P. Yang, J. L. Wang, J. H. Wang, J. Q. Hu, *Adv. Mater.* 2011, 23, 3542 with permission from John Wiley and Sons.[107] Copyright © 2011 Wiley-VCH Verlag GmbH & Co. KGaA, Weinheim.

at 980 nm was determined to be ~1.2×10^9 M^{-1} cm^{-1}. Under laser irradiation (980 nm, 0.51 W cm^{-2}), the Cu_9S_5 dispersion (40 ppm) exhibited a temperature elevation of 15.1 °C in 7 min. The PCE of Cu_9S_5 was determined to be ~25.7%, which was higher than that of AuNRs (PCE = 23.7%) under identical conditions. Importantly, cancer cells *in vivo* could be efficiently killed by the photothermal effects which were realized by Cu_9S_5 dispersion at a very low concentration (40 ppm) under 980 nm laser irradiation with a low and safe power density of 0.51 W cm^{-2}.

Other types of Cu-based photothermal agents are CuSe and CuTe.[110,111] In 2011, Korgel *et al.* utilized a colloidal hot injection method to synthesize $Cu_{2-x}Se$ nanocrystals with an average diameter of ~16 nm, exhibiting a PCE of 22%—higher than that of the commercial AuNSs (13%) and AuNRs (21%).[110] Cabot *et al.* reported highly monodispersed CuTe nanocubes, nanoplates and nanorods based on the lithium bis(trimethylsilyl) amide controlled reaction.[111] Those CuTe nanomaterials could absorb NIR light strongly and convert it to heat.

9.6.2   W-Based Photothermal Agents

Metal oxides were thought to have no photothermal effect under most conditions. However, in 2012, Karthish and coworkers found that tungsten oxide nanocrystals (WO_{3-x}) had a tunable LSPR effect, leading to the intense absorption in visible and NIR regions.[112] This report suggested the application of WO_{3-x} for cancer PAT. In 2013, our group prepared ultrathin PEGylated $W_{18}O_{49}$ nanowires (Figure 9.19A–C) by solvothermal treatment of an ethanol/PEG mixture solution containing WCl_6 at 180 °C for 24 h.[113] The $W_{18}O_{49}$ nanowires had a thickness ~0.9 nm, width ~4 nm, and length ~50 nm. An aqueous dispersion of the $W_{18}O_{49}$ nanowires exhibited a strong blue color, and it also had a short-wavelength absorption edge at approximately 420 nm and a minimum value at around 510 nm (Figure 9.19D). Importantly, it exhibited enhanced photoabsorption as the wavelength increased from 510 to 1100 nm. Under laser irradiation with a safe power density (980 nm, 0.72 W cm^{-2}), the nanowire aqueous dispersions (0.25–3.0 g L^{-1}) were heated up, the maximum temperature elevations can be 12.2–41.2 °C in 5 min. Subsequently, a $W_{18}O_{49}$ dispersion (100 μL, 2 g L^{-1}) was injected into tumors in mice. The tumor surface temperature increased rapidly to reach 44.5 °C at 30 s and 49.1 °C at 60 s under 980 nm laser irradiation (Figure 9.19E), resulting in the efficient photothermal ablation (PTA) of cancer cells in 10 min (Figure 9.19G). The control group was injected with saline solution, and the tumor surface temperature remained below 27.5 °C during the entire irradiation process, without obvious death of cancer cells (Figure 9.19F). PEGylated $W_{18}O_{49}$ nanowires exhibited several advantages as a new NIR-induced photothermal agent, such as small size, high PCE, low cost, and low cytotoxicity. More significantly, this work shows the possibility of using other transition metal oxide nanocrystals as novel photothermal agents for cancer PAT.

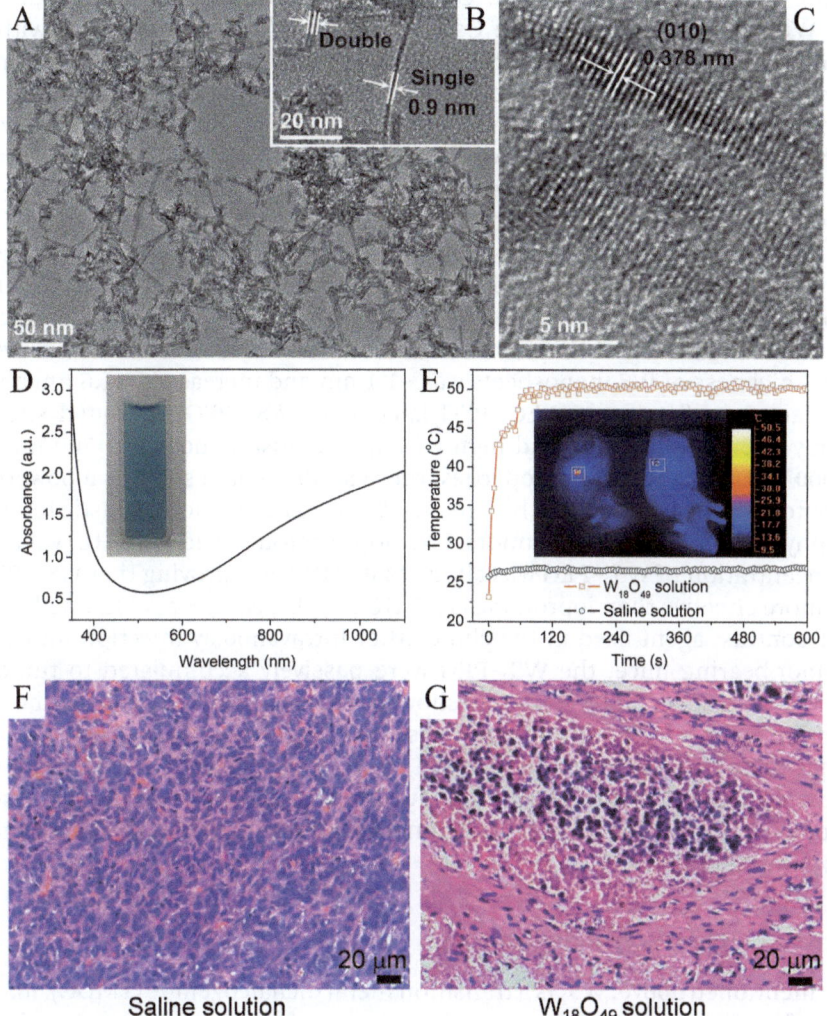

Figure 9.19 TEM (A and B), HRTEM (C) images, and UV-Vis spectrum (D) patterns of $W_{18}O_{49}$ nanowires. (E) Plots of temperature within the irradiated tumor area in two mice treated with saline solution and 2 g L^{-1} $W_{18}O_{49}$ nanowire solution respectively, as a function of irradiation time. Inset: Corresponding full-body thermal image of mice treated with saline solution (right) and $W_{18}O_{49}$ nanowire solution (left) at 180 s. H&E-stained histological images of *in vivo* tumor sections after PTA of saline solution (F) and $W_{18}O_{49}$ nanowires (G). Reproduced from Z. Chen, Q.Wang, H. Wang, L. Zhang, G. Song, L. Song, J. Hu, H. Wang, J. Liu, M. Zhu, D. Zhao, *Adv. Mater.* 2013, **25**, 2095 with permission from John Wiley and Sons.[113] Copyright © 2013 Wiley-VCH Verlag GmbH & Co. KGaA, Weinheim.

However, the photothermal stability is still unsatisfactory. To improve the stability of $W_{18}O_{49}$, we utilized Cs^+ to stabilize the oxygen-induced defect of WO_{3-x}, thus forming stable $CsWO_3$ crystals.[114] $CsWO_3$ nanorods with a diameter of ~11 nm and length ~50 nm were synthesized by a solvothermal synthesis–PEGylation two-step method. They exhibited more stable NIR photothermal performance than PEGylated $W_{18}O_{49}$ nanowires. Moreover, when the mice were injected with the PEGylated $CsWO_3$ nanorods, *in vivo* cancer cells were efficiently destroyed by the photothermal effects of $CsWO_3$ nanorods under laser irradiation (915 nm, 0.72 W cm^{-2}, 10 min).

Apart from tungsten oxides, another kind of layer-like semiconductor, WS_2, has also been investigated as a photothermal conversion agent. Liu and coworkers synthesized hexagonal WS_2 by breaking the weak interlayer forces in bulk WS_2 through lithium ion insertion and ultrasonication.[115] The average thickness of WS_2 nanosheets was ~1.1 nm and increased to 1.6 nm after surface modification by PEG. PEGylated WS_2 (WS_2–PEG) exhibited strong X-ray attenuation ability and high NIR optical absorbance from 700 nm to 1000 nm. Utilizing these properties, the WS_2–PEG could serve as a powerful photothermal agent guided by enhanced X-ray CT and photoacoustic tomography bimodal imaging of tumors. The slope of Hounsfield unit (HU) value to concentration of WS_2–PEG was about 22.01 HU L g^{-1}, proving that WS_2–PEG is more effective than iopromide (15.9 HU L g^{-1}), a commercial iodine-based CT contrast agent used in the clinic. After intravenously injection into 4T1 tumor-bearing mice, the WS_2–PEG were passively accumulated to tumors by the EPR effect and the tumors showed a strong contrast signal in the CT image. Furthermore, tumors could also be imaged by photoacoustic signals after intravenously injection with WS_2–PEG. Under laser irradiation of (808 nm, 0.8 W cm^{-2}, 5 min), the tumor surface temperatures rapidly increased from ~30 °C to ~65 °C. Finally, the tumors were completely eliminated by the two-mode guided PAT.

9.6.3   Other Semiconductors

As mentioned above, WS_2 is a transition metal dichalcogenide (TMDC), made up of one layer of metal atoms and two layers of chalcogenide atoms with a sandwich-like spatial structure. Besides WS_2, there are also other kinds of NIR-absorbing TMDCs as photothermal agent, typically including MoS_2[116] and Bi_2S_3.[117] For example, Chou *et al.* prepared MoS_2 by breaking the weak interlayer forces in bulk MoS_2 through ultrasonication and the formation of H_2, and showed an approximately 7.8-fold absorption increase at 800 nm compared to nano-GO.[116] An aqueous dispersion of MoS_2 (150 ppm) can be heated up rapidly to 60 °C under laser irradiation (800 nm, 0.8 W cm^{-2}).

Recently, Gu *et al.* synthesized oleic acid (OA)-coated Bi_2S_3 nanorods (Bi_2S_3 NRs) by a facile solvothermal method at 150 °C and then functionalized the Bi_2S_3 NRs with Tween 20 in order to disperse them in water and prolong the residence time *in vivo* (Figure 9.20A and B).[117] Those as-prepared Bi_2S_3 NRs had a broad absorption at 700–1100 nm (Figure 9.20C) and showed good

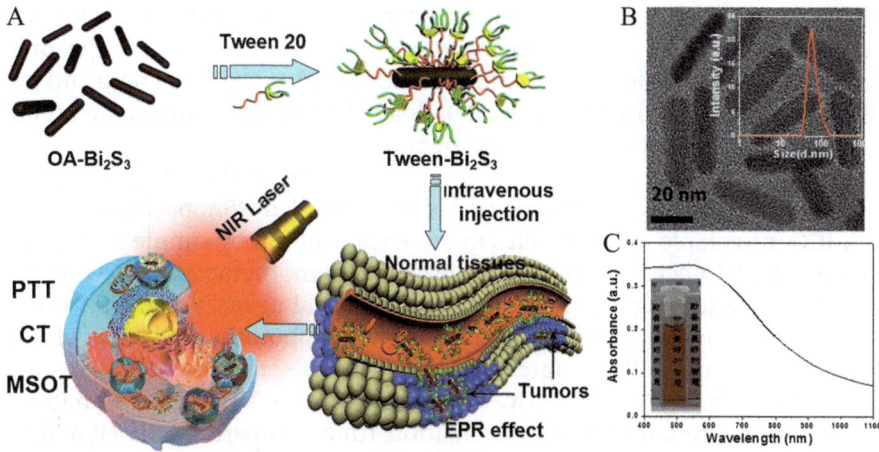

Figure 9.20 (A) Scheme of the theranostic principle based on the unique properties of Bi_2S_3 NRs. (B) TEM image and inset dynamic light scattering figure of as-prepared Bi_2S_3 NRs. (C) Absorbance spectrum of Bi_2S_3 NRs and inset: as-prepared Bi_2S_3 NRs solution in deionized water. Reproduced with permission from J. Liu, X. Zheng, L. Yan, L. Zhou, G. Tian, W. Yin, L. Wang, Y. Liu, Z. Hu, Z. Gu, C. Chen, Y. Zhao, *ACS Nano*, 2015, **9**, 696. Copyright (2015) American Chemical Society.[117]

photothermal ability; PCE was calculated as 28.1%. The 1 mg mL^{-1} sample could cause a temperature increase by ~50 °C within 10 min under laser irradiation (808 nm, 1 W cm^{-2}). As for further biomedical applications, Bi_2S_3 NRs were passively targeted to the tumor site efficiently after intravenous injection and showed satisfactory residence time in tumor. Meanwhile, Bi_2S_3 NRs exhibited enhanced contrast in multispectral optoacoustic tomography and CT imaging, which enabled the simultaneous visual guiding, destruction, and metastasis tracing of cancer cells. Their report showed the potential promise of Bi_2S_3 NRs for precise cancer therapy and therapeutic monitoring.

9.7 Multifunctional Photothermal Agents

For the four main types of photothermal agents, research interests focus more and more on functionalization. Generally, there are two types of multifunctional PAs: synergetic therapy, combining PAT with chemotherapies, or combining PAs with other imaging abilities. Here we give a brief introduction to them both.

9.7.1 Synergetic Therapy

Synergetic therapy always has a better effect on cancer, compared to single PAT or chemotherapy alone. One of the earliest photothermal-chemo synergetic agents was AuNSs. Wu *et al.*[118] utilized nanoliposomes as templates to synthesize a liposome/SiO$_2$/Au nanocomposite. Doxorubicin (DOX) can be incorporated into this nanocomposite as a result of the hollow liposome cores. Under 808 nm laser irradiation, the spherical AuNSs heated up and the DOX was rapidly released from the DOX-loaded AuNSs, providing a therapeutic effect combining PAT and chemotherapy.

There are also synergetic therapeutic agents composed entirely of organic compounds. Recently, Liu and coworkers[11] assembled HSA, paclitaxel (PTX), and ICG into stable nanoparticles in aqueous solution by simple mixing. In this system, PTX acted as an effective drug for chemotherapy, and ICG served as a fluorescent imaging probe and photothermal agent, while HSA was a biocompatible carrier platform. Interestingly, the photothermal effect of ICG promotes the intracellular uptake of HSA–ICG–PTX and thus improves the therapeutic effect. The HSA–ICG–PTX nanoparticles showed an enhanced stability and prolonged blood circulation time compared with HSA–ICG. Moreover, the HSA–ICG–PTX was entirely composed of FDA-approved agents, which was promising for clinical applications. Photothermal-chemo synergetic agents based on semiconductor PAs have also been developed. Liu *et al.*[119] synthesized two-dimensional PEG functionalized MoS$_2$ nanosheets by a chemical exfoliation method. The atomic force microscopy (AFM) result showed the average thickness of MoS$_2$ was ~1 nm. The PEGylated MoS$_2$ was observed to have the highest drug loading ratios (weight ratios between the drug and MoS$_2$) of ~239%, ~39%, and ~118% for DOX, Ce6, and SN38 respectively. After functionalizing PEG–MoS$_2$ with FA, the MoS$_2$–PEG–FA with drugs exhibited an increased specificity to cancer cells *in vitro* and an excellent synergistic anti-cancer effect. In addition, our group prepared the hydrophilic Cu$_9$S$_5$@mSiO$_2$–PEG core–shell nanocomposites, and they exhibited strong NIR photoabsorption, high photothermal conversion characteristics, and excellent biocompatibility *in vitro*, ensuring the effective photothermal ablation of cancer cells and infrared thermal imaging *in vivo*. Moreover, due to their biocompatible mesoporous silica shell, Cu$_9$S$_5$@mSiO$_2$–PEG nanocomposites could serve as a drug carrier for loading DOX for chemotherapy of cancer cells *in vitro* and *in vivo*, and importantly showed an enhanced inhibiting rate when combined with photothermal therapy.[120]

9.7.2   Imaging-Guided PAT

Imaging-guided PAs can be used for confirming the location, size, and metastasis of cancer, and the photothermal ablation of cancer cells can then be achieved by irradiating the corresponding tumor area. This imaging-guided cancer PAT has great significance for clinical applications and attracts much attention. One representative example is CNTs. In 2009, Kim and coworkers[121] reported Au-plated CNTs, which could be used as both photoacoustic contrast agents and photothermal contrast agents. In further research, Liu *et al.*[122] grew a layer of AuNPs on the surface of DNA-coated SWNTs (SWNT–Au), then conjugating FA-modified PEG to the SWNT–Au nanocomposite. The final nanocomposite showed an excellent concentration and excitation-source dependent surface-enhanced Raman scattering (SERS) effect for the detection of FA-receptor-positive cells. The highly enhanced NIR absorbance also allows the agent to be used for targeted photothermal ablation of cancer cells, which was guided by the SERS effect.

Besides CNTs, graphene photothermal agents were also coupled with a number of inorganic nanoparticles to obtain imaging abilities, such as Au nanoclusters[123] and quantum dots.[124] For example, Chen *et al.*[124] synthesized a quantum dot-tagged rGO nanocomposite (QD–rGO), which combined PAT with bioimaging. Remarkably, since QD–rGO absorbed NIR irradiation and converted it into heat, the QD brightness exhibited a marked decrease, providing a means for *in situ* heat/temperature sensing and a real-time indicator of the PAT progress. Besides, superparamagnetic iron oxide nanoparticles (IONP) were also coupled to graphene-based nanosheets for MRI-guided PAT. In 2011, Liu *et al.*[125] synthesized a rGO–IONP nanocomposite and non-covalently functionalized it with PEG in order to give it great stability in physiological environments. The rGO–IONP–PEG could be used as a multifunctional nanoprobe for photoacoustic, MRI and fluorescence imaging because of the high NIR absorbance, strong superparamagnetic property, and an extra fluorescent label, respectively. The rGO–IONP–PEG exhibited a high uptake in 4T1 tumors, which was revealed by the triple-modal tumor probes through intravenous injection after labeling by Cy5. Guided by that, rGO–IONP–PEG achieved highly efficient tumor ablation under NIR laser irradiation (808 nm, 0.5 W cm^{-2}, 5 min). The 2D structure and large surface area gave plentiful room for graphene functionalization, and graphene-based nanocomposites with highly enriched functionalities have become a promising agent for cancer treatment.

Multifunctional semiconductor photothermal agents have also been developed. In 2013, our group reported the first successful design and synthesis of $Fe_3O_4@Cu_{2-x}S$ core–shell nanoparticles, offering both high photothermal stability and superparamagnetic properties.[109] Those properties made these nanoparticles an excellent multifunctional probe for MRI and infrared thermal imaging. They showed low cytotoxicity and highly efficient photothermal effect *in vitro* and *in vivo*. Recently, hexagonal Cu_3BiS_3 nanocrystals were prepared by a one-pot solvothermal method.[126] Owing to the large X-ray attenuation coefficient of Bi and the unusual defects, the Cu_3BiS_3 nanocrystals exhibited both strong NIR absorption and CT imaging response, which can be used as a CT-guided NIR-responsive photothermal agent.

The multifunctional photothermal agents move us one step further toward clinical applications. These results also provide improved recognition of the synergistic effect, which is important for developing multifunctional nanoparticles for biomedical applications.

9.8 Conclusions and Outlook

Despite the exciting achievements of PAT that have appeared in the past few years, there are still challenges for all types of photothermal agents. For the low-cost organic photothermal agents, photobleaching makes them easy to degrade and thus lose the property of photothermal conversion. Classical and widely investigated metal-based photothermal agents, such as Au nanostructures, might melt and be reshaped during irradiation and heating; the high price of noble metals also constrains their applications. Carbon-based photothermal agents have excellent photostability but relatively low NIR absorbance and PCE. Semiconductors seems to be a neutral aggregation of high PCE, low cost, and a slightly lower photostability compared to carbon-based ones. However, they have drawbacks: lack of biocompatibility and cytotoxicity caused by the heavy element content of semiconductor photothermal agents require further modification.

There is still a long way to go for the clinical applications of PAT. The first issue we need to be concerned with is the effective penetration depth of the NIR laser, which is limited to a few centimeters. This may be a sufficient depth for the treatment of some types of disease, such as skin cancer, esophageal cancer and oral cancer, which can be irradiated by NIR light directly, or *via* gastroscopy and endoscopy. However, for internal tumors, such as liver, lung, and kidney tumors, irradiation with stronger penetrating abilities, such as X-rays, is needed.

Another challenge is the metabolism and bioaccumulation of photothermal nanoparticles. For inorganic photothermal nanoparticles that are not biodegradable in the human body, whether they do long-term harm still needs further investigation. Although much research has demonstrated no noticeably toxicity *in vitro* or *in vivo* by appropriate surface modification, it is still extremely hard for such materials to be finally approved by the FDA for clinical applications.

Under the local irradiation, photothermal agents convert light into heat and thus ablate cancer cells. For precise PAT it is therefore a prerequisite to know the location, size, and shape of tumors. If a nanoplatform combines both photothermal and imaging abilities, it will be of great significance for clinical applications. The design and development of novel imaging-guided photothermal agents is another challenge.

Furthermore, the future treatment of cancer relies more and more on synergetic therapy, which is the combination of different treatment approaches, including surgery, chemotherapy, radiotherapy, PDT, and PAT. Those synergetic therapies show the potential of maintaining the advantages of each individual method and overcoming their limitations. Multifunctional nanocarriers will shed new light on the next generation of synergetic therapies for tumors. Nevertheless, there is no doubt that functionalized photothermal agents will play an important role in the future treatment of cancer.

References

1. http://www.who.int/mediacentre/factsheets/fs297/en/.
2. J. R. Wilson, D. M. Mancini, K. McCully, N. Ferraro, V. Lanoce and B. Chance, *Circulation*, 1989, **80**, 1668.
3. B. Chance, M. T. Dait, C. Zhang, T. Hamaoka and F. Hagerman, *Am. J. Physiol.: Cell Physiol.*, 1992, **262**, C766.

4. Z. G. Chen, L. S. Zhang, Y. G. Sun, J. Q. Hu and D. Y. Wang, *Adv. Funct. Mater.*, 2009, **19**, 3815.
5. L. Cheng, C. Wang, L. Feng, K. Yang and Z. Liu, *Chem. Rev.*, 2014, **114**, 10869.
6. A. O. Govorov and H. H. Richardson, *Nano Today*, 2007, **2**, 30.
7. A. B. Taylor, A. M. Siddiquee and J. W. M. Chon, *ACS Nano*, 2014, **8**, 12071.
8. D. K. Roper, W. Ahn and M. Hoepfner, *J. Phys. Chem. C*, 2007, **111**, 3636.
9. X. Zheng, D. Xing, F. Zhou, B. Wu and W. R. Chen, *Mol. Pharmaceutics*, 2011, **8**, 447.
10. J. Yu, D. Javier, M. A. Yaseen, N. Nitin, R. Richards-Kortum, B. Anvari and M. S. Wong, *J. Am. Chem. Soc.*, 2010, **132**, 1929.
11. Q. Chen, C. Liang, C. Wang and Z. Liu, *Adv. Mater.*, 2015, **27**, 903.
12. M. Zheng, P. Zhao, Z. Luo, P. Gong, C. Zheng, P. Zhang, C. Yue, D. Gao, Y. Ma and L. Cai, *ACS Appl. Mater. Inter.*, 2014, **6**, 6709.
13. C. Zheng, M. Zheng, P. Gong, D. Jia, P. Zhang, B. Shi, Z. Sheng, Y. Ma and L. Cai, *Biomaterials*, 2012, **33**, 5603.
14. S. Luo, E. Zhang, Y. Su, T. Cheng and C. Shi, *Biomaterials*, 2011, **32**, 7127.
15. C. Zhang, S. Wang, J. Xiao, X. Tan, Y. Zhu, Y. Su, T. Cheng and C. Shi, *Biomaterials*, 2010, **31**, 1911.
16. C. Zhang, T. Liu, Y. Su, S. Luo, Y. Zhu, X. Tan, S. Fan, L. Zhang, Y. Zhou, T. Cheng and C. Shi, *Biomaterials*, 2010, **31**, 6612.
17. S. Luo, X. Tan, Q. Qi, Q. Guo, X. Ran, L. Zhang, E. Zhang, Y. Liang, L. Weng, H. Zheng, T. Cheng, Y. Su and C. Shi, *Biomaterials*, 2013, **34**, 2244.
18. C.-L. Peng, Y.-H. Shih, P.-C. Lee, T. M.-H. Hsieh, T.-Y. Luo and M.-J. Shieh, *ACS Nano*, 2011, **5**, 5594.
19. L. Cheng, W. He, H. Gong, C. Wang, Q. Chen, Z. Cheng and Z. Liu, *Adv. Funct. Mater.*, 2013, **23**, 5893.
20. Q. Chen, C. Wang, Z. Zhan, W. He, Z. Cheng, Y. Li and Z. Liu, *Biomaterials*, 2014, **35**, 8206.
21. J. Yang, J. Choi, D. Bang, E. Kim, E.-K. Lim, H. Park, J.-S. Suh, K. Lee, K.-H. Yoo, E.-K. Kim, Y.-M. Huh and S. Haam, *Angew. Chem.*, 2011, **123**, 461.
22. E. Ju, K. Dong, Z. Liu, F. Pu, J. Ren and X. Qu, *Adv. Funct. Mater.*, 2015, **25**, 1574.
23. K. Yang, H. Xu, L. Cheng, C. Sun, J. Wang and Z. Liu, *Adv. Mater.*, 2012, **24**, 5586.
24. P. M. George, A. W. Lyckman, D. A. LaVan, A. Hegde, Y. Leung, R. Avasare, C. Testa, P. M. Alexander, R. Langer and M. Sur, *Biomaterials*, 2005, **26**, 3511.
25. A. Ramanaviciene, A. Kausaite, S. Tautkus and A. Ramanavicius, *J. Pharm. Pharmacol.*, 2007, **59**, 311.
26. K. M. Au, Z. Lu, S. J. Matcher and S. P. Armes, *Adv. Mater.*, 2011, **23**, 5792.
27. Z. Zha, X. Yue, Q. Ren and Z. Dai, *Adv. Mater.*, 2013, **25**, 777.
28. M. Chen, X. Fang, S. Tang and N. Zheng, *Chem. Commun.*, 2012, **48**, 8934.

29. L. Cheng, K. Yang, Q. Chen and Z. Liu, *ACS Nano*, 2012, **6**, 5605.
30. J. F. Lovell, C. S. Jin, E. Huynh, H. Jin, C. Kim, J. L. Rubinstein, W. C. W. Chan, W. Cao, L. V. Wang and G. Zheng, *Nat. Mater.*, 2011, **10**, 324.
31. K. K. Ng, J. F. Lovell, A. Vedadi, T. Hajian and G. Zheng, *ACS Nano*, 2013, **7**, 3484.
32. C. S. Jin, J. F. Lovell, J. Chen and G. Zheng, *ACS Nano*, 2013, **7**, 2541.
33. T. W. Liu, T. D. MacDonald, J. Shi, B. C. Wilson and G. Zheng, *Angew. Chem., Int. Ed.*, 2012, **51**, 13128.
34. E. Huynh, Y. C. LeungBen, B. L. Helfield, M. Shakiba, J.-A. Gandier, C. S. Jin, E. R. Master, B. C. Wilson, D. E. Goertz and G. Zheng, *Nat Nano*, 2015, **10**, 325.
35. J. D. Simon, *Acc. Chem. Res.*, 2000, **33**, 307.
36. Y. Liu, K. Ai, J. Liu, M. Deng, Y. He and L. Lu, *Adv. Mater.*, 2013, **25**, 1353.
37. I. H. El-Sayed, X. Huang and M. A. El-Sayed, *Cancer Lett.*, 2006, **239**, 129.
38. J.-L. Li, L. Wang, X.-Y. Liu, Z.-P. Zhang, H.-C. Guo, W.-M. Liu and S.-H. Tang, *Cancer Lett.*, 2009, **274**, 319.
39. C.-P. Liu, F.-S. Lin, C.-T. Chien, S.-Y. Tseng, C.-W. Luo, C.-H. Chen, J.-K. Chen, F.-G. Tseng, Y. Hwu, L.-W. Lo, C.-S. Yang and S.-Y. Lin, *Macromol. Biosci.*, 2013, **13**, 1314.
40. J. Lin, S. Wang, P. Huang, Z. Wang, S. Chen, G. Niu, W. Li, J. He, D. Cui, G. Lu, X. Chen and Z. Nie, *ACS Nano*, 2013, **7**, 5320.
41. J. He, P. Zhang, T. Babu, Y. Liu, J. Gong and Z. Nie, *Chem. Commun.*, 2013, **49**, 576.
42. J. He, Y. Liu, T. Babu, Z. Wei and Z. Nie, *J. Am. Chem. Soc.*, 2012, **134**, 11342.
43. J. Nam, W.-G. La, S. Hwang, Y. S. Ha, N. Park, N. Won, S. Jung, S. H. Bhang, Y.-J. Ma, Y.-M. Cho, M. Jin, J. Han, J.-Y. Shin, E. K. Wang, S. G. Kim, S.-H. Cho, J. Yoo, B.-S. Kim and S. Kim, *ACS Nano*, 2013, **7**, 3388.
44. J. Nam, N. Won, H. Jin, H. Chung and S. Kim, *J. Am. Chem. Soc.*, 2009, **131**, 13639.
45. X. Liu, Y. Chen, H. Li, N. Huang, Q. Jin, K. Ren and J. Ji, *ACS Nano*, 2013, **7**, 6244.
46. J. He, X. Huang, Y.-C. Li, Y. Liu, T. Babu, M. A. Aronova, S. Wang, Z. Lu, X. Chen and Z. Nie, *J. Am. Chem. Soc.*, 2013, **135**, 7974.
47. N. R. Jana, L. Gearheart and C. J. Murphy, *Adv. Mater.*, 2001, **13**, 1389.
48. J. Pérez-Juste, I. Pastoriza-Santos, L. M. Liz-Marzán and P. Mulvaney, *Coord. Chem. Rev.*, 2005, **249**, 1870.
49. H. Chen, L. Shao, Q. Li and J. Wang, *Chem. Soc. Rev.*, 2013, **42**, 2679.
50. K.-S. Lee and M. A. El-Sayed, *J. Phys. Chem. B*, 2005, **109**, 20331.
51. B. Nikoobakht and M. A. El-Sayed, *Chem. Mater.*, 2003, **15**, 1957.
52. X. Huang and M. A. El-Sayed, *J. Adv. Res.*, 2010, **1**, 13.
53. G. Von Maltzahn, J.-H. Park, A. Agrawal, N. K. Bandaru, S. K. Das, M. J. Sailor and S. N. Bhatia, *Cancer Res.*, 2009, **69**, 3892.
54. E. B. Dickerson, E. C. Dreaden, X. Huang, I. H. El-Sayed, H. Chu, S. Pushpanketh, J. F. McDonald and M. A. El-Sayed, *Cancer Lett.*, 2008, **269**, 57.

55. X. Huang, I. H. El-Sayed, W. Qian and M. A. El-Sayed, *J. Am. Chem. Soc.*, 2006, **128**, 2115.
56. L. Tong, Q. Wei, A. Wei and J.-X. Cheng, *Photochem. Photobiol.*, 2009, **85**, 21.
57. L. Tong, Y. Zhao, T. B. Huff, M. N. Hansen, A. Wei and J. X. Cheng, *Adv. Mater.*, 2007, **19**, 3136.
58. Z. Li, P. Huang, X. Zhang, J. Lin, S. Yang, B. Liu, F. Gao, P. Xi, Q. Ren and D. Cui, *Mol. Pharmaceutics*, 2010, **7**, 94.
59. R. S. Norman, J. W. Stone, A. Gole, C. J. Murphy and T. L. Sabo-Attwood, *Nano Lett.*, 2008, **8**, 302.
60. W. I. Choi, J.-Y. Kim, C. Kang, C. C. Byeon, Y. H. Kim and G. Tae, *ACS Nano*, 2011, **5**, 1995.
61. J. L. Li, D. Day and M. Gu, *Adv. Mater.*, 2008, **20**, 3866.
62. R. Bardhan, S. Lal, A. Joshi and N. J. Halas, *Acc. Chem. Res.*, 2011, **44**, 936.
63. A. M. Gobin, M. H. Lee, N. J. Halas, W. D. James, R. A. Drezek and J. L. West, *Nano Lett.*, 2007, **7**, 1929.
64. D. P. O'Neal, L. R. Hirsch, N. J. Halas, J. D. Payne and J. L. West, *Cancer Lett.*, 2004, **209**, 171.
65. C. Loo, A. Lowery, N. Halas, J. West and R. Drezek, *Nano Lett.*, 2005, **5**, 709.
66. L. Huiyu, C. Dong, T. Fangqiong, D. Gangjun, L. Linlin, M. Xianwei, L. Wei, Z. Yangde, T. Xu and L. Yi, *Nanotechnology*, 2008, **19**, 455101.
67. H. Ke, J. Wang, Z. Dai, Y. Jin, E. Qu, Z. Xing, C. Guo, X. Yue and J. Liu, *Angew. Chem.*, 2011, **123**, 3073.
68. R. Fekrazad, N. Hakimiha, E. Farokhi, M. J. Rasaee, M. S. Ardestani, K. A. M. Kalhori and F. Sheikholeslami, *Int. J. Nanomed.*, 2011, **6**, 2749.
69. M. P. Melancon, A. M. Elliott, A. Shetty, Q. Huang, R. J. Stafford and C. Li, *J. Controlled Release*, 2011, **156**, 265.
70. W. Lu, C. Xiong, G. Zhang, Q. Huang, R. Zhang, J. Z. Zhang and C. Li, *Clin. Cancer Res.*, 2009, **15**, 876.
71. W. Lu, M. P. Melancon, C. Xiong, Q. Huang, A. Elliott, S. Song, R. Zhang, L. G. Flores, J. G. Gelovani, L. V. Wang, G. Ku, R. J. Stafford and C. Li, *Cancer Res.*, 2011, **71**, 6116.
72. S. E. Skrabalak, J. Chen, Y. Sun, X. Lu, L. Au, C. M. Cobley and Y. Xia, *Acc. Chem. Res.*, 2008, **41**, 1587.
73. Y. Sun and Y. Xia, *Science*, 2002, **298**, 2176.
74. Y. Sun, B. T. Mayers and Y. Xia, *Nano Lett.*, 2002, **2**, 481.
75. Y. Sun and Y. Xia, *J. Am. Chem. Soc.*, 2004, **126**, 3892.
76. Y. Sun, B. Mayers, T. Herricks and Y. Xia, *Nano Lett.*, 2003, **3**, 955.
77. J. Chen, J. M. McLellan, A. Siekkinen, Y. Xiong, Z.-Y. Li and Y. Xia, *J. Am. Chem. Soc.*, 2006, **128**, 14776.
78. J. Chen, D. Wang, J. Xi, L. Au, A. Siekkinen, A. Warsen, Z.-Y. Li, H. Zhang, Y. Xia and X. Li, *Nano Lett.*, 2007, **7**, 1318.
79. X. Huang, S. Tang, X. Mu, Y. Dai, G. Chen, Z. Zhou, F. Ruan, Z. Yang and N. Zheng, *Nat. Nano*, 2011, **6**, 28.
80. S. Tang, X. Huang and N. Zheng, *Chem. Commun.*, 2011, **47**, 3948.

81. X. Huang, S. Tang, B. Liu, B. Ren and N. Zheng, *Adv. Mater.*, 2011, **23**, 3420.
82. H. W. Kroto, J. R. Heath, S. C. O'Brien, R. F. Curl and R. E. Smalley, *Nature*, 1985, **318**, 162.
83. S. Iijima, *Nature*, 1991, **354**, 56.
84. K. S. Novoselov, A. K. Geim, S. V. Morozov, D. Jiang, Y. Zhang, S. V. Dubonos, I. V. Grigorieva and A. A. Firsov, *Science*, 2004, **306**, 666.
85. N. W. S. Kam, M. O'Connell, J. A. Wisdom and H. Dai, *Proc. Natl. Acad. Sci. U. S. A.*, 2005, **102**, 11600.
86. P. Chakravarty, R. Marches, N. S. Zimmerman, A. D.-E. Swafford, P. Bajaj, I. H. Musselman, P. Pantano, R. K. Draper and E. S. Vitetta, *Proc. Natl. Acad. Sci. U. S. A.*, 2008, **105**, 8697.
87. W. Chung-Hao, H. Yao-Jhang, C. Chia-Wei, H. Wen-Ming and P. Ching-An, *Nanotechnology*, 2009, **20**, 315101.
88. S. Ning, L. Shaoxin, W. Eric and P. Balaji, *Nanotechnology*, 2007, **18**, 315101.
89. H. K. Moon, S. H. Lee and H. C. Choi, *ACS Nano*, 2009, **3**, 3707.
90. A. Burke, X. Ding, R. Singh, R. A. Kraft, N. Levi-Polyachenko, M. N. Rylander, C. Szot, C. Buchanan, J. Whitney, J. Fisher, H. C. Hatcher, R. D'Agostino, N. D. Kock, P. M. Ajayan, D. L. Carroll, S. Akman, F. M. Torti and S. V. Torti, *Proc. Natl. Acad. Sci. U. S. A.*, 2009, **106**, 12897.
91. S. Ghosh, S. Dutta, E. Gomes, D. Carroll, R. D'Agostino, J. Olson, M. Guthold and W. H. Gmeiner, *ACS Nano*, 2009, **3**, 2667.
92. X. Liu, H. Tao, K. Yang, S. Zhang, S.-T. Lee and Z. Liu, *Biomaterials*, 2011, **32**, 144.
93. J. Robinson, K. Welsher, S. Tabakman, S. Sherlock, H. Wang, R. Luong and H. Dai, *Nano. Res.*, 2010, **3**, 779.
94. C. Liang, S. Diao, C. Wang, H. Gong, T. Liu, G. Hong, X. Shi, H. Dai and Z. Liu, *Adv. Mater.*, 2014, **26**, 5646.
95. A. K. Geim, *Science*, 2009, **324**, 1530.
96. G. Hong, S. Diao, A. L. Antaris and H. Dai, *Chem. Rev.*, 2015, **115**, 10816.
97. K. Yang, S. Zhang, G. Zhang, X. Sun, S.-T. Lee and Z. Liu, *Nano Lett.*, 2010, **10**, 3318.
98. W. Zhang, Z. Guo, D. Huang, Z. Liu, X. Guo and H. Zhong, *Biomaterials*, 2011, **32**, 8555.
99. M. Li, X. Yang, J. Ren, K. Qu and X. Qu, *Adv. Mater.*, 2012, **24**, 1722.
100. O. Akhavan, E. Ghaderi and H. Emamy, *J. Mater. Chem.*, 2012, **22**, 20626.
101. J. T. Robinson, S. M. Tabakman, Y. Liang, H. Wang, H. Sanchez Casalongue, D. Vinh and H. Dai, *J. Am. Chem. Soc.*, 2011, **133**, 6825.
102. K. Yang, J. Wan, S. Zhang, B. Tian, Y. Zhang and Z. Liu, *Biomaterials*, 2012, **33**, 2206.
103. O. Akhavan and E. Ghaderi, *Small*, 2013, **9**, 3593.
104. Y. Zhao, H. Pan, Y. Lou, X. Qiu, J. Zhu and C. Burda, *J. Am. Chem. Soc.*, 2009, **131**, 4253.
105. J. M. Luther, P. K. Jain, T. Ewers and A. P. Alivisatos, *Nat. Mater.*, 2011, **10**, 361.

106. Y. Li, W. Lu, Q. Huang, C. Li and W. Chen, *Nanomedicine*, 2010, **5**, 1161.
107. Q. W. Tian, M. H. Tang, Y. G. Sun, R. J. Zou, Z. G. Chen, M. F. Zhu, S. P. Yang, J. L. Wang, J. H. Wang and J. Q. Hu, *Adv. Mater.*, 2011, **23**, 3542.
108. Q. W. Tian, F. R. Jiang, R. J. Zou, Q. Liu, Z. G. Chen, M. F. Zhu, S. P. Yang, J. L. Wang, J. H. Wang and J. Q. Hu, *ACS Nano*, 2011, **5**, 9761.
109. Q. Tian, J. Hu, Y. Zhu, R. Zou, Z. Chen, S. Yang, R. Li, Q. Su, Y. Han and X. Liu, *J. Am. Chem. Soc.*, 2013, **135**, 8571.
110. C. M. Hessel, V. P. Pattani, M. Rasch, M. G. Panthani, B. Koo, J. W. Tunnell and B. A. Korgel, *Nano Lett.*, 2011, **11**, 2560.
111. W. Li, R. Zamani, P. Rivera Gil, B. Pelaz, M. Ibáñez, D. Cadavid, A. Shavel, R. A. Alvarez-Puebla, W. J. Parak, J. Arbiol and A. Cabot, *J. Am. Chem. Soc.*, 2013, **135**, 7098.
112. K. Manthiram and A. P. Alivisatos, *J. Am. Chem. Soc.*, 2012, **134**, 3995.
113. Z. Chen, Q. Wang, H. Wang, L. Zhang, G. Song, L. Song, J. Hu, H. Wang, J. Liu, M. Zhu and D. Zhao, *Adv. Mater.*, 2013, **25**, 2095.
114. W. Xu, Z. Meng, N. Yu, Z. Chen, B. Sun, X. Jiang and M. Zhu, *RSC Adv.*, 2015, **5**, 7074.
115. L. Cheng, J. Liu, X. Gu, H. Gong, X. Shi, T. Liu, C. Wang, X. Wang, G. Liu, H. Xing, W. Bu, B. Sun and Z. Liu, *Adv. Mater.*, 2014, **26**, 1886.
116. S. S. Chou, B. Kaehr, J. Kim, B. M. Foley, M. De, P. E. Hopkins, J. Huang, C. J. Brinker and V. P. Dravid, *Angew. Chem.*, 2013, **125**, 4254.
117. J. Liu, X. Zheng, L. Yan, L. Zhou, G. Tian, W. Yin, L. Wang, Y. Liu, Z. Hu, Z. Gu, C. Chen and Y. Zhao, *ACS Nano*, 2015, **9**, 696.
118. C. Wu, C. Yu and M. Chu, *Int. J. Nanomed.*, 2011, **6**, 807.
119. T. Liu, C. Wang, X. Gu, H. Gong, L. Cheng, X. Shi, L. Feng, B. Sun and Z. Liu, *Adv. Mater.*, 2014, **26**, 3433.
120. G. Song, Q. Wang, Y. Wang, G. Lv, C. Li, R. Zou, Z. Chen, Z. Qin, K. Huo, R. Hu and J. Hu, *Adv. Funct. Mater.*, 2013, **23**, 4281.
121. J.-W. Kim, E. I. Galanzha, E. V. Shashkov, H.-M. Moon and V. P. Zharov, *Nat. Nano*, 2009, **4**, 688.
122. X. Wang, C. Wang, L. Cheng, S.-T. Lee and Z. Liu, *J. Am. Chem. Soc.*, 2012, **134**, 7414.
123. C. Wang, J. Li, C. Amatore, Y. Chen, H. Jiang and X.-M. Wang, *Angew. Chem., Int. Ed.*, 2011, **50**, 11644.
124. S.-H. Hu, Y.-W. Chen, W.-T. Hung, I. W. Chen and S.-Y. Chen, *Adv. Mater.*, 2012, **24**, 1748.
125. K. Yang, L. Hu, X. Ma, S. Ye, L. Cheng, X. Shi, C. Li, Y. Li and Z. Liu, *Adv. Mater.*, 2012, **24**, 1868.
126. B. Li, K. Ye, Y. Zhang, J. Qin, R. Zou, K. Xu, X. Huang, Z. Xiao, W. Zhang, X. Lu and J. Hu, *Adv. Mater.*, 2015, **27**, 1339.

CHAPTER 10

Near Infrared-Triggered Synergetic Cancer Therapy Using Multifunctional Nanotheranostics

JIA-NAN LIU[a], JIAN-LIN SHI[a], AND WEN-BO BU*[a]

[a]State Key Laboratory of High Performance Ceramics and Superfine Microstructures, Shanghai Institute of Ceramics, Chinese Academy of Sciences, Shanghai, 200050, P. R. China
*E-mail: wbbu@mail.sic.ac.cn

10.1 Introduction

Cancer remains one of the world's most devastating diseases, with more than 10 million new cases every year. However, mortality has decreased in recent years thanks to improved diagnostic devices and treatment strategies. Current cancer treatments include surgical intervention, radiation, chemotherapeutic drugs, photodynamic therapy (PDT),[1] and thermotherapy.[2] Though proven to be effective, these strategies often also kill healthy cells and thus cause undesired toxicity to the patient. Recent achievements have provided perspective on the use of nanotechnology as a fundamental tool in cancer research and nanomedicine.[3] After either passively or actively targeting

RSC Nanoscience & Nanotechnology No. 40
Near Infrared Nanomaterials: Preparation, Bioimaging and Therapy Applications
Edited by Fan Zhang
© The Royal Society of Chemistry 2016
Published by the Royal Society of Chemistry, www.rsc.org

cancerous cells, these systems can specifically enhance the accumulation of antitumor agents at desired tumor sites to improve the therapeutic efficacy with minimized side effects.[4]

Although promising, cancer therapy relying on a single therapeutic strategy generally remains suboptimal because each therapeutic approach has its own advantages and limitations. There is almost no doubt that future cancer therapies will very likely rely on the combination of different treatment approaches, which may include surgery, chemotherapy, radiotherapy (RT), PDT, and photothermal therapy (PTT).[5] Such combined therapies with different therapeutic mechanisms have been considered as a promising and compelling alternative strategy to improve therapeutic efficiency and minimize side effects.[6] It has been suggested that, due to the differences in cell-killing mechanisms, synergetic tumor responses may be achieved if the two modalities are combined in appropriate sequence. Therefore, multifunctional nanocarriers that enable cancer combination therapy with different therapeutic mechanisms in one system may play increasingly important roles in the fight against cancer due to their unique advantages such as minimal side effects and high efficacy.

On-demand cancer therapy is becoming feasible through the design of systems that can deliver toxic agents to living organisms in response to specific stimuli, either endogenous (changes in pH, enzyme concentration, or redox gradients) or exogenous (variations in temperature, magnetic field, ultrasound intensity, light, or electric pulses).[7] Of these, exogenous stimuli are more promising because endogenous triggers like pH may vary from person to person. Owing to their non-invasiveness and the possibility of remote spatiotemporal control, a large variety of photoresponsive systems have been engineered in the past few years to achieve on-demand cancer therapy in response to illumination of a specific wavelength (in the ultraviolet (UV), visible, or near infrared (NIR) regions).[8] However, activation by UV light is not only primarily limited to regions of the body that can be directly illuminated (*i.e.*, the teeth, skin, or eyes), but can also promote angiotropism and metastasis.[9] To expand the scope of tissues that can be accessed by light, either photoresponsive moieties that respond to longer wavelengths of light or two-photon excitation is preferred in clinical applications.[10] This makes NIR-responsive nanotheranostics which can trigger a synergistic effect extremely promising for clinical applications in cancer therapy.

In this review, we discuss the most significant progress made in the past 5 years in the field of NIR-responsive nanotheranostics for synergetic cancer therapy. The nanoplatforms discussed include mesoporous silica nanoparticles (mSiO$_2$ NPs),[11] upconversion nanoparticles (UCNPs),[12] Au-related nanomaterials,[13] Fe$_3$O$_4$-related systems,[14] and carbon-related materials. The aim of this review is to provide a summary of various nanoplatforms, their surface modifications, and their efficiency in producing a synergetic therapeutic effect to offer a platform for fresh insight and perspective on the field.

10.2   NIR-Triggered Drug Delivery

Drug delivery systems, which can deliver precise quantities of therapeutic drugs to the targeted cells or tissues in a tailored release manner to enhance drug efficiency and reduce toxicity to the normal tissues, have been one of the most promising applications for human health care and represent a promising field for biomedical materials science.[15] An ideal drug carrier for cancer therapy should encapsulate anticancer drugs with high loading amount and efficiency, and have a "zero release" effect prior to reaching the targeted cells to protect healthy organs from the toxic drugs and prevent the decomposition/denaturing of the drugs.[16] External stimuli such as light, magnetic field, and ultrasound have been widely used for remotely controlled cancer therapy.[17] Among them, light-responsive nanoplatforms based on photoactivatable molecules, photolysis linkers or photo-induced therapies have proved to be one of the most elegant strategies due to the advantages in spatial and temporal control of drug release.[18] In particular, NIR (λ = 700–1100 nm) light has recently become an attractive stimulus because of its minimal absorbance by tissue, allowing for non-invasive and deep tissue penetration.

To achieve NIR-triggered drug delivery, Xing *et al.*[19] and Lin *et al.*[20] use UCNPs to convert the absorbed NIR light into UV to activate the *trans*-platinum(IV) prodrug. Very recently, our group developed a NIR light-triggered anticancer drug release system by integrating $NaYF_4$:Yb/Tm UCNP@$mSiO_2$ and photoresponsive azobenzene molecules into one system (Figure 10.1a).[21] Upon NIR light irradiation, UCNPs emit upconverted UV and visible luminescence, which matches well with the absorption of azobenzene molecules, thereby leading to the continuous rotation–inversion movements of the linked photoresponsive azobenzene molecules in the mesopore channels (Figure 10.1b). These back-and-forth wagging motions of the azobenzene molecules propel the release of the anticancer drugs. The precise control of drug release can be achieved by varying the dosage of NIR light.

Most recently, combined therapy with dual drugs of different therapeutic effects shows excellent performance in the treatment of diseases, especially for drug-resistant cancer.[22] To realize the best therapeutic effect, the species and doses of drugs should be optimized at different clinical manifestations and periods in the treatment. One of the main challenges of combined therapy is to control the release of each drug independently. To solve this problem, Zhao *et al.* developed and synthesized a multifunctional dual-compartment Janus mesoporous silica nanocomposites of UCNP@SiO_2@$mSiO_2$&PMO (PMO = periodic mesoporous organosilica) *via* an anisotropic island nucleation and growth approach with the ordered mesostructure.[23] As shown in Figure 10.1c, each Janus nanocomposite consists of four parts: (1) the NIR-to-UV-vis UCNP core, (2) concentric condensed SiO_2 shells, (3) ordered $mSiO_2$ shells with radial mesopore channels, and (4) the single-crystal PMO nanocubes formed by the novel anisotropic growth approach of the ordered mesostructures on the surface of the core@shell@shell UCNP@SiO_2@$mSiO_2$ NPs. These Janus nanocomposites possess unique dual independent mesopores

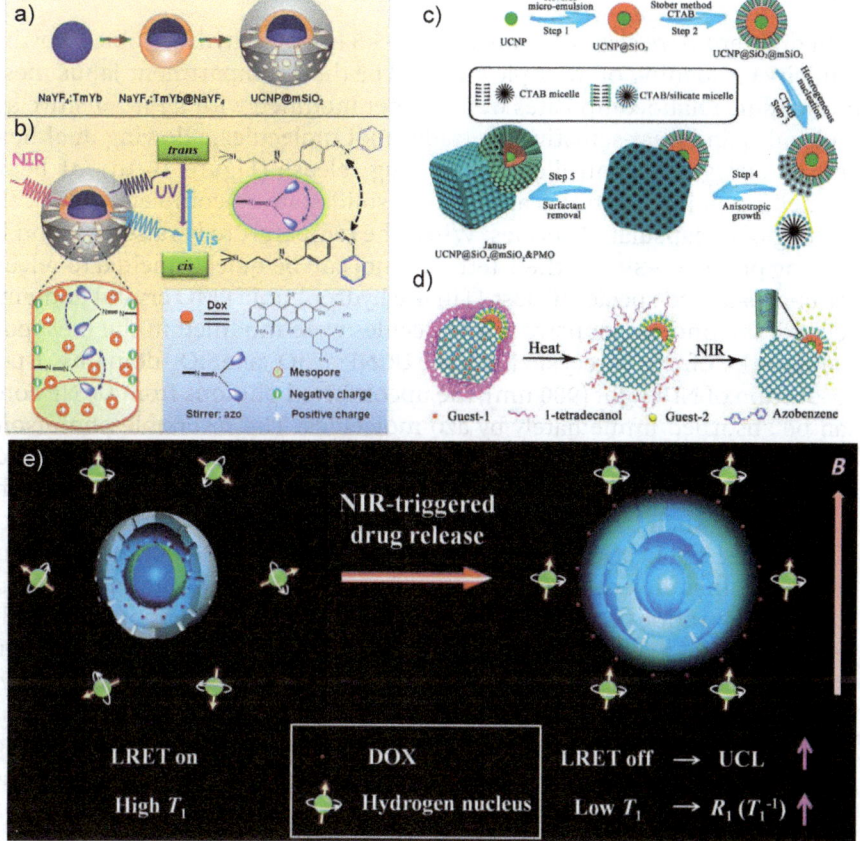

Figure 10.1 (a) Synthetic procedure for upconversion nanoparticles (UCNPs) coated with a mesoporous silica outer layer. (b) NIR light-triggered DOX release by making use of the upconversion property of UCNPs and *trans-cis* photoisomerization of azo molecules grafted into the mesopore network of a mesoporous silica layer. (Reproduced from ref. 21 with permission from John Wiley and Sons. Copyright © 2013 Wiley-VCH Verlag GmbH & Co. KGaA, Weinheim.) (c) Synthetic procedure for the dual-compartment Janus mesoporous silica nanocomposites UCNP@SiO$_2$@mSiO$_2$&PMO by the anisotropic island nucleation and growth method. UCNP = NaGdF$_4$:Yb,Tm@NaGdF$_4$, mSiO$_2$ = mesoporous silica shell, PMO = periodic mesoporous organosilica. (d) Schematic presentation for dual-control drug release systems by using the dual-compartment mesoporous Janus nanocomposites. (Reproduced with permission from X. Li, L. Zhou, Y. Wei, A. M. El-Toni, F. Zhang and D. Zhao, *J. Am. Chem. Soc.*, 2014, **136**, 15086. Copyright (2014) American Chemical Society.[23]) (e) With anticancer drugs (DOX) fully loaded into the mesopores and hollow cavity, UCL signals are quenched through LRET between UCNPs and DOX under 980 nm excitation; the T1-MR signals are almost undetectable due to the low probability of water molecules bonding to the Gd^{3+} ions. Upon NIR-triggered drug release from the nanosensors, both the UCL and T1-MR signals will be restored accordingly. As a result, drug release can be detected by the designed UCL/T1-MRI dual-mode nanosensor. (Reproduced from ref. 24 with permission from John Wiley and Sons. Copyright © 2014 Wiley-VCH Verlag GmbH & Co. KGaA, Weinheim.)

with different pore sizes (2.1 nm and 3.5–5.5 nm) and hydrophobicity/hydro-philicity for loading of multiple guests. The dual-compartment Janus meso-porous silica nanocomposites were further modified with light-sensitive azo molecules and heat-sensitive 1-tetradecanol molecules, allowing dual heat- and NIR-triggered controllable dual drug release (Figure 10.1d). At below its melting point, 1-tetradecanol is in a solid state to completely block the passing of encapsulated species. When the temperature is raised beyond its melting point (38–39 °C), the 1-tetradecanol can be quickly melted to release the encapsulated species (Guest 1) in the hydrophobic PMO crystal domains. Second, the photoresponsive azo molecules were modified in the mesopore frameworks of the core@shell@shell UCNP@SiO_2@$mSiO_2$ domains. Upon absorption of NIR light (980 nm), the upconverted photons from UCNP cores can be absorbed immediately by azo molecules. The reversible photoisom-erization of azo will create continuous rotation–inversion movement, which result in the release of Guest 2, located in the mesopore frameworks of the core@shell@shell UCNP@SiO_2@$mSiO_2$ domains.

In clinical applications, one of the main causes of chemotherapy failure lies in the insufficient or excess drug dosages due to the unknown actual drug concentrations at the focus. In this view, the real-time monitoring of the release of anticancer drugs from drug delivery systems is very important. Our group proposed a UCNP@$hmSiO_2$ nanocomposite and developed a novel concept that monitored NIR-triggered drug release *in vitro* and *in vivo* using simultaneous upconverted luminescence (UCL) and magnetic resonance imaging (MRI) in real time.[24] As shown in Figure 10.1e, the developed nano-sensors contain three essential components: (1) to control the drug release by NIR, the mesopore surfaces of UCNP@$hmSiO_2$ were modified with azo molecules; (2) luminescent resonance energy transfer (LRET) system consist-ing of UCNPs (donor) and DOX (doxorubicin, acceptor) was designed to mon-itor NIR-controlled drug release in real time by detecting UCL signals; (3) the amount of water molecules in the cavity and their bonding probability to the Gd^{3+} ions, and thus the T1-MRI contrast, depends strongly on the amount of the anticancer drug residing in the hollow cavity. Along with the release of DOX molecules, the amount of DOX in the hollow cavity was decreased, which results in the increased probability of water molecule bonding to Gd^{3+} ions, consequently a shortened T1 relaxation time, and finally the increased r_1 value of the nanosensors. Such a monitoring strategy features the high sensitivity of UCL, and the high resolution, non-invasiveness, and tissue depth-independence of MRI. This dual-mode real-time and quantitative monitoring of the drug release can be applied *in vivo* to determine the drug concentrations online in the tissue regions of interest.

10.3 Combined Chemotherapy with PTT

If PTT is combined with chemotherapy, the therapeutic efficacy is expected to be improved since the cytotoxicity of some chemotherapeutic agents is enhanced at elevated temperatures.[25] In addition, the undesired cytotoxicity of free anticancer drugs to the normal cells can be minimized when drugs can be released from carriers triggered by heat at regions of interest. Based on the attractive photothermal property of the nanomaterials to optimize cancer therapy and achieve enhanced antitumor efficacy, the combination of hyperthermia and chemotherapeutic agents is an encouraging approach, which can result in synergistic effects that are greater than either treatment alone.[26] In the past decade, thermoresponsive drug delivery systems have been widely explored in oncology.[27] Some inorganic nanomaterials with a high absorption cross-section for conversion of an extrinsic energy source (*e.g.*, magnetic field, laser, radiofrequency, microwaves, ultrasound) into heat are being intensively investigated for tumor targeted, minimally invasive, and uniform hyperthermia. In the present review, we focus on NIR-triggered thermoresponsive drug delivery systems.

10.3.1 Plasmonic Nanoparticles

Plasmonic gold nanoparticles (AuNPs) exhibit unique size- and shape-dependent optical and photothermal properties due to localized surface plasmon resonance (LSPR), an electromagnetic mode associated with the collective oscillation of free electrons in conduction bands. Following excitation of LSPR by a NIR laser, photon–electron and electron–electron interactions generate heat, which can be used for hyperthermia therapy, or to trigger thermosensitive release in drug delivery systems. The synthesis of AuNPs with controlled size and morphology has led to the production of various nanostructures such as Au nanoshells,[28] Au nanorods (AuNRs), hollow Au nanospheres, and Au nanocages (AuNCs), which process well-defined surface plasmon resonance (SPR) absorption features that can be finely tuned from the visible to the NIR region.

Among various Au nanostructures, AuNRs that can be easily synthesized by seeded growth methods have been extensively explored in PTT due to their strong optical extinction in the visible and NIR region.[29] Our group designed AuNRs-capped magnetic core/mesoporous silica shell nanoellipsoids by coating a uniform layer of AuNRs on the outer surface of a magnetic core/mesoporous silica shell nanostructure, based on a two-step chemical self-assembly process (Figure 10.2a–f).[30] This multifunctional nanocomposite integrated simultaneous chemotherapy, photothermotherapy, *in vivo* MRI, infrared thermal and optical imaging into one single system. Importantly, the prepared multifunctional nanoellipsoids showed high DOX loading capacity and pH value-responsive release mainly due to the electrostatic interaction between DOX molecules and mesoporous silica surface. In addition, a synergistic effect of combined chemo- and photo-thermotherapy was found at moderate power intensity of NIR irradiation based on the DOX release and the photothermal effect of AuNRs.

AuNCs represent another novel class of nanostructures with hollow interiors and porous walls, which were first developed by Xia and coworkers.[31] They can have strong absorption (for the photothermal effect) in the NIR

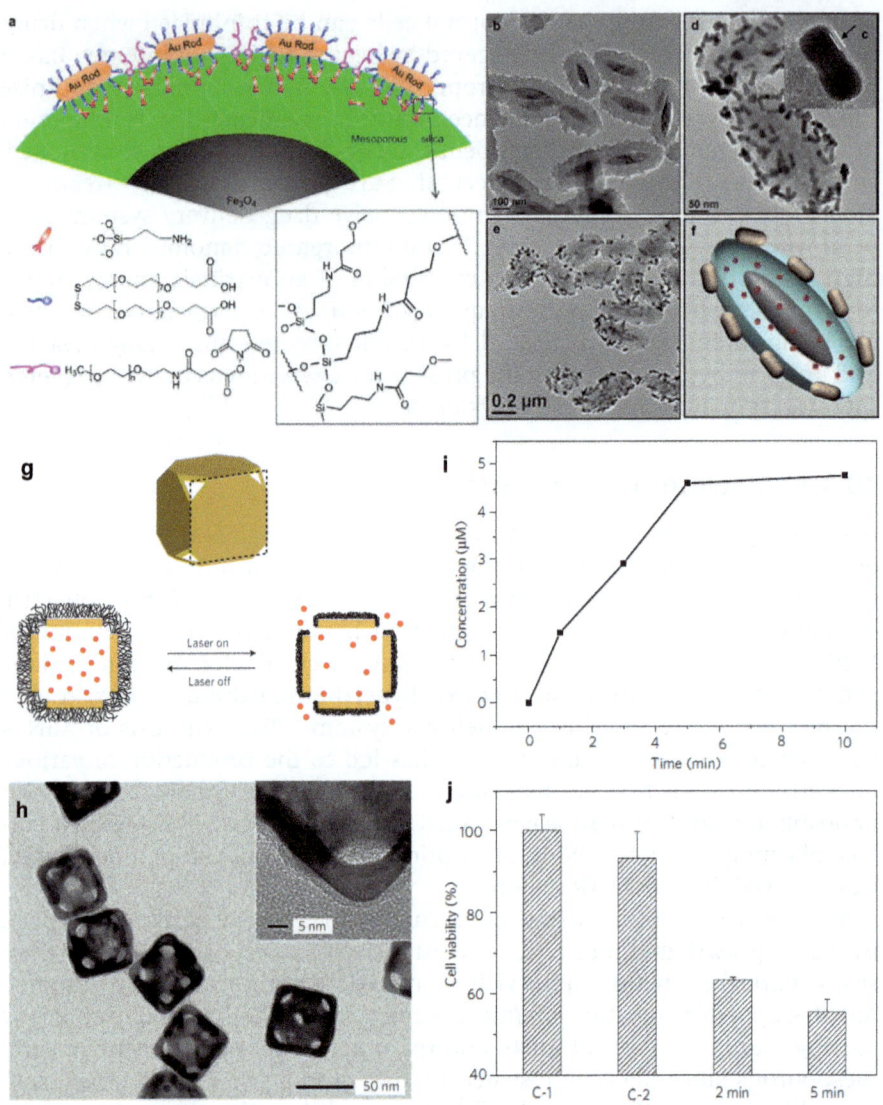

Figure 10.2 (a) Schematic microscopic structure of AuNRs–MMSNEs:Numbers of AuNRs being capped on the outer surface of a magnetite core/mesoporous silica shell nanoellipsoid *via* covalent bonding; (b–f) TEM micrographs of (b) $Fe_2O_3@SiO_2@mSiO_2$, (c–e) AuNRs–MMSNEs and (f) the designed structure model. The arrow in (c) shows the PEG layer on the surface of an AuNR. (Reproduced from M. Ma *et al.*, Au capped magnetic core/mesoporous silica shell nanoparticles for combined photothermo-/chemo-therapy and multimodal imaging, *Biomaterials*, 2012, **33**, 989–998, Copyright (2012) with permission from Elsevier.[30]) (g) Schematic illustrating how the system works. A side view of the Au nanocage is used for the illustration. On exposure to a NIR laser, the light is absorbed by the nanocage and converted into heat, triggering the smart polymer to collapse and thus release the preloaded

while maintaining a compact size. When the surface of an AuNC is covered with a smart polymer, the preloaded effector can be released in a controllable fashion using a NIR laser.[32] This system works well with various effectors without involving sophisticated syntheses, and is well suited for *in vivo* studies owing to the high transparency of soft tissue in the NIR region (Figure 10.2g–j).

10.3.2 Carbon Nanomaterials

Typical sp^2 carbon nanomaterials such as carbon nanotubes (CNTs)[33] and graphenes[34] are also NIR-responsive photothermal agents for cancer PTT treatment *in vitro* and *in vivo* due to their strong light absorbance in the NIR wavelength window in the 700–1300 nm range. On a per mass basis, CNTs and graphenes both exhibit a larger extinction coefficient of NIR light absorption than AuNRs, and consequently higher photothermal conversion efficiency. The ultrahigh surface area of carbon nanomaterials also permits efficient loading of multiple molecules through covalent conjugation. Hence, Guo *et al.* first developed doxorubicin-loaded PEGylated nanographene oxide to facilitate combined chemotherapy and PTT in one system.[35]

However, graphene nanosheet mainly absorbs insoluble molecules *via* non-covalent binding such as π–π interaction, with low drug loading efficiency. As is widely known, mSiO$_2$ NPs themselves are good vectors for insoluble chemotherapeutic drugs.[11,36] Recently, a mSiO$_2$-coated graphene nanosheet was successfully synthesized by Huang *et al.* for the combined therapy of glioma (Figure 10.3).[37] A vertical coating of mSiO$_2$ on a graphene nanosheet could improve the interfacial properties of graphene and integrate the advantages of both materials as drug delivery vectors. These advantages include (1) adsorption of the antitumor drug DOX with a high loading efficiency for chemotherapy *via* both π–π stacking and pore adsorption, (2) high absorption in the NIR window and efficient heat transformation for PTT, and (3) being more easily covalently functionalized.

---

effector. When the laser is turned off, the polymer chains will relax back to the extended conformation and terminate the release. (h) TEM images of Au nanocages for which the surface was covered by a pNIPAAm–co-pAAm copolymer with an LSCT at 39 °C. The inset shows a magnified TEM image of the corner of such a nanocage. (i) A plot of the concentrations of DOX released from the Au nanocages on heating at 45 °C for different periods of time. (j) Cell viability for samples after going through different treatments: (C-1) cells irradiated with a pulsed NIR laser for 2 min in the absence of Au nanocages; (C-2) cells irradiated with the laser for 2 min in the presence of DOX-free Au nanocages; and (2/5 min) cells irradiated with the laser for 2 and 5 min in the presence of DOX-loaded Au nanocages. A power density of 20 mW cm^{-2} was used for all of these studies. (Reproduced with permission by permission from Macmillan Publishers Ltd.:*Nat. Mater.*,[32] Copyright (2009).)

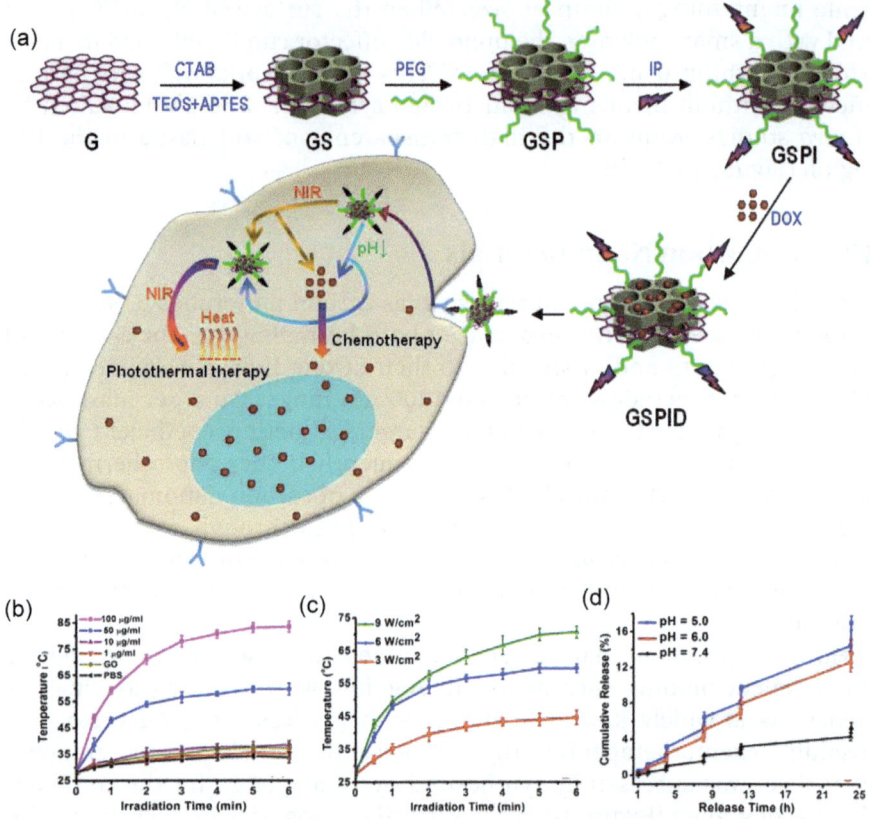

Figure 10.3   (a) Design of GSPID as a multifunctional drug delivery system for combined chemo-photothermal targeted therapy of glioma. (b) Photothermal heating curves of GSPI solution at various GS concentrations at power intensity of 6 W cm^{-2}. The GO solution has the same graphene concentration to GSPI solution with GS concentration at 50 µg mL^{-1}. (c) Photothermal heating curves of GSPI at various power intensities with GS concentration at 50 µg mL^{-1}. (d) Cumulative release profiles of DOX from GSPID at different pHs with 6 W cm^{-2} NIR irradiation. Data are expressed as mean ± SEM (n = 3). (Reproduced with permission from Y. Wang, K. Wang, J. Zhao, X. Liu, J. Bu, X. Yan and R. Huang, *J. Am. Chem. Soc.*, 2013, **135**, 4799. Copyright (2013) American Chemical Society.[37])

10.3.3   Other Inorganic Nanomaterials

10.3.3.1   *Single-Layer MoS$_2$*

Single-layer MoS$_2$, as one of the typical layered metal dichalcogenides, has been demonstrated as a novel NIR absorbing agent, which showed higher absorbance in the NIR region than either graphene or AuNRs.[38] Zhao *et al.* report a simple, high-yield yet low-cost approach to the design of single-layer MoS$_2$ nanosheets with controllable size *via* an improved oleum treatment exfoliation process (Figure 10.4).[39] MoS$_2$ nanosheets functionalized by

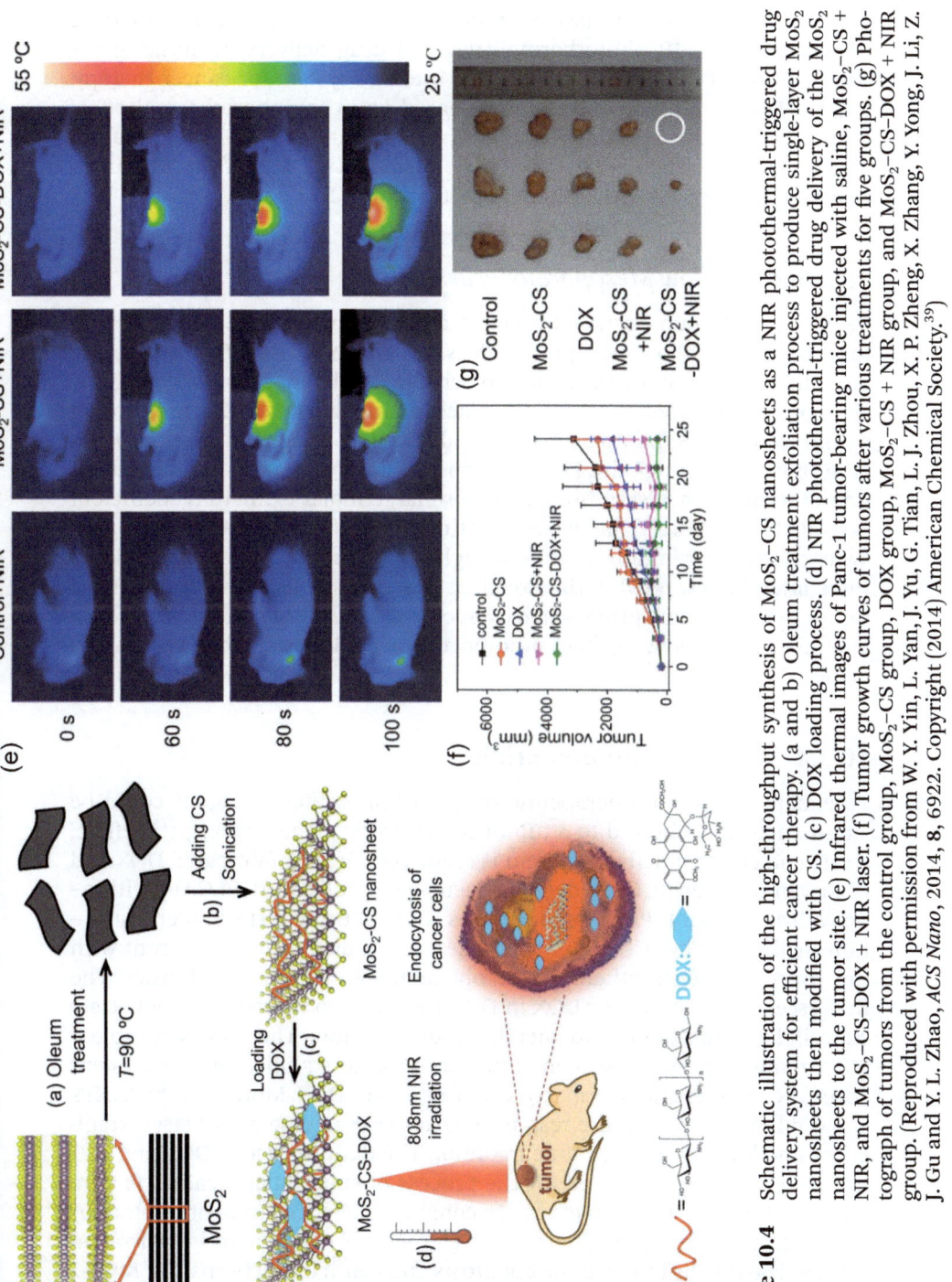

Figure 10.4 Schematic illustration of the high-throughput synthesis of MoS$_2$–CS nanosheets as a NIR photothermal-triggered drug delivery system for efficient cancer therapy. (a and b) Oleum treatment exfoliation process to produce single-layer MoS$_2$ nanosheets then modified with CS. (c) DOX loading process. (d) NIR photothermal-triggered drug delivery of the MoS$_2$ nanosheets to the tumor site. (e) Infrared thermal images of Panc-1 tumor-bearing mice injected with saline, MoS$_2$–CS + NIR, and MoS$_2$–CS-DOX + NIR laser. (f) Tumor growth curves of tumors after various treatments for five groups. (g) Photograph of tumors from the control group, MoS$_2$–CS group, DOX group, MoS$_2$–CS + NIR group, and MoS$_2$–CS-DOX + NIR group. (Reproduced with permission from W. Y. Yin, L. Yan, J. Yu, G. Tian, L. J. Zhou, X. P. Zheng, X. Zhang, Y. Yong, J. Li, Z. J. Gu and Y. L. Zhao, *ACS Nano*, 2014, **8**, 6922. Copyright (2014) American Chemical Society.[39])

decorating with chitosan have been developed as a chemotherapeutic drug nanocarrier for NIR photothermal-triggered drug delivery, facilitating the combination of chemotherapy and PTT into one system for cancer therapy. Loaded DOX could be controllably released upon the photothermal effect induced by 808 nm NIR laser irradiation. *In vitro* and *in vivo* tumor ablation studies demonstrate a better synergistic therapeutic effect of the combined treatment, compared with either chemotherapy or PTT alone.

10.3.3.2 *Hollow Mesoporous Prussian Blue Nanoparticles*

Recently, our group designed a smart and versatile nanotheranostic platform based on hollow mesoporous Prussian blue NPs (HMPBs) with perfluoropentane (PFP) and DOX inside to achieve the distinct *in vivo* synergistic chemothermal tumor therapy and synchronous diagnosis and monitoring by ultrasound (US)/photoacoustic (PA) dual-mode imaging.[40] The prepared HMPBs show ultrahigh drug loading capacity up to 1782 mg g^{-1} and excellent photothermal conversion properties with large molar extinction coefficient ($\sim 1.2 \times 10^{11}$ M^{-1} cm^{-1}). The *in vivo* experiments confirmed that the measured synergistic tumor inhibition ratio is higher than the theoretically calculated value after 5 days, proving the excellent synergistic effect of chemothermal tumor therapy, which promises to overcome the inevitable tumor recurrence and metastasis resulting from inhomogeneous ablation by thermal therapy alone.

10.3.4 Organic Nanomaterials

To ensure that chemotherapeutic drug and photothermal agent could be simultaneously delivered to a tumor region to exert their synergistic effect, a safe and efficient delivery system is required. Cai *et al.* fabricated DOX and indocyanine green (ICG) loaded poly(lactic-co-glycolic acid) (PLGA)–lecithin–polyethylene glycol (PEG) nanoparticles (DINPs) using a single-step sonication method.[41] The ICG (DOX) could serve as a dual-functional agent with integrated PTT (chemotherapy) and optical imaging probe capabilities. The fluorescence (FL) of ICG or DOX in DINPs can be monitored to demonstrate subcellular localization and metabolic distribution. The DINPs exhibited good monodispersity, excellent fluorescence/size stability, and consistent spectral characteristics compared with free ICG or DOX. Moreover, the DINPs showed higher temperature response, faster DOX release under laser irradiation, and longer retention time in tumor. The fluorescence of DOX and ICG in DINPs was also visualized for the process of subcellular location *in vitro* and metabolic distribution *in vivo*. In comparison with chemo- or photothermal treatment alone, the combined treatment of DINPs with laser irradiation synergistically induced the apoptosis and death of DOX-sensitive MCF-7 and DOX-resistant MCF-7/ADR cells, and suppressed MCF-7 and MCF-7/ADR tumor growth *in vivo*. Notably, no tumor recurrence was observed after only a single dose of DINPs with laser irradiation. Hence, the well-defined DINPs exhibited great potential in targeting cancer imaging and chemotherapy-PTT.

10.4   Combined Chemotherapy with PDT

The combination of PDT with chemotherapy is helpful in overcoming limitations encountered by each therapy when used alone. Such combination has the potential to induce antitumor immunity[6] or circumvent multidrug resistance.[42]

Based on this background, Koh *et al.* developed a new type of hollow nanocapsule for use in combining PDT with chemotherapy.[43] Figure 10.5a shows the preparation of hollow nanocapsule by alternating deposition of poly(allylamine hydrochloride) (PAH) and dendritic porphyrin (DP) onto a negatively charged polystyrene nanoparticle (PSNP) and the subsequent removal of the template PSNP. The (PAH/DP)$_n$ multilayer nanocapsules were filled with DOX in order to implement chemotherapy. Moreover, when the DP forms self-assembled nanostructures such as polymeric micelles, large numbers of DPs can effectively generate a high concentration of singlet oxygen at local sites in order to overcome the threshold concentration for oxidative damage.

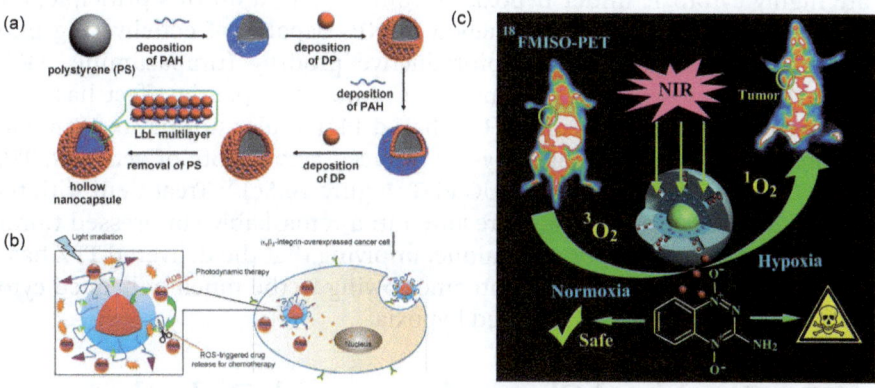

Figure 10.5   (a) Procedure for the preparation of multilayer hollow nanocapsules. (Reproduced from ref. 43 with permission from John Wiley and Sons. Copyright © 2011 Wiley-VCH Verlag GmbH & Co. KGaA, Weinheim.) (b) Schematic illustration of the light-regulated ROS-activated on-demand drug release and the combined chemo-photodynamic therapy. (Reproduced from ref. 44 with permission from John Wiley and Sons. Copyright © 2014 Wiley-VCH Verlag GmbH & Co. KGaA, Weinheim.) (c) Strong hypoxia created by upconversion photodynamic therapy (UC-PDT) activates bioreductive prodrugs codelivered to form cytotoxic species, thereby potentiating the synergetic anticancer efficacy of UC-PDT. This process was accomplished by using upconversion-based nanoparticles designed to simultaneously deliver photosensitizer molecules and bioreductive prodrugs in silica layers. (Reproduced from ref. 45 with permission from John Wiley and Sons. Copyright © 2015 Wiley-VCH Verlga GmbH & Co. KGaA, Weinheim.)

While most nanocapsule (NC) shells used in drug delivery systems are pre-pared from linear polyelectrolytes that lack any function other than that of drug container, this system employs DP not only as a polyelectrolyte for the formation of NC shells but also as photosensitizing units for photodynamic therapy. Cell viability studies showed that combined treatment resulted in higher toxicity than either chemotherapy or PDT alone.

As mentioned above, light-responsive nanocarriers are much preferred because they provide an orthogonal external stimulus that endows the advan-tages of a non-contact mode and precise controllability, as well as the con-trolled release of encapsulated substances both spatially and temporally. Liu *et al.* developed a nanoplatform that combines PDT and chemotherapy with on-demand drug release regulated by one light source (Figure 10.5b).[44] It is widely known that PDT relies on the generation of reactive oxygen species (ROS) to destroy cancer cells upon illumination by photosensitizers. Upon illumination, such a system makes use of PDT-generated ROS to cleave the linker between the PS and the drug.

Typically, PDT processes consume a large amount of oxygen, resulting in local hypoxia in tumors. This is an undesirable consequence of PDT which leads to greatly reduced effectiveness of this therapeutic protocol. Fortu-nately, bioreductive prodrugs that can be activated at low-oxygen conditions are highly cytotoxic under hypoxia in tumors. Based on this principle, our group designed double silica-shelled UCNPs capable of codelivering pho-tosensitizer molecules and a bioreductive prodrug (tirapazamine, TPZ). With these nanostructures a synergetic tumor therapeutic effect has been achieved first by upconversion (UC)-based PDT under normal oxygen con-ditions, immediately followed by the induced cytotoxicity of activated TPZ when oxygen is depleted by the UC-PDT (Figure 10.5c).[45] Treatment with the designed agents plus NIR laser resulted in a remarkably suppressed tumor growth as compared to UC-PDT alone, implying that the delivered TPZ has a profound effect on treatment outcomes owing to the much enhanced cyto-toxicity of TPZ under PDT-induced hypoxia.

10.5   Combined Chemotherapy with Radiotherapy

Although combining chemotherapy with other therapy modalities (ther-motherapy or PDT) under NIR excitation is promising, it has still failed to make satisfactory progress against deep-seated cancers. Fortunately, RT, a conventional and commonly used treatment modality which employs ioniz-ing radiation to control or kill malignant cells, is virtually immune from the restriction of penetration depth.

Tour *et al.* prepared extremely small (<40 nm long and 1 nm wide) hydro-philic carbon clusters (HCCs) that are PEG functionalized (PEG–HCCs), as shown in Figure 10.6a.[46] By simple mixing, this nanovector can be loaded with hydrophobic drugs (paclitaxel, PTX) and functionalized with targeting anti-bodies (cetuximab, Cet), which is an IgG monoclonal antibody that exclusively binds to epidermal growth factor receptor (EGFR) with high affinity and blocks

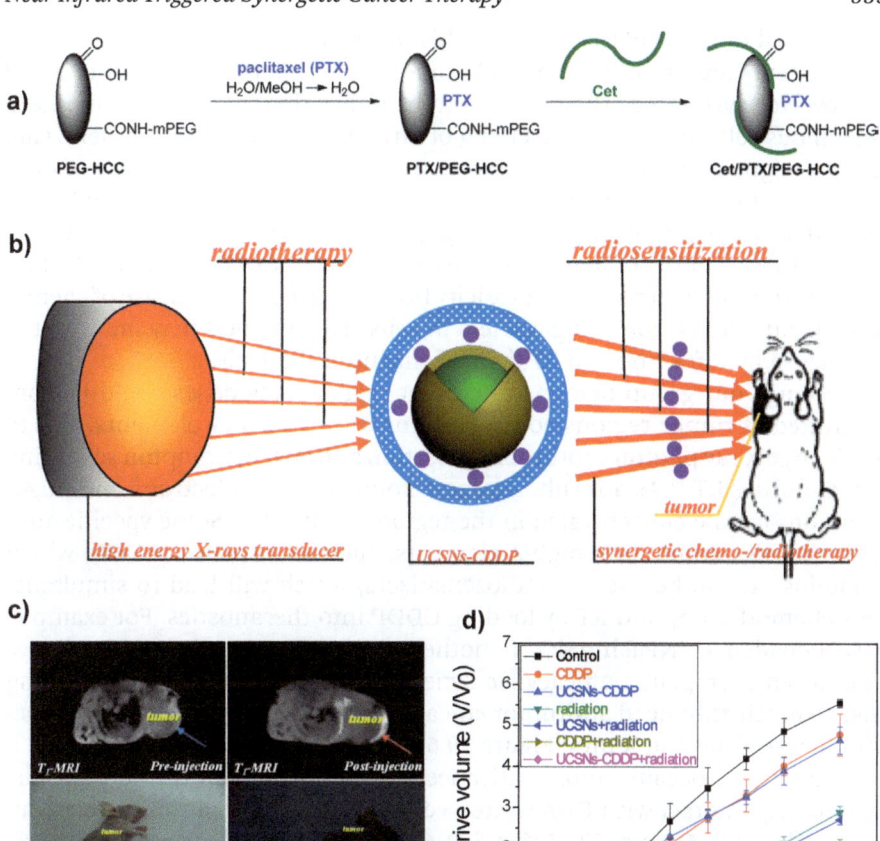

Figure 10.6 (a) Development of Cet/PTX/PEG–HCCs. PEG–HCCs have carbon cores approximately functionalized with various oxygen-containing functional groups. PEG is conjugated to the core *via* an amide bond. By simple mixing protocols, the PEG–HCCs can be loaded with PTX and wrapped with Cet. (Reproduced with permission from D. Sano, J. M. Berlin, T. T. Pham, D. C. Marcano, D. R. Valdecanas, G. Zhou, L. Milas, J. N. Myers and J. M. Tour, *ACS Nano*, 2012, **6**, 2497. Copyright (2012) American Chemical Society.[46]) (b) Schematic illustration of radiosensitization by UCSNs–CDDP. When exposed to high-energy X-ray radiation, CDDP released from UCSNs can enhance the sensitivity of hypoxic tumors to radiation, thus imposing synergetic chemo/radiotherapeutic effects on tumors growth. (c) *In vivo* T1-MR (top) and UCL (bottom) images of a HeLa tumor-bearing mouse before and after the intratumoral injection of UCNP@hmSiO$_2$ (2 mg mL^{-1}, 100 μL). (d) Tumor growth curves of HeLa tumor xenografts following different treatments. (Reproduced with permission from W. Fan, B. Shen, W. Bu, F. Chen, K. Zhao, S. Zhang, L. Zhou, W. Peng, Q. Xiao, H. Xing, J. Liu, D. Ni, Q. He and J. Shi, *J. Am. Chem. Soc.*, 2013, **135**, 6494. Copyright (2013) American Chemical Society.[49])

the normal function of the receptor. This construct is unusual in that all three components are assembled through non-covalent interactions. This targeted nanovector system has the potential to be a new therapy for head and neck squamous cell carcinomas, deserving of further preclinical development. Guo *et al.* designed a special nanocarrier that can allow triggered release of drug molecules in cancer cells under X-ray irradiation.[47] They first prepared DNA-coated AuNPs, which were further conjugated with DOX. Upon hard X-ray radiation, cleavage of DNA can result in the effective release of DOX. This is the first demonstration of increased cytotoxicity from X-ray-triggered release of chemotherapeutic drugs from drug carriers in cells. This method may increase the efficacy of hypofractionated RT that is being used clinically.

Recently, our group demonstrated that higher X-ray doses could be concentrated at tumor regions containing constitutive heavy elements, due to their large X-ray photon capture cross-section and strong Compton scattering effect during RT.[48] As a result, a large amount of photoelectrons and Auger electrons can be concentrated in the regions of interest. Some specific anticancer drugs containing high-Z elements, such as cisplatin (CDDP) which contains Pt, can be used as radiosensitizers, which will lead to simultaneous chemotherapy and RT by loading CDDP into theranostics. For example, CDDP-loaded UCNP@hmSiO$_2$ nanotheranostics were designed to achieve optimized therapeutic efficacy *via* synergetic chemotherapy and RT, giving rise to much enhanced antitumor efficacy with reduced dosages of both chemotherapy drug and X-rays (Figure 10.6b–d).[49]

In addition, because most anticancer drugs must reach the cell nucleus where they interact with DNA to stop cell growth,[50,51] our group further constructed a sub-50 nm UCNP@mSiO$_2$ nanotheranostic system to directly deliver the radiosensitizing drug mitomycin C into the nucleus. The results indicate that the substantially enhanced synergetic chemotherapy and RT can lead to efficient cancer treatment as well as circumvention of multidrug resistance *in vitro* and *in vivo*.[52]

Oxygen deficiency (hypoxia) occurs in the core of most solid tumors and is believed to be one of the major causes for the failure of RT. In order to survive hypoxia, cancer cells develop resistance to proapoptotic signals; these cells become radioresistant and intrinsically prometastatic. To overcome this drawback, our group constructed a core–shell-structured multifunctional nanoradiosensitizer with UCNP as core, mesoporous silica as the shell and a cavity in between. The UCNP core serves as a radiation dose amplifier, the bioreductive prodrug TPZ loaded in the cavity is an hypoxia-selective cytotoxin, and the silica shell provides the protection and diffusion path for TPZ.[53] Such a nanoradiosensitizer has been employed to inhibit the hypoxia, reoxygenation, and subsequent replication of cancer cells that often occurs after a single exposure to low-dose RT alone, and to silence the expression of transcription factors that support the progression of malignancy in cancer. This study confirms the radiotherapeutic benefits of utilizing a nanoradiosensitizer as adjuvant to low-dose RT, and the results demonstrate the highly efficient hypoxia-specific killing in oxygen-dependent antitumor therapies.

10.6   Combined PDT with PTT

PDT and PTT are two major phototherapeutic approaches which require absorption of incoming light by a photosensitizer/reagent to generate respectively ROS and heat for killing cancer cells. To treat deep tissue tumors by phototherapeutic modalities, it is necessary to develop molecules or nanomaterials able to absorb NIR light in the biological windows I (650–950 nm) and II (1000–1350 nm), where the biological components have minimal absorbance.

10.6.1   Use of Two Different Light Sources

The first generation of NP platforms for combined PDT/PTT used two different light sources to excite photosensitizers and photothermal nanomaterials separately, due to their absorption mismatch.

10.6.1.1   Combination of Two Functional Agents into One System

Many research groups used AuNRs along with different kinds of photosensitizers for this combined phototherapy. Yeh and coworkers demonstrated that AuNRs coated with ICG could be used for combined PDT/PTT and biological imaging simultaneously.[54,55] In these two studies, ICG molecules were served as both photosensitizing and NIR-imaging agents. Choi and coworkers reported an AuNR-photosensitizer complex for combined PDT/PTT of cancer.[56] In this work, negatively charged AlPcS4 molecules were attached to the positively charged surface of AuNRs by electrostatic interaction, and the photodynamic effect of the AlPcS4 photosensitizer was temporarily suppressed on the Au surface. Once the photosensitizer was released from the nanocomplex in the intracellular environment, it could finally be optically activated for phototherapeutic effect. Two different light sources were used to separately excite the AuNRs (810 nm laser) and the AlPcS4 photosensitizer (675 nm laser). More recently, AuNRs and photosensitizer were linked using an aptamer switch probe (ASP) in order to achieve combined PDT/PTT.[57] A 812 nm NIR laser and white light were used to activate the AuNRs and the Chlorin e6 (Ce6) photosensitizer, respectively. The ASP was composed of a hairpin-shaped leukemia-specific aptamer with a Ce6 molecule conjugated to the end of the aptamer. When the ASP-modified nanocomplex interacted with the target cancer cells, the ASP allowed the photosensitizer to be released from the surface of the AuNRs, thereby generating both cytotoxic singlet oxygen for PDT and heating for PTT upon dual light irradiation.

Besides Au nanostructures, Liu and coworkers developed polyethylene glycol (PEG)-functionalized nano-graphene oxide (GO) incorporated the Ce6 photosensitizer for synergistic PDT/PTT.[58] Ce6 molecules were loaded on the surface of PEGylated GO *via* supramolecular π–π stacking. The GO–PEG–Ce6 nanocomplex maintained excellent water solubility and efficiently generated singlet oxygen species for PDT. Because of enhanced cellular uptake of the Ce6-loaded nanocomplex compared to free Ce6 molecules, the photodynamic effect was dramatically increased in the tested cancer cells. In this work, the photothermal heating of GO was utilized to further improve intracellular delivery of the nanocomplex, thus enhancing the overall therapeutic effect. The GO and Ce6 in the nanocomplex were irradiated respectively with 808 nm and 660 nm lasers for combined PDT/PTT. However, although such a combined approach using two different light sources showed some improvement in cancer phototherapy, using a single laser source to simultaneously activate two different phototherapies would be a better way to achieve a synergistic phototherapeutic effect.

10.6.1.2   Single Nanoparticles Excited by Two Light Sources

Hwang *et al.* developed PEGylated $W_{18}O_{49}$ nanowires (Figure 10.7a).[59] In this nanosystem, irradiation with 808 nm light results in a pure PTT effect, whereas irradiation with 980 nm light mostly creates a PDT effect, along with

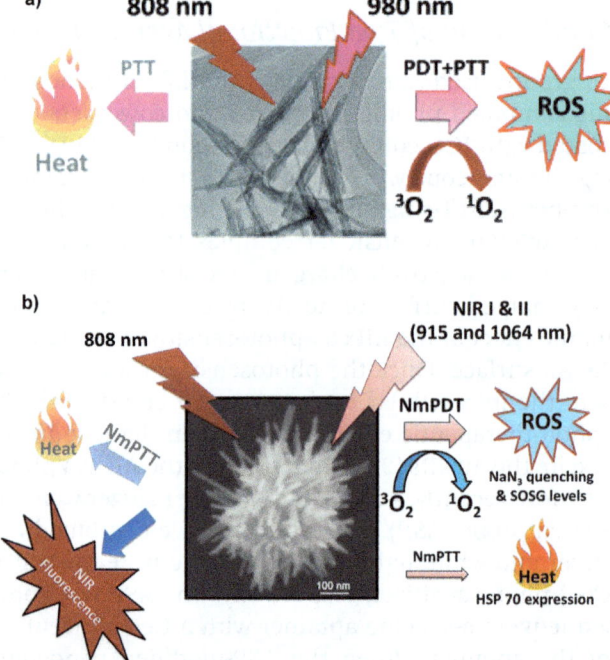

Figure 10.7   (a) Upon excitation with NIR light (980 nm), PEGylated $W_{18}O_{49}$ nanowires can sensitize the formation of singlet oxygen and thus reactive oxygen species (ROS). The resulting photodynamic therapy (PDT) effect can cause the destruction of tumors in the absence of organic photosensitizers. PEG = poly(ethylene glycol). (Reproduced from ref. 59 with permission from John Wiley and Sons. Copyright © 2013 Wiley-VCH Verlag GmbH & Co. KGaA, Weinheim.) (b) Schematic representation of the working mechanisms of NIR-induced fluorescence emission as well as phototherapeutic effects exerted by AuNEs. (Reproduced from ref. 60 with permission from John Wiley and Sons. Copyright © 2014 Wiley-VCH Verlag GmbH & Co. KGaA, Weinheim.)

a very small PTT effect. Under low-power laser irradiation (980 nm, 200 mW cm^{-2}), the major pathway responsible for cell death is PDT-initiated apoptosis with a very small contribution from the PTT effect (53% PDT *vs.* 2.5% PTT in 55.5% total cell death). When 808 nm light was used, cell death was solely a result of the PTT effect. The existence of the PDT and PTT pathways was proven by the detection of ROS and a heat-shock protein (HSP 70). The ability to be excited by NIR light in order to exert both PDT and PTT effects is a very important and unique feature of the PEGW$_{18}$O$_{49}$ nanowires, because organic photosensitizers that can be activated by NIR light are very rare.

Hwang *et al.* further reported the preparation of a unique Au nanoechinus (AuNE) structure with exceptionally high extinction coefficients of ~10^{12} M^{-1} cm^{-1} in the NIR region (800–1700 nm) and extendable NIR absorption covering both the first and second biological windows (Figure 10.7b).[60] Upon 808 nm light photoirradiation of AuNEs, most of the photon energy is converted to thermal heat or emitted fluorescence. In contrast, AuNEs can sensitize the formation of singlet O$_2$ upon photoirradiation of 915 and 1064 nm excitation light and subsequently exert mostly nanomaterial-mediated (Nm) PDT effects, but accompanied by an appreciable NmPTT effect, causing serious damage to the cellular and subcellular components. Therefore, AuNEs can act as a multifunctional theranostic probe for tumor tissue imaging as well as NIR-activated dual-modal NmPDT and NmPTT reagent to exert phototherapeutic effects in the first and second biological windows.

10.6.2 Nanomaterials Using a Single-Wavelength Light Source

In the above-mentioned studies, two different lasers are used independently to trigger PDT and PTT separately. For operational convenience and patient comfort, it would be helpful to find new strategies to realize combined PDT/PTT under single-laser irradiation.

For single-laser-based combined phototherapy, Yudasaka did pioneering work which combines a zinc phthalocyanine (ZnPc) photosensitizer with bovine serum albumin (BSA)-coated CNT derivatives, carbon nanohorns.[61] In this work, carbon nanohorns served as both photosensitizer carrier and photothermal agent. A single 670 nm laser was used to simultaneously activate ZnPc photosensitizer and carbon nanohorn for synergistic phototherapy. This combined phototherapy exhibited a superior anticancer effect both *in vitro* and *in vivo* compared with PDT or PTT alone.

Similarly, Chen and coworkers demonstrated that photosensitizer-coated Au nanostars could induce a synergistic photodynamic/photothermal effect under single-laser irradiation (Figure 10.8).[62] The Au nanostars was precisely tuned to have an absorption peak ~670 nm to match with absorption of photosensitizer Ce6. Ce6-incorporated Au nanostars efficiently destroyed cancer cells *in vitro* under single-laser irradiation (671 nm) compared to unimodal phototherapy. For *in vivo* experiments, combined phototherapy with the Ce6-incorporated Au nanostars significantly reduced tumor growth

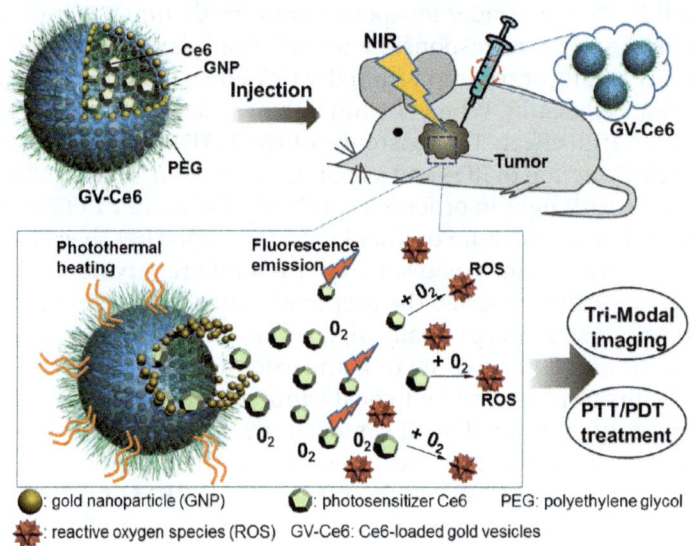

Figure 10.8   Photosensitizer (Ce6)-loaded plasmonic gold vesicles (GVs) for trimodality fluorescence/thermal/photoacoustic imaging guided synergistic photothermal/photodynamic cancer therapy. (Reproduced with permission from S. Wang, P. Huang, L. Nie, R. Xing, D. Liu, Z. Wang, J. Lin, S. Chen, G. Niu, G. Lu and X. Chen, *Adv. Mater.*, 2013, **25**, 3055. Copyright (2013) American Chemical Society.[62])

compared to unimodal phototherapies. It was also demonstrated in this study that the synergistic effect of the combined phototherapeutic system could be modulated by adjusting the irradiation times due to the differing photostability of Au nanostars and photosensitizers.

Nie *et al.* report the design of multifunctional photosensitizer Ce6-loaded plasmonic Au vesicles (GVs) for trimodality fluorescence/thermal/photoacoustic (PA) imaging-guided synergistic PDT/PTT cancer treatment.[63] The GVs, composed of a monolayer of assembled AuNPs, show a strong absorbance in the NIR range of 650–800 nm, as a result of the plasmonic coupling between neighboring AuNPs in the vesicular membranes. This enables the use of 671 nm laser irradiation to simultaneously excite both GVs and Ce6 to produce heat and singlet oxygen, killing the cancer cells. The heating effect upon laser irradiation dissociates the Ce6-loaded GVs (GV–Ce6) to release the encapsulated Ce6 molecules. The efficient loading of Ce6 in GVs significantly increases the accumulation of Ce6 in cancer cells. The tumor tissues visualized by the fluorescence, thermal, and PA signals from GV–Ce6 can be selectively destroyed in a non-invasive manner by 671 nm laser illumination. Both *in vitro* and *in vivo* therapeutic efficacy of GV–Ce6 were enhanced compared to either PTT or PDT alone, or the sum of PTT/PDT due to the synergistic effect.

Lam *et al.* report on a powerful organic NP that can be constructed by using a single organic building block, a porphyrin–cholic acid (CA) hybrid polymer

(Figure 10.9).[64] This NP platform integrates a variety of imaging and therapeutic functions that include near infrared fluorescent imaging (NIRFI), positron emission tomography (PET), MRI, dual-mode PET–MRI, PTT, and PDT, as well as targeted drug delivery. The NPs were formed by the self-assembly of a novel class of hybrid amphiphilic polymers (called telodendrimers) comprising linear PEG and dendritic oligomers of pyropheophorbide-a (Por, a porphyrin analog) and CA. This carefully designed NP platform possesses several properties that are unique and favorable as theranostic agents: (1) intrinsic ability to chelate imaging agents such as ^{64}Cu(II) and Gd(III), (2) excellent efficiency for drug loading, (3) unique architecture-dependent fluorescent, photothermal, photodynamic and magnetic resonance properties, (4) biocompatibility, monodispersivity and relatively small size (20–30 nm) and (5) reversible stabilization *via* disulfide bonds. They have demonstrated in both an ovarian cancer xenograft model and a murine transgenic breast cancer model that these novel NPs could be used as: (1) amplifiable nanoprobes to increase the sensitivity of multimodal imaging for tumor detection, (2) nanotransducers that can be activated to generate heat and ROS efficiently at tumor sites for PTT/PDT dual therapy *via* a single-wavelength light and (3) nanocarriers with programmable releasing property that can minimize the premature drug release in blood and allow efficient release upon light irradiation and/or triggered by the endogenous reducing agent present at the tumor site or in cancer cells.

In addition, two-photon excitation can be another effective way to achieve NIR-triggered PDT/PTT synergistic therapy. For example, Li and coworkers also prepared hypocrellin-loaded AuNCs for two-photon PDT/PTT of cancer.[65] The AuNCs were optimized to have efficient NIR absorption and further coated with lipids containing hydrophobic photosensitizer hypocrellin B (HB). The photodynamic effect of the photosensitizer was quenched in the Au nanostructure, thereby reducing side effects of the photosensitizer in unintended locations. For *in vitro* experiments, cancer cells were treated with the nanocomplex and irradiated with a 790 nm pulsed NIR laser. The photosensitizer HB was released from the AuNCs in the intracellular region, which induced dramatic phototoxicity due to the synergistic effect of PDT and PTT under two-photon irradiation.

Very recently, Cai *et al.* developed a theranostic nanoplatform for imaging-guided synergetic cancer phototherapy, which is expected to have great potential in clinical translation (Figure 10.10).[66] ICG is a NIR dye approved for clinical use by the United States Food and Drug Administration (FDA). It not only can be used for NIR FL and PA imaging, but also can convert the absorbed light energy to ROS and local hyperthermia for PDT and PTT, respectively.[67] ICG could therefore be considered as a kind of ideal theranostic platform for biomedical applications. Smart human serum albumin (HSA)–ICG NPs have some unique advantages: (1) Biosafety: HSA, an endogenous protein approved by the FDA for intravenous administration, is a non-toxic, non-antigenic, and biodegradable ICG delivery system. (2) Tumor targeting: HSA NPs possess both passive and active tumor-targeting abilities *via* the enhanced

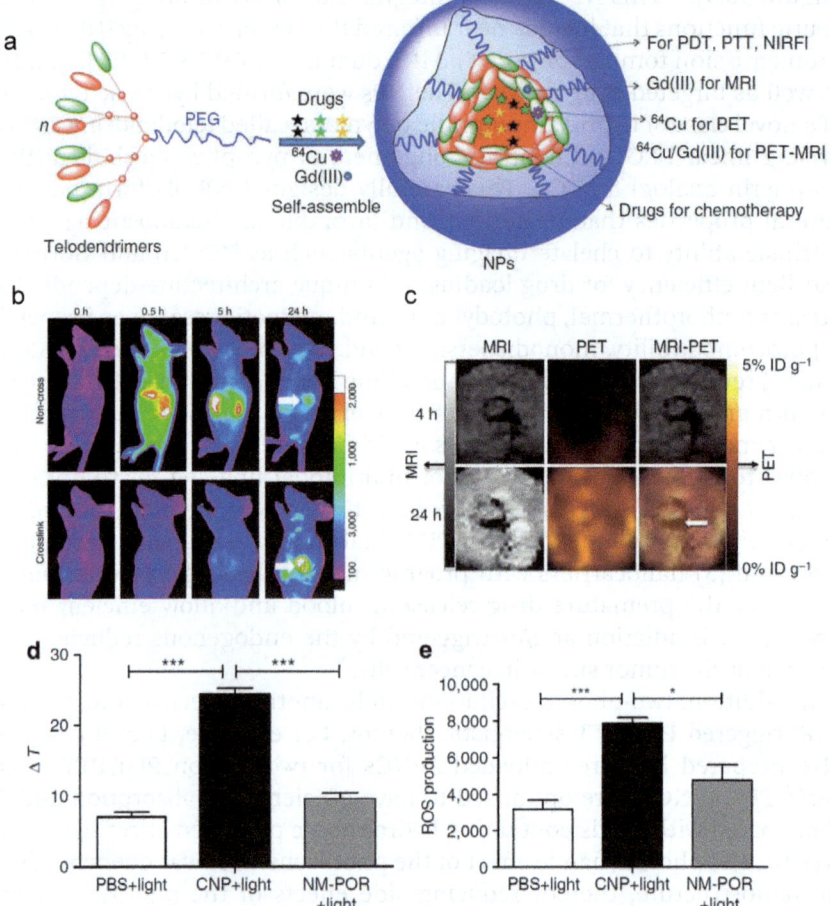

Figure 10.9 (a) Schematic illustration of a multifunctional NP self-assembled by a representative porphyrin–telodendrimer, PEG5k–Por4–CA4, composed of four pyropheophorbide-a molecules and four cholic acids attached to the end of a linear PEG chain. (b) Representative *in vivo* NIRF imaging of nude mice bearing SKOV3 ovarian cancer xenograft following intravenous injection of NPs and CNPs (NP dose 25 mg kg^{-1}). The white arrow points to the tumor site. (c) PET-MR images of tumor slices of nude mice bearing A549 lung cancer xenograft at 4 or 24 h post injection of dual-labelled NPs. White arrow points to the necrotic area in the center of the tumor. (d) The temperature changes (DT) at tumors of nude mice bearing implanted SKOV3 tumor xenografts 24 h post injection of PBS, CNPs, and NM-POR (POR dose 5 mg kg^{-1}) after light exposure ($n = 5$). Light dose 1.25 W cm^{-2} for 120 s. The temperature was monitored using a FLIR thermal camera. The results were expressed as the mean ± s.d. ***$P < 0.001$, one-way ANOVA. (e) ROS production at tumors of nude mice bearing implanted tumor xenografts 24 h post injection of PBS, CNPs, and NM-POR (POR dose 5 mg kg^{-1}) after light exposure ($n = 5$). Light dose 1.25 W cm^{-2} for 120 s. Measured by using DCF-DA as a ROS indicator. The results were expressed as the mean ± S.D. *$P < 0.01$, ***$P < 0.001$, one-way ANOVA. (Reproduced by permission from Macmillan Publishers Ltd.:*Nat. Commun.*,[64] copyright (2014).)

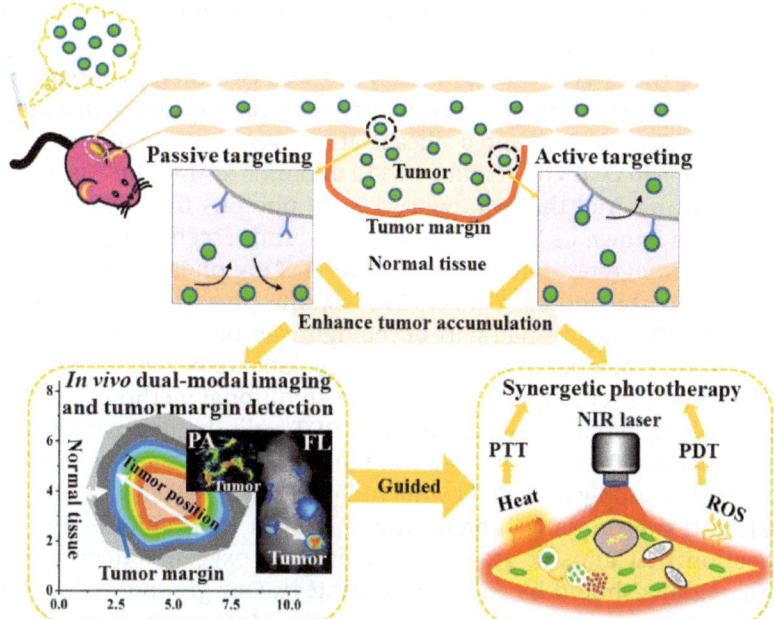

Figure 10.10   Schematic illustration of HSA–ICG NPs for *in vivo* dual-mode imaging, tumor margin detection, and simultaneous PDT/PTT treatments. Uptake of HSA-ICG NPs is presumably mediated by EPR effect (passive targeting) and the gp60 transcytosis pathway (active targeting) and subsequent binding to SPARC (secreted protein acidic and rich in cysteine) in the tumor cells. The tumor, tumor margin, and normal tissue could be detected using *in vivo* NIR and PA dual-mode imaging and spectrum-resolved technology. Upon single NIR laser irradiation, the HSA–ICG NPs can simultaneously convert the absorbed light energy to ROS and heat for synergistic PDT/PTT treatments. (Reproduced with permission from Z. H. Sheng, D. H. Hu, M. B. Zheng, P. F. Zhao, H. L. Liu, D. Y. Gao, P. Gong, G. H. Gao, P. F. Zhang, Y. F. Ma and L. T. Cai, *ACS Nano*, 2014, **8**, 12310. Copyright (2014) American Chemical Society.[66])

permeability and retention (EPR) effect and gp60 and SPARC receptor-mediated transcytosis, respectively. (3) Stimuli response: HSA NPs are prepared with intermolecular cross-linking by disulfide bonds, which exhibit excellent reduction–sensitive activity. (4) Theranostics: HSA-ICG NPs can be used for tumor imaging and margin detection, and synergistic PDT/PTT treatments.

10.7   Combined PDT with Radiotherapy

PDT and radiotherapy, two representative non-invasive treatments, can both direct light/ionizing radiation precisely on targeted tumors to efficiently induce cell death by generating a great deal of ROS. It has been reported that combined PDT and radiotherapy treatments could be more useful since they allow the reduction of the ionizing radiation dose to obtain the same effect as that obtainable by radiotherapy alone.[68]

To achieve the effective combination of these two therapy modes in one nanosystem, Xie *et al.* report a novel integrated nanosystem consisting of a core made of $SrAl_2O_4:Eu^{2+}$ and a silica coating loaded with a photosensitizer, merocyanine 540 (MC540) (Figure 10.11a and b).[69] $SrAl_2O_4:Eu^{2+}$ is a strongly luminescent material that can convert X-ray photons to visible photons, a phenomenon known as X-ray excited optical luminescence. Upon X-ray irradiation, the integrated nanosystem converts X-ray photons to visible photons to activate the photosensitizers MC540, thus producing cytotoxic $^1O_2$. In traditional PDT, only a small part of the light can penetrate into the tissue. The excitation efficiency in this report is calculated to be about 60–125 times greater than the excitation efficiency in the traditional method.

It is well known that oxygen-dependent PDT/radiotherapy usually has limited therapeutic effects on hypoxic solid tumors mainly because of the inadequate oxygen supply provided by the tumor vascular system.[70] To overcome hypoxia and substantially enhance the efficacy of PDT/radiotherapy, tumor oxygenation that aims at greatly increasing the oxygen concentrations in hypoxic regions should be an effective strategy. Our group developed intelligent theranostic nanomaterials based on the MnO_2 nanosheets anchored with UCNPs which can "make" oxygen directly in solid tumors during synergetic PDT/radiotherapy.[71] Through the decomposition of MnO_2 into Mn^{2+} by acidic H_2O_2 in solid tumors, the MnO_2–H_2O_2 redox reaction can generate large quantities of oxygen *in situ* for significantly improving the synergetic PDT/radiotherapy effects on solid tumors upon NIR/X-ray irradiation. Simultaneously, the quenched upconversion luminescence of MnO_2 can be recovered and enhanced for monitoring the therapy process.

Another possible solution relies on the development of PDT and radiotherapy agents which are not dependent on oxygen concentration. For example, our group reported the integration of a scintillator and a semiconductor as an ionizing-radiation-induced PDT agent (Figure 10.11c and d), achieving synchronous radiotherapy and depth-insensitive PDT with diminished oxygen dependence.[72] Aiming at the combination of a nanoscintillator and a semiconductor, octahedral Ce(III)-doped $LiYF_4$ scintillating nanoparticles (SCNP) were first prepared as a nanoconverter. The bulk crystal shows an unusual ultraviolet emission with high luminescence efficiency under high-energy radiation. After coating with SiO_2 and incorporation of thiol groups, ultrafine ZnO semiconductor NPs were attached firmly to the surface of $SCNP@SiO_2$ NPs by strong metal–sulfur bonds. In brief, the SCNP seed is excited using ionizing radiation and emits numerous low-energy photons that match the bandgap of surface-bound ZnO NPs. The subsequent excitons formed (the electron–hole (e^-–h^+) pairs) interact with water molecules to form free radicals. Notably, similar to type I PDT for enhanced antitumor therapeutic efficacy, the highly reactive hydroxyl radicals ($^\bullet OH$) are derived from the reaction between the hole (h^+) and the absorbed water instead of O_2, which essentially minimizes the oxygen-tension dependency for the generation of ROS.

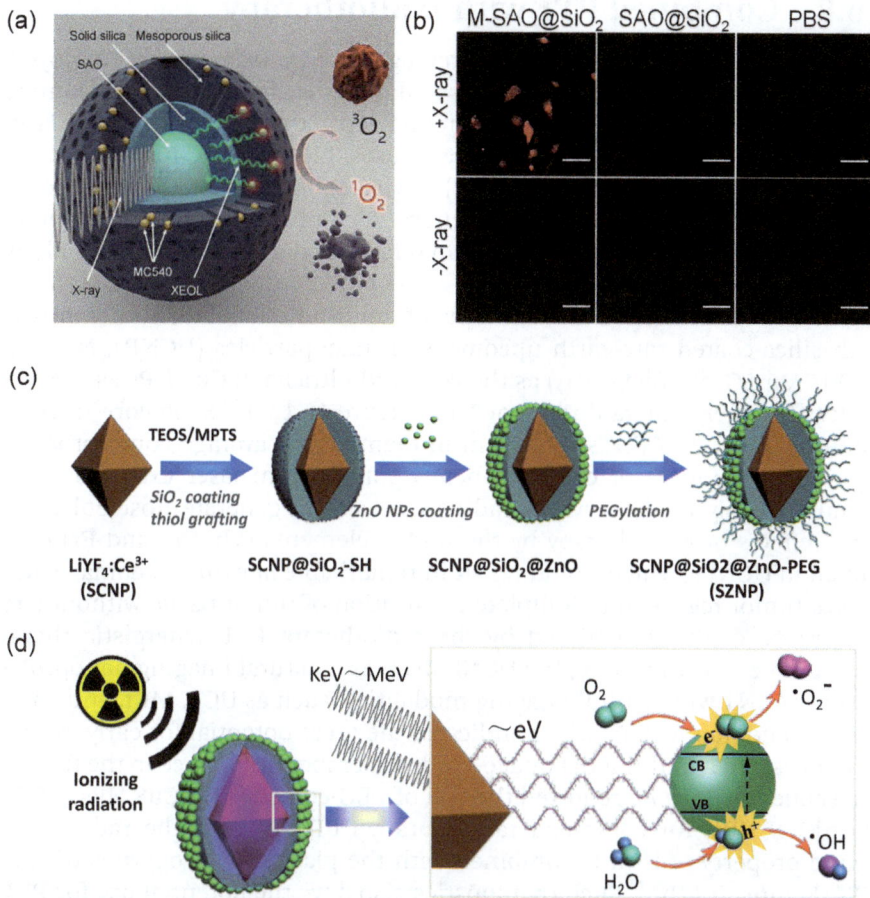

Figure 10.11 (a) Schematic illustration of the working mechanism of X-PDT. A nanoscintillator core made of SAO is coated with two layers of silica, an inner solid layer and an outer mesoporous layer. Into the mesoporous silica layer, a photosensitizer, MC540, is loaded. Under X-ray irradiation, SAO converts X-rays to visible light photons (XEOL). The visible light photons, in turn, activate near-by MC540 molecules to produce cytotoxic $^1O_2$ that destroys nearby cancer cells. $^3O_2$ is the ground-state oxygen. (b) Cytotoxicity studies using ethidium homodimer-1 as a dead cell marker (a.k.a. dead assay). (Reproduced with permission from H. Chen, G. D. Wang, Y.-J. Chuang, Z. Zhen, X. Chen, P. Biddinger, Z. Hao, F. Liu, B. Shen, Z. Pan and J. Xie, *Nano Lett.*, 2015, **15**, 2249. Copyright (2015) American Chemical Society.[69]) (c) Schematic illustration of the synthetic route to monodisperse SZNPs and (d) the mechanism of ionizing radiation-induced photodynamic therapy. The electron–hole (e^-–h^+) pair is formed after exposure to ionizing radiation. TEOS = tetraethyl orthosilicate; MPTS = (3-mercaptopropyl) trimethoxysilane. (Reproduced from ref. 72 with permission from John Wiley and Sons. Copyright © 2015 Wiley-VCH Verlag GmbH & Co. KGaA, Weinheim.)

10.8   Combined PTT with Radiotherapy

Although highly efficient photothermal agents have been widely employed, the difficulty controlling deep tumors still remains an intrinsic shortcoming of optical therapy because of the inevitable depth-dependent decline of laser intensity. By comparison, radiotherapy uses high-energy, highly focused radiation (generally X-rays and γ-rays) as virtual "knives" to kill cancer cells with no depth restriction or invasiveness. A combination of nanomaterial-enhanced radiotherapy with PTT could offset the disadvantage of PTT alone on deep-seated tumors.

Our group developed a new type of multifunctional nanotheranostic with silica-coated rare-earth upconversion nanoparticles (UCNPs, $NaYbF_4$: $2\%Er^{3+}/20\%Gd^{3+}@SiO_2–NH_2$) as the core and ultrasmall CuS NPs as the satellites for synergistic radiotherapy/PTT (Figure 10.12a).[73] Such core-satellite nanotheranostics (CSNTs) of 45 nm in diameter on average could produce significant amounts of cytotoxic heat upon 980 nm laser excitation and simultaneously could serve as radiosensitizers to generate dose-enhancement effects of radiotherapy by the high-Z elements (Yb, Gd, and Er) contained in UCNPs, which would result in remarkable *in vitro* cell damage and *in vivo* tumor regression. Complete eradication of tumor tissue without late recurrence could be achieved by the radiotherapy/PTT synergistic therapeutic effect. Furthermore, the UCNP itself as a natural imaging nanoprobe endows CSNT with several imaging modalities, such as UCL, MRI, and computed tomography (CT), which indicates the great potential for early cancer diagnosis and a multimodal image-guided therapeutic alliance in the future. Subsequently, Li *et al.* reported the use of PEG-coated [^{64}Cu]CuS NPs (PEG–[^{64}Cu]CuS NPs) for combined radiotherapy/PTT, in which the radiotherapeutic property of ^{64}Cu is combined with the plasmonic properties of CuS NPs (Figure 10.12b).[74] Such theranostics also have the potential use for PET image-guided combined therapy due to the presence of ^{64}Cu.[75]

Such a combination serves as an effective strategy for radiotherapy sensitization and brings a strong synergistic effect to conquer the inherent drawbacks of radiotherapy:

- A high radiotherapeutic effect cannot be achieved on cells in the S-phase, which is considered to be the least radiation-sensitive phase in the cell replication cycle, or on hypoxic cells, seriously affecting the tumor's response to radiotherapy. Encouragingly, all these radioresistant cells are very sensitive to the lethal effects of hyperthermia.
- An appropriate level of hyperthermia could increase intratumoral blood flow and subsequently improve oxygenation status in the tumor, resulting in a considerably increased cell sensitivity to radiotherapy.
- The repair of non-lethal damage from radiotherapy might be effectively suppressed by hyperthermia.

Therefore, the integration of PTT and nanomaterial-enhanced radiotherapy could effectively combine their advantages while compensating for their

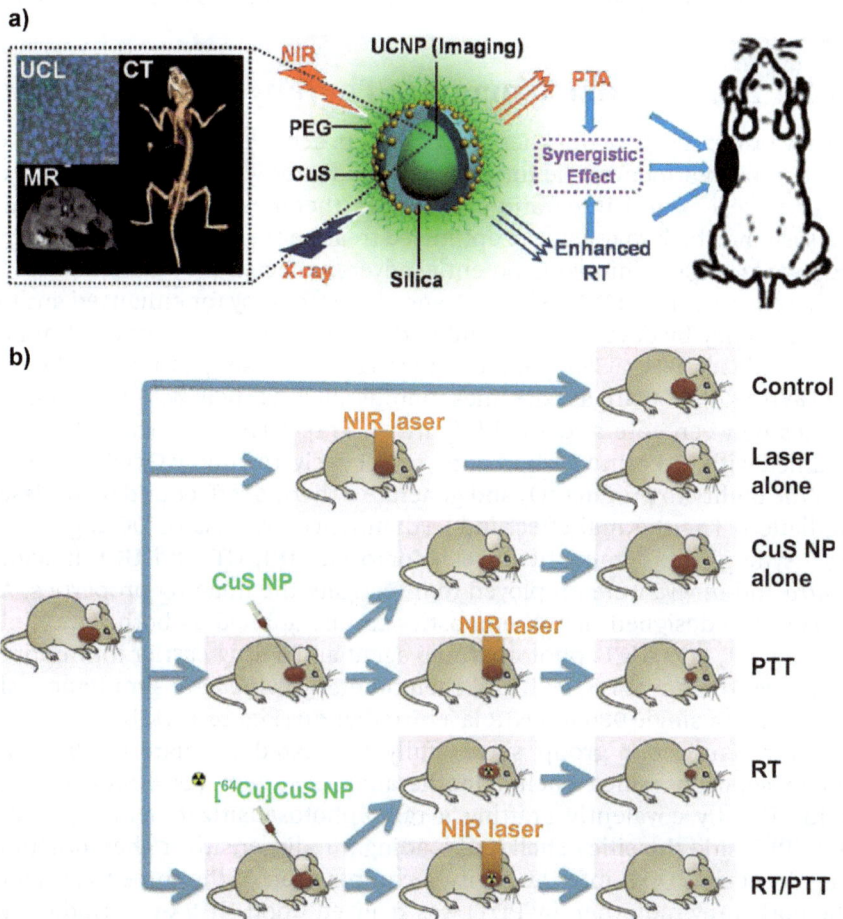

Figure 10.12 (a) Schematic of a CSNT for enhanced RT/PTA synergistic therapy. UCNP cores are used to enlarge the local radiation dose for the enhanced RT, and CuS satellites are responsible for converting the 980 nm laser into heat for PTA. The combination of PTA and CSNT-enhanced RT could give rise to a strong synergistic effect and then construct a RT/PTA synergistic system. (Reproduced with permission from Q. Xiao, X. Zheng, W. Bu, W. Ge, S. Zhang, F. Chen, H. Xing, Q. Ren, W. Fan, K. Zhao, Y. Hua and J. Shi, *J. Am. Chem. Soc.*, 2013, **135**, 13041. Copyright (2013) American Chemical Society.[73]) (b) Experimental design of the antitumor activity study in nude mice bearing subcutaneous ATC tumors (*n* = 7/group). (Reproduced from M. Zhou *et al.*, Single agent nanoparticle for radiotherapy and radio-photothermal therapy in anaplastic thyroid cancer, *Biomaterials*, 2015, **57**, 41–49, Copyright (2015) with permission from Elsevier.[74])

corresponding disadvantages, generating a strong synergistic effect and giving rise to much greater antitumor efficacy than the two treatments alone.

10.9   Multimodal Synergetic Therapy

In view of the much better therapeutic effects achieved by the combination of different treatments, one future direction of study should be the smarter material design and integration of various therapeutic modalities within one system, which may achieve optimized treatment efficacy at relatively low drug doses while minimizing potential adverse effects.

Lin *et al.* integrated PDT with PTT and chemotherapy for enhanced antitumor efficiency by developing a mild and rational route to synthesize multifunctional GdOF:Ln@SiO$_2$ yolk-like microcapsules using UCL GdOF:Ln (Ln = 10%Yb/1%Er/4%Mn) as core, mesoporous silica as shell, with large hollow cavities between core and shell (Figure 10.13a).[76] The microcapsules were modified with ZnPc and carbon dots, respectively, to endow GdOF:Ln@SiO$_2$ with the ability to product $^1O_2$ and generate a thermal effect under NIR laser irradiation. The thermal effect induced enhanced release of DOX, generating a synergistic therapeutic effect. Moreover, MRI, CT, and UCL imaging *in vitro* and *in vivo* were employed to investigate the imaging properties. As a result, the designed nanotheranostics can be applied as both a multiple imaging (CT, MRI, UCL, photothermal) agent and a drug carrier for multiple anticancer therapy (PDT, PTT, and chemotherapy) which are simultaneously triggered by a single 980 nm NIR laser irradiation (Figure 10.13b).

Very recently, our group successfully fabricated a smart Gd–0UCNPs core-mesoporous silica shell nanotheranostic system (UCMSNs) (Figure 10.13c–i).[77] By covalently grafting a radio/photosensitizer hematoporphyrin (HP) inside the silica shell and loading a radiosensitizer/chemotherapy drug docetaxel (Dtxl) into the cavity, a combination of the three treatments (chemotherapy/radiotherapy/PDT) was realized upon NIR/X-ray irradiation, which could ultimately lead to satisfactory treatment efficacy for complete tumor elimination. Again, there are significant synergetic effects among the three treatments. For example, cancerous cells usually reproduce again after radiotherapy *via* the self-repair of the DNA helix structure, while chemotherapy and PDT can efficiently inhibit DNA repair. Furthermore, with the use of MR/UCL bimodal imaging, this nanotheranostic system may be developed as an "all-in-one" approach to accurate, highly efficient, imaging-guided trimodal therapy.

10.10   Summary and Outlook

In this chapter we have systematically reviewed recent advances in the exploration of multifunctional nanomaterials for various kinds of combined cancer therapies. Based on the studies reviewed, NP-assisted NIR-triggered synergetic cancer therapy has been proven to effectively induce site-specific cell death in both *in vitro* and *in vivo* treatments, and to prevent side effects. However,

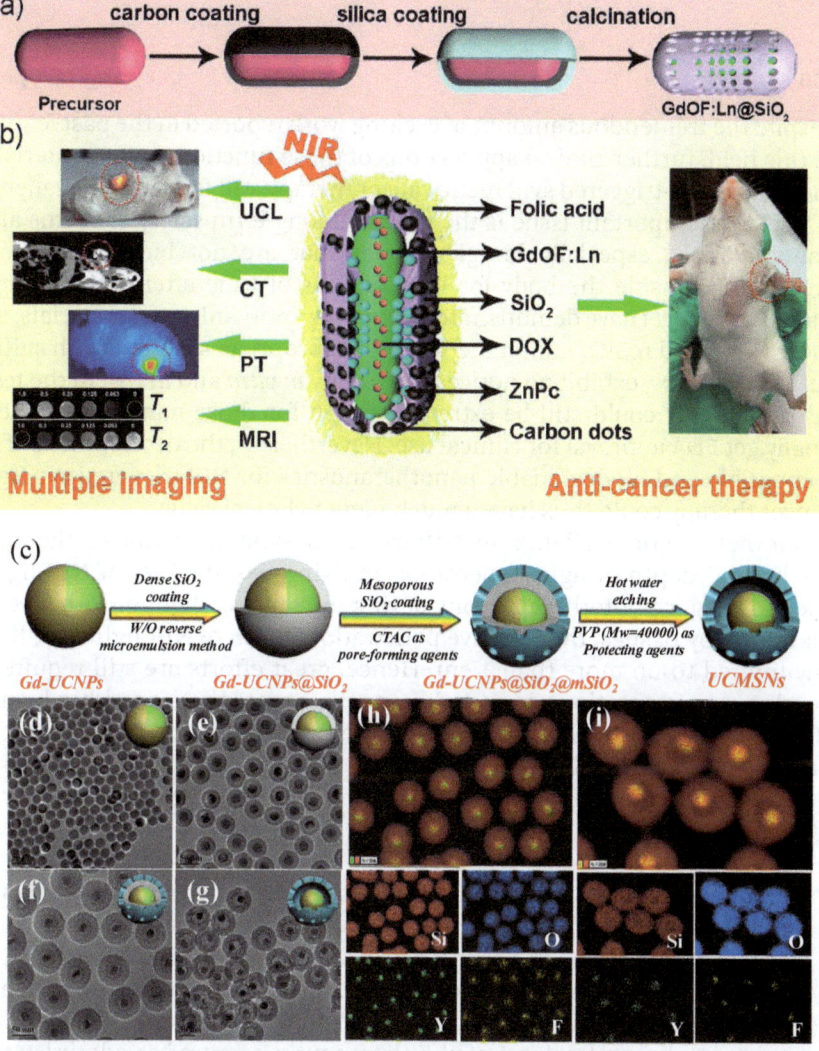

Figure 10.13 Schematic illustration for the synthesis of GdOF:Ln@SiO$_2$_ZnPc-CDs microcapsule (a) and bioapplication for multiple imaging and antitumor therapy (b). (Reproduced with permission from R. Lv, P. Yang, F. He, S. Gai, C. Li, Y. Dai, G. Yang and J. Lin, *ACS Nano*, 2015, **9**, 1630. Copyright (2015) American Chemical Society.[76]) (c) Schematic illustration of the synthetic procedure of UCMSNs. Gd–UCNPs were prepared by epitaxial growth NaGdF$_4$ layer on NaYF$_4$:Yb/Er/Tm through a typical thermal decomposition process. Then a dense silica layer was coated on Gd–UCNPs by a reverse microemulsion method, designed as Gd–UCNPs@SiO$_2$. Subsequently, a mesoporous silica shell was deposited on Gd–UCNPs@SiO$_2$ *via* the template of CTAC, designed as Gd–UCNPs@SiO$_2$@mSiO$_2$. Finally, UCMSNs were successfully fabricated, based on a "surface-protected hot water etching" strategy. (d–g) Transmission electron microscopic (TEM) images of (d) Gd–UCNPs (NaYF$_4$:Yb/Er/Tm@NaGdF$_4$), (e) Gd–UCNPs@SiO$_2$, (f) Gd–UCNPs@SiO$_2$@mSiO$_2$, and (g) UCMSNs (scale bar = 50 nm). (h,i) STEM image and the corresponding element (Si, O, Y, F) mappings of (f) Gd–UCNPs@SiO$_2$@mSiO$_2$ and (g) UCMSNs. (Reproduced from W. Fan *et al.*, A smart upconversion-based mesoporous silica nanotheranostic system for synergetic chemo-/radio-/photodynamic therapy and simultaneous MR/UCL imaging, *Biomaterials*, 2014, **35**, 8992–9002, Copyright (2014) with permission from Elsevier.[77])

despite the tremendous amount of exciting work reported in the past few years in this field, further clinical applications of those functional nanomaterials in the area of NIR-triggered synergetic cancer therapy still faces many challenges.

The most important issue is the potential long-term safety concerns about nanomaterials, especially inorganic ones that are not biodegradable and are retained inside the body for long periods of time after administration. Although reports have demonstrated that many inorganic nanomaterials, such as UCNPs[78] and mSiO$_2$ nanomaterials,[79] of appropriate sizes and with suitable surface coatings, exhibit no noticeable toxicity *in vitro* and *in vivo* in the tested dose ranges, it could still be extremely tough for those nanotheranostics to finally get FDA approval for clinical use. Nevertheless, the development of biocompatible and biodegradable nanotheranostics for NIR-triggered synergetic cancer therapy could thus have a much higher clinical value.

Another major challenge in NIR-triggered synergetic cancer therapy is the limited depth of light penetration in living tissues. Even if NIR light is used to trigger phototherapy, considering the reduced light absorbance and scattering by tissues, the effective penetration depth of NIR light is still usually limited to no more than 1 cm. Hence, great efforts are still required to develop new generations of nanotheranostic agents that can either be more effectively excited by the NIR light or be excited by light within the second biological window (1000–1350 nm).

Ideally, a therapeutic agent should also have imaging capability, for the following reasons:

- Before cancer therapy, careful whole-body imaging can be carried out to find out the exact location, size, and shape of a tumor.
- When the therapeutic agent reaches its highest accumulation in the tumor, direct light irradiation to the tumor area is preferred to achieve the highest therapeutic efficacy. As tumor homing is a dynamic process, real-time tracking of the therapeutic agent after administration by imaging can be very important to achieve the optimized therapeutic outcome.
- After the treatment is finished, advanced imaging techniques are required to determine the therapeutic response as soon as possible.

Therefore, a well-engineered nanoplatform that offers both therapeutic and imaging functionalities could have important clinical value and be of great interest to the future development of combined cancer therapies.

Acknowledgements

This work has been financially supported by the National Natural Science Foundation of China (Grant No. 51372260, 51132009, 51402338), the Development Foundation for Talents of Shanghai (Grant No.2012035), the Shanghai Yangfan Program (Grant No. 14YF1406400), and the Youth Innovation Promotion Association CAS (Grant No. 2015201).

References

1. S. S. Lucky, K. C. Soo and Y. Zhang, *Chem. Rev.*, 2015, **115**, 1990.
2. L. Cheng, C. Wang, L. Feng, K. Yang and Z. Liu, *Chem. Rev.*, 2014, **114**, 10869.
3. E.-K. Lim, T. Kim, S. Paik, S. Haam, Y.-M. Huh and K. Lee, *Chem. Rev.*, 2015, **115**, 327.
4. H. Koo, M. S. Huh, I.-C. Sun, S. H. Yuk, K. Choi, K. Kim and I. C. Kwon, *Acc. Chem. Res.*, 2011, **44**, 1018.
5. N. L. Komarova and C. R. Boland, *Nature*, 2013, **499**, 291.
6. A. P. Castano, P. Mroz, M. X. Wu and M. R. Hamblin, *Proc. Natl. Acad. Sci. U. S. A.*, 2008, **105**, 5495.
7. A. P. Blum, J. K. Kammeyer, A. M. Rush, C. E. Callmann, M. E. Hahn and N. C. Gianneschi, *J. Am. Chem. Soc.*, 2015, **137**, 2140.
8. V. Shanmugam, S. Selvakumar and C.-S. Yeh, *Chem. Soc. Rev.*, 2014, **43**, 6254.
9. T. Bald, T. Quast, J. Landsberg, M. Rogava, N. Glodde, D. Lopez-Ramos, J. Kohlmeyer, S. Riesenberg, D. van den Boorn-Konijnenberg, C. Homig-Holzel, R. Reuten, B. Schadow, H. Weighardt, D. Wenzel, I. Helfrich, D. Schadendorf, W. Bloch, M. E. Bianchi, C. Lugassy, R. L. Barnhill, M. Koch, B. K. Fleischmann, I. Forster, W. Kastenmuller, W. Kolanus, M. Holzel, E. Gaffal and T. Tuting, *Nature*, 2014, **507**, 109.
10. R. Weissleder, *Nat. Biotechnol.*, 2001, **19**, 316.
11. P. Yang, S. Gai and J. Lin, *Chem. Soc. Rev.*, 2012, **41**, 3679.
12. X. Li, F. Zhang and D. Zhao, *Chem. Soc. Rev.*, 2015, **44**, 1346.
13. E. C. Dreaden, A. M. Alkilany, X. Huang, C. J. Murphy and M. A. El-Sayed, *Chem. Soc. Rev.*, 2012, **41**, 2740.
14. D. Ling, N. Lee and T. Hyeon, *Acc. Chem. Res.*, 2015, **48**, 1276.
15. T. M. Allen and P. R. Cullis, *Science*, 2004, **303**, 1818.
16. P. T. Wong and S. K. Choi, *Chem. Rev.*, 2015, **115**, 3388.
17. S. Mura, J. Nicolas and P. Couvreur, *Nat. Mater.*, 2013, **12**, 991.
18. N. Fomina, J. Sankaranarayanan and A. Almutairi, *Adv. Drug Delivery Rev.*, 2012, **64**, 1005.
19. Y. Min, J. Li, F. Liu, E. K. L. Yeow and B. Xing, *Angew. Chem., Int. Ed.*, 2014, **53**, 1012.
20. Y. Dai, H. Xiao, J. Liu, Q. Yuan, P. A. Ma, D. Yang, C. Li, Z. Cheng, Z. Hou, P. Yang and J. Lin, *J. Am. Chem. Soc.*, 2013, **135**, 18920.
21. J. Liu, W. Bu, L. Pan and J. Shi, *Angew. Chem., Int. Ed.*, 2013, **52**, 4375.
22. Q. He, Y. Gao, L. Zhang, Z. Zhang, F. Gao, X. Ji, Y. Li and J. Shi, *Biomaterials*, 2011, **32**, 7711.
23. X. Li, L. Zhou, Y. Wei, A. M. El-Toni, F. Zhang and D. Zhao, *J. Am. Chem. Soc.*, 2014, **136**, 15086.
24. J. Liu, J. Bu, W. Bu, S. Zhang, L. Pan, W. Fan, F. Chen, L. Zhou, W. Peng, K. Zhao, J. Du and J. Shi, *Angew. Chem., Int. Ed.*, 2014, **53**, 4551.
25. T. S. Hauck, T. L. Jennings, T. Yatsenko, J. C. Kumaradas and W. C. W. Chan, *Adv. Mater.*, 2008, **20**, 3832.

26. H. Liu, D. Chen, L. Li, T. Liu, L. Tan, X. Wu and F. Tang, *Angew. Chem., Int. Ed.*, 2011, **50**, 891.

27. L. Klouda and A. G. Mikos, *Eur. J. Pharm. Biopharm.*, 2008, **68**, 34.

28. Y. Jin, *Acc. Chem. Res.*, 2014, **47**, 138.

29. B. Nikoobakht and M. A. El-Sayed, *Chem. Mater.*, 2003, **15**, 1957.

30. M. Ma, H. Chen, Y. Chen, X. Wang, F. Chen, X. Cui and J. Shi, *Biomaterials*, 2012, **33**, 989.

31. S. E. Skrabalak, J. Chen, Y. Sun, X. Lu, L. Au, C. M. Cobley and Y. Xia, *Acc. Chem. Res.*, 2008, **41**, 1587.

32. M. S. Yavuz, Y. Y. Cheng, J. Y. Chen, C. M. Cobley, Q. Zhang, M. Rycenga, J. W. Xie, C. Kim, K. H. Song, A. G. Schwartz, L. H. V. Wang and Y. N. Xia, *Nat. Mater.*, 2009, **8**, 935.

33. N. W. S. Kam, M. O'Connell, J. A. Wisdom and H. Dai, *Proc. Natl. Acad. Sci. U. S. A.*, 2005, **102**, 11600.

34. K. Yang, S. Zhang, G. Zhang, X. Sun, S.-T. Lee and Z. Liu, *Nano Lett.*, 2010, **10**, 3318.

35. W. Zhang, Z. Guo, D. Huang, Z. Liu, X. Guo and H. Zhong, *Biomaterials*, 2011, **32**, 8555.

36. J. E. Lee, N. Lee, T. Kim, J. Kim and T. Hyeon, *Acc. Chem. Res.*, 2011, **44**, 893.

37. Y. Wang, K. Wang, J. Zhao, X. Liu, J. Bu, X. Yan and R. Huang, *J. Am. Chem. Soc.*, 2013, **135**, 4799.

38. S. S. Chou, B. Kaehr, J. Kim, B. M. Foley, M. De, P. E. Hopkins, J. Huang, C. J. Brinker and V. P. Dravid, *Angew. Chem., Int. Ed.*, 2013, **125**, 4254.

39. W. Y. Yin, L. Yan, J. Yu, G. Tian, L. J. Zhou, X. P. Zheng, X. Zhang, Y. Yong, J. Li, Z. J. Gu and Y. L. Zhao, *ACS Nano*, 2014, **8**, 6922.

40. X. Cai, X. Jia, W. Gao, K. Zhang, M. Ma, S. Wang, Y. Zheng, J. Shi and H. Chen, *Adv. Funct. Mater.*, 2015, **25**, 2520.

41. M. Zheng, C. Yue, Y. Ma, P. Gong, P. Zhao, C. Zheng, Z. Sheng, P. Zhang, Z. Wang and L. Cai, *ACS Nano*, 2013, **7**, 2056.

42. A. Khdair, C. Di, Y. Patil, L. Ma, Q. P. Dou, M. P. V. Shekhar and J. Panyam, *J. Controlled Release*, 2010, **141**, 137.

43. K. J. Son, H.-J. Yoon, J.-H. Kim, W.-D. Jang, Y. Lee and W.-G. Koh, *Angew. Chem., Int. Ed.*, 2011, **50**, 11968.

44. Y. Yuan, J. Liu and B. Liu, *Angew. Chem., Int. Ed.*, 2014, **53**, 7163.

45. Y. Liu, Y. Liu, W. Bu, C. Cheng, C. Zuo, Q. Xiao, Y. Sun, D. Ni, C. Zhang, J. Liu and J. Shi, *Angew. Chem., Int. Ed.*, 2015, **54**, 8105.

46. D. Sano, J. M. Berlin, T. T. Pham, D. C. Marcano, D. R. Valdecanas, G. Zhou, L. Milas, J. N. Myers and J. M. Tour, *ACS Nano*, 2012, **6**, 2497.

47. Z. B. Starkewolf, L. Miyachi, J. Wong and T. Guo, *Chem. Commun.*, 2013, **49**, 2545.

48. H. Xing, X. Zheng, Q. Ren, W. Bu, W. Ge, Q. Xiao, S. Zhang, C. Wei, H. Qu, Z. Wang, Y. Hua, L. Zhou, W. Peng, K. Zhao and J. Shi, *Sci. Rep.*, 2013, **3**, 1751.

49. W. Fan, B. Shen, W. Bu, F. Chen, K. Zhao, S. Zhang, L. Zhou, W. Peng, Q. Xiao, H. Xing, J. Liu, D. Ni, Q. He and J. Shi, *J. Am. Chem. Soc.*, 2013, **135**, 6494.

50. L. Pan, Q. He, J. Liu, Y. Chen, M. Ma, L. Zhang and J. Shi, *J. Am. Chem. Soc.*, 2012, **134**, 5722.
51. J.-n. Liu, W. Bu, L.-m. Pan, S. Zhang, F. Chen, L. Zhou, K.-l. Zhao, W. Peng and J. Shi, *Biomaterials*, 2012, **33**, 7282.
52. W. Fan, B. Shen, W. Bu, X. Zheng, Q. He, Z. Cui, K. Zhao, S. Zhang and J. Shi, *Chem. Sci.*, 2015, **6**, 1747.
53. Y. Liu, Y. Liu, W. Bu, Q. Xiao, Y. Sun, K. Zhao, W. Fan, J. Liu and J. Shi, *Biomaterials*, 2015, **49**, 1.
54. W.-S. Kuo, C.-N. Chang, Y.-T. Chang, M.-H. Yang, Y.-H. Chien, S.-J. Chen and C.-S. Yeh, *Angew. Chem., Int. Ed.*, 2010, **49**, 2711.
55. W.-S. Kuo, Y.-T. Chang, K.-C. Cho, K.-C. Chiu, C.-H. Lien, C.-S. Yeh and S.-J. Chen, *Biomaterials*, 2012, **33**, 3270.
56. B. Jang, J.-Y. Park, C.-H. Tung, I.-H. Kim and Y. Choi, *ACS Nano*, 2011, **5**, 1086.
57. J. Wang, G. Zhu, M. You, E. Song, M. I. Shukoor, K. Zhang, M. B. Altman, Y. Chen, Z. Zhu, C. Z. Huang and W. Tan, *ACS Nano*, 2012, **6**, 5070.
58. B. Tian, C. Wang, S. Zhang, L. Feng and Z. Liu, *ACS Nano*, 2011, **5**, 7000.
59. P. Kalluru, R. Vankayala, C.-S. Chiang and K. C. Hwang, *Angew. Chem., Int. Ed.*, 2013, **52**, 12332.
60. P. Vijayaraghavan, C.-H. Liu, R. Vankayala, C.-S. Chiang and K. C. Hwang, *Adv. Mater.*, 2014, **26**, 6689.
61. M. Zhang, T. Murakami, K. Ajima, K. Tsuchida, A. S. D. Sandanayaka, O. Ito, S. Iijima and M. Yudasaka, *Proc. Natl. Acad. Sci. U. S. A.*, 2008, **105**, 14773.
62. S. Wang, P. Huang, L. Nie, R. Xing, D. Liu, Z. Wang, J. Lin, S. Chen, G. Niu, G. Lu and X. Chen, *Adv. Mater.*, 2013, **25**, 3055.
63. J. Lin, S. Wang, P. Huang, Z. Wang, S. Chen, G. Niu, W. Li, J. He, D. Cui, G. Lu, X. Chen and Z. Nie, *ACS Nano*, 2013, **7**, 5320.
64. Y. Li, T.-y. Lin, Y. Luo, Q. Liu, W. Xiao, W. Guo, D. Lac, H. Zhang, C. Feng, S. Wachsmann-Hogiu, J. H. Walton, S. R. Cherry, D. J. Rowland, D. Kukis, C. Pan and K. S. Lam, *Nat. Commun.*, 2014, **5**, 5712.
65. L. Gao, J. Fei, J. Zhao, H. Li, Y. Cui and J. Li, *ACS Nano*, 2012, **6**, 8030.
66. Z. H. Sheng, D. H. Hu, M. B. Zheng, P. F. Zhao, H. L. Liu, D. Y. Gao, P. Gong, G. H. Gao, P. F. Zhang, Y. F. Ma and L. T. Cai, *ACS Nano*, 2014, **8**, 12310.
67. Z. Sheng, D. Hu, M. Xue, M. He, P. Gong and L. Cai, *Nano-Micro Lett.*, 2013, **5**, 145.
68. A. Colasanti, A. Kisslinger, M. Quarto and P. Riccio, *Acta Biochim. Pol.*, 2004, **51**, 1039.
69. H. Chen, G. D. Wang, Y.-J. Chuang, Z. Zhen, X. Chen, P. Biddinger, Z. Hao, F. Liu, B. Shen, Z. Pan and J. Xie, *Nano Lett.*, 2015, **15**, 2249.
70. C. S. Jin, J. F. Lovell, J. Chen and G. Zheng, *ACS Nano*, 2013, **7**, 2541.
71. W. Fan, W. Bu, B. Shen, Q. He, Z. Cui, Y. Liu, X. Zheng, K. Zhao and J. Shi, *Adv. Mater.*, 2015, **27**, 4155.
72. C. Zhang, K. Zhao, W. Bu, D. Ni, Y. Liu, J. Feng and J. Shi, *Angew. Chem., Int. Ed.*, 2015, **54**, 1770.

73. Q. Xiao, X. Zheng, W. Bu, W. Ge, S. Zhang, F. Chen, H. Xing, Q. Ren, W. Fan, K. Zhao, Y. Hua and J. Shi, *J. Am. Chem. Soc.*, 2013, **135**, 13041.
74. M. Zhou, Y. Chen, M. Adachi, X. Wen, B. Erwin, O. Mawlawi, S. Y. Lai and C. Li, *Biomaterials*, 2015, **57**, 41.
75. M. Zhou, R. Zhang, M. Huang, W. Lu, S. Song, M. P. Melancon, M. Tian, D. Liang and C. Li, *J. Am. Chem. Soc.*, 2010, **132**, 15351.
76. R. Lv, P. Yang, F. He, S. Gai, C. Li, Y. Dai, G. Yang and J. Lin, *ACS Nano*, 2015, **9**, 1630.
77. W. Fan, B. Shen, W. Bu, F. Chen, Q. He, K. Zhao, S. Zhang, L. Zhou, W. Peng, Q. Xiao, D. Ni, J. Liu and J. Shi, *Biomaterials*, 2014, **35**, 8992.
78. Y. Sun, W. Feng, P. Yang, C. Huang and F. Li, *Chem. Soc. Rev.*, 2015, **44**, 1509.
79. Y. Chen, H. Chen and J. Shi, *Adv. Mater.*, 2013, **25**(23), 3144.

CHAPTER 11

Nanotoxicity of Near Infrared Nanomaterials

L. YAN[a], Y. L. ZHAO[a,b], AND Z. J. GU*[a]

[a]CAS Key Laboratory for Biomedical Effects of Nanomaterials and Nanosafety, Institute of High Energy Physics and National Center for Nanoscience Technology of China, Chinese Academy of Sciences, Beijing, 100049, P. R. China; [b]Collaborative Innovation Center of Radiation Medicine of Jiangsu Higher Education Institutions, and Jiangsu Provincial Key Lab of Radiation Medicine and Protection, School of Radiation Medicine and Protection, School for Radiological and Interdisciplinary Sciences, Soochow University, Suzhou, 215123, P. R. China
*E-mail: zjgu@ihep.ac.cn

11.1 Introduction

Due to the extraordinary physical and chemical properties endowed by various surface effects and size effects, nanomaterials (NMs) (*e.g.*, noble metal-based nanoparticles, narrow-bandgap semiconductors, and carbon-based materials) have recently demonstrated their applicability to a wide range of biological and biomedical applications including bioimaging, biosensing, drug delivery, and cancer diagnosis and therapy.[1-11] One of the most promising types is the near infrared (NIR) NMs.[12-22] NIR NMs are usually defined as substances that either have the ability to interact with NIR light, such as absorption and reflection, or are capable of emitting NIR light under external stimulations including photoexcitation, chemical reactions, and electric fields. It is well known

RSC Nanoscience & Nanotechnology No. 40
Near Infrared Nanomaterials: Preparation, Bioimaging and Therapy Applications
Edited by Fan Zhang
© The Royal Society of Chemistry 2016
Published by the Royal Society of Chemistry, www.rsc.org

that, in the wavelength range of 650–1450 nm, Rayleigh scattering of biological tissues gradually decreases as the wavelength increases; at the same time, biological samples (such as hemoglobin and water) have the lowest absorption of light in this range (Figure 11.1).[17,23] This means that NIR radiation can penetrate more deeply into biological tissues than either ultraviolet (UV) or visible radiation. Therefore, compared with the traditional fluorescent materials with excitation/emission maxima falling in the UV/visible regions,[24] NIR NMs have great potentials in non-invasive *in vitro/in vivo* imaging of biological tissues with high sensitivity as well as providing a high signal-to-noise ratio. On the

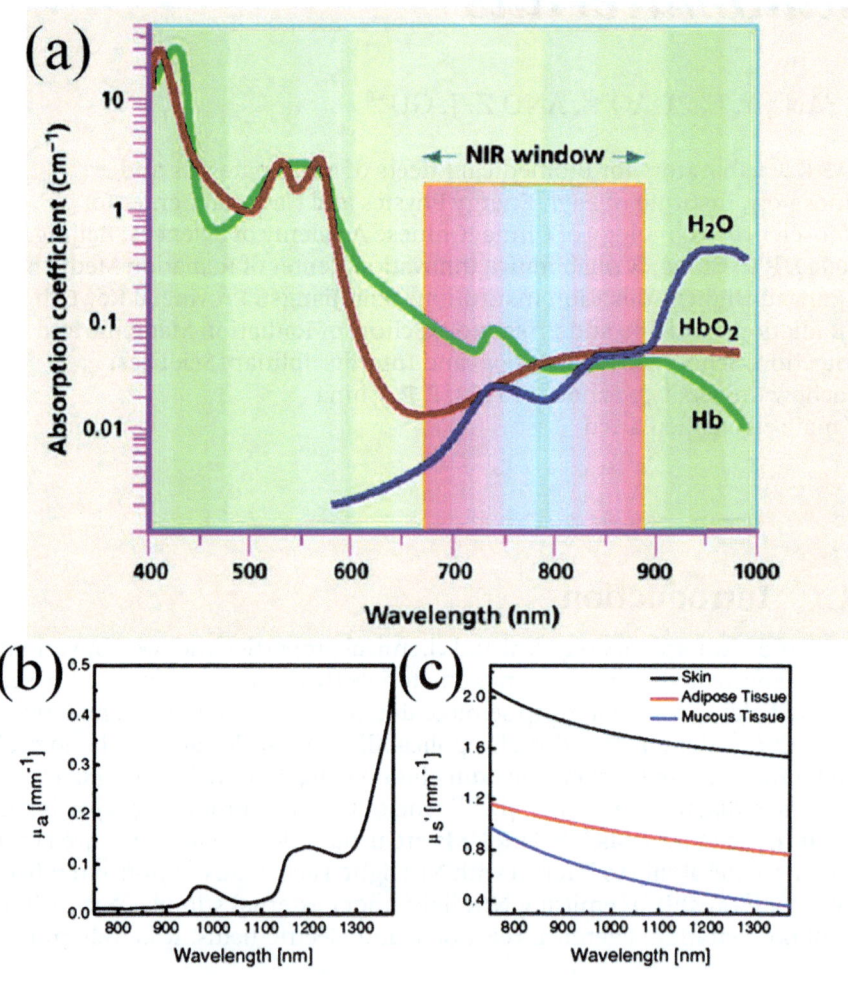

Figure 11.1   (a) Absorption coefficient of both haemoglobin and water in the 400–1000 nm range. (b) Absorption coefficient of water (μ_a). (c) Reduced scattering coefficient (μ_s') for skin, adipose tissue, and mucous tissue in the 800–1400 nm range. Reproduced from ref. 22 with permission from the Royal Society of Chemistry.

other hand, taking advantage of the strong absorption of NMs in the NIR wave-length range, penetrating NIR light can be transduced into local heat by a NIR MN as a thermal transducer. In this way, NIR light can be used as an energy source to provide local heating, consequently resulting in effective tissue dam-age which could result in detrimental consequences, *e.g.*, for tumors. Based on the above-mentioned advantages, many studies have focused on the synthesis and development of various NIR NMs for biomedical imaging and targeted therapy, and the number of NIR NM-based studies has increased exponentially over the years;[22] however, there is little information on the toxicological prop-erties of NIR NMs and their long-term toxicity to human health. Due to their very small size NMs are capable of entering the human body by inhalation, ingestion, skin penetration or injections, and then have the potential to inter-act with intracellular structures and macromolecules for long periods of time. Although the toxicity of many corresponding bulk materials is well understood based on previous studies, for NIR NMs, it is still not known at what concentra-tion or size they can begin to induce toxicity due to their nanoscale dimensions and the common mechanisms of toxicity at the levels of organism, organ, cell, and biomacromolecule.[25,26] In addition, not enough data is available on the toxicity of NIR NMs to properly evaluate their effects on human and environ-mental health and safety. At present, there is a considerable gap between the available data on the production of NIR NMs and their toxicity evaluation. In order to reach conclusions about their toxicity, systematic research is needed on these NIR NMs that show great promise in biomedicine.

In this chapter, we first divide NIR NMs into five species: carbon-based materials, quantum dots (QDs), noble metal-based nanoparticles (NPs), upconversion nanoparticles (UCNPs), and narrow-bandgap semiconductors. Then, we focus primarily on the progress of their toxicity studies in the past several years, discuss in detail how the biophysicochemical properties of NIR NMs influence their *in vitro* and *in vivo* toxicity, present a broad overview of the available *in vitro* and *in vivo* toxicity assessments of NIR NMs, and finally frame the future outlook for NIR NMs by highlighting areas of exceptional promise and challenges. Our emphasis here is mainly on discussion that could offer future opportunities to design and create NIR NMs with good biocompatibility as well as excellent functionalities, rather than attempting to provide a complete historical survey.

11.2 Properties and Applications of Near Infrared Nanomaterials

11.2.1 Physical Properties of Near Infrared Nanomaterials

11.2.1.1 *Fluorescent Properties*

NIR MNs usually have unique electronic properties, which endow them with intrinsic photoluminescence in the NIR region.[27] When absorbing light with the corresponding wavelength, the electrons of NIR NMs (such as QDs,

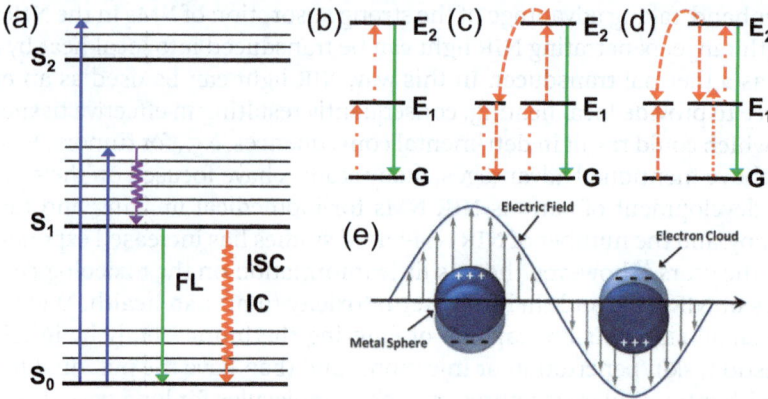

Figure 11.2 (a) Physical process of fluorescence. (b–d) Principal upconversion processes for lanthanide-doped crystals. (e) Schematic diagrams illustrating a localized surface plasmon on a metal particle. Reproduced from ref. 29 with permission from the Royal Society of Chemistry.

carbon-based materials, and narrow-bandgap semiconductors) will excite from the ground state (S_0) to an excited state at a higher energy level (mostly S_1 and sometimes S_2; Figure 11.2a). However, these excited electrons can stay in this excited state for only a very short time before relaxing and returning to the ground state or another state with a lower energy level *via* a radiation pathway (*i.e.*, energy is released as a photon). Because some of the energy is lost during this process, the energy of emission is often lower than that of excitation, which results in a longer wavelength than that of the excitation light. In general, the emission wavelengths are mainly associated with the bandgap and a narrower bandgap is always needed for NIR NMs; this limits the choice of composition elements.

For UCNPs, different mechanisms have been proposed for the occurrence of upconversion luminescence, either alone or in combinations of absorption and non-radiative energy transfer steps (Figure 11.2b–d).[28,30] The first mechanism is excited state absorption (ESA). This mechanism involves multistep excitation by absorption of one or more photons. As shown in Figure 11.2b, when one electron in the ground state (G) absorbs one photon, it first jumps to the intermediate reservoir state (E_1); after sequentially absorbing the second photon, it can be further excited to the excited state (E_2), and then emits higher-energy photons when coming back to the ground state. In the above process, non-radiative energy transfer may take place between either like energy levels or unlike energy levels.

Two other mechanisms are energy transfer upconversion (ETU) and photon avalanche (PA) which are caused by energy transfer between unlike levels. ETU is the most efficient upconversion process in rare earth doped NMs. In this process, the low-lying neighbour (the sensitizer) donates its excitation energy to a neighbouring activator because their excited energies are nearly equal and their distance is near enough, resulting in the excitation of the activator from its ground state to an excited state the before sensitizer emits photons (Figure 11.2c). The excited ion is then promoted to a higher excited state and finally emits higher-energy photons when coming back to the ground state. In fact, ETU is a pairwise or multicentre effect which strongly depends upon the concentration of ions.

PA is an unconventional mechanism as it could lead to strong upconversion emission without any resonance with the absorption transition from the ground state to the intermediate states. In PA, an ion is partially deactivated through energy transfer from the E_2 state to the E_0 state, with both ions at lower excitation levels, as shown in Figure 11.2d.

11.2.1.2 Photothermal Properties

After the electrons of NIR NMs are excited from the ground state of the atoms, the cascade of electrons to the S_0 state also follows a radiationless pathway. In this process, electrons move rapidly from the S_1 energy level to S_0 *via* internal conversion or intersystem crossing, consequently producing strong local heat (Figure 11.2a). Thus, in the case NMs that exhibit strong NIR absorbance, the NIR irradiation can be efficiently transduced into local heat by the NMs. For noble metal-based NPs, strong surface fields are induced due to the coherent excitation of the irradiated electrons to produce electromagnetic radiation (Figure 11.2e).[31,32] The rapid relaxation of these excited electrons is also capable of producing strong localized heat. When NMs are internalized by cancer cells, this heat can be envisaged as a means of destroying the surrounding targeted cancer cells *via* hyperthermia or other thermal-based effects. This is photothermal therapy, a minimally invasive therapeutic strategy. NMs with strong NIR absorption efficiency can therefore be employed as photothermal therapeutic agents for the treatment of cancer.

11.2.2 Applications of Near Infrared Nanomaterials

It is well known that scattering and autofluorescence are the major factors that lead to attenuation of signal proportional to the depth of the feature of interest in living systems.[33] To overcome this problem, it is imperative to tune the mapping parameters away from potential interference from endogenous fluorescence factors and into the NIR spectral region. NIR fluorescence imaging offers a considerable advantage over imaging in the UV/visible range. Not only does it help to enhance the signal-to-noise ratio, but fluorescent labels associated with the NIR emission wavelengths also penetrate deeper into tissue than those associated with visible light emission wavelengths. Owing to its excellent sensitivity and temporal resolution, optical imaging in the NIR window provides enormous potential for the non-invasive detection of diseases and monitoring of therapy. Furthermore, it allows for the assessment of treatment efficacy and thereby facilitates the potential to adjust and customize treatment *in vivo*. For example, Zhang *et al.*[34] demonstrated the effectiveness using the PEI-coated NaYF$_4$:Yb/Er NPs for animal imaging *in vivo* and exhibited the deeper-tissue imaging advantages over QDs. Nyk *et al.*[35] also reported the upconversion bioimaging of cells and mice based on NaYF$_4$:Yb,Tm NPs (20–30 nm) which have an upconversion emission ~800 nm. Their result indicated that upon excitation at 980 nm high-contrast photoluminescence imaging was possible due to the better tissue penetration depths as both the excitation and emission are in the NIR region. Besides these, the idea of upconversion bioimaging was realized by employing some other oxysulfides,[36] oxide,[37–41] and fluoride NMs,[42–47] such as Y$_2$O$_3$:Yb/Er, Gd$_2$O$_3$:Yb/Er, and NaYbF$_4$:Er. In 2009, the group of Dai[48] demonstrated that the exchange-SWCNTs coated with phospholipid–polyethylene glycol (PEG) could be used for whole-animal NIR imaging *in vivo* in the NIR II window. More importantly, the tumour vessels could be resolved within a few micrometres, approaching the diffraction limit. Chen *et al.*[49] also reported their work on InAs/InP/ZnSe NIR QDs in *in vivo* imaging. At 1 h after injection, the fluorescence signal mainly appeared in tumours and gradually increased over time, while the signal from other organs was very low. Alternatively, due to their strong absorbance in the NIR window, many NIR NMs such as graphene-based nanocomposites,[50–53] narrow-bandgap semiconductors,[54–57] and noble metal-based NPs[58–62] also have been used as light-activated nanoscopic heaters for cancer therapy through the selective localized photothermal heating of cancer cells. Gold NPs functionalized with antibodies are capable of selectively targeting the cancer cells, and then selectively heating and damaging tumour tissues upon NIR irradiation.[63]

11.3   Analysis of Toxicity of Near Infrared Nanomaterials

11.3.1   Nanotoxicity Mechanisms of Near Infrared Nanomaterials

So far, mechanisms by which NIR NMs can cause toxicity in the body are still not well understood. Based on previously reported results, several different toxicity mechanisms have been proposed, as summarized in Figure 11.3.[64,65] The first mechanism is attributed to the production of excess reactive oxygen species (ROS). NIR NMs, which could be regarded as foreign materials by cells and tissues, are capable of generating ROS by different routes:[66] (1) direct production of ROS as a result of exposure to cells; (2) interaction of NIR NMs with cellular organelles such as mitochondria; (3) interaction of NIR NMs with redox active proteins such as NADPH oxidase; and (4) interaction of NIR NMs with cell surface receptors and activation of intracellular signalling pathways. For example, after cell internalization iron-based NPs can catalyse ROS generation and the formation of OOH$^{\bullet}$ and OH$^{\bullet}$ radicals from intracellular H$_2$O$_2$.[67,68] Most cells have defence mechanisms to buffer a certain amount of ROS, which plays specific roles in the modulation of several

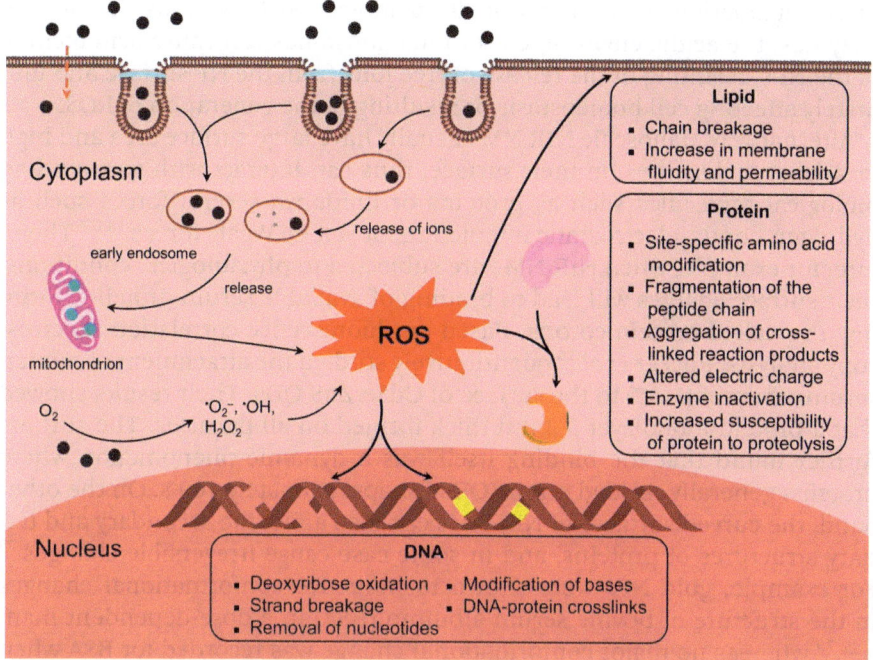

Figure 11.3 ROS generated by internalized NMs may induce oxidative damage to lipids, proteins, and DNA. These reactions can alter the intrinsic properties of the membrane, such as ion transport, and permeability; reduce enzyme activity; inhibit protein synthesis, damage DNA, and so forth, eventually resulting in cell death. Reproduced from ref. 65 with permission from John Wiley and Sons. Copyright © 2013 Wiley-VCH Verlag GmbG & Co. KGaA, Weinheim.

cellular events, such as signal transduction and protein redox regulation.[69,70] However, when the ROS level is too high, cells can undergo a vary of negative effects, including oxidative stress and damage by peroxidizing lipids, interference with signalling functions, and disruption of DNA.[71] ROS have the ability to steal electrons from lipids, resulting in decreased physiological function of cell membrane. For example, exposure of human keratinocytes to insoluble carbon NPs was associated with oxidative stress and apoptosis.[72] It is well documented that the generation and elimination of ROS is dynamically balanced inside cells, and disturbing the balance may induce intracellular protein inactivation, lipid peroxidation, dysfunction of mitochondria, and eventually apoptosis or necrosis.

The second mechanism is often attributed to the release of metal ions from intracellular NIR NMs.[73] When NIR NMs are internalized by cultured cells, the pH values of the surroundings will change from 7.40 in the extracellular medium to 6.00 in the early endosomes and then 4.50 in the lysosomes. Because of the high surface charge density of NIR NMs, the local pH values of their surface can often be lower or higher than those of their surroundings.

After degradation of the functionalized molecules by various degradative enzymes, the acidic environment of the endosomes can cause acid etching of the NPs, resulting in the release of free ions from the NP surface and ultimately affecting cell homeostasis or resulting in the generation of ROS.

Alternatively, nanoscale NIR NMs usually have large surface area and high local charge densities on their surface, thus can interact with surrounding biological molecules such as proteins or lipids *via* several forces such as hydrogen bonds, electrostatic interaction, or van de Waals forces.[74–77] Therefore, for example, when NIR NMs are subjected to physiological conditions, their surface charges will lead to binding of available serum proteins, forming a so-called protein corona. Based on fluorescence correlation spectroscopy analysis, Rocker *et al.*[76] quantitatively studied the attachment of human serum albumin (HAS) to the surface of CdSe/ZnS QDs. Their results showed that a protein monolayer 3.3 nm thick formed on all particles. The authors further found that the binding itself was a dynamic phenomenon where proteins generally resided on the QDs for approximately 100 s. On the other hand, the curved surfaces of NIR NMs can also affect the secondary and tertiary structures of proteins, and in some case cause irreversible changes.[78] For example, gold NPs were shown to influence conformational changes in the structure of bovine serum albumin (BSA) in a dose-dependent manner,[79] whereas no major conformational change was recorded for BSA when adsorbed to carbon C_{60} fullerene NPs.[80] NP-induced protein conformational changes may affect the downstream protein–protein interactions, cellular signalling, and also DNA transcription, which is particularly important for enzymes.

11.3.2 *In vitro vs. In vivo* Assays

In vitro methods, such as the lactate dehydrogenase (LDH) assay of cell membrane integrity, immunochemistry markers for apoptosis and necrosis, and the MTT assay of mitochondrial function, can produce reproducible results rapidly and inexpensively without the use of animals (Figure 11.4).[81] In addition, they also achieve specific and quantitative measurements of toxicity. These methods have therefore been widely employed in the initial evaluation of the expected biocompatibility of NIR NMs. However, one of the disadvantages is that these methods cannot provide enough information on the mechanism of cellular toxicity. For instance, many colorimetric assays such as Live/Dead, trypan blue, and neutral have been used to quantitatively assess the toxicity of NIR NMs; unfortunately, these methods simply discriminate live cells from dead ones, and cannot provide any information regarding the mechanisms of cell toxicity. At the same time, in some cases, the presence of NIR NMs may affect the accuracy and precision of colorimetric assays for *in vitro* toxicity *via* the interaction between NIR NMs and the colour-generating dyes. For example, due to the fluorescence-quenching ability of carbon nanotubes (CNTs), intracellular CNTs can quench the fluorescence of MTT formazan crystals or other dyes such as neutral red or Alamar Blue, and achieve

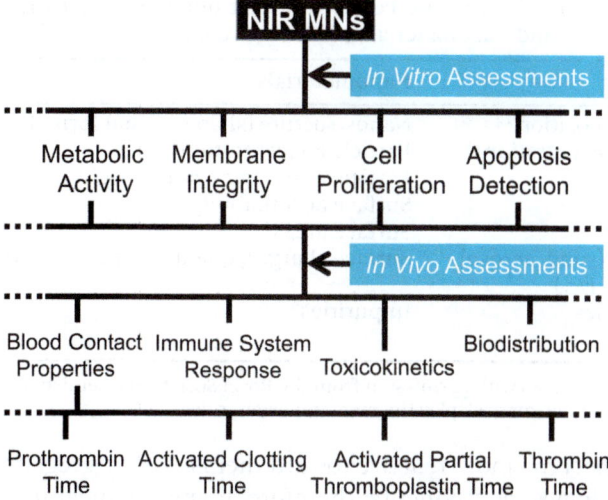

Figure 11.4   *In vivo* and *in vitro* studies for nanotoxicity research.

conflicting results.[82,83] Thus, the *in vitro* methods are mainly used to identify specific characteristics of NMs that can be used as indicators of toxicity and in order to establish a ranking of NP toxicity for mechanistic studies.

On the other hand, *in vivo* methods are expensive and time-consuming. However, these tests not only provide rich information on the carcinogenicity, dermal, pulmonary, and gastrointestinal toxicities induced by NIR NMs, but also can evaluate the immunological, reproductive, and developmental toxicity to determine the chronic systemic toxicity and related mechanism of NIR NMs.[26,84] In addition, animal models are particularly useful to study absorption, distribution, metabolism, and elimination of NIR NMs in the body. Thus, *in vivo* test results are often suitable for use as a prognostic of long-term physiological effects.

11.3.3   Effects of Physicochemical Properties on Nanotoxicity

Compared to the corresponding bulk material, NMs have unique properties and characteristics such as size, shape, and surface area. These can greatly affect the biological interaction of NIR NMs with biomolecules, and endow NIR NMs with unique mechanisms of toxicity from xenobiotics (Table 11.1).[85]

11.3.3.1   Effect of Size and Surface Area

From a toxicological point of view, the interaction of NIR NMs with biological organisms typically takes place at the surface of NIR NMs. As NM size decreases, the surface area exponentially increases and more atoms are displayed on the surface rather than within the bulk. These can make the surface

Table 11.1 Comparison of the key factors that dominate the toxic responses of NMs and bulk materials of identical chemical composition.[a]

Bulk materials	Nanomaterials
Chemical composition*[b]	Nanostructure (shape, crystal forms)*
Dose (mass concentration)*	Particle concentration*
Exposure route*	Particle size*, aggregation/agglomeration*
Reactivity	Surface absorbability*
Conductivity	Surface areas*
Physical form (solid, aerosol, suspension, *etc.*)	Surface charge*, quantum effects*, crystal face
Purity/impurities	Impurities*
Solubility	

[a]Reproduced from ref. 85 with permission from the Royal Society of Chemistry.
[b]Those marked with an asterisk play the most important roles.

of NIR NMs become more reactive toward themselves (aggregation) and surrounding biological components; therefore, many chemical reactions, such as catalytic reactions and redox reactions, may take place on the NM surface, causing the formation of ROS (such as superoxide anions or hydrogen peroxide) which subsequently oxidize other biomolecules. Recently, many clinical and experimental studies have indicated that small size can enhance the production of ROS compared to the corresponding bulk material. On the other hand, size and surface area also play a role in how the body responds to the distribution and elimination of NIR NMs,[86] and influence the mode of cellular uptake and the efficiency of particle processing in the endocytic pathway.[87] *In vitro* studies demonstrated slower uptake and processing of large latex spheres (>200 nm) relative to small ones (50 and 100 nm).[87] More importantly, increased uptake into certain tissues may lead to the accumulation of NIR NMs, where they may interfere with critical biological functions.[88]

11.3.3.2 Effect of Composition

Chemical composition at the surface of NIR NMs largely defines their chemical interactions.[64] Thus, it is also relevant in relation to cell molecular chemistry and oxidative stress. Harper *et al.*[89] studied the effect of NM composition on toxicity using an embryonic zebrafish model. The selected commercially available dispersions of NPs with similar size included aluminium oxide, holmium oxide, titanium oxide, gadolinium oxide, zirconium oxide, dysprosium oxide, samarium oxide and erbium oxide, negatively charged yttrium oxide, silicon dioxide, and alumina-doped cerium oxide. Remarkable mortality was observed after a 5 day continuous waterborne exposure to erbium oxide and samarium oxide with a concentration of 50 ppm, and to holmium oxide and dysprosium oxide with a concentration of 250 ppm. After treatment with yttrium oxide, samarium oxide, and dysprosium oxide at concentrations of 10, 50, and 250 ppm, significant morphological malformations were observed in embryonic zebrafish. In contrast, no significant morbidity or mortality was observed for the other NPs.

Moreover, after functionalization of NIR NMs, the functional groups added to the surface can potentially interact with biological components, alter biological function, and allow passage of NIR NMs that would not normally be taken up by certain cells.

11.3.3.3 Effect of Degradability

The degradability of NIR MNs in cells and in the body is an important factor in acute and long-term toxicity. On one hand, non-degradable NIR NMs can be deposited in organs where they may result in detrimental effects to the cells. On the other hand, biodegradable NIR NMs can lead to unpredicted toxicity due to unexpected toxic degradation products.[90] NIR NMs may contain transition metals (*e.g.* QDs) or other compounds with known toxicity. Thus, degradation of these NMs may release the metal ions and other toxic compounds into the biological environment, leading to generation of ROS, interaction with biomolecules, or reaction with the cellular redox potentials and finally resulting in cell damage.[91]

11.4 *In vitro* and *In vivo* Nanotoxicity of Near Infrared Nanomaterials

11.4.1 Nanotoxicity of Carbon-Based Nanomaterials

Owing to their unique chemical and physical structure, carbon-based NMs are potential candidates for a variety of biomedical applications, including early imaging, NIR photothermal therapy, photoacoustic imaging, and drug delivery.[3,92,93] When these NIR NMs are introduced into complex biological systems, the related risk assessment is of considerable importance to both human and environmental health. Based on current studies, we mainly discuss the *in vitro* and *in vivo* toxicity of CNTs and graphene.

11.4.1.1 Nanotoxicity of Carbon Nanotubes

11.4.1.1.1 *In vitro* Studies. Many groups have investigated the toxic effects of CNTs on cells, including cell apoptosis, overt toxic reactivity, ROS production, membrane perturbations, and cell signalling (Table 11.2). Zhao's group systematically studied the toxicity of unmodified CNT samples prepared by ultrasonication of CNTs in a culture medium without adding surfactants.[94] Their results indicated that the CNTs could induce the loss of phagocytic ability and ultrastructural injury to alveolar macrophages. Moreover, single-walled CNTs (SWCNTs) showed higher toxicity than three other types of carbon-based NMs. The sequence order on a mass basis was SWCNTs > MWCNTs > quartz > C_{60} (Figure 11.5). After exposure to human epidermal keratinocytes, SWCNTs could cause oxidative stress and cellular toxicity, changing the ultrastructure and morphology of the cultured skin cells.[95,96] In

Table 11.2 Summary of *in vitro* toxic effects of CNTs on various typical cell lines.[a]

CNTs	Cells	Dose/exposure time	Toxicity	Ref.
SWCNTs	Human lung epithelial cells	0.25–100 µg mL^{-1} 24 h	Cytotoxicity	98
SWCNTs	Rat lung epithelial cells	2.5, 5.0, and 10.0 µg mL^{-1} 18 h	Cytotoxicity	99
SWCNTs	Rat neuronal PC12 cells	0.01, 0.1, 1, 10, and 100 µg mL^{-1} 24 h	Cytotoxicity	100
SWCNTs	Human epithelial-like HeLa cells	1 mg mL^{-1} 96 h	Cytotoxicity	101
MWCNTs	Human epithelial-like HeLa cells	100 µg mL^{-1} 24 h	Extent of toxicity attenuation increased with serum proteins adsorbed on CNTs increasing	102
MWCNTs	Human lung epithelial cells	1–40 µg mL^{-1} 2, 4, and 24 h	Apoptosis and oxidative DNA damage	103
PEG–SWCNTs	Rat neuronal PC12 cells	0.1, 1, 10, and 100 µg mL^{-1} 24 h	Less toxic than pristine SWCNTs	104
MWCNTs	Rat glioma cells	0, 25, 50, 100, 200, and 400 g mL^{-1} 24 h	Decreased cell viability, cell apoptosis, and G1 cell cycle arrest; increased level of oxidative stress	105
MWCNTs	LPS-primed human macrophage	100 µg mL^{-1} 3, 6, and 9 h	Length-dependent induction of IL-1b and IL-1a secretion proinflammatory response	106
MWCNTs	Mouse macrophages CHO–K1 cells	0–100 g mL^{-1} 16, 24, and 32 h	Cytotoxicity	107

[a]CNTs, carbon nanotubes; IL, interleukin; MWCNTs, multiwalled carbon nanotubes; PEG, poly(ethylene glycol); SWCNTs, single-walled carbon nanotubes.

another dermatological trial, CNTs induced insignificant side effects.[97] The difference may be attributed to the different properties of the CNTs prepared by different processes.

Up to now, many experimental results have demonstrated that CNTs can be internalized by various cells and show cell line-dependent toxicity. For example, cytotoxicity could be induced by oxidative stress when rat macrophages were exposed to MWCNTs.[108] MWCNTs could cause remarkably toxic effects on human T cells in a concentration-dependent manner.[109] Choi *et al.*[110]

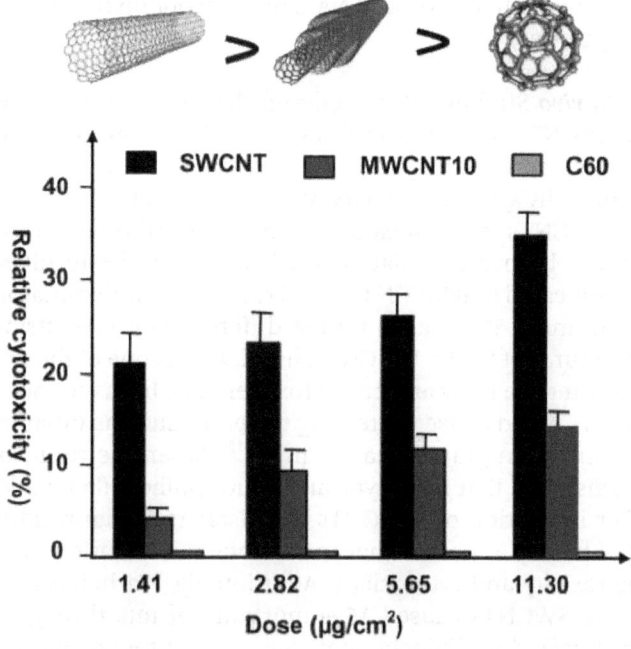

Figure 11.5 The ranking of cytotoxicity of carbon NMs on a mass basis is: SWCNTs > MWCNTs > C_{60}. Reproduced from ref. 94 with permission from the Royal Society of Chemistry.

reported that exposure of A549 cells to SWCNTs could result in inflammatory responses with oxidative stress and cell membrane damage. Cui's group reported that exposure of human embryo kidney cells to SWCNTs could cause the inhibition of cell adhesion and cell proliferation.[111] Kisin *et al.*[112] reported that SWCNTs could cause loss of viability of lung fibroblast cells and result in DNA damage in a concentration- and time-dependent manner. Simon-Deckers *et al.*[113] reported that the proliferation of human lung tumour cells could be inhibited by MWCNTs, finally resulting in cell death. Additionally, *in vitro* studies by Asakura *et al.*[114] indicated that cytotoxic and genotoxic responses could be triggered in Chinese hamster lung cells when they were exposed to MWCNTs.

It is very important to note that the metallic impurities in CNTs are the important factor in inducing significant toxic responses.[115,116] Therefore, a quantitative measurement of the concentration of metal impurities in CNTs is key, although this is very difficult. Recently, neutron activation analysis (NAA) technique as a non-destructive standard method has been used to quantitatively analyse the metal impurities in CNTs, and ICP-MS is regarded as a practical analytical method.[117] In the absence of a true reference material for CNTs, the NAA method can provide the best estimate of the true value of metallic impurities in CNTs, while ICP-MS is a desktop instrumental method. By using both ICP-MS and NAA, one can obtain the absolute concentration of impurities.

11.4.1.1.2 *In vivo* Studies.   It is believed that *in vivo* toxicology study can better evaluate CNT-induced toxicities such as the immunological, reproductive, cardiovascular, neurological, and developmental toxicity, and investigate the chronically systemic toxicity of NMs (Table 11.3). After intratracheal instillation of CNTs to guinea-pigs, no measurable inflammation was observed in the bronchoalveolar space.[118] However, the result reported by Lam *et al.*[119] indicated that SWCNTs could result in an inflammatory response in the lungs of mice. The reason for the difference may be attributed to the different structures of CNTs. For CNTs, inhalation is one of the most popular pathways to enter the body and cause toxic effects. It was found that inhalation of SWCNTs could cause acute lung toxicities such as inflammation and lung injury, and form granulomas in mice.[120] Later, the result of Ma-Hock *et al.*[121] demonstrated that histiocytic and neutrophilic inflammation could be induced after inhalation of MWCNTs by Wistar rats (Figure 11.6). Carrero-Sanchez *et al.*[122] also reported that intratracheal instillation of CNTs could induce lung toxicity and cytotoxicity. Additionally, Warheit *et al.*[123] reported that high-dose SWCNTs caused 15% mortality of rats through instillation and the effect was dose-independent. Bai *et al.*[124] found that the repeated administration of CNTs by intravenous injection could cause reversible testis damage in male mice.

Based on the intrinsic NIR fluorescence, the biodistribution of the unmodified SWCNTs *via* intravenous injection in rabbits was studied.[134] At 24 h after intravenous administration, SWCNTs were found at significant concentrations only in the liver (Figure 11.7). Because of the limited sensitivity of NIR fluorescence method, a more sensitive and quantitative method for measuring the isotope ratios of ^{13}C-labelled SWCNTs was later developed to determine the biodistribution of SWCNTs *in vivo* (Figure 11.7a).[133] The unmodified SWCNTs were found to be distributed in the entire body, and over an extended period of time they mainly accumulated in the liver, lungs, and spleen. Zhao's group[135] further investigated the biodistribution of CNTs functionalized by a covalent hydroxylation reaction and then labelled them with ^{125}I. Their results indicated that functionalized SWCNTs mainly accumulated in the bone, kidney, and stomach of mice, and were finally excreted by the renal route. This result was consist with the biodistribution of ^{131}I-labelled SWCNTs (Figure 11.7b).[136] Notably, the unmodified SWCNTs accumulate mainly in the liver, lungs, and spleen, while the modified SWCNTs had changed their biodistribution, accumulation, and metabolism kinetics. Alternatively, PEGylated SWCNTs labelled with ^{13}C could distribute throughout most organs within 1 h except brain, intestine, and muscle.[137] After 7 days, they accumulated mainly in the liver, spleen, and skin.

Recently, several reports have studied the toxicity and behaviour of functionalized CNTs. After intravenous administration of ^{111}In-labeled diethylene-tri-aminepentaacetic-coated SWCNTs, it was found that the functionalized

Table 11.3 Summary of *in vivo* toxic effects of CNTs on various animals.[a]

CNTs	Animal exposed	Dose/exposure time	Toxicity	Ref.
SWCNTs	8- to 12-week-old nude mice	1 μM (100 μL)/animal 1, 2, 3, and 4 months	No evidence of toxicity after 4 months	125
MWCNTs	Female C57Bl/6 mice	50 μg/animal 24 h and 7 days	Inflammation and granuloma formation	126
MWCNTs	Sprague-Dawley rats	0.5, 2 or 5 mg/animal 60 days	Induced inflammatory and fibrotic responses	127
SWCNTs	Immune-competent hairless SKH-1 mice	40, 80, and 160 μg/animal 5 days	Oxidative stress and increased number of dermal cells	115
MWCNTs	Male C57BL/6 mice	0.3, 1, and 5 mg m^{-3} for 6 h per day 14 days	Suppression of systemic immune function	128
MWCNTs	Male Sprague-Dawley rats	1, 10, and 100 μg/animal 1, 7, 30, 90, and 180 days	Only evidence of apoptosis of alveolar macrophages	129
MWCNTs	C57BL/6J mice	10, 20, 40 and 80 μg/animal 1, 7, 28 and 56 days	Penetrations of alveolar macrophages, the alveolar wall, and visceral pleura are both frequent and sustained	130
MWCNTs	Kunming mice	90 min a time for 4 times a day with initial concentration 80 mg m^{-3} 8, 16, and 24 days	Proliferation and thickening of alveolar walls	131
SWCNTs	Normal New Zealand rabbits	20 μg kg^{-1} 24 h	Liver only significant site of accumulation. Absence of acute toxicity	132
SWCNTs	Male Kunming mice	0.6 and 2 mg/animal 1, 7, and 28 days	No signs of acute toxicity	133

[a]CNTs, carbon nanotubes; MWCNTs, multiwalled carbon nanotubes; SWCNTs, single-walled carbon nanotubes.

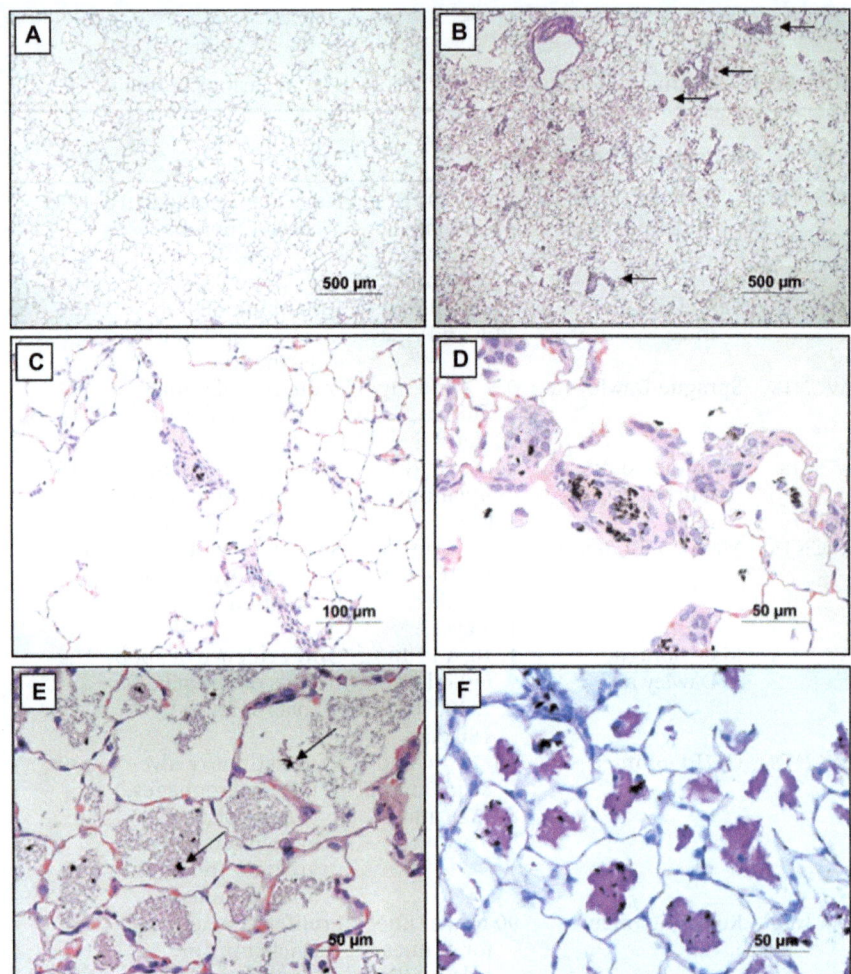

Figure 11.6 Overview of the lung section of (a) control rat and (b) rat exposed to 2.5 mg m^{-3} MWCNT. (c) is a detail view of focal granulomatous inflammation observed in a rat exposed to 0.1 mg m^{-3} MWCNT, and (d) in a rat exposed to 2.5 mg m^{-3} MWCNT. (e) Lung section of a rat exposed to 2.5 mg m^{-3} showing intra-alveolar eosinophilic granular material which was diagnosed as lipoproteinosis. (f) This intra-alveolar eosinophilic granular material was PAS-positive. Reproduced from L. Ma-Hock *et al.*, Inhalation toxicity of multiwall carbon nanotubes in rats exposed for 3 months, *Toxicol. Sci.*, 2009, **112**, 2, with permission from Oxford University Press.[121]

SWCNTs were mostly cleared from the systemic blood circulation system through renal excretion at 3 h and they were not found to be retained by the reticuloendothelial organs.[138] Later, Liu *et al.*[139] reported that the PEGylated SWCNTs showed a longer blood circulation time after intravenous injection in mice than the surfactant-suspended pristine SWCNTs in rabbits. Their

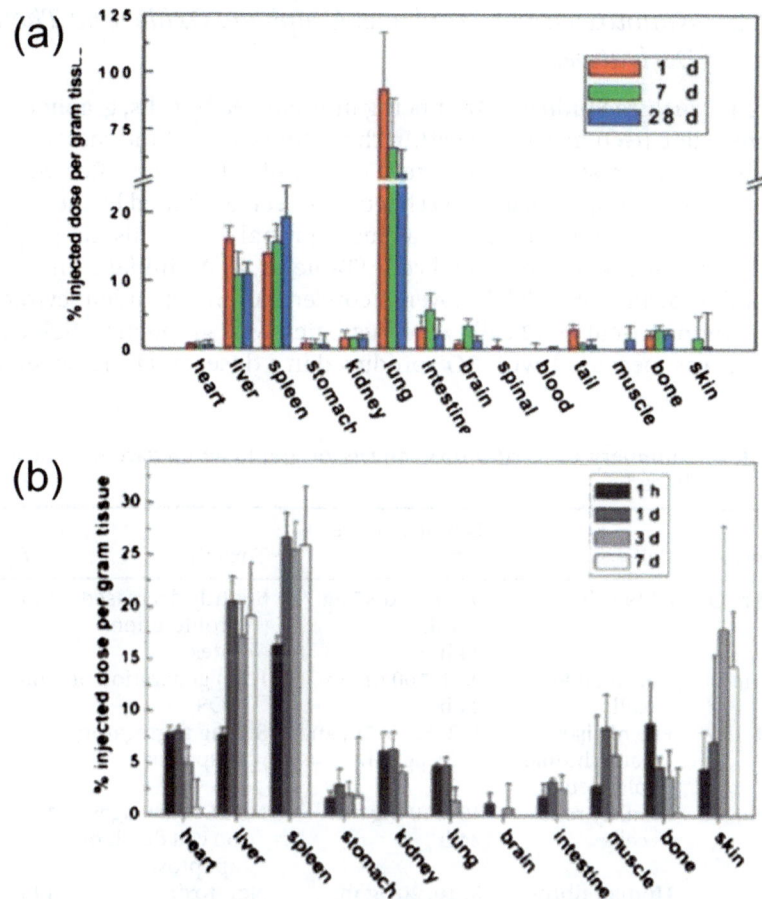

Figure 11.7 (a) Biodistribution histograms of unmodified SWCNTs in mice at different time points post exposure *via* intravenous injection. Reproduced from S.-t. Yang, W. Guo, Y. Lin, X.-y. Deng, H.-f. Wang, H.-f. Sun, Y.-f. Liu, X. Wang, W. Wang, M. Chen, Y.-p. Huang and Y.-P. Sun, *J. Phys. Chem C*, 2007, **111**, 17761. Copyright (2007) American Chemical Society.[133] (b) Biodistribution histograms of PEG-modified SWCNTs in mice at different time points post exposure *via* intravenous injection, quantified by the isotope (^{13}C) labelling technique. Reproduced from ref. 137 with permission from John Wiley and Sons. Copyright © 2008 Wiley-VCH Verlag GmbH & Co. KGaA, Weinheim.

results further demonstrated that CNTs functionalized with PEG–phospholipid with increasingly branched PEG chains exhibited enhanced blood circulation half-life when intravenously injected into animals. The higher degree of surface PEGylation of SWCNTs also improved the excretion of the CNTs from the body after 3 months of exposure. The PEG coatings described in this study enabled low reticuloendothelial system uptake and increased elimination from the body.

11.4.1.2 Nanotoxicity of Graphene, Graphene Oxide, and Their Derivatives

11.4.1.2.1 *In vitro* Studies. After being internalized by cells, graphene and graphene oxide (GO) may travel within the cytoplasm and interact with biomolecules, finally resulting in a certain degree of cytotoxicity. Recently, the *in vitro* toxicity of graphene and GO sheets has been evaluated by the cell survival test using different cells, such as rat neuronal PC12 cells, lung epithelial cells, fibroblasts, and neuronal cells (Table 11.4). An initial comparative examination by Biris *et al.*[100] discovered concentration-dependent cytotoxicity when rat neuronal PC12 cells were incubated with graphene. High oxidative stress was measured, with ROS produced in a dose- and time-dependent

Table 11.4 Summary of *in vitro* toxic effects of graphene on various typical cell lines.[a]

Graphene	Cells	Dose/exposure time	Toxicity	Ref.
GO and rGO	A549 cells	0, 20, and 85 µg mL^{-1} 24 h	Slightly decreased proliferation rates	143
Graphene	Neuronal PC12 cells	0.01–100 mg mL^{-1} 24 h	High generation of ROS	100
GO and rGO	Freshly isolated human platelets	1, 2, 5, 10, 20, and 25 µg mL^{-1}	Strong aggregatory response	144
Carboxylated GO	Monkey renal cells	10–300 mg mL^{-1} 24 h	No LDH leakage; no cell death or apoptosis	142
GO	Human fibroblast cells	5, 10, 20 µg mL^{-1} 5 days	Non-toxic	145
GO	Monkey renal cells	10–300 mg mL^{-1} 24 h	Cell membrane accumulation and dose-dependent oxidative stress	142
GO	Human neuroblastoma SH-SY5Y cell line	<80 µg mL^{-1} 48 and 96 h	No obvious cytotoxicity	146
GO	Human erythrocytes	3.125–200 mg mL^{-1} 3 h	High haemolytic activity	140
GO	Human lung cells	10–100 µg mL^{-1} 24 and 48 h	Significant increase of early- and late-apoptotic cells	147
GO	Bone marrow derived dendritic cells	1–25 µg mL^{-1} 48 h	No alteration of antigen engulfment	148

[a]GO, graphene oxide; LDH, lactate dehydrogenase; rGO, reduced graphene oxide; ROS, reactive oxygen species.

manner. In a comprehensive study, graphene was tested on A549 adenocarcinomic human epithelial cells. At low GO concentrations, graphene induced neither obvious cytotoxicity nor significant cellular uptake. However, high concentrations of GO could induce oxidative stress that slightly reduced cell viability. A large number of other studies indicated that, compared to GO, reduced GO (rGO) and graphene exhibited less dispersibility and were more toxic to cells.[140,141] The results indicated that the production of ROS, which is a critical characteristic of oxidative stress within cells, was an important mechanism leading to the *in vitro* toxicity of graphene-based NMs. In another study, GO and carboxylated GO were tested on monkey renal cells at concentrations between 10 and 300 µg mL^{-1} (Figure 11.8).[142] Their results demonstrated that GO accumulated mainly at the surface of the cell membrane, resulting in significant destabilization of the F-actin alignment; however, carboxylated GO was largely internalized by the cells and accumulated in the perinuclear region without causing damage to the cytoskeletal morphology, even when the concentration was a high as 300 µg mL^{-1}. In this case, no physical damage to the cell membrane was observed, thus supporting the hypothesis that another mechanism of cell death must be attributed to the *in vitro* toxicity of graphene.

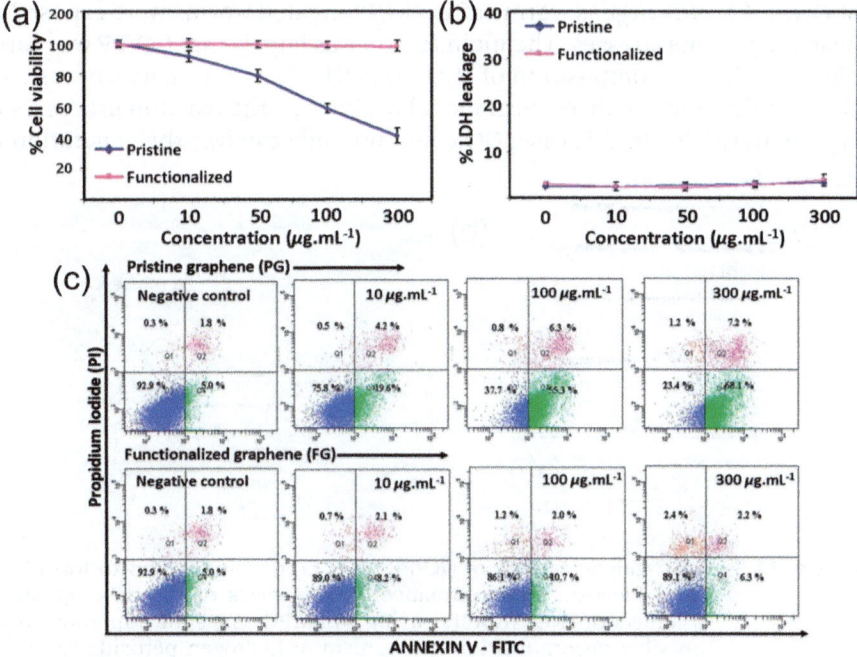

Figure 11.8 (a) Cell viability and (b) LDH leakage of Vero cells treated with p-G and f-G at different concentrations. (c) Flow cytogram showing apoptosis assay based on annexin V-FITC and PI staining of cells. Reproduced from ref. 142 with permission from the Royal Society of Chemistry.

Recently, Li *et al.* studied the toxicity of graphene in RAW 264.7 macrophage cells in detail at the molecular level, and provided an improved molecular mechanism for further understanding the cytotoxicity induced by graphene.[149] After treatment with graphene, the signalling pathways of both mitogen-activated protein kinase (MAPK) and transforming growth factor beta (TGF-β) were activated, further resulting in the downstream Bcl-2 protein family initiating mitochondrial-related apoptosis. Moreover, all three phosphorylated kinases, *i.e.*, p38 mitogen-activated protein kinase (p38), extracellular signal-regulated kinase (ERK), and c-Jun N-terminal kinase (JNK), were significantly upregulated, indicating that the three MAPK signal pathways were activated after graphene treatment. The stimulated transcription activity of the Smads proteins verified that TGF-β, a key component of cell apoptosis, was also found to be activated by graphene. As a result, Bim and Bax were further activated, causing permeabilization of the mitochondrial outer membrane and release of mitochondrial proapoptotic factors into the cytosol. The above results indicate that graphene could induce ROS production and trigger the mitochondrial apoptotic pathway, ultimately leading to mitochondrial damage. Alternatively, Zhang *et al.*[150] reported that GO functionalized with PEGylated poly-L-lysine (GO/PP) could cause ROS production under environmental stress such as heat and juglone stresses. The results indicated that GO/PP significantly impaired the stress-resistance capacity of *Caenorhabditis elegans* worms after GO/PP-treated worms were exposed to heat and juglone stresses. The main reason was largely that GO/PP was capable of catalytic decomposition of H_2O_2 to COH (Figure 11.9), which was consist with the result of theoretical model to simulate the reaction processes of H_2O_2 with GO/PP. In this case, GO could not only catalyse the generation of

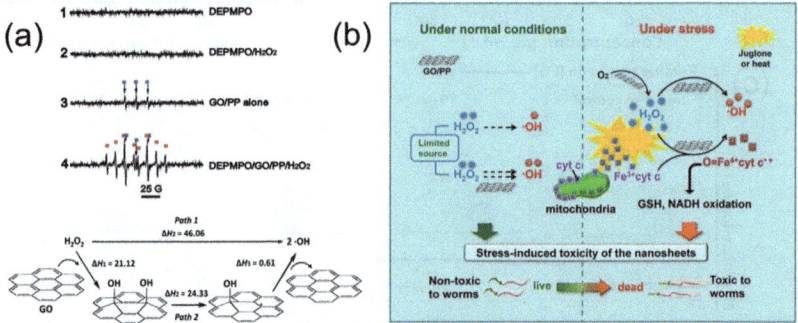

Figure 11.9   (a) Catalytic activity of GO/PP on H_2O_2 decomposition to form ·OH. (top) Electron spin resonance (ESR) spectra of the decomposition products of H_2O_2 in water using DEPMPO as a spin trap. (bottom) A possible decomposition mechanism of hydrogen peroxide catalysed by coronene, a model for the graphitic area of GO using the density functional theory method. (b) Schematic diagram of the proposed mechanism for the stress-induced toxicity of GO/PP on *C. elegans*. Reproduced from ref. 150 with permission from John Wiley and Sons. Copyright © 2012 Wiley-VCH Verlag GmbH & Co. KGaA, Weinheim.

COH, but also facilitated electron transfer from cytochrome c to H_2O_2, finally generating ROS. Therefore, GO/PP in the *C. elegans* body had the ability to accelerate and enhance electron transfer between the interface of juglone/O_2, H_2O_2/COH, or cytochrome c/H_2O_2, which impaired the inherent antioxidant defence system and eventually resulted in dramatic toxicity to the worms.

Many studies have demonstrated that uncoated graphene and GO exhibited concentration-dependent toxicity to various types of cells, and many hydrophobic macromolecules including chitosan, Tween, PEG, dextran, and protein have been employed to significantly attenuate their cytotoxic effects.[151–153] Using both electrostatic and hydrophobic interactions, proteins such as fetal bovine serum could be adsorbed on the surface of pristine graphene and GO, impeding the direct interaction of GO with cells to reduce their cytotoxicity. Liu's group and others have also found that coating pristine graphene and GO with biocompatible polymers such as PEG or Tween-80 could endow them with excellent solubility and stability in physiological solutions, reducing non-specific binding with functional biomolecules, and lowering their *in vitro* cytotoxicity to various cells.[154]

11.4.1.2.2 *In vivo* Studies. To understand the potential *in vivo* toxicity of graphene and graphene-based NMs, their behaviour in animals should be investigated first (Table 11.5). Yang *et al.*[155] first studied the *in vivo* biodistribution of PEGylated GO nanosheets with ultrasmall sizes using fluorescent labelling. Their results indicated that PEGylated nano-GO showed the ability for efficient passive targeting tumour due to the enhanced permeability and retention (EPR) effect. In later studies, the *in vivo* biodistribution and behaviour of graphene and its derivatives in animals have been investigated by the radiolabelling method. Zhang *et al.*[156] found that after intravenous injection [188]Re-labelled GO without surface coating mainly accumulated in the lung for a long time without much excretion, which was in agreement with the results of Cui's group.[145] In recent work, Yang *et al.*[157] studied the long-term *in vivo* biodistribution of nGO–PEG labelled with [125]I. They found that intravenously injected ^{125}I–nGO–PEG predominantly deposited in the reticuloendothelial system (RES) including the liver and spleen. Unlike the uncoated GO, no appreciable lung uptake of nGO–PEG was observed. There results demonstrated that the observed time-dependent decrease in the RES was indeed owing to the excretion of nGO–PEG. The analysis of liver slices stained with haematoxylin and eosin (H&E) exhibited that large numbers of black spots (aggregated nGO–PEG) were observed at early time points after injecting nGO–PEG, but gradually disappeared over time, consistent with the biodistribution data based on radioactivity measurements. Moreover, high radioactivities were found in both urine and faecal samples, suggesting that nGO–PEG with ultrasmall sizes in the range of 10–30 nm (sheet diameter) might be cleared out through both renal and fecal excretion.

In order to further understand the effect of sizes and surface coatings on the *in vivo* behaviour of GO, three GO samples, including ultrasmall nGO–PEG (~23 nm) with covalent PEG coating, ultrasmall nRGO–PEG (~27 nm),

Table 11.5 Summary of *in vivo* toxic effects of graphene on various animals.[a]

Graphene	Animal exposed	Dose/exposure time	Toxicity	Ref.
GO	Japanese white rabbits	100–300 mg/eye 2 and 49 days	No eye changes	158
GO	Male C57BL/6 mice	50 µg/mouse 4, 24 h, 21 days	Severe and persistent lung injury	159
Amino–GO	Swiss male mice	250 µg kg^{-1} 24 h	Absence of thrombotoxicity	144
GO	Kunming mice	0.4 mg/mouse 1, 7, and 30 days	Exhibited chronic toxicity	145
rGO	Swiss male mice	250 µg kg^{-1} 2 h	Less effective in platelet aggregation	160
GO	Japanese white rabbits	0.1, 0.2, and 0.3 mg 2, 7, 28, and 49 days	Few changes in eyeball appearance	85
PEG–GO	Healthy female Balb/c mice	20 mg kg^{-1} 3, 7, 20, 40, and 90 days	RES accumulation and no alteration of biochemical blood parameters	161
GO	Swiss male mice	250 µg kg^{-1} 15 min	Significantly large numbers of lung vessels with platelet thrombi	160
Graphene	Female C57BL/6 strain mice	5 and 50 µg/ animal 7 days	Inflammatory cytokine release; acute pulmonary inflammatory response	162
Dextran–GO	Healthy female Balb/c mice	20 mg kg^{-1} 1, 3, and 7 days	Accumulation in liver and spleen; gradual clearance within 1 week; no short-term toxicity	163

[a]GO, graphene oxide; PEG, poly(ethylene glycol); RES, reticuloendothelial system.

and larger RGO–PEG (65 nm) with non-covalent PEGylation were intravenously injected into mice.[157] The results showed that nRGO–PEG and RGO–PEG with non-covalent PEG coatings showed much longer blood circulation compared to nGO–PEG, leading to a remarkably increased tumour uptake for the first two GO derivatives due to the EPR effect. The size of NMs also played an important role in regulating their *in vivo* behaviour. Compared to larger RGO–PEG NMs, ultrasmall nGO–PEG and nRGO–PEG exhibited significantly reduced RES accumulation.

   In addition, after intravenous injection of pristine GO without a surface coating, both Huang[156] and Cui[145] found that the pristine GO nanosheets mainly accumulated in the lung for a long period of time, inducing granuloma formation and pulmonary oedema. The Dash group also found that, after intravenous injection, GO without surface modification could induce high thrombogenicity in mice and evoke a strong aggregatory response in human platelets, which indicated that pristine GO nanosheets may induce blood clots.[160]

Pulmonary toxicity is a major concern in the industrial production of NIR NMs, because their respirability might cause damage and eventually long-term disease in humans who come into contact with this type of NMs. It has been shown that few-layer graphene nanosheets with diameters up to 25 μm could deposit beyond the ciliated airways after inhalation, and provoked high levels of inflammation in the mouse lung (Figure 11.10).[162] Another study also indicated that, after directly injecting aggregated graphene, Pluronic-dispersed graphene, and GO into the lungs of mice, GO could induce the mitochondrial generation of ROS, activate inflammatory and apoptotic pathways, and cause severe and persistent injury in the lungs.[159] However, the mice treated with aggregated graphene and dispersed graphene showed no obvious lung injury. The authors further measured lung fibrosis in mice treated with GO, aggregated graphene or Pluronic-dispersed graphene, and then examined trichrome-stained lung sections. The results indicated that the aggregated graphene could induce patchy fibrosis in mice, while Pluronic-dispersed graphene did not cause obvious fibrosis.

Therefore, chemical functionalization is critical to modulate the toxicity of graphene. So far, various polymers (*e.g.*, PEG, Dextran, and chitosan) on the surface of graphene have been studied to eliminate the *in vivo* toxicity of graphene and its composites.[160,164-172] PEGylation of GO can, for example, decrease the toxic effect in mice. Liu *et al.* systematically studied the potential long-term toxicity of nGO–PEG to mice. After intravenous injection at a dose of 20 mg kg^{-1}, the blood of treated mice was collected for blood biochemistry and haematology tests. The results indicated that nGO–PEG treated mice at different times after injection were normal compared with the control groups. Moreover, no noticeable organ damage or inflammation was observed, suggesting no obvious toxicity caused by nGO–PEG at the tested dose. In another recent work by Gollavelli *et al.*,[173] polyacrylic acid (PAA) was used to functionalize GO to reduce its toxicity to zebrafish. Therefore, well-designed surface modifications of graphene could effectively decrease its *in vivo* toxicity.

11.4.2 Nanotoxicity of Quantum Dots

QDs, a kind of quasi-zero-dimensional semiconductor NMs, are usually made up of group II–VI or group III–V elements with diameters ranging from 1 nm to 10 nm.[174,175] Due to their unique nanostructure and quantum confinement effects, QDs could be approved as a promising alternative to traditional organic luminescence probe with many advantages such as preferable photostability, wide excitation spectrum, narrow emission spectrum, long fluorescence lifetime, and large Stokes shift.[176] Specifically, by controlling the size and the chemical composition, the fluorescence emission of QDs may be tuned from the near UV, through the visible region, and into the NIR, which is suitable for bioimaging because of the deeper penetration of NIR light. In recent years, NIR QDs have emerged as a promising tool in analytical applications, including biosensor and *in vivo* imaging.[174,177-179] Although they offer potentially invaluable societal benefits such as drug targeting and

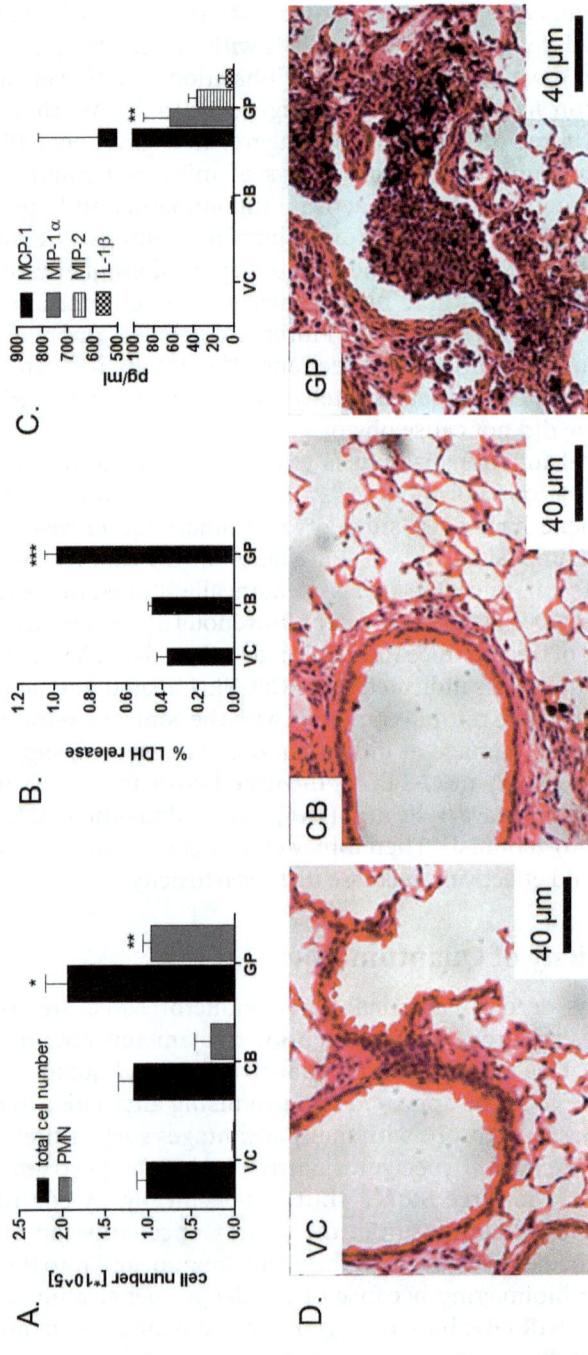

Figure 11.10 Pulmonary inflammatory response to CB and GP 24 h postaspiration. (a) Total cell number and total granulocyte number in the lavage fluid following exposure to CB and GP. (b) Measurement of the membrane leakage as LDH levels in the lavage fluid. (c) Concentration of the chemokines MCP-1 and MIP-1R as well as cytokines MIP-1 and interleukin (IL)-1β in BAL. (d) Lung histology 24 h postaspiration. Reproduced with permission from A. Schinwald, F. A. Murphy, A. Jones, W. MacNee and K. Donaldson, *ACS Nano*, 2012, **6**, 736. Copyright (2012) American Chemical Society.[162]

in vivo biomedical imaging, they may also pose risks to human health and the environment under certain conditions. For example, Cd and Se, two of the most widely used constituent metals in QD core metalloid complexes, are known to cause acute and chronic toxicities in vertebrates and are of considerable concern for human health and the environment. Unfortunately, most of the highly luminescent QDs currently used for biomedical imaging contain toxic elements, including Cd, Hg, Pb, Se, Te, and As.[180,181] Moreover, several studies have demonstrated that these toxic components strongly influence the cytotoxic effect of QDs, due to their eventual release into the cellular environment.

11.4.2.1   In vitro *Studies*

In vitro toxicological studies are important for understanding the *in vivo* data as well as a preliminary to *in vivo* applications. Cell culture models are necessary for preclinical safety assessment during the development of biomedical NPs. Up to now, many studies have demonstrated that QDs exhibit inherent toxicities in cells and living systems, including hepatic, renal, neurologic, and/or genetic toxicities. The main reason is that most of QDs contain Cd, Se, or Te, which may be released from the QDs, finally resulting in heavy metal toxicity. Cd ions have been shown to bind to thiol groups on critical molecules in the mitochondria and cause enough stress and damage to cause significant cell death. QDs derived from CdTe and CdSe were prone to photo and air oxidation, which could potentially promote free-radical formation that can instigate cytotoxicity.[180,182] Derfus *et al.*[183] used hepatic cells to monitor the toxic effects of QDs, as the liver is the primary site for acute damage from Cd and a major accumulation site for NPs. Their results showed that oxidation of the NP surface, either induced by exposure to air before solubilization or catalysed by UV light, could result in oxidation of Se and/or S, exposing free Cd. Exposure to air before solubilization or moderate to prolonged exposure to UV light after incubation increased the amount of Cd in the cells enough to cause observable cell death. Further efforts demonstrated that cells labelled with QDs synthesized under stable, inert conditions showed no toxic effects. By examining the effects of the free core materials in solution, Cd was determined as the primary cause of cytotoxicity, but its levels could be reduced or eliminated by adding additional surface coatings. Rizvi *et al.*[184] evaluated the cytotoxicity of MUA–QDs in three different cell lines, including SK-BR-3, MCF7, and HepG2. Their results indicated that MUA–QDs at concentration of 60 µg mL^{-1} were rather biocompatible with SK-BR-3 and MCF7 cell lines at 1 h and 24 h. However, HepG2 cells showed evidence of toxicity at MUA–QD concentrations of 15 µg mL^{-1} at 24 h. This demonstrated that QDs exhibited differential toxicity in different cell lines and further suggested different effects at the tissue and organ levels. The reason for the difference might be that after uptake by hepatocytes QDs were exposed to metabolic degradation, thus leading to the release of their toxic core components and the generation of ROS. Lovrić *et al.*[185] demonstrated that CdTe QDs could enter the cell and distribute in different subcellular compartments, resulting in changes in cell nuclear morphology as well as decreases in metabolic activity.

QDs can also damage DNA and disrupt normal cell activity. Hoshino *et al.*[186] reported that DNA damage was observed based on interactions with QDs coated with carboxylic acids. The authors further evaluated the genotoxic potential of QDs by comet assay. For 2 h after treating the WTK1 cells with QD–COOH at 2 μM, it was found that the tail length was significantly increased. After treatment for 12 h, the tail length was equal to that of the control cells, suggesting that the induced DNA damage was efficiently repaired during prolonged incubation. Remarkably, crude QD–COOH samples prepared only by membrane filtration, not by ultrafiltration, exhibited stronger DNA damage than purified QD–COOH. There is also evidence that free-radical generation promotes DNA damage in the absence and presence of light photon activation. Photoactivation of QDs by visible or UV light was shown to increase free-radical formation.[187,188] The mechanism involved in these environments is that a photon of light excites the QD, generating an excited electron that migrates to molecular oxygen, thus generating singlet oxygen.

In order to reduce the cytotoxicity of QDs, surface functionalization is widely used. The presence of a shell is capable of slowing down the release of Cd^{2+} ions into the cellular environment and the nature of the organic capping layer is responsible for both the cellular distribution and the protection of the QDs against oxidation.[189,190] Lovric *et al.*[185] found that CdTe QDs functionalized with mercaptopropionic acid and cysteamine were cytotoxic to rat pheochromocytoma cells at a concentration of 10 μg mL^{-1}, while uncoated CdTe QDs were cytotoxic at 1 μg mL^{-1}. These results suggested that surface coatings could enhance biocompatibility and decrease toxicity. Another common surface shell coating for QDs is ZnS. The ZnS shell is capable of protecting the core from oxidation and other environmental factors that contribute to Cd release. Without the disruption of the shell, the environment has no interaction with the Cd of the core, and it therefore cannot be toxic to the system.[182,191]

11.4.2.2   In vivo *Toxicity*

However, data from *in vitro* studies could be misleading and will require verification from animal experiments. *In vivo* systems are extremely complicated and the interactions of QDs with biological components, such as proteins and cells, could lead to unique biodistribution, clearance, immune response, and metabolism.[192,193] More importantly, it could lead to predictive models for assessing toxicity. The *in vivo* research involves experiments performed in the context of the entire system consisting of the body of an experimental animal. The overall behaviour of QDs could be summed up as follow: (1) QDs can enter the body *via* six principal routes—subcutaneous, intravenous, dermal, intraperitoneal, inhalation, and oral; (2) QDs can interact with biological components; (3) subsequently QDs can distribute to various organs; (4) QDs can enter the cells of the organ and reside in the cells.[90]

Two initial studies showed QDs did not excrete and remain intact *in vivo*.[194,195] This, however, has very recently been demonstrated to be dependent on size and surface chemistry.[196] Choi *et al.*[197] showed that cysteine-functionalized QDs <5.5 nm in diameter were excreted in the urine. In another study, at 24 h after intravenous injection of 15 mg kg^{-1} of PEGylated-Ag$_2$S QDs into the mouse tail vein, the NIR II photoluminescence signal suggested high tumour accumulation of QDs due to the non-specific EPR effect.[198] After 1, 2, 3, 7, 14, 28, and 60 days after injection of PEGylated-Ag$_2$S QDs, the RES including liver and spleen retained higher concentrations of PEGylated-Ag$_2$S QDs than other organs, which largely was attributed to the initial high uptake of PEGylated-Ag$_2$S QDs and the slow metabolic process of PEGylated-Ag$_2$S QDs. The concentration of PEGylated-Ag$_2$S QDs in intestine was low at first, and then gradually rose in the third week. It may account for the excretion of PEGylated-Ag$_2$S QDs through biliary pathway. It was worth noting that bone took up a considerable amount of PEGylated-Ag$_2$S QDs and still retained most of them even at 60 days after injection, this result demonstrated that the absolute elimination of QDs from the body may be much more difficult than anticipated. Therefore, the accumulation and retention of QDs in the tissue, whether in the long or short term, are likely to cause acute or chronic toxicity.[199] Lin *et al.*[200] also performed pharmacokinetic and toxicology studies in mice at time points of up to 6 months. According to their results, after intravenously injection into mice, commercially available Qtracker 705 non-targeted QDs (QD705) primarily accumulated within the liver, spleen, and kidney. The authors found no evidence of excretion or metabolism of the QD705 NPs within 28 days. Concerned by the persistence of the QDs, the authors examined the kidneys by TEM at 6 months after dosing, and observed significant renal toxicity in dosed mice but not in control mice. The "subtle but definitive" cytological changes noted in dosed mice consisted of proximal tubular degeneration, with pronounced changes to mitochondria in the proximal convoluted tubules. Based on these results, the authors cautioned that the *in vivo* administration of QD705 may be highly toxic.

Moreover, the *in vivo* fate and physiological behaviour of QDs was studied in *C. elegans* using various advanced techniques including green fluorescent protein (GFP) transfection, fluorescent imaging, synchrotron radiation, and classic toxicological approaches.[174] Contreras *et al.*[201] found that CdSe@ZnS core–shell QDs were degraded, even with ZnS coating, and oxidation of inner cores after digestion in *C. elegans* was verified. The degradation of QDs leads to release of toxic ions (Cd and Se), which are most likely to be the primary sources of QD toxicity.

11.4.3   Nanotoxicity of Noble Metal-Based Nanoparticles

A large number of studies have focused on applications of NIR active noble metal-based NPs.[13,202–207] In the past decades, because of their ultrasmall size, good biocompatibility, high photothermal conversion efficiency, and excellent luminescence properties, noble metal-based NPs have shown excellent potential for multiple applications. The so-called noble metal are metals that are resistant to corrosion and oxidation in moist air, including Au, Ag, Pt, Pd, Rh, Ir, Os, and Ru. Noble metals are traditionally valued because they are rare and precious, but now their biomedical applications are also important.

Recently, NIR techniques have been used extensively in both therapy and clinical diagnosis. NIR NMs have been extensively studies due to the capacity of NIR light to penetrate more deeply into biological tissues than visible light, because living cells and tissues have low light scattering and adsorption in this region. Due to their excellent NIR fluorescence and photothermal conversion efficiency, noble metal-based NPs can be useful for biomedical applications. For biological clinical applications, good biocompatibility and low toxicity are the significant conditions. Here, we mainly describe the toxicity of noble metal NPs which have a NIR response.

11.4.3.1 In vitro *Studies*

The toxicity of noble metal-based NPs is associated with many factors such as the particle's composition, surface modification, size, and shape. For example, when comparing the toxicity of Au, Ag, and Pt NPs, AgNPs were found to be the most toxic and AuNPs were non-toxic.[208] Chanu *et al.*[209] studied the *in vitro* cytotoxicity of ZnO-supported Au clusters and NPs, and reported that the as-made NMs were non-toxic to normal cells. Jang *et al.*[210] synthesized polyvinylpyrrolidone (PVP)-coated spherically clustered porous Au–Ag alloy NP (PVP–SPAN) which exhibited remarkable photothermal conversion efficiency under NIR light (Figure 11.11). They compared the cytotoxicity of several materials by MTT cell viability assay: PVP–HAN (PVP–coated hollow Au@Ag nanoshell without large pores), PVP–AEN(Au embedded NPs), and PVP–SPAN. It was found that PVP–SPAN showed much lower cytotoxicity than PVP–AEN, suggesting that etching process removed cytotoxic Ag and AgCl from PVP–AEN. Cheng *et al.*[211] systemically investigated the biosecurity of bimetallic Au/Ag NMs. They found that moderate concentrations of modified chitosan-capped bimetallic star-shaped NPs were not only non-toxic to normal cells and cancer cells, but also facilitated highly efficient photothermal elimination of cancer cells.

Xiao *et al.*[213] studied the cytotoxicity of PdNPs with a porous structure with increased NIR absorption by a cell viability assay of several cell lines. They found that the modified porous PdNPs showed excellent biocompatibility for A549 cells. The cell viability remained ~100% when the concentration of porous PdNPs was as high as 90 mg mL^{-1}.

Appropriate surface modification of noble metal-based NPs can provide low cytotoxicity, good stability, and excellent biocompatibility.[20,214] Synthetic polymers (*e.g.* PEG or polyoxyethylene) and peptides (*e.g.* heparin, glutathione, cysteine, BSA) are two kinds of popular reagents for surface modification. Liu *et al.*[215] modified Au nanorods (AuNRs) with a polythiol PEG-based copolymer. The obtained AuNR–PTPEGm950 had very low cytotoxicity and showed high efficacy for the ablation of cancer cells *in vitro*. Xu's group synthesized AgNPs with different sizes and then assessed their size-dependent toxicity.[216,217] Their results indicated that the smaller AgNPs (11.6 ± 3.5 nm) were less toxic than the larger AgNPs (41.6 ± 9.1 nm) at the same molarity, and exhibited size-dependent nanotoxicity.

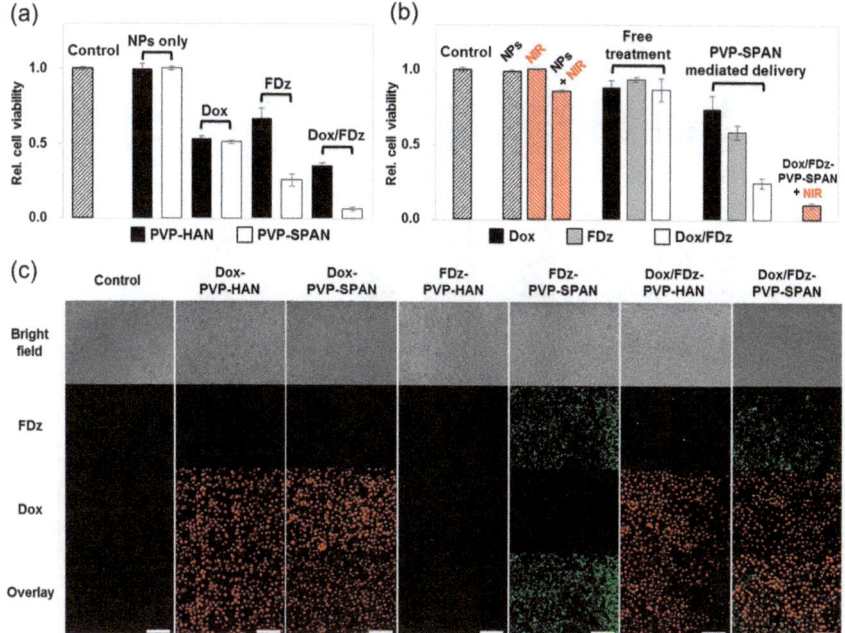

Figure 11.11 Quantitative evaluation of therapeutic efficacy of DOX- and/or FDz-loaded PVP–HAN and PVP–SPAN in NS3 replicon Huh7 cells. (a) Comparison of cell viability of NS3 replicon Huh7 cells treated with PVP–HAN, PVP–SPAN, DOX-loaded PVP–HAN, Dox-loaded PVP–SPAN, FDz-loaded PVP–HAN, FDz-loaded PVP–SPAN, DOX/FDz-loaded PVP–HAN, and DOX/FDz-loaded PVP–SPAN. (b) Synergistic chemo/thermotherapeutic efficacy was observed by using DOX- and FDz-loaded PVP–SPAN. PVP–SPAN mediated FDz/DOX delivery in combination with NIR irradiation showed significantly enhanced therapeutic efficacy. However, in control experiments, cells treated with only PVP–SPAN, NIR irradiation, and free DOX and/or FDz showed no notable decrease in viability. (c) Fluorescence microscopy images of cells treated with the NPs showed the successful delivery of the loaded DOX and/or FDz. Scale bars are 50 μm. Reproduced with permission from H. Jang and D.-H. Min, *ACS Nano*, 2015, **9**, 2696. Copyright (2015) American Chemical Society.[212]

11.4.3.2 In vivo *Studies*

Understanding the *in vivo* toxicity of noble metal-based NPs is crucial for evaluation of their potential health risk. After injection with AuNPs, there was no significant changes in the weight of mice, indicating the non-toxic nature of the AuNPs.[218] Tan *et al.*[219] studied the acute toxicity of the Ag_2S–GSH–SNO NPs with NIR fluorescence imaging capability (Figure 11.12). The histological analysis indicated that the as-made NPs could be safely used as a bioimaging and NO delivery agent. Chandirasekar *et al.*[220] evaluated the toxicity of Na cholate-templated AgNPs using developmental-stage zebrafish

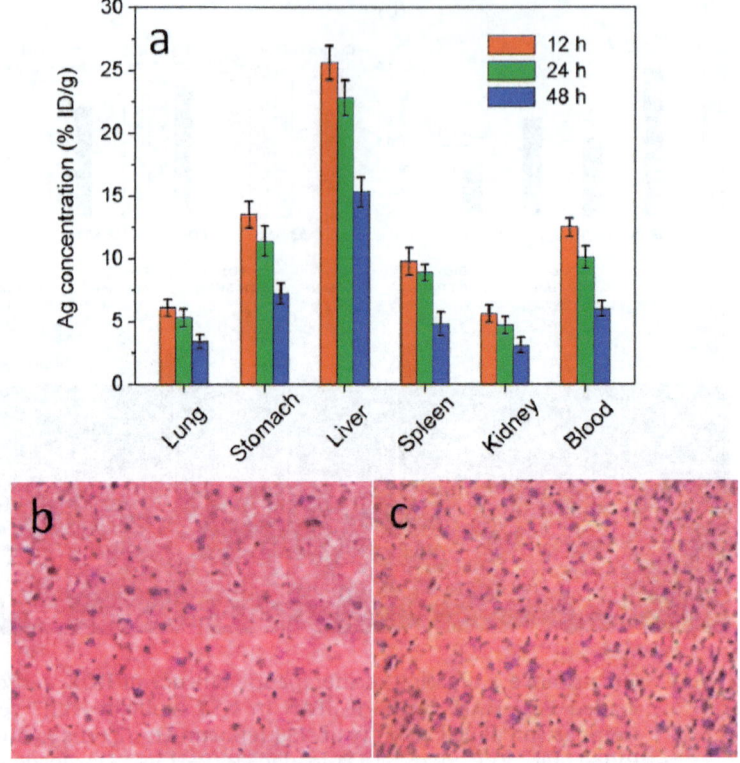

Figure 11.12   (a) *In vivo* biodistribution and toxicity analysis of Ag₂S–GSH–SNO
NPs over a period of 48 h in a nude mouse. The Ag concentration in
the organs was determined at different time points after tail intravenous injection of the Ag₂S–GSH–SNO NPs (5 µg) using ICP-MS. (b)
Without injection and (c) 12 h after injection of Ag₂S–GSH–SNO NPs.
Tissue images are acquired at ×400. Reproduced with permission
from L. Tan, A. Wan and H. Li, *ACS Appl. Mater. Interfaces*, 2013, 5,
11163. Copyright (2013) American Chemical Society.[219]

embryos. In a survival and hatching experiment, no significant toxic effect
was observed at AgNP concentrations up to 200 µg mL⁻¹, and the NP-stained
embryos exhibited blue fluorescence with high intensity for a long period
of time, which showed that AgNPs were relatively stable in the living system. Chen *et al.*[221] examined the potential *in vivo* toxicity of Pd@Au–PEG
which displayed strong NIR absorbance and excellent stability. Their results
showed that there were no obvious signs of toxic side effects for Pd@Au–
PEG-injected mice at a dose of 15 mg kg⁻¹ within 60 days and 1 year.

Due to their strong absorption abilities, tunable extinction spectra, and ease
of surface modification, AuNPs have made a remarkable impact in the world
of nanoscience, and have attracted great interest for biosensing, labelling,
and photothermal therapy. Thus, it is important to evaluate their *in vivo* toxic
effects on human health and environment. Biodistribution of functionalized

AuNPs was widely investigated, and most of the results showed that function-alized AuNPs are non-toxic to living systems, suggesting that AuNPs possess potential applications in biology and nanomedicine.[222-224] The study of Hirn *et al.*[225] indicated that functionalization and surface charge had a significant effect on the *in vivo* biodistribution of AuNPs. The accumulation of TPPMS-coated AuNPs was size-dependent only in liver, increasing from 50% of 1.4 nm AuNPs to >99% of 200 nm AuNPs, with a sharp increase in accumulation between 1.4 nm and 5 nm. Both the negatively (functionalized with TGA) and positively (functionalized with CA) charged 2.8 nm AuNPs accumulated in the liver and spleen. However, accumulation was significantly higher for negative AuNPs than for positive AuNPs in the liver, and the reverse was true in the spleen. Moreover, the hepatobiliary clearance of TPPMS-coated AuNPs showed an inverse linear relationship to the AuNPs diameter over the size range of 5–200 nm. When conjugating AuNPs to tumour-targeting ligands, the obtained conjugate NPs were able to target tumour biomarkers. From the biodistribution study, it was found that targeted AuNPs were more efficiently accumulated in the tumour than non-targeted AuNPs.[198,226,227] AuNPs coated with biocompatible glycoprotein were also more efficiently localized in the liver.[228] Chemical functionalization also has great effects on the pharmaco-kinetics of AuNPs.[229-231] Tong *et al.* reported that PEGylated AuNRs exhibited a biphasic clearance mode, with a significantly prolonged blood residence time for branched PEG in contrast to linear PEG.[232] Similarly, low-molecu-lar-weight PEG-coated AuNPs showed enhanced clearance rates compared to low-molecular-weight PEG-coated particles.[233]

11.4.4 Nanotoxicity of Upconversion Nanoparticles

Owing to their unique photophysical properties, rare earth ion-doped upconversion nanoparticles (UCNPs) have attracted extensive attention in recent years.[234,235] UCNPs have many special merits, such as a long lumines-cence lifetime, narrow emission bandwidths, high quantum yields, and low toxicity, which allows their potential applications in the biomedical field as biological luminescent labels and drug delivery carriers. Compared with traditional phosphors exited by UV, such as organic dyes and QDs, UCNPs are capable of transferring NIR light into UV/visible light (*i.e.* upconversion luminescence, UCL).[236] Utilization of NIR excitation light not only allows for deeper penetration of light and reduces photodamage effects induced by UV/visible light, but also decreases or eliminates autofluorescence, light scattering, and phototoxicity.[45,237] Therefore, UCNPs, particularly lantha-nide-doped UCNPs such as $NaYF_4$:Yb,Er NPs, CaF_2:Yb,Er NPs and $NaYbF_4$:Er NPs, have recently attached enormous interest owing to their unique optical properties; simultaneously, UCNPs have also provided numerous opportunities in the biomedical field with potential applications in bio-imaging, cell and tissue labelling, biodetection, therapy and multiplex analysis.[238]

11.4.4.1   In vitro *Studies*

MTT, MTS, and CCK-8 mitochondrial metabolic activity assays have been utilized to evaluate the potential toxicity of UCNPs in various cells, such as human pancreatic cancer Panc 1 cells, human nasopharyngeal epidermal carcinoma cells, and human glioblastoma U87MG cells. So far, a wide range of UCNP concentrations (0.05–20 000 µg mL^{-1}) and incubation periods (1–336 h) have been studied. For example, Zhang[239] and Li[240] examined the biocompatibility of Yb/Er-doped rare earth fluorides coated with a silica shell or a layer of azelaic acid molecules. They found that these NPs had essentially no effect on the cell viability of human nasopharyngeal epidermal carcinoma KB cells after incubation for 20 h at a particle concentration of 800 µg mL^{-1}. Shan *et al.* also reported that there was limited or no toxicity of carboxyl- and amino-functionalized NPs after incubation for 9 days with human osteosarcoma cells.[241] UCNPs non-covalently functionalized with PEG exerted only a minor negative effect on the proliferation of human nasopharynx carcinoma cells. After incubation for 24 h with 500 µg mL^{-1} of UCNPs, more than 90% of cells survived.[242] Chatterjee *et al.*[34] also investigated the impact of PEI-coated UCNPs on rats and in cell culture. The results demonstrated that bone marrow derived stem cells exposed to UCNPs exhibited no toxic effect. Li's group also investigated the cytotoxicity of mPEG-LaF$_3$:Yb,Ho on KB cells at different concentrations.[243] They found that ~80% of incubated cells were viable even at an incubation time of 12 h and a high incubation concentration of 500 µg mL^{-1}. In another study, Xing *et al.*[42] also demonstrated the low *in vitro* toxicity of UCNPs. When both HL-7702 and RAW 264.7 cells were treated with a high concentration of NaYbF$_4$ for 24 h, the viability of cells was >82.7% for HL-7702 and >88.9% for RAW 264.7 cells. In addition, Li's group studied the interaction of NaYF$_4$ UCNPs with living cells (including HeLa, LO2, and KB).[244] They found that >80% of treated cells survived after incubation with 800 µg mL^{-1} UCNPs for 24 h, proving the low toxicity of UCNPs.

Noticeably, many studies indicated that the viability of cells treated with UCNPs mainly depended on the incubation time and concentration. For example, as reported by Zhang *et al.*,[239] the cell viability of skeletal myoblasts and marrow-derived stem cells significantly decreased with the increase of the incubation concentration of the NaYF$_4$:Yb,Er@SiO$_2$ nanocomposite from 1 to 100 µg mL^{-1}. When the incubation concentration reached 100 µg mL^{-1}, it was found that approximately 63% of these cells were viable.

Furthermore, some studies also investigated the toxic effects of UCNPs on cell behaviour based on their photoluminescent properties. For example, after mesenchymal stem cells (mMSCs) were treated with oligo-arginine-PEG-coated NaYF$_4$:Yb,Er NPs, systematic *in vitro* tests revealed that the proliferation and differentiation of mMSCs were not notably affected, suggesting that the cells labelled with oligo-arginine-PEG-UCNPs were able to maintain their stem cell potency.[245] Zhang *et al.*[246] used SiO$_2$-coated NaYF$_4$:Yb,Er NPs (UCNP@SiO$_2$) as a luminescent probe to dynamically track live myoblast cells *in vitro* using confocal microscopy. The UCL signal of UCNP@

SiO$_2$ tracking transplanted cells in mouse limb muscle was followed over 4 h, and suggested subtle migratory activity of the transplanted cells. Han et al.[247] also reported that mesenchymal stem cells labelled with PEI covalently conjugated NaYbF$_4$:Tm@CaF$_2$ were able to undergo osteogenic and adipogenic differentiation upon *in vitro* induction. However, the osteogenesis of labelled rat mesenchymal stem cells appeared to be less potent than that of the unlabelled MSCs.

11.4.4.2 In vivo *Studies*

In vivo toxicity is one of the most important characteristics for theranostic applications of UCNPs. So far, *in vivo* toxicity of hydrophilic UCNPs has been systematically investigated in mice.[248,249] Most of the individual studies have indicated that lanthanide-based UCNPs are safe at the dosage used for imaging. In 2010, Li's group systematically studied the long-term toxicity of PAA-coated NaYF$_4$:Yb,Er NPs (PAA–UCNPs) by observing the behaviour, body weight, histology and hematology, and serum biochemistry of the animals.[250] No abnormal indicators were observed at 115 days after intravenous injection, except a small weight difference compared with the control. Moreover, similar to the control group, the PAA–UCNPs treated mice also had normal eating and drinking behaviour, fur colour, exploratory behaviour, activity, and neurological status. At a dose of 15 mg kg^{-1}, hepatocytes in the liver samples appeared normal, and there were no inflammatory infiltrates; cardiac muscle tissue in the heart samples did not show hydropic degeneration; the glomerular structure was clear and no necrosis was found in any of the groups; and no pulmonary fibrosis was observed in the lung samples. However, the difference compared with the control was that the spleen was affected, and slight hyperplasia was found in the periarteriolar lymphoid sheath (PALS) of the white pulp. In haematological and serological studies, the blood smears from PAA–UCNP-treated mice indicated that the number and shape of red blood cells, platelets, and white blood cells were normal. Alanine aminotransferase, aspartate aminotransferase, and total bilirubin were also similar between the treated and untreated mice. In addition, the toxicity of NaGdF$_4$:Yb,Er,Tm, DTPA–NaLuF$_4$:Gd,Yb,Er/Tm and 6-aminohexanoic acid modified NaLuF$_4$:Sm,Yb,Tm has been investigated and few toxic effects were observed.

Similarly, Liu et al. found no obvious hepatic toxicity induced by intravenous injection of PAA–UCNPs (~35 nm) and PEG–UCNPs (~30 nm), by measuring the levels of alanine aminotransferase, aspartate aminotransferase, alkaline phosphatase, albumin, globulin and total protein. Li et al.[251] synthesized highly water-soluble NaYF$_4$:Yb,Er@NaGdF$_4$ multilayer core–shell UCNPs by a microwave-assisted route for the first time. The results revealed that core–shell UCNPs showed low cytotoxicity and long circulation time *in vivo*. More significantly, multilayer core–shell UCNPs provided a much higher efficacy compared to the monolayer core–shell agent. Besides, other UCNPs including peptide-modified NaYF$_4$:Yb,Er/Ce NPs,[252] BaGdF$_5$:Yb/Tm, NaGdF$_4$:Yb,Er@NaGdF$_4$@SiO$_2$,[253] hyaluronic acid (HA)-modified

$NaYF_4$:Yb,Gd,Tm,[254] and ANG/PEG–UCNPs[255] also exhibited no obvious tissue damage or any other side effect to the treated animals.

Additionally, Chatterjee *et al.*[34] studied the impact of PEI-coated UCNPs on rats. After injection into the animals, UCNPs quickly accumulated in the lungs. After 24 h the concentration of UCNPs in the lungs was reduced while the amount in the spleen increased. After 7 days, the UCNPs were undetectable in the animals. Xiong *et al.*[256] tested the biodistribution of UCNPs conjugated with folic acid. Their result showed that UCNPs mainly accumulated in cancer cells because of the presence of folic acid (FA) which was capable of guiding the UCNPs to FA receptors located on the surface of cancer cells. At the same time, small amounts were found in the liver and spleen, while they were rarely detected in the kidneys. The study of Xing *et al.*[42] showed that $NaYbF_4$ UCNPs administered to mice were excreted with faeces and urine up to 7 days after injection. After 1 month, no fluorescence from $NaYbF_4$ was observed, suggesting that all UCNPs had been removed from the body. Histological analysis proved that no damage or toxic effects were caused to the organs by the long distribution time of UCNPs *in vivo*.

11.4.5   Nanotoxicity of Narrow-Bandgap Semiconductors

In recent years, much attention has increasingly turned to narrow-bandgap semiconductor NMs active in the infrared (IR) range.[257–267] Due to their excellent electrical structures, these narrow-bandgap semiconductors can emit from the NIR out to longer wavelengths. Alternatively, these NMs are capable of translating the NIR light into heat, based on the high NIR absorption ability. Therefore, these NMs have the potential to use as fluorophores in connection with biological tissue studies or as the photothermal agents for photothermal therapy of cancers.[268–270] The chalcogenides include CdE, HgE, PbE, Ag_2E, SnE, Bi_2E_3, $CuInE_2$ (E = S, Se or Te) and alloys and heterostructures of many of these materials.

11.4.5.1   In vitro *Studies*

Turyanska *et al.*[271] reported that apoferritin–PbS nanocomposites could not induce a change in the cell cycle of non-tumorigenic cells at concentrations up to 1 mg mL^{-1}, but cause apoptotic cell death at concentrations >0.2 mg mL^{-1} after exposure to human-derived breast cancer cell lines. Tian *et al.*[54] carefully evaluated the cytotoxicity of the Cu_9S_5 nanocrystals using the MTT assay. The result indicated that exposure of HeLa cells to Cu_9S_5 nanocrystals caused insignificant toxicity even at a Cu_9S_5 nanocrystal concentration up to 100 ppm. Song *et al.*[272] also evaluated the cytotoxicity of the PAA-coated Co_9Se_8 nanoplates on HepG2 cells using a CCK-8 assay. During the concentration of PAA-coated Co_9Se_8 nanoplates ranging from 0 to 120 µg mL^{-1}, all of the cells retained >90% viability, indicating low cell cytotoxicity and good biocompatibility. Liu *et al.*[273] also demonstrated that $Cu_{2-x}Se$@mSiO$_2$–PEG

NPs exhibited low toxicity after incubating HeLa cells with different concentrations (0–500 μg mL^{-1}) for 24 h. Liu *et al.*[274] carefully studied MoS$_2$ nanosheets with or without functionalization. The results showed that pristine MoS$_2$ nanosheets could induce slight cytotoxicity after culturing for 2 days and 3 days, while MoS$_2$ nanosheets functionalized with PEG exhibited insignificant cytotoxicity. They further assessed the intracellular ROS levels in MoS$_2$-treated cells using the dihydroethidine probe. Compared with the MTT assay, no significant increase on the percentage of DHE-positive cells was observed for cells treated with MoS$_2$, suggesting minimal oxidative stress induced by those nanosheets.

11.4.5.2 In vivo *Studies*

The biodistribution of CuS NPs with different surface functionalization was studied by Zhou *et al.*[275] For 24 h after intravenous injection of CuS NPs, it was found that citrate-coated CuS NPs displayed noticeably higher uptake than did PEG-coated CuS NPs in the liver and the spleen. Conversely, PEG-coated CuS NPs were less likely to be captured by RES cells and therefore displayed higher levels in the heart, kidney, lung, stomach, intestine, and bone. Moreover, the mean systemic clearance was significantly slower with PEG than with citrate, suggesting that citrate-coated CuS NPs were cleared faster than PEG-coated CuS NPs. This difference may be attributed to surface functionalization which resulted in higher uptake by the RES. Wang *et al.*[276] systemically studied Cu$_{2-x}$S nanocrystals *in vivo* using the method provided by the Organisation for Economic Cooperation and Development (OECD). Their results indicated that no toxic reactions were observed in the treated mice following intravenous administration of Cu$_{2-x}$S NPs at 25 and 50 mg kg^{-1}. Analysis of H&E-stained samples showed that no abnormalities were found in the liver, spleen, kidney, lung, and brain. However, treatment with Cu$_{2-x}$S NPs at 100 and 150 mg kg^{-1} led to degenerative necrosis and the disappearance of hepatocytes. The result of Qian *et al.*[277] showed that after intravenous injection of PEGylated TiS$_2$ nanosheets no mouse death or any sign of toxic effect was observed within 60 days. Moreover, H&E analysis showed that major organs were not damaged. Liu *et al.*[274] carried out a pilot study to assess the *in vivo* toxicity of MoS$_2$ functionalized with PEG. The blood of female Balb/c mice was collected at 1, 7, 15, and 30 days after injection. The results of both a serum biochemistry assay and a complete blood panel test were rather close to those of untreated healthy mice, suggesting that MoS$_2$–PEG may not be noticeably toxic to mice.

11.5 Conclusions, Remarks, and Perspectives

The toxicity of NIR NMs is a very important issue to address before their potential can be fully utilized for healthcare and medical research applications. A great deal of concern has been raised about the possible health risks of NIR NMs, both in the press and by practicing physicians. Thus a critical assessment of risk *versus* benefit in the use of engineered NIR NMs for diagnostics and therapy is extremely vital for the advancement of these NMs in medicine. This chapter has focused on the nanotoxicity of NIR NMs, including carbon-based materials, QDs, noble metal-based NPs, UCNPs, and narrow-bandgap semiconductors, emphasizing the complexity of this subject. Here we summarize some significant aspects of the challenges for the field.

11.5.1   Challenge 1: The toxicity Mechanisms of NIR NMs

The observed toxic responses result from many different ingredients. Taking GO nanosheets as example: which toxicity, and how much comes from GO, and how much comes from the size, from aggregates of GO nanosheets, from defects, or from surface adsorbability of GO? These uncertainties make the understanding of nanotoxicity mechanisms particularly complicated. Moreover, our knowledge of cellular mechanisms and the molecular mechanisms of toxic responses to NMs are almost non-existent so far.

11.5.2   Challenge 2: Standardized NIR NMs for Toxicity Tests

In the toxicological testing of NMs, at present it is rather difficult to compare them with other xenobiotic substances with different chemical compositions. The main reason is that, at the nanoscale, more than 10 factors can coincidentally influence the toxic responses of NIR NMs (see Table 11.1), thus making any toxicological comparison with different substances less meaningful. Fortunately, when standardized NMs whose toxicity has been well characterized are developed, comparison with other substances becomes possible.

11.5.3   Challenge 3: Theoretical Modelling for Cellular and Molecular Interactions of Nanoparticles

This has recently become a significant research topic. As shown in Table 11.1, more than 10 factors can sensitively influence the biological effects of NIR NMs. Experimental investigation of each factor and their combinations would create an enormous workload, requiring much time to complete, in order to reach an explicit conclusion. In addition, many interaction processes of NMs with biological systems are impossible to explore experimentally because proper techniques are currently lacking. Theoretical modelling studies provide an alternative means of solving these issues.

11.5.4   Challenge 4: Systematic Knowledge Frameworks for Nanotoxicology

In the field of nanotoxicology there are notoriously many disagreements in the experimental data because the factors listed in Table 11.1 are not the same in different studies that outwardly seem to use the same nanomaterial.

Recently, these contradictions have begun to lessen due to better characterization of these properties of NIR NMs. It is important to develop systematic knowledge frameworks for the nanotoxicology of NIR NMs.

Acknowledgements

National Basic Research Programs of China (973 program, No. 2012CB932504), and National Natural Science Foundation of China (No. 21177128, 31751015, and 21320102003).

References

1. L. Moody and A. A. Holder, *Annu. Rep. Prog. Chem., Sect. A: Inorg. Chem.*, 2009, **105**, 505.
2. K.-T. Yong, I. Roy, M. T. Swihart and P. N. Prasad, *J. Mater. Chem.*, 2009, **19**, 4655.
3. W. Yang, K. R. Ratinac, S. P. Ringer, P. Thordarson, J. J. Gooding and F. Braet, *Angew. Chem., Int. Ed.*, 2010, **49**, 2114.
4. R. Bardhan, S. Lal, A. Joshi and N. J. Halas, *Acc. Chem. Res.*, 2011, **44**, 936.
5. S. S. Kelkar and T. M. Reineke, *Bioconjugate Chem.*, 2011, **22**, 1879.
6. M. J. Sailor and J.-H. Park, *Adv. Mater.*, 2012, **24**, 3779.
7. L. Zhu and V. P. Torchilin, *Integr. Biol.*, 2013, **5**, 96.
8. C. Li, *Nat. Mater.*, 2014, **13**, 110.
9. H. Xu, Q. Li, L. Wang, Y. He, J. Shi, B. Tang and C. Fan, *Curr. Top. Med. Chem.*, 2014, **43**, 2650.
10. Y. Chen, C. Tan, H. Zhang and L. Wang, *Chem. Soc. Rev.*, 2015, **44**, 2681.
11. O. S. Wolfbeis, *Chem. Soc. Rev.*, 2015, **44**, 4743.
12. C. Amiot, S. Xu, S. Liang, L. Pan and J. Zhao, *Sensors*, 2008, **8**, 3082.
13. P. K. Jain, X. Huang, I. H. El-Sayed and M. A. El-Sayed, *Acc. Chem. Res.*, 2008, **41**, 1578.
14. J. Gao, X. Chen and Z. Cheng, *Curr. Top. Med. Chem.*, 2010, **10**, 1147.
15. X. He, J. Gao, S. S. Gambhir and Z. Cheng, *Trends Mol. Med.*, 2010, **16**, 574.
16. S. A. Hilderbrand and R. Weissleder, *Curr. Opin. Chem. Biol.*, 2010, **14**, 71.
17. Q. Ma and X. Su, *Analyst*, 2010, **135**, 1867.
18. V. J. Pansare, S. Hejazi, W. J. Faenza and R. K. Prud'homme, *Chem. Mater.*, 2012, **24**, 812.
19. Z. Zhang, J. Wang and C. Chen, *Adv. Mater.*, 2013, **25**, 3869.
20. F. Jabeen, M. Najam-ul-Haq, R. Javeed, C. W. Huck and G. K. Bonn, *Molecules*, 2014, **19**, 20580.
21. V. Shanmugam, S. Selvakumar and C.-S. Yeh, *Curr. Top. Med. Chem.*, 2014, **43**, 6254.
22. R. Wang and F. Zhang, *J. Mater. Chem. B*, 2014, **2**, 2422.
23. H. Kobayashi, M. Ogawa, R. Alford, P. L. Choyke and Y. Urano, *Chem. Rev.*, 2010, **110**, 2620.

24. S. Luo, E. Zhang, Y. Su, T. Cheng and C. Shi, *Biomaterials*, 2011, **32**, 7127.
25. K. L. Aillon, Y. Xie, N. El-Gendy, C. J. Berkland and M. L. Forrest, *Adv. Drug Delivery Rev.*, 2009, **61**, 457.
26. S. Sharifi, S. Behzadi, S. Laurent, M. Laird Forrest, P. Stroeve and M. Mahmoudi, *Curr. Top. Med. Chem.*, 2012, **41**, 2323.
27. X. Huang and M. A. El-Sayed, *J. Adv. Res.*, 2010, **1**, 13.
28. F. Wang and X. Liu, *Curr. Top. Med. Chem.*, 2009, **38**, 976.
29. N. E. Motl, A. F. Smith, C. J. DeSantis and S. E. Skrabalak, *Chem. Soc. Rev.*, 2014, **43**, 3823.
30. X. Huang, S. Han, W. Huang and X. Liu, *Curr. Top. Med. Chem.*, 2013, **42**, 173.
31. Y.-H. Wang, S.-P. Chen, A.-H. Liao, Y.-C. Yang, C.-R. Lee, C.-H. Wu, P.-C. Wu, T.-M. Liu, C.-R. C. Wang and P.-C. Li, *Sci. Rep.*, 2014, **4**, 5685.
32. X. Huang and M. A. El-Sayed, *Alexandria Med. J.*, 2011, **47**, 1.
33. R. Weissleder, *Nat. Biotechnol.*, 2001, **19**, 316.
34. D. K. Chatterjee, A. J. Rufaihah and Y. Zhang, *Biomaterials*, 2008, **29**, 937.
35. M. Nyk, R. Kumar, T. Y. Ohulchanskyy, E. J. Bergey and P. N. Prasad, *Nano Lett.*, 2008, **8**, 3834.
36. Y. Song, Y. Huang, L. Zhang, Y. Zheng, N. Guo and H. You, *RSC Adv.*, 2012, **2**, 4777.
37. Z. Liu, Z. Li, J. Liu, S. Gu, Q. Yuan, J. Ren and X. Qu, *Biomaterials*, 2012, **33**, 6748.
38. Z. Liu, F. Pu, S. Huang, Q. Yuan, J. Ren and X. Qu, *Biomaterials*, 2013, **34**, 1712.
39. W. Ren, G. Tian, L. Zhou, W. Yin, L. Yan, S. Jin, Y. Zu, S. Li, Z. Gu and Y. Zhao, *Nanoscale*, 2012, **4**, 3754.
40. Z. Xu, C. Li, P. A. Ma, Z. Hou, D. Yang, X. Kang and J. Lin, *Nanoscale*, 2011, **3**, 661.
41. X. Chen, Z. Liu, Q. Sun, M. Ye and F. Wang, *Opt. Commun.*, 2011, **284**, 2046.
42. H. Xing, W. Bu, Q. Ren, X. Zheng, M. Li, S. Zhang, H. Qu, Z. Wang, Y. Hua, K. Zhao, L. Zhou, W. Peng and J. Shi, *Biomaterials*, 2012, **33**, 5384.
43. Y. Yang, Y. Sun, T. Cao, J. Peng, Y. Liu, Y. Wu, W. Feng, Y. Zhang and F. Li, *Biomaterials*, 2013, **34**, 774.
44. S. Zeng, M.-K. Tsang, C.-F. Chan, K.-L. Wong and J. Hao, *Biomaterials*, 2012, **33**, 9232.
45. G. Chen, J. Shen, T. Y. Ohulchanskyy, N. J. Patel, A. Kutikov, Z. Li, J. Song, R. K. Pandey, H. Ågren, P. N. Prasad and G. Han, *ACS Nano*, 2012, **6**, 8280.
46. N.-N. Dong, M. Pedroni, F. Piccinelli, G. Conti, A. Sbarbati, J. E. Ramírez-Hernández, L. M. Maestro, M. C. Iglesias-de la Cruz, F. Sanz-Rodriguez, A. Juarranz, F. Chen, F. Vetrone, J. A. Capobianco, J. G. Solé, M. Bettinelli, D. Jaque and A. Speghini, *ACS Nano*, 2011, **5**, 8665.
47. Y. Dai, C. Zhang, Z. Cheng, P. A. Ma, C. Li, X. Kang, D. Yang and J. Lin, *Biomaterials*, 2012, **33**, 2583.

48. K. Welsher, Z. Liu, S. P. Sherlock, J. T. Robinson, Z. Chen, D. Daranciang and H. Dai, *Nat. Nanotechnol.*, 2009, **4**, 773.
49. J. Gao, K. Chen, R. Xie, J. Xie, S. Lee, Z. Cheng, X. Peng and X. Chen, *Small*, 2010, **6**, 256.
50. H. Kim, D. Lee, J. Kim, T.-i. Kim and W. J. Kim, *ACS Nano*, 2013, **7**, 6735.
51. D.-K. Lim, A. Barhoumi, R. G. Wylie, G. Reznor, R. S. Langer and D. S. Kohane, *Nano Lett.*, 2013, **13**, 4075.
52. W. Zhang, Z. Guo, D. Huang, Z. Liu, X. Guo and H. Zhong, *Biomaterials*, 2011, **32**, 8555.
53. M.-C. Wu, A. R. Deokar, J.-H. Liao, P.-Y. Shih and Y.-C. Ling, *ACS Nano*, 2013, **7**, 1281.
54. Q. Tian, F. Jiang, R. Zou, Q. Liu, Z. Chen, M. Zhu, S. Yang, J. Wang, J. Wang and J. Hu, *ACS Nano*, 2011, **5**, 9761.
55. G. Song, Q. Wang, Y. Wang, G. Lv, C. Li, R. Zou, Z. Chen, Z. Qin, K. Huo, R. Hu and J. Hu, *Adv. Funct. Mater.*, 2013, **23**, 4281.
56. Q. Tian, M. Tang, Y. Sun, R. Zou, Z. Chen, M. Zhu, S. Yang, J. Wang, J. Wang and J. Hu, *Adv. Mater.*, 2011, **23**, 3542.
57. Q. Tian, J. Hu, Y. Zhu, R. Zou, Z. Chen, S. Yang, R. Li, Q. Su, Y. Han and X. Liu, *J. Am. Chem. Soc.*, 2013, **135**, 8571.
58. P. Taladriz-Blanco, V. Pastoriza-Santos, J. Pérez-Juste and P. Hervés, *Langmuir*, 2013, **29**, 8061.
59. B. Kang, M. A. Mackey and M. A. El-Sayed, *J. Am. Chem. Soc.*, 2010, **132**, 1517.
60. J. Chen, Z. Guo, H.-B. Wang, M. Gong, X.-K. Kong, P. Xia and Q.-W. Chen, *Biomaterials*, 2013, **34**, 571.
61. B. Dong, S. Xu, J. Sun, S. Bi, D. Li, X. Bai, Y. Wang, L. Wang and H. Song, *J. Mater. Chem.*, 2011, **21**, 6193.
62. R. Jiang, S. Cheng, L. Shao, Q. Ruan and J. Wang, *J. Phys. Chem. C*, 2013, **117**, 8909.
63. X. Huang, P. K. Jain, I. H. El-Sayed and M. A. El-Sayed, *Photochem. Photobiol.*, 2006, **82**, 412.
64. S. J. Soenen, P. Rivera-Gil, J.-M. Montenegro, W. J. Parak, S. C. De Smedt and K. Braeckmans, *Nano Today*, 2011, **6**, 446.
65. L. Yan, Z. Gu and Y. Zhao, *Chem.–Asian J.*, 2013, **8**, 2342.
66. T. R. Pisanic, S. Jin and V. I. Shubayev, in *Nanotoxicity: From in vivo and in vitro Models to Health Risks*, ed. S. C. Sahu and D. A. Casciano, John Wiley and Sons, Ltd., London, 2009, pp. 397.
67. A. Stroh, C. Zimmer, C. Gutzeit, M. Jakstadt, F. Marschinke, T. Jung, H. Pilgrimm and T. Grune, *Free Radical Biol. Med.*, 2004, **36**, 976.
68. L. Gao, J. Zhuang, L. Nie, J. Zhang, Y. Zhang, N. Gu, T. Wang, J. Feng, D. Yang, S. Perrett and X. Yan, *Nat. Nanotechnol.*, 2007, **2**, 577.
69. H. Meng, T. Xia, S. George and A. E. Nel, *ACS Nano*, 2009, **3**, 1620.
70. A. Nel, T. Xia, L. Mädler and N. Li, *Science*, 2006, **311**, 622.
71. G. Oberdörster, E. Oberdörster and J. Oberdörster, *Environ. Health Perspect.*, 2005, **113**, 823.
72. R. K. Srivastava, A. B. Pant, M. P. Kashyap, V. Kumar, M. Lohani, L. Jonas and Q. Rahman, *Nanotoxicology*, 2011, **5**, 195.

73. Y.-N. Chang, M. Zhang, L. Xia, J. Zhang and G. Xing, *Materials*, 2012, **5**, 2850.
74. X.-R. Xia, N. A. Monteiro-Riviere and J. E. Riviere, *Nat. Nanotechnol.*, 2010, **5**, 671.
75. Y. Tu, M. Lv, P. Xiu, T. Huynh, M. Zhang, M. Castelli, Z. Liu, Q. Huang, C. Fan, H. Fang and R. Zhou, *Nat. Nanotechnol.*, 2013, **8**, 594.
76. C. Rocker, M. Potzl, F. Zhang, W. J. Parak and G. U. Nienhaus, *Nat. Nanotechnol.*, 2009, **4**, 577.
77. M. P. Monopoli, C. Aberg, A. Salvati and K. A. Dawson, *Nat. Nanotechnol.*, 2012, **7**, 779.
78. J. W. E. Worrall, A. Verma, H. Yan and V. M. Rotello, *Chem. Commun.*, 2006, 2338.
79. N. Wangoo, C. R. Suri and G. Shekhawat, *Appl. Phys. Lett.*, 2008, **92**, 133104.
80. S. Liu, Y. Sui, K. Guo, Z. Yin and X. Gao, *Nanoscale Res. Lett.*, 2012, **7**, 1.
81. M. Sabbioni, *Proceedings of the Workshop: Research Needs on Nanoparticles*, Brussels, 2005, p. 22.
82. L. Guo, A. Von Dem Bussche, M. Buechner, A. Yan, A. B. Kane and R. H. Hurt, *Small*, 2008, **4**, 721.
83. J. M. Wörle-Knirsch, K. Pulskamp and H. F. Krug, *Nano Lett.*, 2006, **6**, 1261.
84. Environmental Protection Agency, 2008.
85. L. Yan, F. Zhao, S. Li, Z. Hu and Y. Zhao, *Nanoscale*, 2011, **3**, 362.
86. K. W. Powers, M. Palazuelos, B. M. Moudgil and S. M. Roberts, *Nanotoxicology*, 2007, **1**, 42.
87. J. Rejman, V. Oberle, I. S. Zuhorn and D. Hoekstra, *Biochem. J.*, 2004, **377**, 159.
88. W. Kreyling, M. Semmler-Behnke and W. Möller, *J. Nanopart. Res.*, 2006, **8**, 543.
89. S. Harper, C. Usenko, J. E. Hutchison, B. L. S. Maddux and R. L. Tanguay, *J. Exp. Nanosci.*, 2008, **3**, 195.
90. H. C. Fischer and W. C. W. Chan, *Curr. Opin. Biotechnol.*, 2007, **18**, 565.
91. S. Lanone and J. Boczkowski, *Curr. Mol. Med.*, 2006, **6**, 651.
92. C. Fisher, A. E. Rider, Z. J. Han, S. Kumar, I. Levchenko and K. Ostrikov, *J. Nanomater.*, 2012, **2012**, 2817.
93. M. Adeli, R. Soleyman, Z. Beiranvand and F. Madani, *Curr. Top. Med. Chem.*, 2013, **42**, 5231.
94. G. Jia, H. Wang, L. Yan, X. Wang, R. Pei, T. Yan, Y. Zhao and X. Guo, *Environ. Sci. Technol.*, 2005, **39**, 1378.
95. A. Shvedova, V. Castranova, E. Kisin, D. Schwegler-Berry, A. Murray, V. Gandelsman, A. Maynard and P. Baron, *J. Toxicol. Environ. Health, Part A*, 2003, **66**, 1909.
96. A. D. Maynard, P. A. Baron, M. Foley, A. A. Shvedova, E. R. Kisin and V. Castranova, *J. Toxicol. Environ. Health, Part A*, 2004, **67**, 87.
97. A. Huczko and H. Lange, *Fullerene Sci. Technol.*, 2001, **9**, 247.
98. K.-K. Liu, C.-L. Chang, C.-C. Chang and J. I. Chao, *Nanotechnology*, 2007, **18**, 325102.

99. C. S. Sharma, S. Sarkar, A. Periyakaruppan, J. Barr, K. Wise, R. Thomas, B. L. Wilson and G. T. Ramesh, *J. Nanosci. Nanotechnol.*, 2007, **7**, 2466.

100. Y. Zhang, S. F. Ali, E. Dervishi, Y. Xu, Z. Li, D. Casciano and A. S. Biris, *ACS Nano*, 2010, **4**, 3181.

101. H. N. Yehia, R. K. Draper, C. Mikoryak, E. K. Walker, P. Bajaj, I. H. Musselman, M. C. Daigrepont, G. R. Dieckmann and P. Pantano, *J. Nanobiotechnol.*, 2007, **5**, 8.

102. Y. Zhu, W. Li, Q. Li, Y. Li, Y. Li, X. Zhang and Q. Huang, *Carbon*, 2009, **47**, 1351.

103. C. L. Ursini, D. Cavallo, A. M. Fresegna, A. Ciervo, R. Maiello, G. Buresti, S. Casciardi, F. Tombolini, S. Bellucci and S. Iavicoli, *Toxicol. In Vitro*, 2012, **26**, 831.

104. Y. Zhang, Y. Xu, Z. Li, T. Chen, S. M. Lantz, P. C. Howard, M. G. Paule, W. Slikker, F. Watanabe, T. Mustafa, A. S. Biris and S. F. Ali, *ACS Nano*, 2011, **5**, 7020.

105. Y.-g. Han, J. Xu, Z.-g. Li, G.-g. Ren and Z. Yang, *NeuroToxicology*, 2012, **33**, 1128.

106. J. Palomäki, E. Välimäki, J. Sund, M. Vippola, P. A. Clausen, K. A. Jensen, K. Savolainen, S. Matikainen and H. Alenius, *ACS Nano*, 2011, **5**, 6861.

107. S. Hirano, S. Kanno and A. Furuyama, *Toxicol. Appl. Pharmacol.*, 2008, **232**, 244.

108. K. Pulskamp, S. Diabaté and H. F. Krug, *Toxicol. Lett.*, 2007, **168**, 58.

109. M. Bottini, S. Bruckner, K. Nika, N. Bottini, S. Bellucci, A. Magrini, A. Bergamaschi and T. Mustelin, *Toxicol. Lett.*, 2006, **160**, 121.

110. S.-J. Choi, J.-M. Oh and J.-H. Choy, *J. Inorg. Biochem.*, 2009, **103**, 463.

111. D. Cui, F. Tian, C. S. Ozkan, M. Wang and H. Gao, *Toxicol. Lett.*, 2005, **155**, 73.

112. E. R. Kisin, A. R. Murray, M. J. Keane, X.-C. Shi, D. Schwegler-Berry, O. Gorelik, S. Arepalli, V. Castranova, W. E. Wallace, V. E. Kagan and A. A. Shvedova, *J. Toxicol. Environ. Health, Part A*, 2007, **70**, 2071.

113. A. Simon-Deckers, B. Gouget, M. Mayne-L'Hermite, N. Herlin-Boime, C. Reynaud and M. Carrière, *Toxicology*, 2008, **253**, 137.

114. M. Asakura, T. Sasaki, T. Sugiyama, M. Takaya, S. Koda, K. Nagano, H. Arito and S. Fukushima, *J. Occup. Health*, 2010, **52**, 155.

115. A. R. Murray, E. Kisin, S. S. Leonard, S. H. Young, C. Kommineni, V. E. Kagan, V. Castranova and A. A. Shvedova, *Toxicology*, 2009, **257**, 161.

116. S. Koyama, Y. A. Kim, T. Hayashi, K. Takeuchi, C. Fujii, N. Kuroiwa, H. Koyama, T. Tsukahara and M. Endo, *Carbon*, 2009, **47**, 1365.

117. C. Ge, F. Lao, W. Li, Y. Li, C. Chen, Y. Qiu, X. Mao, B. Li, Z. Chai and Y. Zhao, *Anal. Chem.*, 2008, **80**, 9426.

118. A. Huczko, H. Lange, E. Całko, H. Grubek-Jaworska and P. Droszcz, *Fullerene Sci. Technol.*, 2001, **9**, 251.

119. C.-W. Lam, J. T. James, R. McCluskey and R. L. Hunter, *Toxicol. Sci.*, 2004, **77**, 126.

120. A. A. Shvedova, E. R. Kisin, R. Mercer, A. R. Murray, V. J. Johnson, A. I. Potapovich, Y. Y. Tyurina, O. Gorelik, S. Arepalli, D. Schwegler-Berry, A. F. Hubbs, J. Antonini, D. E. Evans, B.-K. Ku, D. Ramsey, A. Maynard, V. E. Kagan, V. Castranova and P. Baron, *Am. J. Physiol.: Lung Cell. Mol. Physiol.*, 2005, **289**, L698.

121. L. Ma-Hock, S. Treumann, V. Strauss, S. Brill, F. Luizi, M. Mertler, K. Wiench, A. O. Gamer, B. van Ravenzwaay and R. Landsiedel, *Toxicol. Sci.*, 2009, **112**, 468.

122. J. C. Carrero-Sánchez, A. L. Elías, R. Mancilla, G. Arrellín, H. Terrones, J. P. Laclette and M. Terrones, *Nano Lett.*, 2006, **6**, 1609.

123. D. B. Warheit, B. R. Laurence, K. L. Reed, D. H. Roach, G. A. Reynolds and T. R. Webb, *Toxicol. Sci.*, 2004, **77**, 117.

124. Y. Bai, Y. Zhang, J. Zhang, Q. Mu, W. Zhang, E. R. Butch, S. E. Snyder and B. Yan, *Nat. Nanotechnol.*, 2010, **5**, 683.

125. M. L. Schipper, N. Nakayama-Ratchford, C. R. Davis, N. W. S. Kam, P. Chu, Z. Liu, X. Sun, H. Dai and S. S. Gambhir, *Nat. Nanotechnol.*, 2008, **3**, 216.

126. C. A. Poland, R. Duffin, I. Kinloch, A. Maynard, W. A. H. Wallace, A. Seaton, V. Stone, S. Brown, W. MacNee and K. Donaldson, *Nat. Nanotechnol.*, 2008, **3**, 423.

127. J. Muller, F. Huaux, N. Moreau, P. Misson, J.-F. Heilier, M. Delos, M. Arras, A. Fonseca, J. B. Nagy and D. Lison, *Toxicol. Appl. Pharmacol.*, 2005, **207**, 221.

128. L. A. Mitchell, F. T. Lauer, S. W. Burchiel and J. D. McDonald, *Nat. Nanotechnol.*, 2009, **4**, 451.

129. D. Elgrabli, S. Abella-Gallart, F. Robidel, F. Rogerieux, J. Boczkowski and G. Lacroix, *Toxicology*, 2008, **253**, 131.

130. R. Mercer, A. Hubbs, J. Scabilloni, L. Wang, L. Battelli, D. Schwegler-Berry, V. Castranova and D. Porter, *Part. Fibre Toxicol.*, 2010, **7**, 1.

131. J.-G. Li, W.-X. Li, J.-Y. Xu, X.-Q. Cai, R.-L. Liu, Y.-J. Li, Q.-F. Zhao and Q.-N. Li, *Environ. Toxicol.*, 2007, **22**, 415.

132. P. Cherukuri, C. J. Gannon, T. K. Leeuw, H. K. Schmidt, R. E. Smalley, S. A. Curley and R. B. Weisman, *Proc. Natl. Acad. Sci. U. S. A.*, 2006, **103**, 18882.

133. S.-t. Yang, W. Guo, Y. Lin, X.-y. Deng, H.-f. Wang, H.-f. Sun, Y.-f. Liu, X. Wang, W. Wang, M. Chen, Y.-p. Huang and Y.-P. Sun, *J. Phys. Chem. C*, 2007, **111**, 17761.

134. P. Cherukuri, C. J. Gannon, T. K. Leeuw, H. K. Schmidt, R. E. Smalley, S. A. Curley and R. B. Weisman, *Proc. Natl. Acad. Sci. U. S. A.*, 2006, **103**, 18882.

135. H. Wang, J. Wang, X. Deng, H. Sun, Z. Shi, Z. Gu, Y. Liu and Y. Zhao, *J. Nanosci. Nanotechnol.*, 2004, **4**, 1019.

136. J. Wang, X. Deng, S. Yang, H. Wang, Y. Zhao and Y. Liu, *Nanotoxicology*, 2008, **2**, 28.

137. S.-T. Yang, K. A. S. Fernando, J.-H. Liu, J. Wang, H.-F. Sun, Y. Liu, M. Chen, Y. Huang, X. Wang, H. Wang and Y.-P. Sun, *Small*, 2008, **4**, 940.

138. R. Singh, D. Pantarotto, L. Lacerda, G. Pastorin, C. Klumpp, M. Prato, A. Bianco and K. Kostarelos, *Proc. Natl. Acad. Sci. U. S. A.*, 2006, **103**, 3357.

139. G. Prencipe, S. M. Tabakman, K. Welsher, Z. Liu, A. P. Goodwin, L. Zhang, J. Henry and H. Dai, *J. Am. Chem. Soc.*, 2009, **131**, 4783.

140. K.-H. Liao, Y.-S. Lin, C. W. Macosko and C. L. Haynes, *ACS Appl. Mater. Interfaces*, 2011, **3**, 2607.

141. X. Zhang, W. Hu, J. Li, L. Tao and Y. Wei, *Toxicol. Res.*, 2012, **1**, 62.

142. A. Sasidharan, L. S. Panchakarla, P. Chandran, D. Menon, S. Nair, C. N. R. Rao and M. Koyakutty, *Nanoscale*, 2011, **3**, 2461.

143. W. Hu, C. Peng, W. Luo, M. Lv, X. Li, D. Li, Q. Huang and C. Fan, *ACS Nano*, 2010, **4**, 4317.

144. S. K. Singh, M. K. Singh, M. K. Nayak, S. Kumari, S. Shrivastava, J. J. A. Grácio and D. Dash, *ACS Nano*, 2011, **5**, 4987.

145. K. Wang, J. Ruan, H. Song, J. Zhang, Y. Wo, S. Guo and D. Cui, *Nanoscale Res. Lett.*, 2010, **6**, 1.

146. M. Lv, Y. Zhang, L. Liang, M. Wei, W. Hu, X. Li and Q. Huang, *Nanoscale*, 2012, **4**, 3861.

147. N. V. S. Vallabani, S. Mittal, R. K. Shukla, A. K. Pandey, S. R. Dhakate, R. Pasricha and A. Dhawan, *J. Biomed. Nanotechnol.*, 2011, **7**, 106.

148. A. V. Tkach, N. Yanamala, S. Stanley, M. R. Shurin, G. V. Shurin, E. R. Kisin, A. R. Murray, S. Pareso, T. Khaliullin, G. P. Kotchey, V. Castranova, S. Mathur, B. Fadeel, A. Star, V. E. Kagan and A. A. Shvedova, *Small*, 2013, **9**, 1686.

149. Y. Li, Y. Liu, Y. Fu, T. Wei, L. Le Guyader, G. Gao, R.-S. Liu, Y.-Z. Chang and C. Chen, *Biomaterials*, 2012, **33**, 402.

150. W. Zhang, C. Wang, Z. Li, Z. Lu, Y. Li, J.-J. Yin, Y.-T. Zhou, X. Gao, Y. Fang, G. Nie and Y. Zhao, *Adv. Mater.*, 2012, **24**, 5391.

151. Z. Liu, J. T. Robinson, X. Sun and H. Dai, *J. Am. Chem. Soc.*, 2008, **130**, 10876.

152. X. Sun, Z. Liu, K. Welsher, J. Robinson, A. Goodwin, S. Zaric and H. Dai, *Nano Res.*, 2008, **1**, 203.

153. K. Yang, L. Feng, X. Shi and Z. Liu, *Curr. Top. Med. Chem.*, 2013, **42**, 530.

154. K. Yang, Y. Li, X. Tan, R. Peng and Z. Liu, *Small*, 2013, **9**, 1492.

155. K. Yang, S. Zhang, G. Zhang, X. Sun, S.-T. Lee and Z. Liu, *Nano Lett.*, 2010, **10**, 3318.

156. X. Zhang, J. Yin, C. Peng, W. Hu, Z. Zhu, W. Li, C. Fan and Q. Huang, *Carbon*, 2011, **49**, 986.

157. K. Yang, J. Wan, S. Zhang, B. Tian, Y. Zhang and Z. Liu, *Biomaterials*, 2012, **33**, 2206.

158. L. Yan, Y. Wang, X. Xu, C. Zeng, J. Hou, M. Lin, J. Xu, F. Sun, X. Huang, L. Dai, F. Lu and Y. Liu, *Chem. Res. Toxicol.*, 2012, **25**, 1265.

159. M. C. Duch, G. R. S. Budinger, Y. T. Liang, S. Soberanes, D. Urich, S. E. Chiarella, L. A. Campochiaro, A. Gonzalez, N. S. Chandel, M. C. Hersam and G. M. Mutlu, *Nano Lett.*, 2011, **11**, 5201.

160. S. K. Singh, M. K. Singh, P. P. Kulkarni, V. K. Sonkar, J. J. A. Grácio and D. Dash, *ACS Nano*, 2012, **6**, 2731.

161. K. Yang, J. Wan, S. Zhang, Y. Zhang, S.-T. Lee and Z. Liu, *ACS Nano*, 2011, **5**, 516.

162. A. Schinwald, F. A. Murphy, A. Jones, W. MacNee and K. Donaldson, *ACS Nano*, 2012, **6**, 736.
163. S. Zhang, K. Yang, L. Feng and Z. Liu, *Carbon*, 2011, **49**, 4040.
164. Y. Wang, Z. Li, J. Wang, J. Li and Y. Lin, *Trends Biotechnol.*, 2011, **29**, 205.
165. H. Shen, L. Zhang, M. Liu and Z. Zhang, *Theranostics*, 2012, **2**, 283.
166. T. Kuila, S. Bose, A. K. Mishra, P. Khanra, N. H. Kim and J. H. Lee, *Prog. Mater. Sci.*, 2012, **57**, 1061.
167. L. Yin, C. He, C. Huang, W. Zhu, X. Wang, Y. Xu and X. Qian, *Chem. Commun.*, 2012, **48**, 4486.
168. D. Bitounis, H. Ali-Boucetta, B. H. Hong, D.-H. Min and K. Kostarelos, *Adv. Mater.*, 2013, **25**, 2258.
169. C. Chung, Y.-K. Kim, D. Shin, S.-R. Ryoo, B. H. Hong and D.-H. Min, *Acc. Chem. Res.*, 2013, **46**, 2211.
170. L. Feng, L. Wu and X. Qu, *Adv. Mater.*, 2013, **25**, 168.
171. S. Kim and C. B. Park, *Adv. Funct. Mater.*, 2013, **23**, 10.
172. R. Romero-Aburto, T. N. Narayanan, Y. Nagaoka, T. Hasumura, T. M. Mitcham, T. Fukuda, P. J. Cox, R. R. Bouchard, T. Maekawa, D. S. Kumar, S. V. Torti, S. A. Mani and P. M. Ajayan, *Adv. Mater.*, 2013, **25**, 5632.
173. G. Gollavelli and Y.-C. Ling, *Biomaterials*, 2012, **33**, 2532.
174. A. M. Smith, H. Duan, A. M. Mohs and S. Nie, *Adv. Drug Delivery Rev.*, 2008, **60**, 1226.
175. R. Hardman, *Environ. Health Perspect.*, 2006, **114**, 165.
176. J. Li, X. Chang, X. Chen, Z. Gu, F. Zhao, Z. Chai and Y. Zhao, *Biotechnol. Adv.*, 2014, **32**, 727.
177. A. Bernardin, A. L. Cazet, L. Guyon, P. Delannoy, F. O. Vinet, D. Bonnaffé and I. Texier, *Bioconjugate Chem.*, 2010, **21**, 583.
178. J. M. Mauro, H. Mattoussi, I. L. Medintz, E. R. Goldman, P. T. Tran and G. P. Anderson, in *Defense Applications of Nanomaterials*, American Chemical Society, 2005, ch. 2, vol. 891, ACS Symposium Series, pp. 16–30.
179. W. H. D. Jong and P. J. Borm, *Int. J. Nanomed.*, 2008, **3**, 133.
180. N. Chen, Y. He, Y. Su, X. Li, Q. Huang, H. Wang, X. Zhang, R. Tai and C. Fan, *Biomaterials*, 2012, **33**, 1238.
181. K. C. Nguyen, V. L. Seligy and A. F. Tayabali, *Nanotoxicology*, 2013, **7**, 202.
182. S. J. Soenen, B. B. Manshian, T. Aubert, U. Himmelreich, J. Demeester, S. C. De Smedt, Z. Hens and K. Braeckmans, *Chem. Res. Toxicol.*, 2014, **27**, 1050.
183. A. M. Derfus, W. C. W. Chan and S. N. Bhatia, *Nano Lett.*, 2004, **4**, 11.
184. S. B. Rizvi, S. Rouhi, S. Taniguchi, S. Y. Yang, M. Green, M. Keshtgar and A. M. Seifalian, *Int. J. Nanomed.*, 2014, **9**, 1323.
185. J. Lovrić, H. Bazzi, Y. Cuie, G. A. Fortin, F. Winnik and D. Maysinger, *J. Mol. Med.*, 2005, **83**, 377.
186. A. Hoshino, K. Fujioka, T. Oku, M. Suga, Y. F. Sasaki, T. Ohta, M. Yasuhara, K. Suzuki and K. Yamamoto, *Nano Lett.*, 2004, **4**, 2163.
187. W. E. Smith, J. Brownell, C. C. White, Z. Afsharinejad, J. Tsai, X. Hu, S. J. Polyak, X. Gao, T. J. Kavanagh and D. L. Eaton, *ACS Nano*, 2012, **6**, 9475.
188. C.-W. Peng and Y. Li, *J. Nanomater.*, 2010, **2010**, 1.

189. J. Liu, W.-C. Law, J. Liu, R. Hu, L. Liu, J. Zhu, H. Chen, J. Wang, Y. Hu, L. Ye and K.-T. Yong, *RSC Adv.*, 2013, **3**, 1768.

190. M. L. Schipper, G. Iyer, A. L. Koh, Z. Cheng, Y. Ebenstein, A. Aharoni, S. Keren, L. A. Bentolila, J. Li, J. Rao, X. Chen, U. Banin, A. M. Wu, R. Sinclair, S. Weiss and S. S. Gambhir, *Small*, 2009, **5**, 126.

191. T. Pons, E. Pic, N. Lequeux, E. Cassette, L. Bezdetnaya, F. Guillemin, F. Marchal and B. Dubertret, *ACS Nano*, 2010, **4**, 2531.

192. C. Wang, X. Gao and X. Su, *Anal. Bioanal. Chem.*, 2010, **397**, 1397.

193. T. S. Hauck, R. E. Anderson, H. C. Fischer, S. Newbigging and W. C. W. Chan, *Small*, 2010, **6**, 138.

194. D. Marsolais,É. Duchesne, C. H. Côté and J. Frenette, *Inflammatory cells do not decrease the ultimate tensile strength of intact tendons in vivo and in vitro: protective role of mechanical loading*, 2007, vol. 102.

195. M. A. Walling, J. A. Novak and J. R. E. Shepard, *Quantum dots for live cell and in vivo imaging*, 2009, vol. 10.

196. H. Sarin, A. Kanevsky, H. Wu, K. Brimacombe, S. Fung, A. Sousa, S. Auh, C. Wilson, K. Sharma, M. Aronova, R. Leapman, G. Griffiths and M. Hall, *J. Transl. Med.*, 2008, **6**, 1.

197. H. S. Choi, W. Liu, P. Misra, E. Tanaka, J. P. Zimmer, B. I. Ipe, M. G. Bawendi and J. V. Frangioni, *Nat. Biotechnol.*, 2007, **25**, 1165.

198. Y. Zhang, Y. Zhang, G. Hong, W. He, K. Zhou, K. Yang, F. Li, G. Chen, Z. Liu, H. Dai and Q. Wang, *Biomaterials*, 2013, **34**, 3639.

199. Z. Chen, H. Chen, H. Meng, G. Xing, X. Gao, B. Sun, X. Shi, H. Yuan, C. Zhang, R. Liu, F. Zhao, Y. Zhao and X. Fang, *Toxicol. Appl. Pharmacol.*, 2008, **230**, 364.

200. G. Lin, Q. Ouyang, R. Hu, Z. Ding, J. Tian, F. Yin, G. Xu, Q. Chen, X. Wang and K.-T. Yong, *Nanomed. Nanotechnol. Biol. Med.*, 2015, **11**, 341.

201. E. Q. Contreras, M. Cho, H. Zhu, H. L. Puppala, G. Escalera, W. Zhong and V. L. Colvin, *Environ. Sci. Technol.*, 2013, **47**, 1148.

202. R. R. Arvizo, S. Bhattacharyya, R. A. Kudgus, K. Giri, R. Bhattacharya and P. Mukherjee, *Curr. Top. Med. Chem.*, 2012, **41**, 2943.

203. X. Yuan, Z. Luo, Y. Yu, Q. Yao and J. Xie, *Chem.–Asian J.*, 2013, **8**, 858.

204. L. Polavarapu, S. Mourdikoudis, I. Pastoriza-Santos and J. Perez-Juste, *CrystEngComm*, 2015, **17**, 3727.

205. Y. Xiurong, W. Xiaolei, Z. Hui and X. Xiaowen, in *Functional Nanoparticles for Bioanalysis, Nanomedicine, and Bioelectronic Devices Volume 1*, American Chemical Society, 2012, ch. 9, vol. 1112, ACS Symposium Series, pp. 241–279.

206. Y. Liu, Y. Zhang, H. Ding, S. Xu, M. Li, F. Kong, Y. Luo and G. Li, *J. Mater. Chem. A*, 2013, **1**, 3362.

207. N. Meir, I. Jen-La Plante, K. Flomin, E. Chockler, B. Moshofsky, M. Diab, M. Volokh and T. Mokari, *J. Mater. Chem. A*, 2013, **1**, 1763.

208. P. V. Asharani, Y. lianwu, Z. Gong and S. Valiyaveettil, *Nanotoxicology*, 2011, **5**, 43.

209. T. I. Chanu, T. Muthukumar and P. T. Manoharan, *Phys. Chem. Chem. Phys.*, 2014, **16**, 23686.

210. H. Jang and D.-H. Min, *ACS Nano*, 2015, **9**, 2696.
211. L.-C. Cheng, J.-H. Huang, H. M. Chen, T.-C. Lai, K.-Y. Yang, R.-S. Liu, M. Hsiao, C.-H. Chen, L.-J. Her and D. P. Tsai, *J. Mater. Chem.*, 2012, **22**, 2244.
212. H. Jang and D.-H. Min, *ACS Nano*, 2015, **9**, 2696.
213. J.-W. Xiao, S.-X. Fan, F. Wang, L.-D. Sun, X.-Y. Zheng and C.-H. Yan, *Nanoscale*, 2014, **6**, 4345.
214. L. Dykman and N. Khlebtsov, *Curr. Top. Med. Chem.*, 2012, **41**, 2256.
215. X. Liu, N. Huang, H. Li, H. Wang, Q. Jin and J. Ji, *ACS Appl. Mater. Interfaces*, 2014, **6**, 5657.
216. C. K. Ng, K. Sivakumar, X. Liu, M. Madhaiyan, L. Ji, L. Yang, C. Tang, H. Song, S. Kjelleberg and B. Cao, *Biotechnol. Bioeng.*, 2013, **110**, 1831.
217. L. M. Browning, K. J. Lee, P. D. Nallathamby and X.-H. N. Xu, *Chem. Res. Toxicol.*, 2013, **26**, 1503.
218. X. Wu, X. He, K. Wang, C. Xie, B. Zhou and Z. Qing, *Nanoscale*, 2010, **2**, 2244.
219. L. Tan, A. Wan and H. Li, *ACS Appl. Mater. Interfaces*, 2013, **5**, 11163.
220. S. Chandirasekar, C. Chandrasekaran, T. Muthukumarasamyvel, G. Sudhandiran and N. Rajendiran, *ACS Appl. Mater. Interfaces*, 2015, **7**, 1422.
221. M. Chen, S. Tang, Z. Guo, X. Wang, S. Mo, X. Huang, G. Liu and N. Zheng, *Adv. Mater.*, 2014, **26**, 8210.
222. L. Balogh, S. S. Nigavekar, B. M. Nair, W. Lesniak, C. Zhang, L. Y. Sung, M. S. T. Kariapper, A. El-Jawahri, M. Llanes, B. Bolton, F. Mamou, W. Tan, A. Hutson, L. Minc and M. K. Khan, *Nanomed. Nanotechnol. Biol. Med.*, 2007, **3**, 281.
223. C. A. Simpson, B. J. Huffman, A. E. Gerdon and D. E. Cliffel, *Chem. Res. Toxicol.*, 2010, **23**, 1608.
224. C. Sun, H. Yang, Y. Yuan, X. Tian, L. Wang, Y. Guo, L. Xu, J. Lei, N. Gao, G. J. Anderson, X.-J. Liang, C. Chen, Y. Zhao and G. Nie, *J. Am. Chem. Soc.*, 2011, **133**, 8617.
225. S. Hirn, M. Semmler-Behnke, C. Schleh, A. Wenk, J. Lipka, M. Schäffler, S. Takenaka, W. Möller, G. Schmid, U. Simon and W. G. Kreyling, *Eur. J. Pharm. Biopharm.*, 2011, **77**, 407.
226. E. Morales-Avila, G. Ferro-Flores, B. E. Ocampo-García, L. M. De León-Rodríguez, C. L. Santos-Cuevas, R. O. García-Becerra, L. A. Medina and L. Gómez-Oliván, *Bioconjugate Chem.*, 2011, **22**, 913.
227. X. Qian, X.-H. Peng, D. O. Ansari, Q. Yin-Goen, G. Z. Chen, D. M. Shin, L. Yang, A. N. Young, M. D. Wang and S. Nie, *Nat. Biotechnol.*, 2008, **26**, 83.
228. R. Kannan, V. Rahing, C. Cutler, R. Pandrapragada, K. K. Katti, V. Kattumuri, J. D. Robertson, S. J. Casteel, S. Jurisson, C. Smith, E. Boote and K. V. Katti, *J. Am. Chem. Soc.*, 2006, **128**, 11342.
229. G. S. Terentyuk, G. N. Maslyakova, L. V. Suleymanova, B. N. Khlebtsov, B. Y. Kogan, G. G. Akchurin, A. V. Shantrocha, I. L. Maksimova, N. G. Khlebtsov and V. V. Tuchin, *J. Biophotonics*, 2009, **2**, 292.
230. H. Xie, Z. J. Wang, A. Bao, B. Goins and W. T. Phillips, *Int. J. Pharm.*, 2010, **395**, 324.

231. G. Zhang, Z. Yang, W. Lu, R. Zhang, Q. Huang, M. Tian, L. Li, D. Liang and C. Li, *Biomaterials*, 2009, **30**, 1928.
232. L. Tong, W. He, Y. Zhang, W. Zheng and J.-X. Cheng, *Langmuir*, 2009, **25**, 12454.
233. C. A. Simpson, A. C. Agrawal, A. Balinski, K. M. Harkness and D. E. Cliffel, *ACS Nano*, 2011, **5**, 3577.
234. M. Haase and H. Schäfer, *Angew. Chem., Int. Ed.*, 2011, **50**, 5808.
235. J. Zhou, Z. Liu and F. Li, *Chem. Soc. Rev.*, 2012, **41**, 1323.
236. P. Qiu, N. Zhou, H. Chen, C. Zhang, G. Gao and D. Cui, *Nanoscale*, 2013, **5**, 11512.
237. G. Chen, H. Agren, T. Y. Ohulchanskyy and P. N. Prasad, *Curr. Top. Med. Chem.*, 2015, **44**, 1680.
238. Z. Gu, L. Yan, G. Tian, S. Li, Z. Chai and Y. Zhao, *Adv. Mater.*, 2013, **25**, 3758.
239. R. Abdul Jalil and Y. Zhang, *Biomaterials*, 2008, **29**, 4122.
240. J. Zhou, Y. Sun, X. Du, L. Xiong, H. Hu and F. Li, *Biomaterials*, 2010, **31**, 3287.
241. J. Shan, J. Chen, J. Meng, J. Collins, W. Soboyejo, J. S. Friedberg and Y. Ju, *J. Appl. Phys.*, 2008, **104**, 094308.
242. T. Cao, Y. Yang, Y. Sun, Y. Wu, Y. Gao, W. Feng and F. Li, *Biomaterials*, 2013, **34**, 7127.
243. H. Hu, M. Yu, F. Li, Z. Chen, X. Gao, L. Xiong and C. Huang, *Chem. Mater.*, 2008, **20**, 7003.
244. Q. Liu, M. Chen, Y. Sun, G. Chen, T. Yang, Y. Gao, X. Zhang and F. Li, *Biomaterials*, 2011, **32**, 8243.
245. C. Wang, L. Cheng, H. Xu and Z. Liu, *Biomaterials*, 2012, **33**, 4872.
246. N. M. Idris, Z. Li, L. Ye, E. K. Wei Sim, R. Mahendran, P. C.-L. Ho and Y. Zhang, *Biomaterials*, 2009, **30**, 5104.
247. L. Zhao, A. Kutikov, J. Shen, C. Duan, J. Song and G. Han, *Theranostics*, 2013, **3**, 249.
248. Y. Sun, W. Feng, P. Yang, C. Huang and F. Li, *Curr. Top. Med. Chem.*, 2015, **44**, 1509.
249. A. Gnach, T. Lipinski, A. Bednarkiewicz, J. Rybka and J. A. Capobianco, *Curr. Top. Med. Chem.*, 2015, **44**, 1561.
250. L. Xiong, T. Yang, Y. Yang, C. Xu and F. Li, *Biomaterials*, 2010, **31**, 7078.
251. F. Li, C. Li, J. Liu, X. Liu, L. Zhao, T. Bai, Q. Yuan, X. Kong, Y. Han, Z. Shi and S. Feng, *Nanoscale*, 2013, **5**, 6950.
252. X.-F. Yu, Z. Sun, M. Li, Y. Xiang, Q.-Q. Wang, F. Tang, Y. Wu, Z. Cao and W. Li, *Biomaterials*, 2010, **31**, 8724.
253. D. Yang, Y. Dai, J. Liu, Y. Zhou, Y. Chen, C. Li, P. a. Ma and J. Lin, *Biomaterials*, 2014, **35**, 2011.
254. F. Liu, X. He, L. Liu, H. You, H. Zhang and Z. Wang, *Biomaterials*, 2013, **34**, 5218.
255. D. Ni, J. Zhang, W. Bu, H. Xing, F. Han, Q. Xiao, Z. Yao, F. Chen, Q. He, J. Liu, S. Zhang, W. Fan, L. Zhou, W. Peng and J. Shi, *ACS Nano*, 2014, **8**, 1231.
256. L.-Q. Xiong, Z.-G. Chen, M.-X. Yu, F.-Y. Li, C. Liu and C.-H. Huang, *Biomaterials*, 2009, **30**, 5592.

257. C.-N. Zhu, P. Jiang, Z.-L. Zhang, D.-L. Zhu, Z.-Q. Tian and D.-W. Pang, *ACS Appl. Mater. Interfaces*, 2013, **5**, 1186.

258. Y. Zhang, G. Hong, Y. Zhang, G. Chen, F. Li, H. Dai and Q. Wang, *ACS Nano*, 2012, **6**, 3695.

259. L. Tan, A. Wan, T. Zhao, R. Huang and H. Li, *ACS Appl. Mater. Interfaces*, 2014, **6**, 6217.

260. L. Tan, A. Wan and H. Li, *Langmuir*, 2013, **29**, 15032.

261. W. Li, R. Zamani, P. Rivera Gil, B. Pelaz, M. Ibáñez, D. Cadavid, A. Shavel, R. A. Alvarez-Puebla, W. J. Parak, J. Arbiol and A. Cabot, *J. Am. Chem. Soc.*, 2013, **135**, 7098.

262. J. Kolny-Olesiak and H. Weller, *ACS Appl. Mater. Interfaces*, 2013, **5**, 12221.

263. I. Hocaoglu, F. Demir, O. Birer, A. Kiraz, C. Sevrin, C. Grandfils and H. Yagci Acar, *Nanoscale*, 2014, **6**, 11921.

264. M. Gong, A. Kirkeminde and S. Ren, *Sci. Rep.*, 2013, **3**, 2092.

265. H. Chen, B. Li, M. Zhang, K. Sun, Y. Wang, K. Peng, M. Ao, Y. Guo and Y. Gu, *Nanoscale*, 2014, **6**, 12580.

266. Q. Cao, R. Che and N. Chen, *Chem. Commun.*, 2014, **50**, 4931.

267. Q. Cao and R. Che, *RSC Adv.*, 2014, **4**, 16641.

268. S. S. Chou, B. Kaehr, J. Kim, B. M. Foley, M. De, P. E. Hopkins, J. Huang, C. J. Brinker and V. P. Dravid, *Angew. Chem., Int. Ed.*, 2013, **52**, 4160.

269. L. Cheng, J. Liu, X. Gu, H. Gong, X. Shi, T. Liu, C. Wang, X. Wang, G. Liu, H. Xing, W. Bu, B. Sun and Z. Liu, *Adv. Mater.*, 2014, **26**, 1886.

270. V. C. Sanchez, A. Jachak, R. H. Hurt and A. B. Kane, *Chem. Res. Toxicol.*, 2012, **25**, 15.

271. L. Turyanska, T. D. Bradshaw, M. Li, P. Bardelang, W. C. Drewe, M. W. Fay, S. Mann, A. Patane and N. R. Thomas, *J. Mater. Chem.*, 2012, **22**, 660.

272. X.-R. Song, X. Wang, S.-X. Yu, J. Cao, S.-H. Li, J. Li, G. Liu, H.-H. Yang and X. Chen, *Adv. Mater.*, 2015, **27**, 3285.

273. X. Liu, Q. Wang, C. Li, R. Zou, B. Li, G. Song, K. Xu, Y. Zheng and J. Hu, *Nanoscale*, 2014, **6**, 4361.

274. T. Liu, C. Wang, X. Gu, H. Gong, L. Cheng, X. Shi, L. Feng, B. Sun and Z. Liu, *Adv. Mater.*, 2014, **26**, 3433.

275. M. Zhou, R. Zhang, M. Huang, W. Lu, S. Song, M. P. Melancon, M. Tian, D. Liang and C. Li, *J. Am. Chem. Soc.*, 2010, **132**, 15351.

276. S. Wang, A. Riedinger, H. Li, C. Fu, H. Liu, L. Li, T. Liu, L. Tan, M. J. Barthel, G. Pugliese, F. De Donato, M. Scotto D'Abbusco, X. Meng, L. Manna, H. Meng and T. Pellegrino, *ACS Nano*, 2015, **9**, 1788.

277. X. Qian, S. Shen, T. Liu, L. Cheng and Z. Liu, *Nanoscale*, 2015, **7**, 6380.

Subject Index